METHODS IN MOLECULAR BIOLOGY™

For other titles published in this series, go to
www.springer.com/series/7651

Free Radicals and Antioxidant Protocols

Second Edition

Edited by

Rao M. Uppu

Southern University and A & M College, Baton Rouge, LA, USA

Subramanyam N. Murthy

Tulane University School of Medicine, New Orleans, LA, USA

William A. Pryor

Louisiana State University, Baton Rouge, LA, USA

Narasimham L. Parinandi

The Ohio State University Medical Center, Columbus, OH, USA

Humana Press

Editors
Rao M. Uppu
Department of Environmental Toxicology and the Health Research Center
Southern University and A & M College
108 Fisher Hall
James L. Hunt Street
Baton Rouge, LA 70813
USA
rao_uppu@subr.edu

William A. Pryor
Department of Chemistry
Louisiana State University
Baton Rouge, LA 70808
USA
wapryor@att.net

Subramanyam N. Murthy
Department of Medicine and Pharmacology
Tulane University
School of Medicine
1430 Tulane Ave.
New Orleans, LA 70112
USA
smurthy@tulane.edu

Narasimham L. Parinandi
Department of Internal Medicine and the Davis Heart & Lung Research Institute
Ohio State University
473 West 12th Ave.
Columbus, OH 43210
USA
narasimham.parinandi@osumc.edu

ISSN 1064-3745 e-ISSN 1940-6029
ISBN 978-1-58829-710-5 e-ISBN 978-1-60327-029-8
DOI 10.1007/978-1-60327-029-8

Library of Congress Control Number: 2009940921

Printed on acid-free paper

springer.com

Preface

Reactive oxygen and nitrogen species (RONs) affect normal physiological processes and pathological conditions. In spite of the development of analytical and technological advancements in the detection and determination of RONs, there is a growing demand to achieve more rapid and accurate detection and determination of RONs and also of redox stress. In order to satisfy such a demand, we chose to prepare this current volume which comprises 27 chapters contributed by world-renowned experts on the state-of-the-art analytical and technological aspects of the detection and determination of RONs, and oxidative, nitrative, nitrosative, and redox stresses in biological systems, in vitro and in vivo.

There is a need to establish biomarkers for the oxidative stress-induced genetic damage in cellular and animal models. This need is addressed herein by offering protocols for (1) the generation of stable oxidative stress-resistant phenotypes of Chinese hamsters fibroblasts, (2) in vivo detection and measurement of free radicals and oxygen, (3) in vivo determination of tissue and DNA damage as a result of radical exposure, and (4) in vivo and in vitro monitoring of footprints of free-radical and antioxidant reactions. In the sections dedicated to the analysis of antioxidants and their metabolites, we describe methods for the analysis of phenolic acids and flavonoids, eugenol antioxidants, and the recycling of ascorbic and lipoic acids. We also report on cellular reductive capacity, determination of glutathione, mitochondrial transmembrane potential, and cytotoxicity in cardiomyocytes under RONs stresses.

Lipids are at the epicenter of oxidative stress. However, the analysis of lipid peroxidation, either in vitro or in vivo, is still complex, and novel methods and technologies are always in great demand to accurately analyze the peroxidized lipids. This volume presents the recent advances in the soft (electro spray) ionization mass spectrometry (MS) of phospholipid hydroperoxides for cellular and tissue oxidative lipidomics, the simultaneous analysis of multiple lipid oxidation products in vivo by LC-MS, and protocols for enzyme immunoassay of isoprostanes and chemiluminescence determination of nitrite in plasma. We also discuss the preparation, purification, and characterization of lipoxygenase-catalyzed phosphatidylinositol peroxides, and the biology and chemistry of oxidized low-density lipoprotein.

Gene transfer has become a useful therapeutic strategy in the treatment of certain diseases/disorders. Along those lines, the current volume describes techniques for gene therapy involving endothelial nitric oxide synthase delivery to the lung in pulmonary hypertension and the techniques for delivery of the extracellular superoxide dismutase (ecSOD) gene for erectile dysfunction therapy. Drug delivery to the target areas has been identified as a preferred therapeutic approach to treat certain diseases, and this pharmacological strategy is being actively explored to combat certain oxidative stress-mediated diseases. In view of that, this volume describes the preparation of drug-loaded polymeric nanoparticles and evaluation of the antioxidant activity against lipid peroxidation, design, synthesis, and action of antiatherogenic antioxidants, synthesis and characterization of polymer nanocarriers for the targeted delivery of therapeutic enzymes in vitro, nanoparticle and iron chelators as a potential novel Alzheimer therapy, and a simple method for effective and safe removal of membrane cholesterol from lipid rafts in endothelial cells with implications in the oxidant-mediated lipid signaling.

Molecular oxygen is the key life-supporting species in aerobes, and methods for the accurate measurement of oxygen levels in the biological systems are just emerging. For that, electron paramagnetic resonance spectroscopy (EPR) appears to be a novel analytical technique of choice, and the current volume presents the methods of EPR imaging of free radicals and oxygen in vivo and the EPR spectroscopic determination of tissue oxygen in vivo with the aid of oxygen-sensitive paramagnetic lithium phthalocyanine particles.

This book would not have been possible without the contribution of chapters by several experts in the field of oxidative stress, and therefore, the editors deem it a distinct privilege to gratefully acknowledge every individual author of the chapters that has made this book a reality. We thank Drs. Achuthan Raghavamenon, Michelle Fletcher Claville, and Deidra Atkins-Ball for help in editing and organizing the chapters. The patience shown by Professor Walker and the team of Humana Press is beyond words and cannot be adequately articulated. The editors express profound gratitude to Professor John Walker for the confidence shown and consider it an honor that he gave us this assignment. We also thank the following organizations and universities for financial support and use of institutional facilities: Departments of Chemistry and Environmental Toxicology, Southern University and A&M College, Baton Rouge, LA; Departments of Medicine and Pharmacology, Tulane University School of Medicine, New Orleans, LA; Department of Internal Medicine and the Davis Heart and Lung Research Institute, The Ohio State University Medical College, Columbus, OH; National Institutes of Health, Bethesda, MD; National Science Foundation, Washington, DC; and US Department of Education, Washington, DC. Finally, we would be remiss if we did not express our heartfelt appreciation to our families for their unparalleled support.

Rao M. Uppu
Subramanyam N. Murthy
William A. Pryor
Narasimham L. Parinandi

Contents

Contributors

ANTONYSUNIL ADAIKALAKOTESWARI • *Department of Cell and Molecular Biology, Madras Diabetes Research Foundation & Dr. Mohan's Diabetes Specialties, Gopalapuram, Chennai, India*

ANDREW A. AMOSCATO • *Department Pathology, University of Pittsburgh, Pittsburgh, PA, USA*

MUTHUSWAMY BALASUBRAMANYAM • *Department of Cell and Molecular Biology, Madras Diabetes Research Foundation & Dr Mohan's Diabetes Specialties Centre, Gopalapuram, Chennai, India*

HÜLYA BAYIR • *Center for Free Radical and Antioxidant Health, Departments of Environmental & Occupational Health and Critical Care Medicine, University of Pittsburgh, Pittsburgh, PA, USA*

T.J. BIVALACQUA • *Department of Pharmacology, Tulane University Health Sciences Center, New Orleans, LA, USA. Present address: Department of Urology, Johns Hopkins Hospital, Baltimore, MD, USA*

SESIKERAN BOINDALA • *National Institute of Nutrition, Indian Council of Medical Research, Jamai-Osmania, Hyderabad, India*

ALLAN BUTTERFIELD • *Department of Chemistry, Sanders-Brown Center on Aging and Center of Membrane Sciences, University of Kentucky, Lexington, Kentucky, USA*

E.S. O'CONNOR BUTLER • *Division of Pulmonary, Allergy, Critical Care, and Sleep Medicine, Department of Internal Medicine, The Ohio State University Medical Center, Columbus, OH, USA*

VITTORIO CALABRESE • *Department of Chemistry, Biochemistry & Molecular Biology Section, Faculty of Medicine, University of Catania, Catania, Italy*

DENIS M. CALLEWAERT • *Department of Chemistry and Center for Biomedical Research, Oakland University, Rochester, MI, USA*

CAROLIN CORNELIUS • *Department of Chemistry, Biochemistry and Molecular Biology Section, Faculty of Medicine, University of Catania, Catania, Italy*

H.C. CHAMPION • *Department of Medicine University, Pittsburg, PA, USA*

MARCUS S. COOKE • *Radiation & Oxidative Stress Group, Department of Genetics, University of Leicester, Leicester, UK*

TÂNIA B. CRECZYNSKI-PASA • *Departamento de Ciências Farmacêuticas, Universidade Federal de Santa Catarina, UFSC, Porto Alegre, Brazil*

JASON P. CRUFF • *Division of Pulmonary, Allergy, Critical Care, and Sleep Medicine, Department of Internal Medicine, The Ohio State University, College of Medicine, Columbus OH, USA*

OSAMU CYNSHI • *Fuji-gotemba Research Laboratories, Chugai Pharmaceutical Co., Ltd., Shizuoka, Japan*

TODD DAVIS • *Department of Chemistry, Idaho State University, Pocatello, ID 83209, USA*

PETER C. DEDON • *Biological Engineering Division and Center for Environmental Health Science, Massachusetts Institute of Technology, Cambridge, MA, USA*

W. DENG • *Department of Pharmacology, Tulane University Health Sciences Center, New Orleans, LA, USA*

STEVEN T. DEKOSKY • *Department of Neurology, University of Pittsburgh, Pittsburgh, PA, USA*

THOMAS D. DZIUBLA • *Department of Pharmacology and Institute for Environmental Medicine, University of Pennsylvania School of Medicine, Philadelphia, PA; Department of Chemical and Materials Engineering, University of Kentucky, Lexington, KY, USA*

MARK D. EVANS • *Radiation & Oxidative Stress Group, Department of Cancer Studies and Molecular Medicine, University of Leicester, Leicester, UK*

PETER B. FARMER • *Cancer Biomarkers and Prevention Group, Department of Cancer Studies and Molecular Medicine, University of Leicester, Leicester, UK*

KARUNAKARAN GANESHKUMAR • *Experimental Obesity Laboratory, Pennington Biomedical Research Center, Louisiana State University, Baton Rouge, LA, USA*

XUELI GAO • *Department of Environmental Toxicology and the Health Research Center, Southern University and A&M College, Baton Rouge, LA, USA*

MAHDI OMAR GARELNABI • *Division of Cardiothoracic Surgery, The Ohio State University Medical Center Columbus, OH, USA*

JOEL S. GREENBERGER • *Departments of Radiation Oncology University of Pittsburgh, Pittsburgh, PA, USA*

SÍLVIA S. GUTERRES • *Faculdade de Farmácia da Universidade Federal do Rio Grande do Sul, UFRGS, Porto Alegre, Brazil*

RONALD HAMILTON • *Department of Pathology, University of Pittsburgh, Pittsburgh, PA, USA*

W.J. HELLSTROM • *Department of Urology, Tulane University Health Sciences Center, New Orleans, LA, USA*

ADAM HINZEY • *Division of Pulmonary, Allergy, Critical Care, and Sleep Medicine, Department of Internal Medicine, The Ohio State University, Columbus, OH, USA*

F. HYODO • *Biophysical Spectroscopy Section, Radiation Biology Branch, Center for Cancer Research, National Cancer Institute, Bethesda, MD, USA*

PHILIP J. KADOWITZ • *Department of Pharmacology, Tulane University Health Sciences Center, New Orleans, LA, USA*

VALERIAN E. KAGAN • *Center for Free Radical and Antioxidant Health, Department of Environmental & Occupational Health, University of Pittsburgh, Pittsburgh, PA, USA*

MICHELLE A. KLINE • *Division of Pulmonary, Allergy, Critical Care, and Sleep Medicine, Department of Internal Medicine, The Ohio State University, College of Medicine, Columbus, OH, USA*

PATRICK, M. KOCHANEK • *Department Critical Care Medicine, University of Pittsburgh, Pittsburgh, PA, USA*

SAINATH R. KOTHA • *Lipidomics, Division of Pulmonary, Allergy, Critical Care, and Sleep Medicine, Lipid Signaling and Vasculotoxicity Laboratory, Dorothy M. Davis Heart and Long Research Institute, The Ohio State University College of Medicine, Columbus, OH, USA*

M.C. KRISHNA • *Biophysical Spectroscopy Section, Radiation Biology Branch, Center for Cancer Research, National Cancer Institute, Bethesda, MD, USA*

PERIANNAN KUPPUSAMY • *Center for Biomedical EPR Spectroscopy and Imaging, Comprehensive Cancer Center, Davis Heart and Lung Research Institute, Department of Internal Medicine, The Ohio State University Medical Center, Columbus, OH, USA*

RICCARDO LENTILE • *Department of Biochemical, Physiological and Nutritional Sciences, University of Messina, Messina, Italy*

GANG LIU • *Department of Radiology, University of Utah, Salt Lake City, Utah, USA*

HEATHER LORD • *Department of Chemistry, University of Waterloo, Waterloo, ON, Canada*

ANDRIY LYSYTSYA • *Center for Free Radical and Antioxidant Health, Department of Environmental & Occupational Health, University of Pittsburgh, Pittsburgh, PA, USA*

VICTOR MA • *Department of Chemistry, University of Waterloo, Waterloo, ON, Canada*

CESARE MANCUSO • *Institute of Pharmacology, Catholic University School of Medicine, Rome, Italy*

YELENA MARGOLIN • *Biological Engineering Division and Center for Environmental Health Science, Massachusetts Institute of Technology, Cambridge, MA, USA*

CLAY B. MARSH • *Division of Pulmonary, Allergy, Critical Care, and Sleep Medicine, Department of Internal Medicine, Ohio State University Medical Center, Columbus OH, USA*

A. MATSUMOTO • *Biophysical Spectroscopy Section, Radiation Biology Branch, Center for Cancer Research, National Cancer Institute, Bethesda, MD, USA*

K. MATSUMOTO • *Biophysical Spectroscopy Section, Radiation Biology Branch, Center for Cancer Research, National Cancer Institute, Bethesda, MD, USA*

S. MATSUMOTO • *Biophysical Spectroscopy Section, Radiation Biology Branch, Center for Cancer Research, National Cancer Institute, Bethesda, MD, USA*

JAMES M. MAY • *Departments of Medicine and Molecular Physiology and Biophysics, Vanderbilt University Medical Center, Nashville, TN, USA*

JESSICA N. MAZERIK • *Division of Pulmonary, Allergy, Critical Care, and Sleep Medicine, Department of Internal Medicine, The Ohio State University Medical Center, Columbus OH, USA*

PING MEN • *Department of Radiology, University of Utah, Salt Lake City, Utah, USA*

VILAS MISTRY • *Radiation & Oxidative Stress Group, University of Leicester, Leicester, UK*

VISWANATHAN MOHAN • *Madras Diabetes Research Foundation & Dr. Mohan's Diabetes Specialties Centre, Gopalapuram, Chennai, India*

PAULA I. MOREIRA • *Center for Neuroscience and Cell Biology of Coimbra, University of Coimbra, Coimbra, Portugal*

MELISSA MORLEY • *Department of Chemistry, University of Waterloo, Waterloo, ON, Canada*

SUBRAMANYAM N. MURTHY • *Departments of Medicine and Pharmacology, Tulane University School of Medicine, New Orleans, LA, USA*

VLADIMIR R. MUZYKANTOV • *Department of Pharmacology, Institute for Environmental Medicine, and Institute for Translational Medicine and Therapeutics, University of Pennsylvania School of Medicine, Philadelphia, PA, USA*

ENIKA NAGABABU • *Molecular Dynamics Section, National Institute on Aging, National Institutes of Health, Baltimore, MD, USA*

LAKSHMAIAH NAKKA • *National Institute of Nutrition, Indian Council of Medical Research, Jamai-Osmania, Hyderabad, India*

ETSUO NIKI • *Human Stress Signal Research Center, National Institute of Advanced Industrial Science and Technology, Osaka, Japan*

AKIHIKO NUNOMURA • *Department of Psychiatry and Neurology, Asahikawa Medical College, Asahikawa, Japan*

NARASIMHAM L. PARINANDI • *Department of Internal Medicine and the Davis Heart and Lung Research Institute, The Ohio State University Medical Center, Columbus, OH, USA*

SAMPATH PARTHASARATHY • *Division of Cardiothoracic Surgery, The Ohio State University Medical Center, Columbus, OH, USA*

JANUSZ PAWLISZYN • *Department of Chemistry, University of Waterloo, Waterloo, ON, Canada*

GEORGE PERRY • *Department of Pathology, Case Western Reserve University, Cleveland, Ohio, USA; and College of Sciences, University of Texas at San Antonio, San Antonio, Texas, USA*

ADRIANA R. POHLMANN • *Instituto de Química da Universidade Federal do Rio Grande do Sul, UFRGS, Porto Alegre, Brazil*

NED A. PORTER • *Department of Chemistry, Center in Molecular Toxicology, and Vanderbilt Institute of Chemical Biology, Vanderbilt University, Nashville, TN, USA*

Achuthan C. Raghavamenon • *Department of Environmental Toxicology and the Health Research Center, Southern University and A&M College, Baton Rouge, LA, USA*

Joseph M. Rifkind • *Molecular Dynamics Section, National Institute on Aging, National Institutes of Health, Baltimore, MD, USA*

Vladimir B. Ritov • *Department of Medicine, University of Pittsburgh, Pittsburgh, PA, USA*

Brian K. Rivera • *Center for Biomedical EPR Spectroscopy and Imaging, Comprehensive Cancer Center, Davis Heart and Lung Research Institute, Department of Internal Medicine, The Ohio State University, Columbus, OH, USA*

Zaheer Sameermahmood • *Dept of Cell and Molecular Biology, Madras Diabetes Research Foundation & Dr Mohan's Diabetes Specialties, Gopalapuram, Chennai, India*

Nalini Santanam • *Department of Pharmacology, Marshall University, Huntington, WV, USA*

K. Sathishkumar • *Department of Environmental Toxicology and the Health Research Center, Southern University and A&M College, Baton Rouge, LA, USA*

Lawrence M. Sayre • *Departments of Chemistry, Case Western Reserve University, Cleveland, OH, USA*

Scheila Rezende Schaffazick • *Pós-Graduação em Ciências Farmacêuticas da Universidade Federal do Rio Grande do Sul, UFRGS, Porto Alegre, Brazil*

Harald H.O. Schmid • *The Hormel Institute, University of Minnesota, Austin, MN, USA*

Shariq I. Sherwani • *Division of Pulmonary, Allergy, Critical Care, and Sleep Medicine, Department of Internal Medicine, The Ohio State University Medical Center, Columbus, OH, USA*

Vladimir V. Shuvaev • *Institute for Environmental Medicine, University of Pennsylvania School of Medicine, Philadelphia, PA, USA*

Eric A. Simone • *Vertex Pharmaceuticals, Inc, Formulation Development, 130 Waverly Street, Cambridge, MA, USA*

Rajinder Singh • *Cancer Biomarkers and Prevention Group, Department of Cancer Studies and Molecular Medicine, University of Leicester, Leicester, UK*

Sean Sliman • *Division of Pulmonary, Allergy, Critical Care, and Sleep Medicine, Department of Internal Medicine, The Ohio State University Medical Center, Columbus, OH, USA*

Charles Sloan • *Oxford Biomedical Research, Rochester Hills, MI, USA*

Mark A. Smith • *Department of Pathology, Case Western Reserve University, Cleveland, Ohio, USA*

Douglas R. Spitz • *Department of Radiation Oncology, Holden Comprehensive Cancer Center, The University of Iowa, Iowa City, IA, USA*

Constantine D. Stalikas • *Laboratory of Analytical Chemistry, Department of Chemistry, University of Ioannina, Ioannina 451 10 Greece*

A.M. GIUFFRIDA STELLA • *Department of Chemistry, Biochemistry & Molecular Biology Section, Faculty of Medicine, University of Catania, Catania, Italy*

SHANNON J. SULLIVAN • *Department of Pediatrics, University of Iowa, Iowa City, IA, USA*

KUNIO TAMURA • *Fuji-gotemba Research Laboratories, Chugai Pharmaceutical Co., Ltd., Shizuoka, Japan*

RAMESHWARI TELAPROLU • *Department of Environmental Toxicology and the Health Research Center, Southern University and A&M College, Baton Rouge, LA, USA*

VLADIMIR A. TYURIN • *Center for Free Radical and Antioxidant Health, Department of Environmental & Occupational Health, University of Pittsburgh, Pittsburgh, PA, USA*

YULIA Y. TYURINA • *Center for Free Radical and Antioxidant Health, Departments of Environmental & Occupational Health, University of Pittsburgh, Pittsburgh, PA, USA*

RAO M. UPPU • *Department of Environmental Toxicology and the Health Research Center, Southern University and A&M College, Baton Rouge, LA, USA*

DEEPTI S. VIKRAM • *Center for Biomedical EPR Spectroscopy and Imaging, Comprehensive Cancer Center, Davis Heart and Lung Research Institute, Department of Internal Medicine, The Ohio State University Medical Center, Columbus, OH, USA*

BARBARA K. WEIS • *The Hormel Institute, University of Minnesota, Austin, MN, USA*

HUIYONG YIN • *Division of Clinical Pharmacology, Departments of Medicine and Chemistry Vanderbilt University, Nashville, TN, USA*

XIONGWEI ZHU • *Departments of Pathology, Case Western Reserve University, Cleveland, OH, USA*

Section I

Reactive Oxygen and Nitrogen Species

Chapter 1

In Vivo Imaging of Free Radicals and Oxygen

Deepti S. Vikram, Brian K. Rivera, and Periannan Kuppusamy

Abstract

Free radicals are highly reactive compounds that play an essential role in many biological processes, both beneficial and deleterious. Detection and quantification of these species is critical to develop a better understanding of normal and pathophysiological functions at the cellular and tissue levels. Electron paramagnetic resonance (EPR) spectroscopy is the technique most commonly used for this purpose through the detection of exogenous probes or spin traps that interact with the free radical species of interest. Over the past several years, the spatial and temporal distribution of free radicals within cells and tissues has been of particular interest. This chapter briefly explains the principles and challenges in the use of EPR for biological samples and introduces the concept of EPR for free radical imaging purposes. In addition, specific examples are given for the use of EPR imaging in four principal areas: free radical probes, nitric oxide ($^{\bullet}NO$), redox state, and oxygen (O_2) concentration.

Key words: Free radicals, EPR spectroscopy, imaging, nitric oxide, nitroxyl, redox state, metabolism, oximetry.

1. Introduction

Free radicals are atomic or molecular species with unpaired electrons. They are highly reactive and unstable as compared to similar ions. Free radicals play an important role in many biological processes including metabolic pathways, cell signaling, immune response, and a variety of pathophysiological conditions (1). Detection and quantification of these species is critical to decipher cellular pathways and mechanisms to understand disease and function. Free radicals are generated in the biological environment as a result of reactions associated with common biochemical pathways involving oxygen metabolism. Thus, their universal presence

R.M. Uppu et al. (eds.), *Free Radicals and Antioxidant Protocols*, Methods in Molecular Biology 610,
DOI 10.1007/978-1-60327-029-8_1, © Humana Press, a part of Springer Science+Business Media, LLC 1998, 2010

and their role as critical mediators of normal and pathophysiology have resulted in considerable development of techniques that can detect these radicals.

Electron paramagnetic resonance (EPR) spectroscopy is the most widely used method for the detection of free radicals. By definition, the paramagnetic molecules contain one or more unpaired electrons, e.g., nitric oxide (•NO), oxygen (O_2), nitroxyls (>N = O), or copper sulfate ($CuSO_4$). Free radicals are paramagnetic, but the term is limited to short-lived fragments or redox intermediates that possess unpaired electrons. "EPR spin trapping" is a technique commonly used for the detection of reactive oxygen and nitrogen species (2).

"EPR imaging" is used to obtain spatial information about the distribution of free radical species in vitro and in vivo. The advantages of using an imaging technique are obvious; since free radicals are not involved in just one process, and are an integral part of the ubiquitous metabolism in organisms, obtaining spatial information helps in understanding in vivo mechanisms of free radical generation and subsequent reactions. Thus, while detection and quantification of free radical species is valuable, the ability to extract information about the spatial distribution is also important. This chapter focuses on EPR imaging of free radicals in vivo. It also includes a discussion on imaging of •NO, oxygen, and the redox environment in tissues as well.

1.1. Principle of EPR Spectroscopy

EPR spectroscopy was first reported in 1945 by Zavoisky (3). The technique involves the detection of molecules with unpaired electrons. In the presence of an external magnetic field, the moments arising from the electron "spins" of these unpaired electrons will be aligned with or in opposition to the applied field. Thus, the unpaired electrons can exist in two spin states that have different energies. The difference in magnitude of these energies is proportional to the applied field strength.

In an EPR experiment, the sample is placed in a magnetic field and exposed to electromagnetic radiation, typically in the microwave frequency range. The resonance condition is satisfied when the magnetic field and microwave energy correspond to the difference in energy of the two spin states of the unpaired electrons. The magnetic field strength is varied, and the absorbance of microwave energy is recorded to obtain an EPR spectrum for the sample being analyzed. Today, because of its unique capability to detect materials with unpaired electrons, EPR spectroscopy is used in applications as diverse as chemistry, physics, biology, and medicine. In addition to the detection of unpaired electrons, EPR data provide information about the local tissue environment, and this aspect makes EPR spectroscopy and imaging (EPRI) very valuable for in vivo applications.

1.2. EPR Instrumentation for Biological Samples

The most common operating frequency for EPR spectrometers is X-band (~9 GHz). However, X-band instrumentation is limited to non-aqueous samples of a few millimeters in size. Biological samples are aqueous in nature and hence cause serious "dielectric loss" problems. A further limitation is associated with the penetration depth of the excitation field at higher microwave frequencies. It has been estimated that for a water-containing tissue, the penetration depth at X-band is about 1 mm. This severely restricts the use of conventional X-band techniques for studying larger biological samples. Hence, low-frequency spectrometers operating in the range 200 MHz to 3 GHz were developed to enable EPR measurements on large aqueous samples (4–6).

The main feature of these low-frequency spectrometers is the resonators capable of accommodating large aqueous samples with minimal dielectric dissipation. However, there is a disadvantage in the use of lower frequencies as the sensitivity of the measurement is directly related to the square of the operating frequency. The decrease in sensitivity at lower frequencies is to some extent compensated by the higher filling factor for the sample. For example, while X-band EPR measurements on lossy samples are limited to a volume of few microliters, L-band (1–2 GHz) measurements are possible on volumes of a few milliliters. Radiofrequency resonators (200–300 MHz) are used for samples of up to 200 mL. Over the last decade, a variety of low-frequency instrumentation suitable for use with lossy biological samples have been developed. This enabled EPR spectroscopic measurements on large biological samples up to and including a whole rat (4–11).

1.3. EPR Imaging

EPR spectroscopy reports free radical concentration in the global object. That is, the spectroscopic measurements provide a summation of spectra over the entire active volume within the object. It does not, however, provide information regarding spatial distribution of spins within the object. Spatial information is provided by imaging. Both qualitative and quantitative information can be obtained at every spatial location within the object.

Unlike spectroscopy, which uses a homogeneous magnetic field in the volume of measurement, CW EPR imaging is performed in the presence of an inhomogeneous magnetic field produced by additional set(s) of field coils. These coils generate a linear field gradient within the sample volume. Measurements of the spin distribution (object) are obtained by performing a conventional field sweep in the presence of these gradients. During the sweep, successive pseudo isofield planes orthogonal to the magnetic field gradient are brought into resonance. The plane integrals of the EPR absorption function, collected as a function of sweep field, give a projection of the object corresponding to the particular gradient. The image is then obtained by reconstruction from backprojections (12). Image reconstructions are commonly performed by filtered-backprojection or Fourier reconstruction techniques. In addition, a variety of projection and image data processing techniques, including spectral deconvolution and hyperfine correction, are utilized to improve the quality and resolution of the reconstructed images (13–16).

The CW EPRI has evolved over the past decade to be an important tool in studying the distribution of free radicals in various branches of science (17–19). In the last few years the potential applications of EPRI to studies of living biological systems have been recognized. However, the broad application of the EPRI technique to obtain high-quality images of lossy biological samples has been limited by several factors including gradient design and accuracy, sensitivity, and speed of acquisition. The difficulties inherent in a successful and useful biological EPRI experiment include the rare occurrence of sufficient concentration of endogenous free radicals, the lack of availability of ideal stable spin probes, short relaxation times, slow data acquisition, drifts in microwave frequency, and the magnitude of the static and gradient fields.

1.4. Imaging Instrumentation

Instrumentation and computer software were developed to enable high-resolution multidimensional imaging of free radicals and paramagnetic species in isolated organs and tissues (13, 20–22). A particular challenge in building this instrumentation was to be able to fit the resonator and three sets of gradient coils into the relatively small magnet gaps available in standard resistive magnets. Three sets of water-cooled gradient coils were built for the *x*, *y*, and *z* gradients, and powered by six power supplies. The resonator and gradient coils were fitted into the gap of a 38 cm pole face iron-core electromagnet whose pole caps and ring shims were machined to yield a 104 mm gap with field homogeneity of greater than 10 mG over a 25 mm diameter sphere. The gradients and power supplies were designed to achieve gradient fields of up to 150 G/cm.

Computer software was developed for IBM compatible PCs for acquiring spatial or spectral–spatial EPR projections via GPIB (IEEE-488) bus control of a Bruker signal channel and field controller (23). Image reconstruction was performed by filtered-backprojection methods (12). Algorithms were developed to remove hyperfine-based image artifacts, further enhancing the image resolution (14, 16). In extensive validation studies on phantoms, isolated hearts, and other tissues, it was observed that high-quality, spatially accurate images of the distributions of free radicals could be obtained with submillimeter resolution.

EPR imaging of biological samples has many technical challenges for instrumentation development in general, and for the sample resonator design in particular, beyond those of simple spectroscopy. The most important challenge is the need for a resonator design of minimum thickness, which makes it possible to achieve higher magnetic field gradients for a given coil's driving power. Thinner resonators also enable multidimensional gradient coils to be placed in the gap.

Loop–gap resonators (LGR) provide straightforward design and high filling factors (24). However, due to the open structure of the inductive loop element, LGRs require a shield. The need for the shield leads to problems in achieving an optimum magnitude of modulation field and a minimum 20% increase of overall resonator thickness. Reentrant resonators (RER) do not require a shield; however, since they were constructed from milled and silver-plated plastic, they exhibit low rigidity and poor thermal stability (25). Ceramics, being a rigid material with high structural strength and stable mechanical and thermal parameters, are thus a good choice for resonator construction. Several RER sample resonators have been designed and fabricated using ceramics (26, 27).

To further decrease the overall thickness, modulation coils were wound as a thin coil, epoxy-impregnated, and mounted onto the side walls of the resonator with adhesive. We observed that for a fixed concentration of free radical sample, the ceramic L-band RERs yield sensitivity similar to that which can be obtained at X-band using standard microwave cavities, assuming that optimal filling of the resonators is performed with suitable cylindrical tubes. The L-band resonator can accommodate much larger volumes of lossy aqueous samples and can thus compensate for the inherently lower sensitivity of L-band measurements. In addition, we have shown that this design can be modified with a piezoelectric actuator to serve as an electronically tunable resonator with the frequency locked to that of a low noise fixed frequency source (28). This latter approach eliminates any frequency drift and is useful in that it maintains the isofrequency condition for a given imaging experiment.

1.5. Measurement and Imaging of Free Radicals

Some of the areas where in vivo EPR plays a major role in the understanding of normal and pathophysiology conditions include measurement of oxygen concentrations in tissue, measurement of free radicals, redox state, and pH. Endogenous paramagnetic materials are present in tissues in very small concentrations ($< \mu M$) and some species are not stable for long periods of time. Most of the reactive oxygen species (ROS) are free radicals and therefore cannot be imaged. Thus, an exogenous probe which can interact with the targeted paramagnetic species needs to be introduced into the tissue region of interest.

EPR spin trapping uses a probe (called the "spin trap") that reacts with free radical species that are normally not detectable under normal conditions. Examples of free radicals detected using this method are hydroxyl radical ($^{\bullet}OH$), superoxide ($O_2^{\bullet -}$), peroxyl ($ROO^{\bullet}$), and $^{\bullet}NO$ (29). Commonly employed spin traps for oxygen radicals include DMPO, DEPMPO, and PBN (29). EPR imaging of free radicals can be performed using spin traps, as well as probes that interact with free radicals. However, in biological tissues the concentration of the spin adduct will not be sufficient for the imaging of these radical adducts.

The following sections focus on the use of EPR imaging for four major applications: (i) imaging of free radical probes, (ii) $^{\bullet}NO$ imaging, (iii) redox imaging, and (iv) oxygen imaging.

2. Imaging of Free Radical Probes – Specific Applications/ Examples

2.1. Imaging of Free Radicals in the Heart

Cardiac ischemia leading to secondary myocardial infarction is among the most common causes of morbidity and mortality. Chemical or surgical interventions allow the recovery of the ischemic myocardium by restoration of blood flow or reperfusion. This reperfusion, however, is known to be associated with ventricular arrhythmias and myocardial dysfunction that can lead to severe cardiac impairment and cell death (30–33). ROS such as $^{\bullet}OH$, $O_2^{\bullet -}$, hydrogen peroxide (H_2O_2), and singlet oxygen ($^1O_2$) have been implicated as important factors in the pathogenesis of cellular injury in the postischemic heart (34–37). These species have been shown to mediate the contractile dysfunction observed during reperfusion and may be implicated in reoxygenation injury (38–40). Most of the evidences for the free radical generation in the heart were indirect and based on the beneficial effects of free radical scavengers in animal models. There was a great need for techniques capable of direct measurement of free radical generation in experimental models of this disease (41).

The isolated, perfused heart model is an important and versatile tool that is commonly used to study normal cardiac physiology and the basic mechanisms of cardiac disease (37). There is extensive evidence that free radical generation and metabolism is greatly altered by ischemia-reperfusion, and there has been a great need to be able to measure and map alterations in myocardial radical generation and metabolism as well as tissue oxygenation in this setting. It has been demonstrated that alterations in $^{\bullet}NO$ generation also occur, and there has been much interest in being able to measure and image this process (42, 43). Thus, EPR spectroscopy and imaging studies of the isolated heart offered the unique potential to provide important insights into the basic mechanisms of injury to the heart during and following a heart attack.

2.2. Imaging of Ischemic Heart Using a Free Radical Probe

Three-dimensional (3D) spatial imaging is necessary to obtain a complete, unambiguous image of an asymmetric object. However, in the past, most of the EPR imaging experiments were performed only in two dimensions (2D), and in some cases making use of the axial symmetry of the object to avoid the need for a third gradient and the extra time required for data acquisition. A 2D image superimposes all of the slices along the third dimension onto the plane of projection. This naturally integrates the information along the third axis and provides only a 2D projection of the image. An example of the 3D image from a rat heart obtained with glucose char (paramagnetic contrast) label is shown in **Fig. 1.1** (13). After 15 min of normal perfusion, an aqueous suspension of glucose char label was infused and the heart subjected to no-flow global ischemia. EPR spectra were continuously measured to monitor the sharpening of the signal due to the decrease in O_2 concentration. The heart was then imaged by collecting 1,024 projections. In the compiled 3D images shown in **Figs. 1.2** and **1.3**, the external shape of the epicardium, large vessels including the aorta and pulmonary arteries are clearly seen and corresponded with the visually observed external surface of the rat heart. The internal endocardial surface of the left ventricle could also be clearly seen, and within the image, the ascending aorta, aortic root, the left main coronary artery, the bifurcation of the left anterior descending coronary, and the circumflex coronary arteries could be seen. The left anterior descending coronary artery could be observed down to a diameter of 0.2 mm.

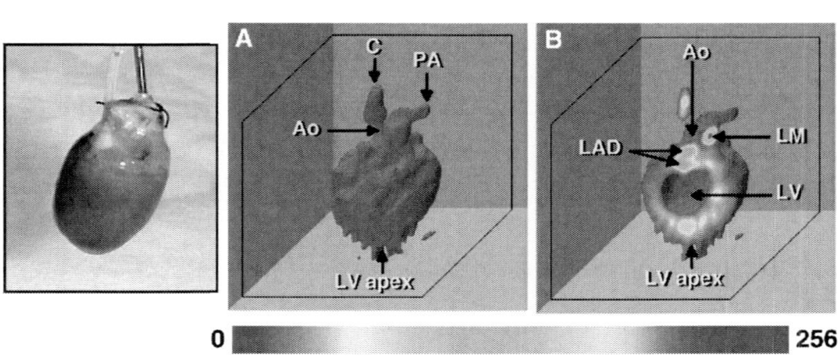

Fig. 1.1. Three-dimensional image of rat heart infused with glucose char suspension. (**a**) Full view of the heart; (**b**) a longitudinal cutout showing the internal structure of the heart; photograph of an isolated perfused rat heart is shown on the *left*. Legends: C, cannula; Ao, aortic root; PA, pulmonary artery; LM, left main coronary artery; LAD, left anterior descending artery; LV, left ventricular cavity. The void seen in the LV cavity is due to the inflated balloon. Image acquisition parameters: projections, 1,024; magnetic field gradient, 50.0 G/cm; acquisition time, 78 min (Reprinted with permission from Kuppusamy et al. (13)).

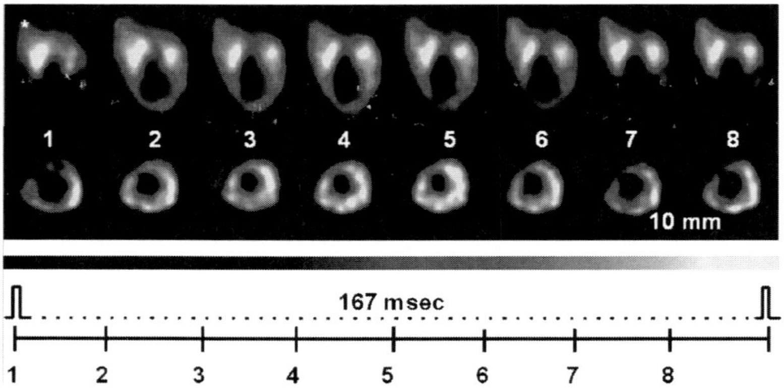

Fig. 1.2. Three-dimensional spatial images of a beating heart. A mid-vertical slice (*top*) and a transverse slice (*bottom*) through the LV cavity are shown for 8 out of the 16 three-dimensional images of the perfused heart as a function of cardiac cycle. The pacing frequency was 6 Hz. The data acquisition parameters were number of gates, 16; number of field points, 64; projections, 144; gradient, 20 G/cm; time constant, 1.2 ms; acquisition time, 64 min (Reprinted with permission from Kuppusamy et al. (8)).

Imaging of nitroxyl probes in hearts subjected to long durations of ischemia has also been reported (11).

2.3. Gated Cardiac Imaging of Beating Heart

EPR imaging of beating hearts is faced with many challenges. Of the constraints that limit or compromise application of EPR imaging, the problems associated with organ movement, such as the contractile motion of the heart or respiratory motion with breathing, have considerably limited applications to living systems where motion occurs during the process of data acquisition. Thus in vivo EPR spectroscopy and imaging studies to date have provided only time-averaged information. This results in a loss of information regarding the temporal and spatial changes. While random motional artifacts are difficult to control, periodic motions such as heart beat can be controlled by pacing at a fixed frequency and synchronizing the data acquisition system to that frequency, a process known as gated-acquisition. We have developed instrumentation capable of performing gated imaging measurements on perfused beating rat hearts (8). The instrumentation is capable of performing gated-acquisitions of up to 256 images per cycle, with rates of up to 16 Hz. Thus, a temporal resolution of 4,096 Hz is possible at this maximum rate. We used 6 Hz (τ = 167 ms) pacing for perfused rat hearts and collected 16 points per cardiac cycle for 64 field steps. The typical data acquisition time was 20–25 s per spectrum.

Female Sprague-Dawley rat hearts were perfused by the Langendorff method with a modified Krebs bicarbonate perfusate. The aluminum support tube of the aortic cannula served as one of the pacing electrodes, while a copper wire connected to the ventricular wall functioned as the other electrode. The heart was

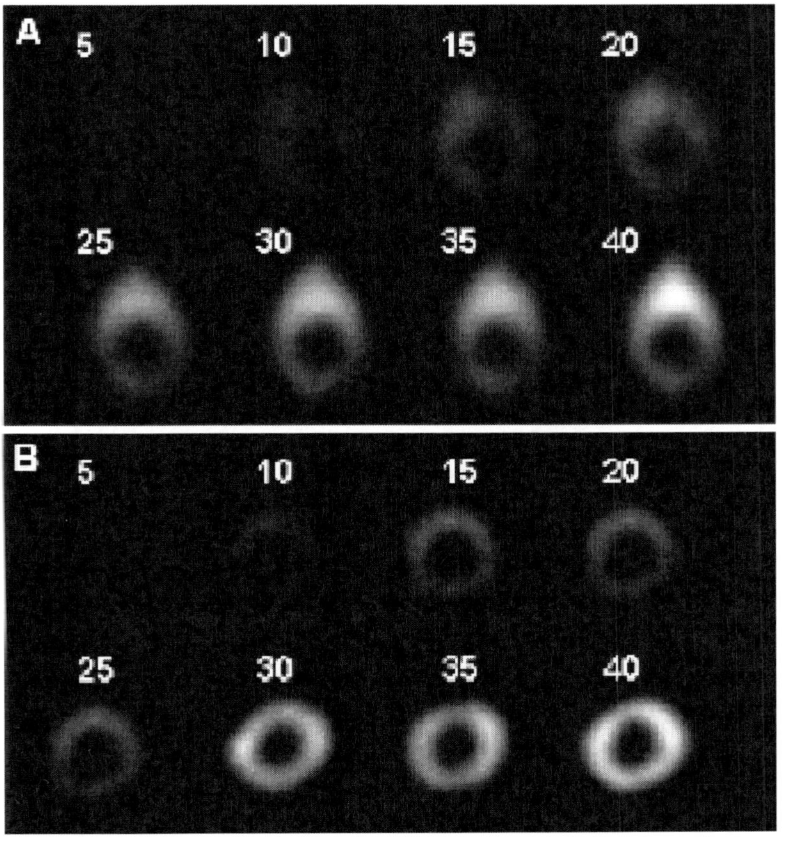

Fig. 1.3. Images of $(MGD)_2$–Fe(II)–NO in the rat heart as a function of ischemic duration. A time-course series (as indicated in min of ischemia) of 25 × 25 mm^2 longitudinal (**a**) and transverse slices obtained from 3D image of a heart loaded with 2 mM $(MGD)_2$–Fe(II) and 10 mM nitrite and subjected to no-flow global ischemia at room temperature are shown. The image acquisition time was 5 min. No generation in the RV myocardium is fourfold lower than in the LV (Reprinted with permission from Kuppusamy et al. (42)).

paced at 360 bpm (6 Hz) with an electrical stimulator using a pulse of 5 V and 7 ms duration. EPR imaging measurements were performed on these hearts, while maintaining continuous pacing and perfusion.

After 15 min of equilibration, the heart was transferred to the resonator and the perfusate solution was switched to that containing 1 mM 4-oxo-2,2,5,5,-tetramethylpiperidine-d$_{16}$-*N*-oxyl or perdeuterated tempone (PDT). Gated projections were acquired using a magnetic field gradient of 20 G/cm. A total of 144 projections were acquired, decomposed into 16 data sets, and the images reconstructed. Vertical and transverse slices of 8 out of 16 images of the heart are shown as a function of cardiac cycle in **Fig. 1.2**. The contraction–relaxation cycle is clearly seen in these images. The systolic and end-diastolic pressures during the cycle were 120 and 8 mmHg, respectively. The aorta is identified at the top-left corner of the vertical slices in **Fig. 1.2**. The aortic cannula was not visible in these images due to the aluminum tubing that was used in the cannula. The LV cavity is clearly seen as the central void and the two bright spots appear to correspond to the proximal coronary arteries. During systole, the LV cavity clearly narrows with vertical elongation and the LV wall also is seen to markedly thicken.

3. Imaging of Nitric Oxide

Nitric oxide is a gaseous paramagnetic molecule that is endogenously produced by a variety of mammalian cells. It regulates blood pressure, smooth muscle relaxation, neuronal signaling, and immune response. Increased generation of $^{\bullet}NO$ can lead to cell injury and cell death by excessive binding to heme proteins or reaction with $O_2^{\bullet -}$, leading to the formation of the potent oxidant peroxynitrite. There is considerable interest in the scientific community for measuring and imaging $^{\bullet}NO$ in order to understand its role in various disease states.

Nitric oxide exhibits a strong EPR signal in the gas phase (44). However, due to fast relaxation of the electron spin, no signal is observed from $^{\bullet}NO$ in solution or under physiological conditions. Detection of $^{\bullet}NO$ by EPR spectroscopy is accomplished using spin traps. The method involves trapping of $^{\bullet}NO$ by iron complexes such as heme in hemoglobin or iron (II) dithiocarbamates (45). The use of dithiocarbamate–Fe(II) complexes was first reported by Mordvintcev et al. (46) using *N,N*-diethyldithiocarbamate (DETC) in the form of the $(DETC)_2$–Fe(II) complex, and then shortly thereafter by Lai and Komarov (47) using the water-soluble *N*-methyl-D-glucamine dithiocarbamate (MGD) as the corresponding $(MGD)_2$–Fe(II) complex. The EPR spectrum of low-spin $(MGD)_2$–Fe(II)–NO complex yields a characteristic three-line spectrum. There are no background contributions from soluble or gaseous $^{\bullet}NO$ or from the diamagnetic $(MGD)_2$–Fe(II) complex.

3.1. Imaging of Nitric Oxide Generation in the Heart

We have shown that rat hearts subjected to global ischemia generate $^{\bullet}NO$ via an enzyme-independent pathway involving direct reduction of nitrite under the acidic and reducing conditions that occur during myocardial ischemia (48, 49). In view of the important implications of this enzyme-independent mechanism of $^{\bullet}NO$ generation on the pathogenesis and treatment of tissue injury, we performed real time isotope tracer measurements of the mechanism of $^{\bullet}NO$ generation in the heart.

Isolated rat hearts were loaded with the •NO trap, $(MGD)_2$–Fe(II), and ^{15}N-nitrite and subjected to global no-flow ischemia. No signal was observed at the onset of ischemia; however, a prominent doublet signal characteristic of the $(MGD)_2$–Fe(II)–NO complex appeared after a period of time. We mapped the spatial distributions of this •NO generation in 3D in the ischemic myocardium using L-band EPR spectroscopy (42). The images (**Fig. 1.3**) clearly showed that •NO is formed throughout the myocardium, enabling visualization of the external shape of the epicardium, right ventricular (RV) myocardium, and internal endocardial surface of the left ventricle and left ventricular chamber. Kinetic experiments showed that maximum •NO generation and trapping occurred at the midmyocardium and spreads out to the endocardium and epicardium of the left ventricle. The magnitude of generation in the right ventricle myocardium was fourfold lower than in the left ventricle. Thus, real time kinetics and 3D mapping of •NO generation could be performed in whole biological organs using EPR imaging.

3.2. Imaging of NO Generation in Mice Subjected to Cardiopulmonary Arrest

We have performed spectroscopic measurements of •NO generation in a whole-body murine model subjected to cardiopulmonary arrest (9). Mice were infused (i.v.) with 70 mg/kg of nitrite (to augment the tissue levels of nitrite) and cardiopulmonary arrest was induced 5 min later with an overdose of pentobarbital. The mice were placed in the L-band resonator and EPR spectra were acquired. Approximately 15 min after the induction of cardiopulmonary arrest, a triplet signal with hyperfine coupling constants 15.9 G and 17.4 G was observed (**Fig. 1.4**). The signal continued to grow with time and reached a plateau in about 2 h. Control mice in which arrest was induced, but received only saline, did not show any characteristic EPR spectrum, suggesting that any •NO produced in the animal in the absence of added nitrite was below the detection level (1 mM) of the L-band spectrometer. To distinguish whether the observed •NO (signal) is derived from the infused nitrite, measurements were performed in mice infused with 70 mg/kg of ^{15}N-labeled nitrite. A doublet spectrum with hyperfine coupling constants 23.3 G was observed in mice labeled with ^{15}N-nitrite (**Fig. 1.4**). Since the ^{15}N has a nuclear spin of $^1/_2$, a doublet spectrum is expected from the ^{15}NO isotope. The doublet signal and the control spectrum confirm that the observed •NO originated from the infused nitrite, and not from endogenous tissue nitrite or from enzyme-mediated reactions.

In order to visualize the distribution of the •NO complex in different organs, we performed spatial EPR imaging measurements using the L-band imaging instrumentation. Mice were labeled with ^{15}N-nitrite and cardiopulmonary arrest was induced 5 min later with an overdose of pentobarbital. The animal was then placed inside the L-band EPR resonator and serial spectral

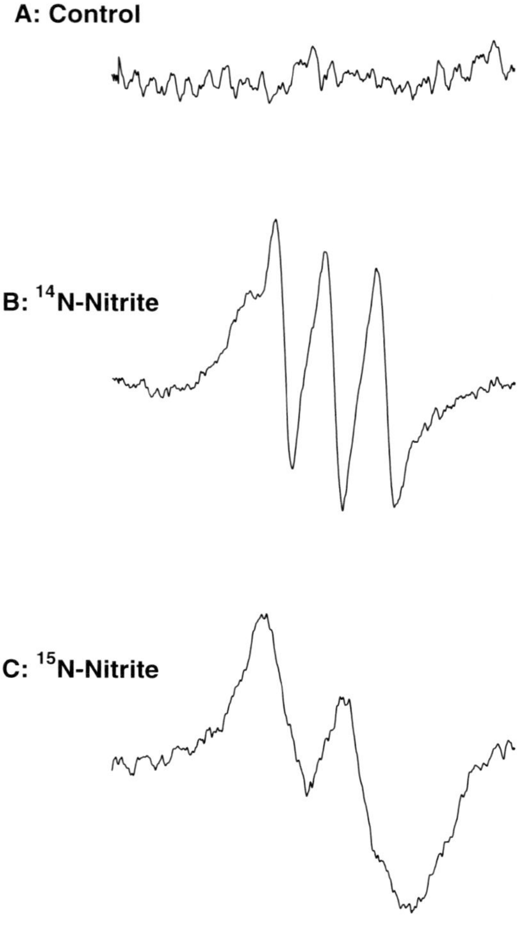

Fig. 1.4. In vivo L-band EPR spectrum of •NO measured in mice subjected to cardiopulmonary arrest. The mice were labeled (i.v.) with saline (**a**), 70 mg/kg (b.w.) of ^{14}N-nitrite (**b**), or 70 mg/kg ^{15}N-nitrite (**c**) and subjected to cardiopulmonary arrest with an overdose of pentobarbital 5 min later. The spectra were measured from the thoracic region of the intact animal after 120 min of cardiopulmonary arrest. The spectrum (**b**) is a triplet with ^{14}N(I = 1) hyperfine coupling constants 15.9 G and 17.4 G from a ^{14}NO–heme complex, while spectrum (**c**) is a doublet with ^{15}N(I = ½) hyperfine coupling constants 23.3 G from a ^{15}NO–heme complex. Spectra were recorded at ambient temperature with the spectrometer settings: microwave frequency, 1.322 GHz; microwave power, 10 mW; modulation field, 2.0 G; modulation frequency, 100 kHz; scan time, 5 min (Reprinted with permission from Kuppusamy et al. (9)).

data were acquired to monitor the growth of the heme–NO signal. Image projection acquisitions were started 2 h after induction of cardiopulmonary arrest.

A total of 144 projections were acquired from a volume of $40 \times 40 \times 40$ mm^3 in the object using a constant magnetic field gradient of 25 G/cm. The projection data were deconvoluted and the image was reconstructed using filtered-backprojection methods. Three-dimensional spatial images of the •NO complex in the animal were acquired in the head, thoracic, and abdomen regions of the intact animal. A representative image obtained in the thoracic region is shown in **Fig. 1.5**. The 3D composite image in the figure clearly shows the accumulation of the heme–NO complex primarily in the lungs and heart, and to a lesser extent in the liver and great vessels.

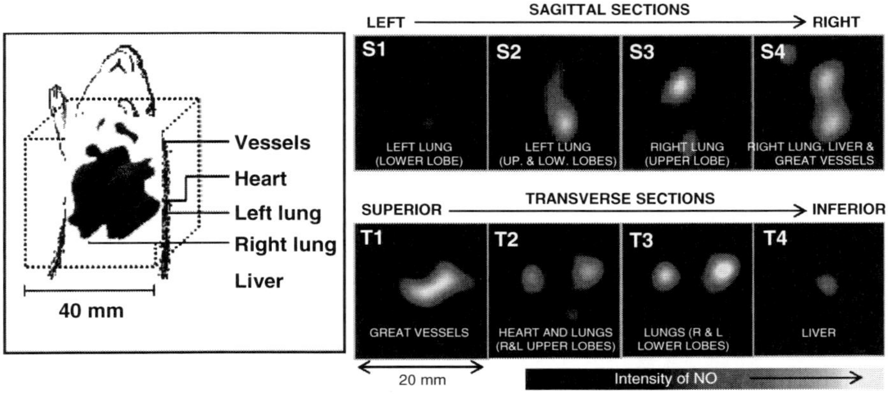

Fig. 1.5. Three-dimensional EPR image of •NO in the thoracic region of a mouse subjected to cardiopulmonary arrest. The mouse was labeled with 70 mg/kg (b.w.) of nitrite (i.v.) and subjected to cardiopulmonary arrest with an overdose of pentobarbital 5 min later. The image was obtained on the whole body intact animal after 2 h of cardiopulmonary arrest. (*Left*) 3D image of •NO in the thoracic region. The sketch shows the position of the image within the body of the mouse. The lungs, heart, great vessels, and liver are seen in the image. Image acquisition parameters: gradient, 25 G/cm; projections, 144. The images are rendered at 20% background transparency. (*Right panes*) Sagittal and transverse sections (20×20 mm^2) of •NO distribution obtained in the thoracic region. S1–S4, sagittal slices from left to right of the thoracic region. T1–T4, transverse slices from superior to inferior of the thoracic region. The following features are observed in the images: S1, left lung (lower lobe); S2, left lung; S3, right lung (upper lobe); S4, great vessels, right lung, and liver; T1, great vessels; T2, heart and lungs (upper lobes); T3, lungs (lower lobes); T4, liver (Reprinted with permission from Kuppusamy et al. (9)).

Figure 1.5 also shows some of the transverse and sagittal slices obtained from the 3D image. The •NO distributions observed in the slices correlate well with the anatomical structure of the major organs in the thoracic region. It should be noted that the images obtained from the head or abdominal regions of these animals did not show any meaningful structure, suggesting that the signal intensity was too small for imaging.

The studies presented show that increased amounts of nitrosoheme complexes are also formed as early as 15 min after the onset of cardiopulmonary arrest. The maximum rate of nitrosoheme complex formation was seen within the first 30 min following the onset of cardiopulmonary arrest. Thus, this mechanism of increased •NO and mononitrosyl-heme formation also occurs in the early minutes following cardiopulmonary arrest and may contribute to the increasing difficulty with resuscitation after long periods of arrest. This mechanism of •NO generation may have important implications in the understanding of the pathogenesis and the treatment of tissue injury in this setting.

3.3. Imaging of •NO Generation in the Brain

By the systemic administration of diethyldithiocarbamate and iron into the rat, •NO produced in the brain during ischemia-hypoxia was trapped. The right hemisphere of the brain was then removed and frozen with liquid nitrogen. 3D imaging of the production of •NO in the brain was performed (43). The results suggested that •NO molecules were produced and trapped in the areas that are known to have high nitric oxide synthase activity, such as cortex, hippocampus, hypothalamus, amygdala, and substantia nigra. In this ischemia-hypoxia model, which did not interrupt the posterior circulation, the production and trapping of •NO in the cerebellum was approximately 30% of that in the cerebrum.

Despite these published reports, there are several problems associated with imaging of •NO in vivo using EPR. One problem is that the spin adduct, for example, the $(MGD)_2$–Fe(II)–NO, has a large linewidth (3.5 G), which results in poor spatial resolution. Other problems include the need to use relatively high concentration of the trap agents (e.g., iron), solubility, stability, and non-specific reactions of the trap in biological systems.

4. Redox Imaging

The term "redox environment" is usually used when dealing with biological systems because they have numerous linked redox couples (50). Normally, the biological environment of an organism is maintained in redox homeostasis regulated by several factors such as level of O_2, thiols, and enzymatic pathways. Alteration in any of these factors due to disease will cause changes in the redox environment of the cells and tissues. It is important to be able to detect and visualize such changes, especially in systems like tumors, where the redox environment plays a role in the efficacy of treatment. The development of low-frequency EPR (1.0 GHz and lower) has enabled mapping and imaging of the tumor redox environment (51–53).

4.1. Nitroxyls as Redox Probes for EPR Imaging

Low-molecular weight stable nitroxyl free radicals are useful as spin probe contrast agents in functional imaging studies (54, 55). EPR imaging, using the nitroxyls, can provide functional information on the global redox status in experimental animal models of pathological conditions, in vivo. Two critical properties of nitroxyls provide such capability: (i) nitroxyl labels participate in redox reactions (**Fig. 1.6**) where the nitroxyl (paramagnetic, EPR detectable) is reduced to the corresponding hydroxylamine (diamagnetic, EPR "silent"), and vice versa,

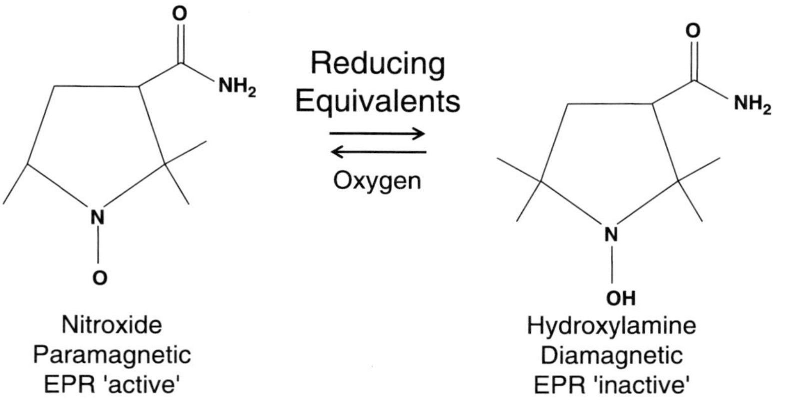

Fig. 1.6. Structure of 3-CP and its tissue metabolite. The "EPR active" 3-CP (3-carbamoyl-2,2,5,5-tetramethylpyrrolidine-*N*-oxyl, or 3-carbamoyl-proxyl) nitroxyl probe undergoes one-electron reduction in tissues to "EPR-silent" hydroxylamine form.

establishing equilibrium between the two species (56, 57); (ii) in cells, the nitroxyls undergo reduction to the corresponding hydroxylamines more efficiently under hypoxic conditions than under aerobic conditions via intracellular enzymatic processes (58). When nitroxyls are administered in vivo, a rapid equilibrium is established between the levels of nitroxyl and hydroxylamine. The level of the nitroxyl, detectable in vivo, is independent of whether the nitroxyl or the hydroxylamine is administered. Based on the above-mentioned properties, several studies were carried out using stable nitroxyls as probes in EPR spectroscopy and imaging experiments to obtain tissue/tumor morphology and redox status.

4.2. Imaging of the Nitroxyl Pharmacokinetics in Tumor

Since tumor tissues are characterized by significant heterogeneities in terms of redox status and oxygenation, it is desirable to obtain spatially resolved images of nitroxyl distribution and clearance simultaneously within the tumor volume. **Figure 1.7** shows a few selected images from a series of 2D spatial maps of nitroxyl content in normal and buthionine sulfoximine (BSO)-treated tumor tissues obtained as a function of time after 3-CP infusion. Each time-course image within the series, as shown in the figure, was normalized with respect to the maximum intensity obtained within that series, usually in the images collected in about 4–6 min after infusion. BSO-treated animals demonstrated a slower rate of reduction of nitroxyl in the tumor compared to untreated mice. The measurements clearly suggest that the nitroxyl is reduced more efficiently in the tumor than in normal tissue. However, the tumor may contain variable regions of redox state, depending on a variety of factors including differences in reducing equivalents, oxygenation, and pH.

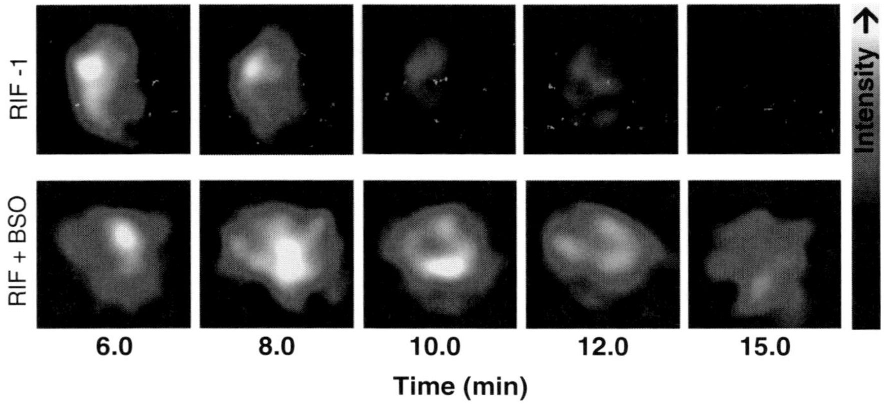

Fig. 1.7. Spatially resolved clearance of nitroxyl in RIF-1 tumor tissue. After tail vein infusion of 3-CP, a series of two-dimensional images of the nitroxyl from tumor (untreated and BSO-treated) were measured using L-band EPRI method. A few selected images and the corresponding approximate time after infusion are shown. The images represent the mean nitroxyl concentration in a two-dimensional projection of the tissue volume (10×10 mm^2; depth, 5 mm) averaged over 1.5–2.0 min. The image data were acquired using a magnetic field gradient of 15 G/cm at 16 orientations in the two-dimensional plane. Each image within a series was normalized with respect to the maximum intensity in that series. The nitroxyl in the tumor of BSO-treated mouse persisted longer, compared with that in the untreated mouse.

4.3. Redox Mapping of Tumor

In order to obtain spatially resolved information regarding the reduction rate constants of nitroxyls in the tumor, we developed algorithms for direct measurement and mapping of the rate constants (redox mapping) (53, 59). The "redox mapping" method is based on the following principle (**Fig. 1.8**). Spatial imaging of

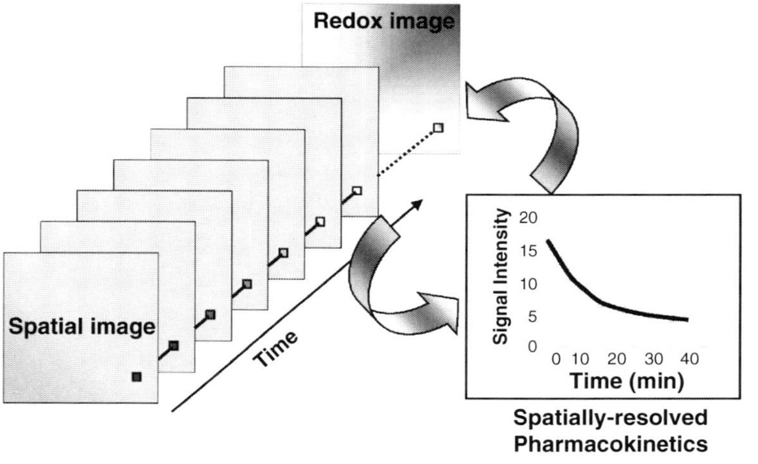

Fig. 1.8. Reconstruction of redox image. The procedure of obtaining redox image from a 2D pharmacokinetic image data is illustrated. Nitroxyl intensity from a given voxel within the image is followed as a function of time to obtain the reduction profile (plot) versus *t*. The profile is modeled with an appropriate kinetic expression, usually a pseudo first-order decay process, to obtain rate constant at each voxel. The computation is repeated over all the voxels to reconstruct the redox image.

nitroxyl in the tumor is performed to obtain time-series images. The time-dependent intensity of the nitroxyl in a given voxel is obtained from the images and fitted to a pseudo first-order kinetic rate equation to compute the rate constant.

The procedure will be repeated for all voxels in the image to obtain rate constants for each of the individual voxels. The resulting array of rate constants is then displayed on a color scale or frequency (histogram) plot. An appropriate threshold of intensity level is defined to exclude those regions where the nitroxyl concentration is too small to make calculations. This means that areas with nitroxyl intensity above the threshold level provide spatially resolved visualization (mapping) of the rate constants in the tissue.

4.4. Redox Mapping of RIF-1 Tumor: Effect of BSO Treatment

The rate constants for the clearance of the probe in each pixel of the pharmacokinetic images in **Fig. 1.7** were computed and displayed in the form a color-coded image and frequency histogram in **Fig. 1.9** (52). The plot shows the presence of a range

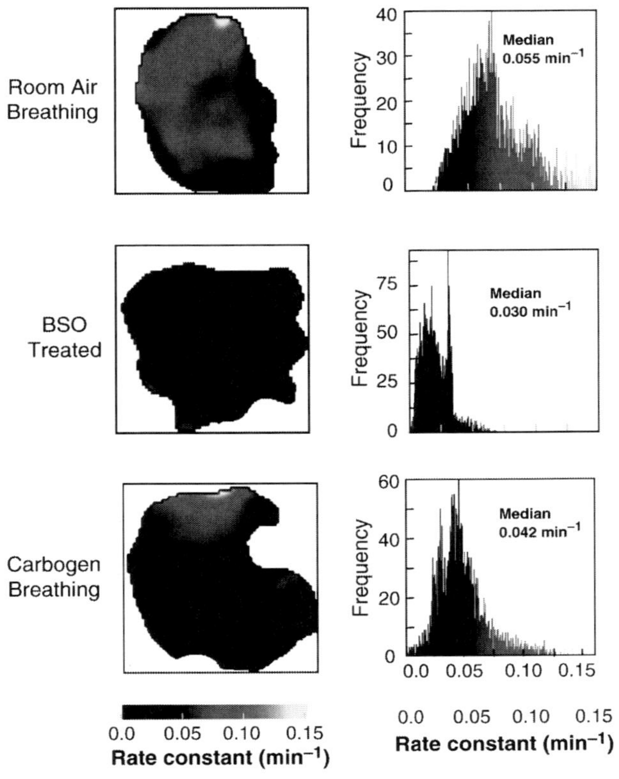

Fig. 1.9. Redox mapping of tumor. Two-dimensional spatial mapping of pseudo first-order rate constants (*left panels*) and frequency plot (*right panels*) of the nitroxide reduction rate constants in the RIF-1 tumors of untreated (air-breathing), BSO-treated, and carbogen-breathing mice were obtained from the time-course image data similar to those shown in **Fig. 1.7**.

of rate constants, which is a measure of heterogeneity in the redox status within each tissue. The redox data show that the BSO treatment significantly decreases the rate constant of nitroxyl reduction in the tumor. It is known that BSO inhibits the γ-glutamylcysteine synthetase enzyme that is responsible for GSH synthesis (60). Studies have shown that tumor tissue levels of GSH can be depleted to very low levels without any toxicity (61). The observed effect of BSO treatment on the redox constants is, therefore, attributed to alterations in the tissue GSH levels. In order to investigate the correlation between the observed rate constants and tissue GSH, the tissue levels of GSH were measured in the tissues of normal (control) and BSO-treated (same dose as for the EPR measurements) mice using EC-HPLC. The BSO-treated tissue data showed a significant reduction in the GSH levels in the tumor tissue when compared to the untreated control (52). On the other hand, no significant differences were observed between treated and untreated normal tissues. Thus, it is evident that BSO causes a differential depletion among normal and tumor tissues.

4.5. Redox Mapping of RIF-1 Tumor: Effect of Carbogen-Breathing

Figure 1.9 also shows redox images obtained from RIF-1 tumor-bearing mice subjected to carbogen (95% O_2 + 5% CO_2 breathing (51)). The redox map indicates the spatially resolved rates of nitroxyl depletion within the tumor and the differences in the intensities noticed in the map indicate the spatial heterogeneity in the tumor redox activity. From this plot, it is clear that the rate constants in the tumor of an air-breathing mouse are distributed widely with a median value of 0.055 min^{-1}. Similar measurements of EPR imaging pharmacokinetics were made when the animal was breathing carbogen. The carbogen-breathing data revealed marked differences in the magnitude and distribution of the rate constants with a median value of 0.042 min^{-1}. The distribution of the rate constants is significantly narrowed and less variable when compared to the air-breathing animal. These results indicate that the redox rate constants are significantly heterogeneous in the case of air-breathing animal. On the other hand, the distribution is significantly narrowed during carbogen-breathing indicating that carbogen-breathing decreases the tumor redox activity with less heterogeneity in the distribution of the redox state.

The heterogeneity in the redox status may arise as a consequence of the heterogeneity in perfusion or O_2 concentration in the tumor. The diminished heterogeneity of the tumor redox status in the case of carbogen-breathing may be due to elimination of diffusion-limited hypoxia, thereby rendering the tumor more homogeneously oxygenated.

4.6. Imaging Nitroxyls Metabolism in Human Skin

While altered cellular free radicals and redox metabolism are critical factors in many human diseases, it has not been previously possible to both measure and image these processes in humans. Using a custom-built 2.2-GHz microwave bridge, we performed noninvasive measurements of the distribution and metabolism of the topically applied nitroxide, ^{15}N-perdeuterated tempone, in human forearm skin (62). The penetration of the label into the skin and its metabolic clearance were modeled using kinetic equations.

It was observed that the penetration process from the skin surface into the dermis and subcutaneous regions, as well as its clearance from these regions, could be described by single exponential functions. Phantom imaging experiments using the nitroxide showed that a spatial resolution of up to 50 μm could be achieved. The skin imaging measurements showed two bands in the distribution of the label along the skin depth. The first band appeared in the outer 400 μm of the skin, the epidermis region, whereas the second band was centered at a depth of 1,000 μm in the subcutaneous region with a thickness about 400 μm. These two bands decayed and merged into a single band with time. The experiments clearly demonstrated that EPR imaging enable noninvasive detection and mapping of free radicals in the human skin. This technique can provide important data regarding redox state and free radical metabolism in the skin and thus has great promise in the study of dermatological disease and the development of skin treatment formulations.

5. Oxygen Imaging

5.1. EPR Oximetry

Measurement of O_2 concentration (oximetry) can be done using EPR spectroscopy (63). EPR oximetry refers to the measurement of oxygen concentration by EPR spectroscopy or imaging. The principle of EPR oximetry is based on the paramagnetic characteristics of molecular oxygen, which in its ground state has two unpaired electrons and undergoes spin exchange interaction with the paramagnetic EPR spin probe. This process is sensitive to oxygen content, with the relaxation rate of spin probe increasing as a function of oxygen content (concentration/pressure). This increased spin–spin relaxation rate results in increased line-broadening. According to Smoluchowsky theory, which is applicable to soluble oximetry probes such as nitroxyls, the oxygen-broadened linewidth is directly proportional to the O_2 concentration. The fact that the linewidths of EPR resonance spectra correlate with O_2 concentration has been used in a variety of biological settings.

5.2. Probes for EPR Oximetry

EPR oximetry requires the presence of a suitable EPR probe. In general, particulate probes such as lithium phthalocyanine (7, 51, 64) are suitable for measurements of O_2 partial pressure (pO_2), while soluble probes such as nitroxyls and trityl molecules measure dissolved O_2 concentration. The soluble oximetry probes are noninvasive; they can be injected into the animal intraperitoneally or intravenously, distributed into a variety of pharmacological compartments, and structurally modified to target a specific distribution compartment. They report the average O_2 concentration as well as the concentration of the probe in the pharmacological compartment. On the other hand, the particulate probes have to be surgically inserted into tissues and report the pO_2 in a tiny region of the total tissue volume. The advantages in using the particulate probes as compared to the soluble probes are the following: (i) higher resolution – resolutions in the range of 0.1 Torr (mmHg) can be obtained; (ii) suitability for repeated measurements in vivo without reintroduction of the probe into the tissue; (iii) noninvasive measurement – it may require one-time introduction of the probe, but subsequent measurements are performed under noninvasive conditions; (iv) accuracy – highly reproducible and correlating closely with measurements by other methods; (v) localized measurements – the measurement is made from a single voxel/region containing the particulate, thus the spatial resolution is the size of the particulate deposit; (vi) insolubility in aqueous solvents; (vii) no effect of the various biological oxidoreductants, pH, temperature, etc.; (viii) nontoxic – the probes are very inert in biological systems; (ix) temporal response is very good, usually less than 1 s; and (x) respond to pO_2, rather than concentration of O_2, which may be quite heterogeneous in cellular/tissue environment making calibration difficult.

5.3. In Vivo Mapping of O_2 Concentration

Mapping of O_2 concentration using EPR imaging involves the use of spectral–spatial (spectroscopic) imaging techniques, where the image contains not only spatial information (the spatial distribution of spin density) but also spectral information (spatial distribution of O_2 concentration). The method requires stepped field gradients, and the resulting image has information on one, two, or three spatial dimensions and one spectral dimension (10). Thus, a 4D spectral–spatial image provides information about the signal amplitude along three spatial dimensions and the linewidth information along the spectral dimension.

5.4. Visualization of the Emergence of Hypoxia in a Growing Tumor

We have developed a novel procedure for in situ monitoring of O_2 concentration in the microenvironment of growing tumors by EPR-based oximetry using embedded paramagnetic particulates (7). Unlike the existing methods of O_2 measurement, wherein the oximetry probes (needle electrodes, optical probes, or EPR implants) are physically inserted during the measurement, the new approach uses spin probes that are permanently embedded or implanted in the tumor. A particular advantage of this procedure is that it is noninvasive, both in terms of implantation of the probe and in oximetry measurements.

The toxicity of LiPc to RIF-1 cells was studied in vitro by cell proliferation and clonogenic assays (7). Treatment of 1×10^6 cells with 1 mg of LiPc for 2 h did not show any significant effect on the membrane integrity of the cells as measured by Trypan blue dye uptake. Cell proliferation studies performed with 1×10^6 cells treated with 1 mg of LiPc for 24 h followed by 4 days of incubation of cells did not show any significant effect. We also performed clonogenic assays on cells treated with LiPc and found that there was no significant change in the colony forming ability of the cells. The in vitro study clearly suggested that the LiPc particles have no apparent toxicity on the RIF-1 tumor cells.

Next, the effect of co-implanted LiPc on tumor growth volume curve was determined. Two groups of mice were implanted with RIF-1 cells with and without LiPc and their tumor size (volume) was measured daily up to 14 days. The tumor volume was observed to increase exponentially in both cases and no statistically significant differences in tumor growth were observed between the two groups (7).

A mixture of RIF-1 tumor cells and microparticulates of LiPc was implanted in the hind leg of C3H mice. This enabled repeated measurements of O_2 concentration from the implanted site for more than 2 weeks during the growth of tumor. The particulates in the tumor were stable and nontoxic. Repeated measurements of EPR imaging were performed to monitor the location and redistribution of the injected particulates during tumor growth. **Figure 1.10a** shows the distribution of the probe in a typical tumor on day 12 (tumor volume $\sim$400 mm^3) after inoculation. **Figure 1.10b and c** shows the O_2 mapping obtained from the tumor. The distribution of O_2 concentration in the tumor in room air-breathing animal (**Fig. 1.10b**) shows that the tumor is significantly hypoxic and heterogeneous with respect to oxygenation. The O_2 distribution was mostly $<$10 mmHg. We further imaged the distribution of O_2 concentration in the tumor under carbogen-breathing conditions (**Fig. 1.10c**). The image showed that O_2 concentration was increased overall, with maximum around 20 mmHg.

5.5. Prospects of Oxygen Imaging Using EPR

EPR oximetry is a minimally invasive method that reports direct and absolute pO_2 or O_2 concentration in intact biological tissue. It has a high sensitivity to pO_2 and can provide localized mapping, if needed. There is very little interference from other sources

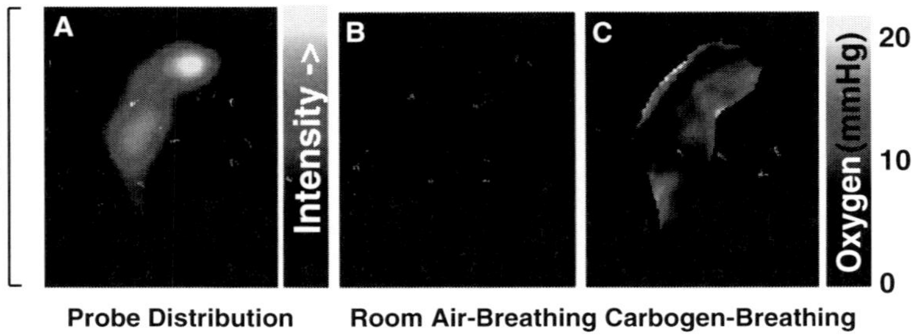

Fig. 1.10. EPR images of LiPc and O_2 in the tumor. (**a**) Distribution of the probe in a RIF-1 tumor. (**b**) Distribution of O_2 concentration in the same tumor under room air-breathing conditions. (**c**) Distribution of O_2 concentration in the same tumor under carbogen-breathing conditions. The tumor size at the time of measurement was $8 \times 10 \times 11$ mm^3 (Reprinted with permission from Ilangovan et al. (7)).

(high specificity), no toxicity, and is not affected by the tissue environment. The probes have reasonable half-lives and adequate distribution within the tissue. Qualitative as well as quantitative pO_2 readings can be obtained using this technique. The most important advantage is that repeated and noninvasive measurements of tissue oxygenation over time are possible. However, this is limited by the duration of probe retention in the tissue (65).

Data acquisition time is a major issue in EPR imaging of O_2. A reasonable 3D spectral–spatial image takes about 30 min to acquire (10). Soluble probes are used for imaging, which have low sensitivity and specificity when compared to particulate probes. However, improvements in image reconstruction algorithms have considerably reduced data acquisition time for spatial imaging and these will be extended to spectral–spatial imaging in the near future (66, 67). In biological tissues, the penetration depth of the microwave energy is limited to a few millimeters at 1.2 GHz (L-band), but is higher at lower frequencies.

6. Conclusions

We have provided a summary of the EPR technology that we have developed for the imaging of free radicals in in vivo biological systems. The technology has progressed to the point of enabling unique information to be obtained with spatial mapping of free radical metabolism, oxygenation, and •NO generation in a number of different applications. In the future, advances in probe development, instrumentation, and data acquisition and processing algorithms will allow us to build upon this work in the hopes of one day bringing EPR imaging capability to the clinical realm.

Acknowledgments

The EPR imaging results illustrated in the review were composed from several publications from our laboratory. We gratefully acknowledge the publishers of the respective journals for permitting us to reproduce the figures used in this article. The work was supported by the NIH grants CA102264, EB004031, and EB005004.

References

1. Ansari, K.N. (1997) The free radicals – the hidden culprits – an update. *Indian J. Med. Sci.* **51**, 319–336.
2. Kuppusamy, P. (2004) EPR spectroscopy in biology and medicine. *Antioxid. Redox Signal* **6**, 583–585.
3. Zavoisky, E. (1945) Spin-magnetic resonance in paramagnetics. *J. Phys. E.* **9**, 245–249.
4. Halpern, H.J., Spencer, D.P., Polen, J.V., Bowman, M.K., Nelson, A.C., Dowey, E.M., and Teicher, B.A. (1989) Imaging radiofrequency electron spin resonance spectrometer with high resolution and sensitivity for in vivo measurements. *Rev. Sci. Instrum.* **60**, 1040–1050.
5. He, G., Samouilov, A., Kuppusamy, P., and Zweier, J.L. (2001) In vivo EPR imaging of the distribution and metabolism of nitroxide radicals in human skin. *J. Magn. Reson.* **148**, 155–164.
6. Zweier, J.L. and Kuppusamy, P. (1988) Electron paramagnetic resonance measurements of free radicals in the intact beating heart: A technique for detection and characterization of free radicals in whole biological tissues. *Proc. Natl. Acad. Sci. U.S.A.* **85**, 5703–5707.
7. Ilangovan, G., Bratasz, A., Li, H., Schmalbrock, P., Zweier, J.L., and Kuppusamy, P. (2004) In vivo measurement and imaging of tumor oxygenation using coembedded paramagnetic particulates. *Magn. Reson. Med.* **52**, 650–657.
8. Kuppusamy, P., Chzhan, M., Wang, P., and Zweier, J.L. (1996) Three-dimensional gated EPR imaging of the beating heart: Time-resolved measurements of free radical distribution during the cardiac contractile cycle. *Magn. Reson. Med.* **35**, 323–328.
9. Kuppusamy, P., Shankar, R.A., Roubaud, V.M., and Zweier, J.L. (2001) Whole body detection and imaging of nitric oxide generation in mice following cardiopulmonary arrest: Detection of intrinsic nitrosoheme complexes. *Magn. Reson Med.* **45**, 700–707.
10. Kuppusamy, P., Shankar, R.A., and Zweier, J.L. (1998) In vivo measurement of arterial and venous oxygenation in the rat using 3D spectral-spatial electron paramagnetic resonance imaging. *Phys. Med. Biol.* **43**, 1837–1844.
11. Kuppusamy, P., Wang, P., Zweier, J.L., Krishna, M.C., Mitchell, J.B., Ma, L., Trimble, C.E., and Hsia, C.J. (1996) Electron paramagnetic resonance imaging of rat heart with nitroxide and polynitroxyl-albumin. *Biochemistry* **35**, 7051–7057.
12. Woods, R.K., Hyslop, W.B., Maar, R.B., and Lauterbur, P.C. (1991) Image reconstruction. In: *EPR Imaging and In vivo EPR* (G.R. Eaton, S.S. Eaton, and K. Ohno, Eds.), CRC Press, Boca Raton, FL, pp. 91–117.
13. Kuppusamy, P., Wang, P., and Zweier, J.L. (1995) Three-dimensional spatial EPR imaging of the rat heart. *Magn. Reson. Med.* **34**, 99–105.
14. Deng, Y., He, G., Kuppusamy, P., and Zweier, J.L. (2003) Deconvolution algorithm based on automatic cutoff frequency selection for EPR imaging. *Magn. Reson. Med.* **50**, 444–448.
15. Kuppusamy, P. and Zweier, J.L. (1996) A forward-subtraction procedure for removing hyperfine artifacts in electron paramagnetic resonance imaging. *Magn. Reson. Med.* **35**, 316–322.
16. Kuppusamy, P. and Zweier, J.L. (1996) Hyperfine artifacts in electron paramagnetic resonance imaging. *Res. Chem. Intermed.* **22**, 593–604.
17. Berliner, J.L. and Fujii, H. (1985) Magnetic resonance imaging of biological specimens by electron paramagnetic resonance of nitroxide spin labels. *Science* **227**, 517–519.
18. Berliner, L.J. (1992) Applications of EPR imaging to materials, agriculture and medicine. In: *Magnetic Resonance Microscopy: Methods and Applications in Material Science, Agriculture and Biomedicine* (B. Blumich

and W. Kuhn, Eds.), VCH Publishers, Weinheim, pp. 151–163.
19. Eaton, G.R., Eaton, S.S., and Ohno, K. (1991) *EPR Imaging and In Vivo EPR*, CRC Press, Boca Raton, FL.
20. Kuppusamy, P., Chzhan, M., Vij, K., Shteynbuk, M., Lefer, D.J., Giannella, E., and Zweier, J.L. (1994) Three-dimensional spectral-spatial EPR imaging of free radicals in the heart: A technique for imaging tissue metabolism and oxygenation. *Proc. Natl. Acad. Sci. U.S.A.* **91**, 3388–3392.
21. Kuppusamy, P., Chzhan, M., Wang, P., and Zweier, J.L. (1996) Three-dimensional gated EPR imaging of the beating heart: Time-resolved measurements of free radical distribution during the cardiac contractile cycle. *Magn. Reson. Med.* **35**, 323–328.
22. Kuppusamy, P., Chzhan, M., and Zweier, J.L. (1995) Development and optimization of three-dimensional spatial EPR imaging for biological organs and tissues. *J. Magn. Reson. B* **106**, 122–130.
23. Kuppusamy, P., Chzhan, M., Samouilov, A., Wang, P., and Zweier, J.L. (1995) Mapping the spin-density and lineshape distribution of free radicals using 4D spectral-spatial EPR imaging. *J. Magn. Reson. B* **107**, 116–125.
24. Froncisz, W. and Hyde, J.S. (1982) The loop gap resonator, a new microwave lumped circuit ESR sample structure. *J. Magn. Reson.* **47**, 515–521.
25. Sotgiu, A. (1985) Resonator design for in vivo EPR spectroscopy. *J. Magn. Reson.* **65**, 206–214.
26. Chzhan, M., Kuppusamy, P., and Zweier, J.L. (1995) Development of an electronically tunable L-band resonator for EPR spectroscopy and imaging of biological samples. *J. Magn. Reson. B* **108**, 67–72.
27. Chzhan, M., Shteynbuk, M., Kuppusamy, P., and Zweier, J.L. (1993) An optimized L-Band resonator for EPR imaging of biological samples. *J. Magn. Reson.* **A105**, 49–53.
28. Chzhan, M., Kuppusamy, P., Samouilov, A., He, G., and Zweier, J.L. (1999) A tunable reentrant resonator with transverse orientation of electric field for in vivo EPR spectroscopy. *J. Magn. Reson.* **137**, 373–378.
29. Villamena, F.A. and Zweier, J.L. (2004) Detection of reactive oxygen and nitrogen species by EPR spin trapping. *Antioxid. Redox. Signal* **6**, 619–629.
30. Forman, M.B., Virmani, R., and Puett, D.W. (1990) Mechanisms and therapy of myocardial reperfusion injury. *Circulation* **81**, IV-69–IV-78.
31. Manning, A.S., Coltart, J.D., and Hearse, D.J. (1984) Ischemia and reperfusion-induced arrhythmias in the rat – effect of xanthine oxidase inhibition with allopurinol. *Circ. Res.* **55**, 545–548.
32. Pogwizd, S.M. and Corr, P.B. (1986) Mechanisms of arrhythmogenesis during myocardial ischemia and reperfusion: A perspective of our current understanding. *J. Mol. Cell. Cardiol.* **18**(suppl 4), 43–47.
33. Yamada, M., Hearse, D.J., and Curtis, M.J. (1990) Reperfusion and readmission of oxygen. *Circ. Res.* **67**, 1211–1224.
34. Kukreja, R.C., Kearns, A.A., Zweier, J.L., Kuppusamy, P., and Hess, M.L. (1991) Singlet oxygen interaction with Ca^{2+}-ATPase of cardiac sarcoplasmic reticulum. *Circ. Res.* **69**, 1003–1014.
35. Halliwell, B., Gutteridge, J.M.C., and Cross, C.E. (1992) Free radicals, antioxidants, and human disease: Where are we now? *J. Lab. Clin. Med.* **119**, 598–620.
36. Esterbauer, H., Wag, G., and Puhl, H. (1993) Lipid peroxidation and its role in atherosclerosis. *Br. Med. Bull.* **49**, 566–576.
37. Zweier, J.L., Kuppusamy, P., Williams, R., Rayburn, B.K., Smith, D., Weisfeldt, M.L., and Flaherty, J.T. (1989) Measurement and characterization of postischemic free radical generation in the isolated perfused heart. *J. Biol. Chem.* **264**, 18890–18895.
38. Hearse, D.J. and Tosaki, A. (1988) Free radicals and calcium: Simultaneous interacting triggers as determinants of vulnerability to reperfusion-induced arrhythmias in the rat heart. *J. Mol. Cell Cardiol.* **20**, 213–223.
39. Horton, J.W. and White, D.J. (1995) Role of xanthine oxidase and leukocytes in postburn cardiac dysfunction. *J. Am. Coll. Surg.* **181**, 129–137.
40. Tosaki, A., Blasig, I.E., Pali, T., and Ebert, B. (1990) Heart protection and radical trapping by DMPO during reperfusion in isolated working rat hearts. *Free Radic. Biol. Med.* **8**, 363–372.
41. Zweier, J.L., Flaherty, J.T., and Weisfeldt, M.L. (1987) Direct measurement of free radical generation following reperfusion of ischemic myocardium. *Proc. Natl. Acad. Sci. U.S.A.* **84**, 1404–1407.
42. Kuppusamy, P., Wang, P., Samouilov, A., and Zweier, J.L. (1996) Spatial mapping of nitric oxide generation in the ischemic heart using electron paramagnetic resonance imaging. *Magn. Reson. Med.* **36**, 212–218.
43. Kuppusamy, P., Ohnishi, S.T., Numagami, Y., Ohnishi, T., and Zweier, J.L. (1995) Three-dimensional imaging of nitric oxide production in the rat brain subjected to

ischemia-hypoxia. *J. Cereb. Blood Flow Metab.* **15**, 899–903.
44. Henry, Y. and Guissani, A. (1994) Nitric oxide: A biological effector. Detection using electron paramagnetic resonance. *Transfus. Clin. Biol.* **1**, 157–164.
45. Berliner, L.J. and Fujii, H. (2004) In vivo spin trapping of nitric oxide. *Antioxid. Redox Signal* **6**, 649–656.
46. Mordvintcev, P., Mulsch, A., Busse, R., and Vanin, A. (1991) On-line detection of nitric oxide formation in liquid aqueous phase by electron paramagnetic resonance spectroscopy. *Anal Biochem.* **199**, 142–146.
47. Lai, C.S. and Komarov, A.M. (1994) Spin trapping of nitric oxide produced in vivo in septic-shock mice. *FEBS Lett.* **345**, 120–124.
48. Zweier, J.L., Wang, P., Samouilov, A., and Kuppusamy, P. (1995) Enzyme-independent formation of nitric oxide in biological tissues. *Nat. Med.* **1**, 804–809.
49. Zweier, J.L., Wang, P., and Kuppusamy, P. (1995) Direct measurement of nitric oxide generation in the ischemic heart using electron paramagnetic resonance spectroscopy. *J Biol. Chem.* **270**, 304–307.
50. Schafer, W.M., Nowak, B., Kaiser, H.J., Block, S., Koch, K.C., vom Dahl, J., and Bull, U. (2001) An approach for comparative quantification of myocardial blood flow (O-15-H2O-PET), perfusion (Tc-99m-tetrofosmin-SPECT) and metabolism (F-18-FDG-PET). *Nuklearmedizin* **40**, 164–171.
51. Ilangovan, G., Li, H., Zweier, J.L., Krishna, M.C., Mitchell, J.B., and Kuppusamy, P. (2002) In vivo measurement of regional oxygenation and imaging of redox status in RIF-1 murine tumor: Effect of carbogen-breathing. *Magn. Reson. Med.* **48**, 723–730.
52. Kuppusamy, P., Li, H., Ilangovan, G., Cardounel, A.J., Zweier, J.L., Yamada, K., Krishna, M.C., and Mitchell, J.B. (2002) Noninvasive imaging of tumor redox status and its modification by tissue glutathione levels. *Cancer Res.* **62**, 307–312.
53. Yamada, K.I., Kuppusamy, P., English, S., Yoo, J., Irie, A., Subramanian, S., Mitchell, J.B., and Krishna, M.C. (2002) Feasibility and assessment of non-invasive in vivo redox status using electron paramagnetic resonance imaging. *Acta Radiol.* **43**, 433–540.
54. Krishna, M.C., Subramanian, S., Kuppusamy, P., and Mitchell, J.B. (2001) Magnetic resonance imaging for in vivo assessment of tissue oxygen concentration. *Semin. Radiat. Oncol.* **11**, 58–69.
55. Mitchell, J.B., Russo, A., Kuppusamy, P., and Krishna, M.C. (2000) Radiation, radicals, and images. *Ann N.Y. Acad. Sci.* **899**, 28–43.
56. Krishna, M.C., Grahame, D.A., Samuni, A., Mitchell, J.B., and Russo, A. (1992) Oxoammonium cation intermediate in the nitroxide-catalyzed dismutation of superoxide. *Proc. Natl. Acad. Sci. U.S.A.* **89**, 5537–5541.
57. Swartz, H.M. (1990) Principles of the metabolism of nitroxides and their implications for spin trapping. *Free Radic. Res. Commun.* **9**, 399–405.
58. Chen, K., Glockner, J.F., Morse, P.D., 2nd, and Swartz, H.M. (1989) Effects of oxygen on the metabolism of nitroxide spin labels in cells. *Biochemistry* **28**, 2496–2501.
59. Kuppusamy, P. and Krishna, M.C. (2002) EPR imaging of issue redox status. *Curr. Topics Biophys.* **26**, 29–34.
60. Griffith, O.W. (1982) Mechanism of action, metabolism, and toxicity of buthionine sulfoximine and its higher homologs, potent inhibitors of glutathione synthesis. *J. Biol. Chem.* **257**, 13704–13712.
61. Yu, N.Y. and Brown, J.M. (1984) Depletion of glutathione in vivo as a method of improving the therapeutic ratio of misonidazole and SR 2508. *Int. J. Radiat. Oncol. Biol. Phys.* **10**, 1265–1269.
62. He, G., Samouilov, A., Kuppusamy, P., and Zweier, J.L. (2001) In vivo EPR imaging of the distribution and metabolism of nitroxide radicals in human skin. *J. Magn. Reson.* **148**, 155–164.
63. Swartz, H.M. (2004) Using EPR to measure a critical but often unmeasured component of oxidative damage: Oxygen. *Antioxid. Redox Signal* **6**, 677–686.
64. Ilangovan, G., Zweier, J.L., and Kuppusamy, P. (2004) Mechanism of oxygen-induced EPR line broadening in lithium phthalocyanine microcrystals. *J. Magn. Reson.* **170**, 42–48.
65. Gallez, B., Baudelet, C., and Jordan, B.F. (2004) Assessment of tumor oxygenation by electron paramagnetic resonance: Principles and applications. *NMR Biomed.* **17**, 240–262.
66. Ahmad, R., Clymer, B., Deng, Y., He, G., Vikram, D., Kuppusamy, P., and Zweier, J.L. (2006) Optimization of data acquisition for EPR imaging. *J. Magn. Reson.* **179**, 263–272.
67. Deng, Y., He, G., Petryakov, S., Kuppusamy, P., and Zweier, J.L. (2004) Fast EPR imaging at 300 MHz using spinning magnetic field gradients. *J. Magn. Reson.* **168**, 220–227.

Chapter 2

In Vivo Measurement of Tissue Oxygen Using Electron Paramagnetic Resonance Spectroscopy with Oxygen-Sensitive Paramagnetic Particle, Lithium Phthalocyanine

F. Hyodo, S. Matsumoto, E. Hyodo, A. Matsumoto, K. Matsumoto, and M.C. Krishna

Abstract

The partial pressure of oxygen (pO_2) plays a determining role in the energy metabolism of aerobic cells. However, low pO_2 level induces pathophysiological conditions such as tumor hypoxia, ischemia or reperfusion injury, and delayed/altered wound healing. Especially, pO_2 level in the tumor is known to be related to tumor progression and effectiveness of radiotherapy. To monitor the pO_2 levels in vivo, continuous wave (CW) and time-domain (TD) electron paramagnetic resonance (EPR) spectroscopy method was used, in which surface coil resonator and Lithium phthalocyanine (LiPc) as oxygen sensor were crucial. Once LiPc particles are embedded in a desired location of organ/tissue, the pO_2 level can be monitored repeatedly and non-invasively. This method is based on the effect of oxygen concentration on the EPR spectra of LiPc which offers several advantages as follows: (1) high sensitivity, (2) minimum invasiveness, (3) repeated measurements, (4) absence of toxicity (non-toxic), and (5) measurement in a local region of the tissue with embedded LiPc. Therefore, in this chapter, we describe the method using CW and TD EPR spectroscopy with oxygen-sensitive particle, LiPc, for in vivo monitoring of oxygen.

Key words: Lithium phthalocyanine (LiPc), oxymetry, in vivo tissue oxygen, tumor, EPR.

1. Introduction

The partial pressure of oxygen (pO_2) plays a determining role in the energy metabolism of aerobic cells. However, low pO_2 level induces pathophysiological conditions such as tumor hypoxia,

R.M. Uppu et al. (eds.), *Free Radicals and Antioxidant Protocols*, Methods in Molecular Biology 610,
DOI 10.1007/978-1-60327-029-8_2, © Humana Press, a part of Springer Science+Business Media, LLC 1998, 2010

ischemia or reperfusion injury, and altered wound healing (1–3). Especially, since hypoxic environment is known to render tumor cells resistant to ionizing irradiation and might be implicated in metastatic ability of tumors, the presence of hypoxic region in tumor is one of the main limiting factors for treatment in clinic (1, 4). Tumor oxygenation is also known to enhance the efficacy of radiotherapy when pO_2 is increased. Some of the early clinical efforts that were designed to overcome chronic hypoxia by breathing higher (concentrated) levels of oxygen were initiated in the 1950s. The breathing of carbogen (95% oxygen and 5% carbon dioxide) to increase the tumor pO_2 has also been reported. The carbon dioxide component is supposed to maintain tumor blood flow via peripheral vasodilation and also increase oxygen delivery by shifting the hemoglobin–oxygen dissociation curve to the right. Several studies have shown the improvement of tumor oxygenation by breathing 100% oxygen or carbogen, resulting in radio-sensitization. Therefore, tracking the levels of fluctuating pO_2 in tissue provides useful information for the treatment of cancer.

Electron paramagnetic resonance spectroscopy/imaging (EPRS/EPRI) oxymetry is being applied to monitor the in vivo oxygen levels in the murine models (5–9). EPR addresses unpaired spins and provides the characteristic spectra of free radical probes such as nitroxyl radical, trityl radical, spin trap agents, and solid free radical materials. There are two approaches for EPRI/EPRS, namely the continuous-wave (CW) modality (constant frequency with a field sweep) or the time-domain (TD; termed often Fourier transform, FT) modality (acquisition of impulse response of the system at constant field, just as in FT NMR) (7, 8). EPR using the CW methodology was developed earlier than TD EPRI. This is because of the ultrafast spin dynamics of the ensemble of unpaired electrons and the requirement for very special electronic components and devices that are necessary to parallel the advances in MRI. There are no stable free radicals of adequate concentration or biological half-life that occurs naturally in vivo. Therefore, EPR oxymetry technique would involve prior implant or infusion of stable non-toxic particulate or stable/slowly metabolizable water-soluble free radicals, into the monitoring tissue/subject as an oxygen sensor.

Oxygen-sensitive paramagnetic materials such as lithium phthalocyanine (LiPc) (10–14), coals (15), and India ink (16) have been successfully utilized for EPR oxymetry (17). Especially, LiPc is widely utilized as an oxygen sensor for the in vivo EPR oxymetry and exhibits a single and narrow-line EPR spectrum with its spectral line width is sensitive to the oxygen concentration surrounding the LiPc particles. Liu et al. demonstrated that LiPc is a useful material for studying the effect of molecular oxygen on the EPR spectrum line width in viable biological systems (12). The EPR oxymetry is based on the Heisenberg spin exchange between paramagnetic oxygen and LiPc spins in the excited state (10). Molecular oxygen is paramagnetic with two unpaired electrons in its ground state, and the unpaired electrons can interact with LiPc electrons. The exchange process is sensitive to the oxygen concentration/levels, with relaxation rate of spin of LiPc increasing as a function of the oxygen concentration. This increased relaxation rate results in broadening EPR line width of LiPc. Therefore oxygen concentration is monitored by determining the EPR line width of the paramagnetic oxygen-sensitive materials (probes). **Figure 2.1** shows a typical X-band EPR spectra of LiPc that partly responds to oxygen, under two different levels of pO_2, 0% O_2, and air (about 21% O_2), respectively (18). The EPR line width of LiPc is fairly narrow under 0% oxygen, however, in ambient air (21% O_2), it exhibits a broadened (wide) spectrum (*see* **Note 1**). Taking advantage of this phenomenon (EPR spectral characteristics), in vivo pO_2 is monitored by using EPR spectroscopy with LiPc as described in **Section 3**.

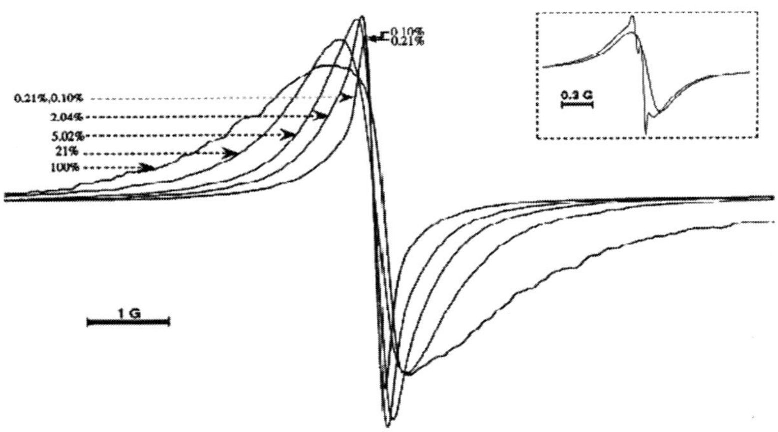

Fig. 2.1. Continuous wave (cw) X-band EPR spectra of an oxygen-dependent LiPc under different pO_2 levels. The oxygen percentages from the narrowest to the broadest spectra were: 0.10, 0.21, 2.04, 5.02, 21, and 100%, and the corresponding receiver gains were 1.25×10^3, 1.25×10^3, 2.5×10^3, 4×10^3, and 2.5×10^4, respectively. The spectrometer settings were: modulation amplitude, MA, 0.8 G; modulation frequency, MF, 25 kHz; microwave power, P, 0.2 mW; sweep width, SW, 20 G; time constant, TC, 1 s; scan time, ST, 4 min. The spectra were severely *over modulated* to avoid multiplets caused due to magnet inhomogeneity, and thus were used only in broad qualitative terms. The inset shows the spectra of an LiPc batch (electrochemically synthesized at a constant potential of 0.15 V using a coiled coil working electrode) that is predominantly composed of oxygen-insensitive crystals of LiPc at 0% O_2 and 21% O_2 (compressed air); the broad and narrow lines correspond to the oxygen-insensitive and oxygen-sensitive crystals, respectively.

2. Materials

2.1. Oxygen-Sensitive Probe

In our studies, LiPc is used as the oxygen-sensitive paramagnetic material. The structure of LiPc is shown in **Fig. 2.2**. LiPc is suspended in phosphate-buffered saline (PBS, pH 7.4) for calibration and in vivo implantation. LiPc particle is stable at room temperature.

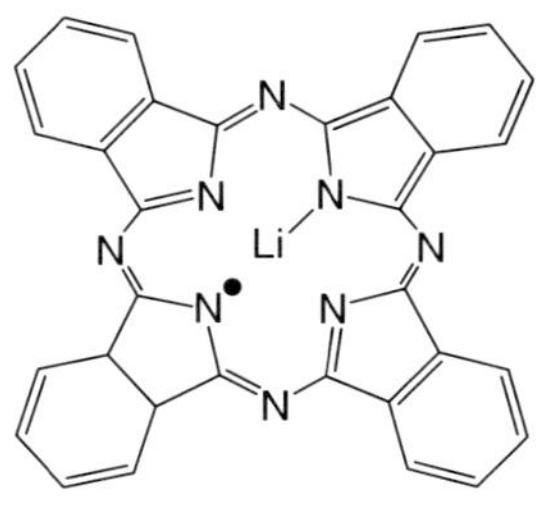

Fig. 2.2. Structure of LiPc.

2.2. Animals

In general, choice of different animal species as experimental models for oxymetry using LiPc particle seldom cause any problems; however, there are certain limitations while monitoring the pO_2 levels in large animals using the in vivo EPR system (*see* **Note 2**). In this chapter, specifically, the EPR technique with LiPc as the oxygen-sensitive probe to measure in vivo pO_2 in mouse model is described.

Female C3H mice are supplied by the Frederick Cancer Research Center, Animal Production (Frederick, MD). Animals, received at 6 wk of age, are housed 5/cage in climate-controlled and circadian rhythm-adjusted rooms and allowed food and water ad libitum. Experiments are carried out in compliance with the Guidelines for the Care and Use of Laboratory Animal Resources (1996), National Research Council, and approved by the National Cancer Institute Animal Care and Use Committee.

2.3. EPR Instruments

2.3.1. CW EPR

The 300 MHz CW EPR system, which was previously described (19), is equipped with a 700 MHz bridge and surface coil type resonator (7.3 mm i.d.), instead of a 300 MHz bridge and a parallel coil type resonator.

2.3.2. Time-Domain EPR

The individual components of the time-domain pulsed EPR spectrometer operating in the quadrature mode of data collection (two receivers with 90° out of phase to each other) for the complex Fourier transformation (FT) are essentially identical as described elsewhere except for the operating frequency at 750 MHz (8). Briefly, the spectrometer consists of a transmitting arm, a receiver arm, and a resonator. The transmitting arm utilizes a 750 MHz frequency source, which is derived from an 800 MHz frequency source. RF pulses delivered from the transmitting arm are amplified by a 200 W amplifier and send to the resonator through a diplexer. The pulse width is appropriately adjusted to give a 90° flip angle. The receiver gate is opened for ~300 ns after the delivery of the power to the resonator. Under those conditions, the major part of ringing diminishes. The receiving arm collects the time evolution response of paramagnetic probe in the resonator. The signal received at 750 MHz is amplified and mixed, resulting in an intermediate frequency signal at 20 MHz containing the FID. The real and imaginary components of FID signal are converted by the complex FT of the absorption spectrum. Then, the absorption spectrum is curve fitted as a mixture of Lorentian or Gaussian spectrum, and the line width information is thus obtained.

3. Methods

3.1. Process Outline

In vivo EPR oxymetry experiment typically includes the following steps:

1. Selection of the target organ to measure pO_2;
2. Establishing the standard pO_2 calibration curve;
3. Implantation of LiPc particle in one or several target organs;
4. Stabilization for at least 4–7 d prior to starting the measurements;
5. Measurement of pO_2 in the target organ using EPR; and
6. Data processing (conversion of line width (mGauss) to pO_2 (mmHg) using the calibration curve).

3.2. pO_2 Calibration (Standard) Curve

In order to obtain the in vivo pO_2 data from the conversion of the acquired EPR line width of LiPc to the corresponding pO_2 level, the establishment of pO_2 calibration (standard) curve is required. An aliquot (1–3 mg) of LiPc particles suspended in PBS (10–20 μL) is taken in a gas-permeable Teflon tube. The Teflon tube is placed on a surface coil type resonator with the LiPc particles at the center of the coil (**Fig. 2.3**). The resonator and sample are covered with a plastic tube and warmed to 36–37°C using a combination of hot air and an IR lamp (*see* **Note 3**).

The EPR line widths of the LiPc particles are measured under several conditions of gas flow, including 21 (medical air), 15, 10, 5, 1.25 (O_2/N_2), and 0% (argon), and then a calibration curve for LiPc oxymetry is obtained. The measurements are done after

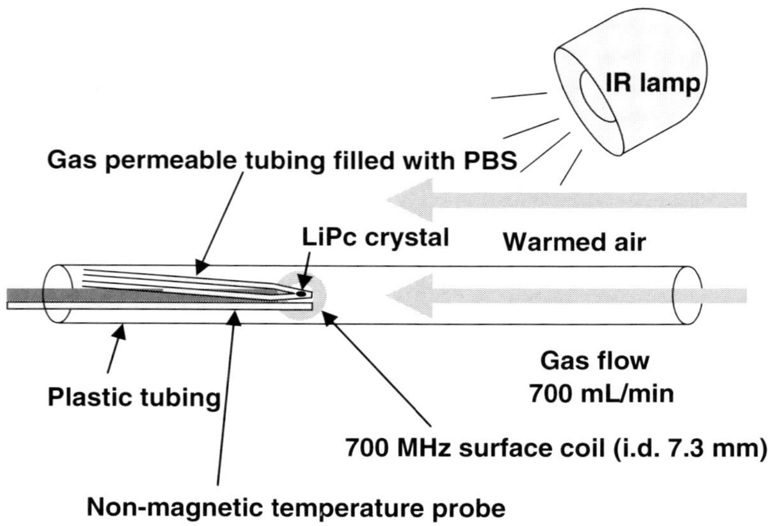

Fig. 2.3. Schematic representation of the LiPc setting for pO_2 calibration (standard) curve. *See* **Section 3.2** and **Note 3** for additional information on the placement of Teflon tube, surface coil type resonator, and the LiPc particles.

at least 30 min equilibration under each gas flow condition at the rate of 700 mL/min. The EPR signal is measured by CW EPR at 700 MHz using a single loop surface coil resonator.

3.3. Implantation of LiPc Particles

For the in vivo EPR oxymetry with oxygen-sensitive probe, the LiPc particles can be implanted in several target organs. As the surface coil is placed on one preferred site of target organ/tissue, we can repeatedly measure the pO_2 level in a specific/localized area (*see* **Note 4**).

The mouse is anesthetized by administering 1.5% isoflurane in medical air (flow rate was 700 mL/min) through a nose cone. The fur of lower abdomen and femoral region is removed by shaving. A portion of LiPc particles (5–10 mg) is suspended with an appropriate volume (10–20 μL) of physiologic saline and kneaded until a slurry paste is obtained. Then, the LiPc slurry is placed in the tip of a 20-gauge injection needle to make a soft pellet 2–3 mm long. The needle is injected into the desired region of the animal, and the pellet is pushed out using a smooth-fitting piston. **Fig. 2.4** shows a diagram of the LiPc implantation scheme. The average dry weight of the pellet is 0.8 ± 0.1 mg (mean ± SD), which is estimated from 7 randomly selected pilot runs. For example, implantations of LiPc are made into both the right femoral muscle and the left lower abdominal mammary gland pad of the mouse (*see* **Note 5**) The quantitation of pO_2 is achieved by the calibration curve shown in **Fig. 2.5**.

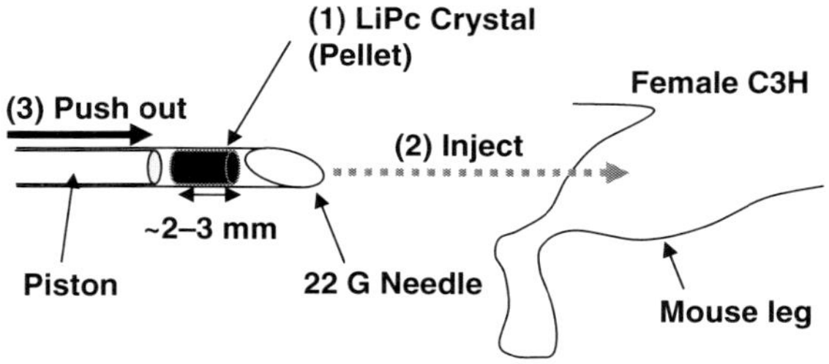

Fig. 2.4. Schematic drawing of the implantation of LiPc crystals (1). A portion of LiPc crystals (5–10 mg) is suspended with an appropriate volume (10–20 μL) of physiologic saline and kneaded until a slurry paste is obtained. Then, the LiPc slurry was placed in the tip of a 20-gauge needle to make a 2–3 mm long, soft pellet (2). The needle is injected into the region of interest of the animal, and (3) the pellet was pushed out using a smooth piston. The average dry weight of the pellet was 0.8 ± 0.1 mg (mean ± SD).

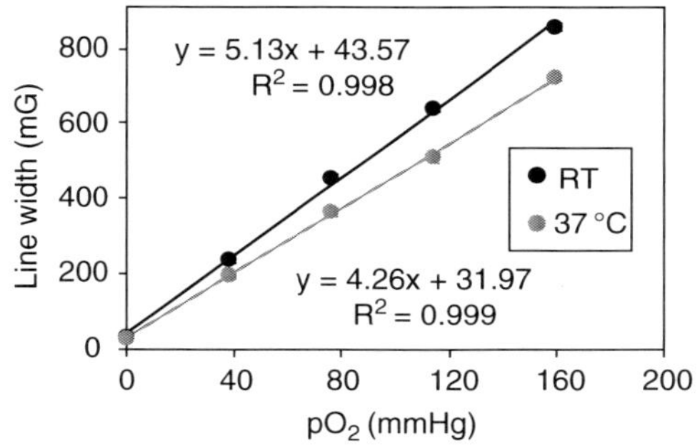

Fig. 2.5. Calibration curve of the EPR linewidth of LiPc versus pO_2 under various gas conditions, including 21 (room air), 15, 10, 5, 1.25, and 0% (Ar) oxygen, each with the constant gas flow rate of 700 mL/min. The EPR measurement was done after at least 30 min equilibration under each gas flow condition at 37°C (*gray circle*) or at room temperature (*black circle*). The EPR signal was measured using a 700 MHz CW EPR spectrometer. The EPR conditions were as follows: microwave frequency: 700 MHz; scan rate: 0.25 mT/s; time constant: 0.003 s; field modulation frequency: 13.5 kHz. The microwave power (0.005–0.3 mW) and field modulation width (0.01–0.07 mT) were adjusted to avoid signal saturation and line broadening.

3.4. In Vivo EPR Measurement

3.4.1. The Animal Setting

The animal setting for in vivo CW and TD EPR measurement is basically same.

1. Mice are anesthetized with 1.5% isoflurane in medical air (flow rate 700 mL/min) and placed in a mouse holder (*see* **Note 6**).
2. The mouse legs and lower abdomen are restrained by adhesive tape to the holder. (*see* **Note 7**).
3. Before inserting the temperature probe into the rectum, the tip of non-magnetic temperature probe (FISO; FOT-L-SD, Fiber Optic sensor) is coated with gel.
4. The surface coil is placed on the region where the LiPc crystal is implanted.

5. Animal is positioned to resonator in the EPR magnet.
6. EPR measurement is initiated using the minimum modulation amplitude and power. (*see* **Note 8**)
7. Data processing is done with the calibration curve.

3.4.2. CW EPR Conditions

Microwave frequency = 700 MHz, scan rate = 0.25 mT/s, time constant = 0.003 s, field modulation frequency: 13.5 kHz. The microwave power = 0.005–0.3 mW and field modulation width (0.01–0.07 mT).

Figure 2.6 shows an example of CW EPR data of time course of the pO_2 levels in mouse femoral and mammary pad during and after breathing carbogen (95% O_2, 5% CO_2; *see* **Note 9**). The pO_2 levels in both the studied tissues increase gradually during carbogen breathing in mice, and decrease when the animals are switched back to breathing room air.

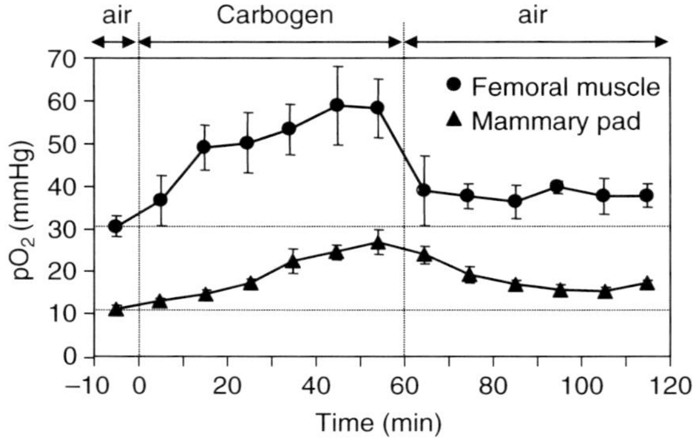

Fig. 2.6. Time course of pO_2 in the femoral muscle and mammary pad before, during, and after the mouse breathed carbogen. Values are indicated as mean ± SE of average values of 3 mice each, every 10 min. The pO_2 levels in both sites increased gradually when the mice were exposed to carbogen. When the breathing gas was switched back to room air, the pO_2 levels went down faster in the femoral muscle compared to the mammary pad. The pO_2 levels in the mammary tissue are lower than in the femoral muscle.

3.4.3. TD EPR

As an oxygen sensor for TD EPR oxymetry, LiPc crystal possesses all the desirable properties including high spin density, single spectrum, very narrow line width (long transversal relaxation time T_2^*), high sensitivity to changes in oxygen concentration, and stability in biological tissues. The line width of EPR spectrum is inversely related to the decay constant of TD EPR signal after irradiation pulse which must be longer than the recovery time of receiver system.

TD EPR conditions: resonant frequency = 750 MHz; excitation pulse width = 125 ns; TR = 7.5 μs; No. of sampling = 1,760. The dead-time of the spectrometer was estimated as 270 ns.

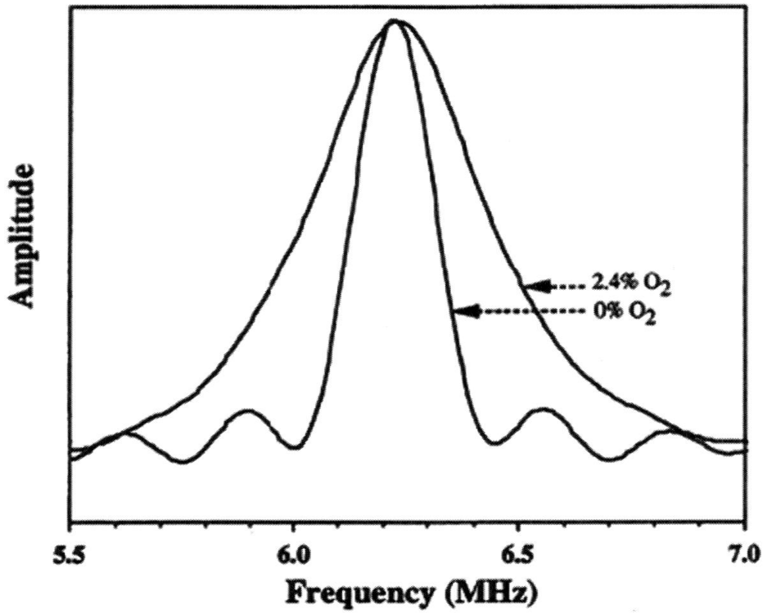

Fig. 2.7. The FT spectra of the oxygen-sensitive LiPc batch, using a time-domain RF EPR spectrometer at two different oxygen percentages, 0% O_2 and 2.04% O_2, respectively. A 1-GHz sampler with a channel limitation of 4 k points, corresponding to 4.096 ms was used to collect the data. The wiggles in the 0% O_2 data are due to truncation of the FID at 4.1 ms, no digital filters are used.

Typical TD EPR spectra are shown in **Fig. 2.7** under two different pO_2 levels (0% O_2 and 2.4% O_2) for the same batch whose CW spectra and the oxygen responses as shown in **Fig. 2.2**. For example, time-course pO_2 change after carbogen breathing in normal muscle and tumor are shown in **Fig. 2.8**. The basal pO_2 values in muscle region are higher than those of tumor region. In addition, the response to carbogen breathing in the tumor region is low.

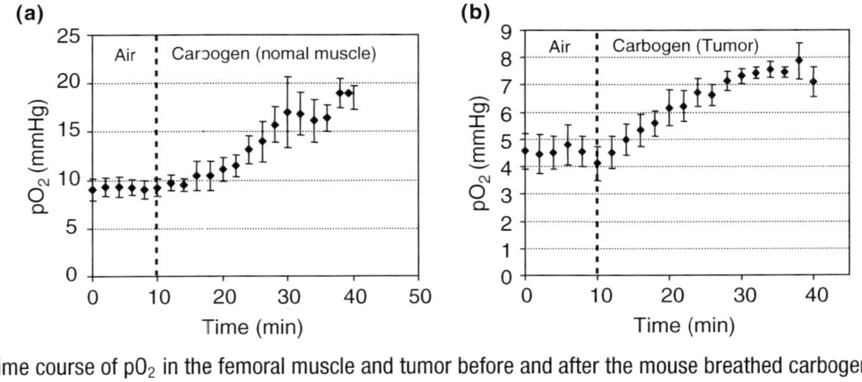

Fig. 2.8. Time course of pO_2 in the femoral muscle and tumor before and after the mouse breathed carbogen.

4. Notes

1. If oxygen level shows variable values in the tissue close to LiPc particles, the line width of LiPc will show average pO_2 levels surrounding the particle.
2. Selection of proper frequency for in vivo measurements is necessary because there is a penetration problem. For example, the penetration depth at 1.2 GHz is 5–10 mm. Therefore, to measure pO_2 in large animals, the frequency should be as low as possible and can be detected.
3. Because body temperature of animals is between 37 and 39°C, the calibration curve of LiPc has to be obtained under 37°C to match the in vivo conditions. In addition, it is necessary to prepare a new calibration curve for every lot of LiPc particle.
4. LiPc particle is needed to be implanted at least 4–7 d prior to start of the experiments for the stabilization of pO_2 of the implanted tissue. It is preferred to measure the tissue pO_2 levels everyday following the LiPC implantation in order to confirm the stabilization of pO_2 levels in the chosen target tissue. It was reported that the implants do not cause any adverse histological actions under appropriate conditions (17).
5. Method of LiPc implantation in brain tissue has been reported by Dun et al. (17).
6. Following anesthesia, body temperature in mouse will drop to ~ 25°C in the absence of body warming. Therefore, anesthetization of mouse and preparation of the animal for in vivo pO_2 EPR measurements should be carried out under proper body warming conditions with the aid of heating pad, RI ramp, and warm air in order to keep the constant physiological normal body core temperature.
7. In the absence of proper animal restraint, the matching and tuning may be erratic during pO_2 measurements.
8. Find out the minimum modulation amplitude and power to avoid the line broadening by saturation.
9. The carbon dioxide component is supposed to maintain tumor blood flow via peripheral vasodilation and also to increase oxygen delivery by shifting the hemoglobin–oxygen dissociation curve to the right. Many studies have succeeded in showing the improvement of tumor oxygenation by 100% oxygen or carbogen breathing, resulting in radiosensitization.

References

1. Gallez, B., Baudelet, C., and Jordan, B.F. (2004) Assessment of tumor oxygenation by electron paramagnetic resonance: Principles and applications. *NMR Biomed.* **17**, 240–262.
2. Liu, S., Liu, W., Ding, W., Miyake, M., Rosenberg, G.A., and Liu, K.J. (2006) Electron paramagnetic resonance-guided normobaric hyperoxia treatment protects the brain by maintaining penumbral oxygenation in a rat model of transient focal cerebral ischemia. *J. Cereb. Blood Flow Metab.* **26**, 1274–1284.
3. Sen, C.K., Khanna, S., Gordillo, G., Bagchi, D., Bagchi, M., and Roy, S. (2002) Oxygen, oxidants, and antioxidants in wound healing: An emerging paradigm. *Ann. NY Acad. Sci.* **957**, 239–249.
4. De Jaeger, K., Kavanagh, M.C., and Hill, R.P. (2001) Relationship of hypoxia to metastatic ability in rodent tumours. *Br. J. Cancer.* **84**, 1280–1285.
5. Elas, M., Ahn, K.H., Parasca, A., Barth, E.D., Lee, D., Haney, C., and Halpern, H.J. (2006) Electron paramagnetic resonance oxygen images correlate spatially and quantitatively with Oxylite oxygen measurements. *Clin. Cancer Res.* **12**, 4209–4217.
6. Ilangovan, G., Zweie, J.L., and Kuppusamy, P. (2004) Mechanism of oxygen-induced EPR line broadening in lithium phthalocyanine microparticles. *J. Magn. Reson.* **170**, 42–48.
7. Matsumoto, K., Subramanian, S., Devasahayam, N., Aravalluvan, T., Murugesan, R., Cook, J.A., Mitchell, J.B., and Krishna, M.C. (2006) Electron paramagnetic resonance imaging of tumor hypoxia: Enhanced spatial and temporal resolution for in vivo pO2 determination. *Magn. Reson. Med.* **55**, 1157–1163.
8. Subramanian, S., Matsumoto, K., Mitchell, J.B., and Krishna, M.C. (2004) Radio frequency continuous-wave and time-domain EPR imaging and Overhauser-enhanced magnetic resonance imaging of small animals: Instrumental developments and comparison of relative merits for functional imaging. *NMR Biomed.* **17**, 263–294.
9. Swartz, H.M. and Clarkson, R.B. (1998) The measurement of oxygen in vivo using EPR techniques. *Phys. Med. Biol.* **43**, 1957–1975.
10. Ilangovan, G., Bratasz, A., Li, H., Schmalbrock, P., Zweier, J.L., and Kuppusamy, P. (2004) In vivo measurement and imaging of tumor oxygenation using coembedded paramagnetic particulates. *Magn. Reson. Med.* **52**, 650–657.
11. Ilangovan, G., Li, H., Zweier, J.L., Krishna, M.C., Mitchell, J.B., and Kuppusamy, P. (2002) In vivo measurement of regional oxygenation and imaging of redox status in RIF-1 murine tumor: Effect of carbogen-breathing. *Magn. Reson. Med*. **48**, 723–730.
12. Liu, K.J., Gast, P., Moussavi, M., Norby, S.W., Vahidi, N., Walczak, T., Wu, M., and Swartz, H.M. (1993) Lithium phthalocyanine: A probe for electron paramagnetic resonance oximetry in viable biological systems. *Proc. Nat. Acad. Sci .USA* **90**, 5438–5442.
13. Matsumoto, A., Matsumoto, S., Sowers, A.L., Koscielniak, J.W., Trigg, N.J., Kuppusamy, P., Mitchell, J.B., Subramanian, S., Krishna, M.C., and Matsumoto, K. (2005) Absolute oxygen tension (pO(2)) in murine fatty and muscle tissue as determined by EPR. *Magn. Reson. Med.* **54**, 1530–1535.
14. Swartz, H.M., Boyer, S., Gast, P., Glockner, J.F., Hu, H., Liu, K.J., Moussavi, M., Norby, S.W., Vahidi, N. et al. (1991) Measurements of pertinent concentrations of oxygen in vivo. *Magn. Reson. Med.* **20**, 333–339.
15. Goda, F., Liu, K.J., Walczak, T., O'Hara, J.A., Jiang, J., and Swartz, H.M. (1995) In vivo oximetry using EPR and India ink. *Magn. Reson .Med.* **33**, 237–245.
16. Swartz, H.M., Liu, K.J., Goda, F., and Walczak, T. (1994) India ink: A potential clinically applicable EPR oximetry probe. *Magn. Reson. Med.* **31**, 229–232.
17. Dunn, J.F. and Swartz, H.M. (2003) In vivo electron paramagnetic resonance oximetry with particulate materials. *Methods* **30**, 159–166.
18. Afeworki, M., Miller, N.R., Devasahayam, N., Cook, J., Mitchell, J.B., Subramanian, S., and Krishna, M.C. (1998) Preparation and EPR studies of lithium phthalocyanine radical as an oxymetric probe. *Free Radic. Biol. Med.* **25**, 72–78.
19. Yamada, K., Murugesan, R., Devasahayam, N., Cook, J.A., Mitchell, J.B., Subramanian, S., and Krishna, M.C. (2002) Evaluation and comparison of pulsed and continuous wave radiofrequency electron paramagnetic resonance techniques for in vivo detection and imaging of free radicals. *J. Magn. Reson.* **154**, 287–297.

Chapter 3

Measurement of Plasma Nitrite by Chemiluminescence

Enika Nagababu and Joseph M. Rifkind

Abstract

Studies have demonstrated that plasma nitrite (NO_2^-) reflects endothelial nitric oxide (NO) production. In addition, NO_2^- has been shown to have biological activities associated with its reduction to NO in blood and tissues. Therefore, determination of plasma NO_2^- has been proposed as a prognostic marker for cardiovascular diseases. Typical concentrations of NO_2^- in the plasma are in the nanomolar range and determination of this NO_2^- poses a challenge in terms of both sensitivity and specificity. Thus, a highly sensitive, chemiluminescence method that is based on the reduction of NO_2^- by potassium iodide and iodine is being used to determine the nitrite in biological fluids. This method has the sensitivity, but also measures other nitric oxide species such as *S*-nitrosothiols and *N*-nitrosamines. We, therefore, developed an alternative method based on the reduction of NO_2^- by ascorbic acid in strongly acidic media. As part of the methodology, glacial acetic acid and ascorbic acid are introduced into the purge vessel of the NO analyzer. Samples containing NO_2^- are injected into the purge vessel and the chemiluminescence signals generated as a result of the formation of NO are then measured. We find that under these conditions NO_2^- is stoichiometrically reduced to NO. Other traditional NO-generating species, such as *S*-nitrosothiols, *N*-nitrosamines, nitrated lipids, and nitrated proteins, did not interfere in the determination of plasma NO_2^-. Using the present method, plasma NO_2^- in fasting human subjects has been determined to be in the range of 56–210 nM (mean $\pm$ SD $= 110 \pm 36$ nM; $n = 8$).

Key words: Plasma, nitrite, nitric oxide, endothelial nitric oxide synthase, chemiluminescence, ascorbic acid.

1. Introduction

Nitric oxide (NO) synthesized by endothelial nitric oxide synthase (eNOS) diffuses randomly into smooth muscle cells and blood. The life span of NO in blood milieu is less than 0.2 ms. The reaction of NO with blood components results in the formation of nitrite (NO_2^-), nitrate (NO_3^-), *S*-nitrosothiols (RSNOs), iron

Grant Support: This research was supported by the Intramural Research Program of the National Institute on Aging, NIH.

R.M. Uppu et al. (eds.), *Free Radicals and Antioxidant Protocols*, Methods in Molecular Biology 610,
DOI 10.1007/978-1-60327-029-8_3, © Humana Press, a part of Springer Science+Business Media, LLC 1998, 2010

nitrosyls, *N*-nitrosamines, nitrated lipids, and nitro-proteins (1). Related to the role of NO in regulating vascular tone and blood flow, decreased NO synthesis in the endothelium is a strong risk factor for the development of atherosclerosis, vascular thrombosis, reperfusion injury, and hypertension (reviewed in refs 2 and 3). The measurement of NO_2^- and NO_3^-, the dominant species formed from NO in plasma is used as a measure of NO synthesis by vascular endothelial cells (4). The reliability of the determination of NO_3^- to assess the eNOS activity has been a subject of debate as the levels of NO_3^- are also influenced by several eNOS-independent mechanisms such as diet, intestinal bacteria, renal function, and breathing of atmospheric NO and NO_2 gases. In addition, due to the presence of high concentrations of NO_3^- in plasma (~25 μM), acute changes of eNOS activity are not easy to determine. Recent studies have demonstrated the possibility of using a determination of the nanomolar concentrations of plasma NO_2^- alone to evaluate the activity of eNOS (5–7). Furthermore, some studies have documented that NO_2^- can serve as an intravascular NO storage molecule (8–11). Despite all the biological significance of this compound, a serious limitation faced by workers in the area has been the lack of an accurate, sensitive, and reproducible method for the determination of plasma NO_2^-.

Currently, for the determination of NO_2^- in human biological fluids the methods employed rely on mass spectroscopy, colorimetry, HPLC, and flow injection analysis (4). One of the most widely used methods for the determination of NO_2^- is the colorimetric method using the Griess reagent (12). This method, however, lacks sensitivity for the determination of NO_2^- in physiological fluids, which is likely to be less than 1 μM. Among others, the ozone-based chemiluminescence method is highly sensitive and can be used to measure nanomolar concentrations of NO. This method relies on the reaction of NO with ozone and the measurement and quantification of the energy released while returning from an excited state (NO_2^*) to the ground state.

$$NO + O_3 \rightarrow NO_2^* + O_2$$

$$NO_2^* \rightarrow NO_2 + h\nu$$

Using the above approach different NO-related species can be detected after being selectively reduced to NO. The potassium iodide- and iodine (KI/I_2)-based reductive chemiluminescence assay has been used to determine plasma NO_2^- (13–15). This reagent measures NO_2^- in the presence of NO_3^-; it does not reduce NO_3^- to NO. This reagent, however, is not specific for NO_2^- as it can also react with plasma RSNOs and *N*-nitrosamines (16, 17).

Therefore, a direct reductive chemiluminescence method has been developed to determine NO_2^- without interference of NO_3^-, RSNOs, and *N*-nitrosamines (18). This method involves a reduction of nitrite to NO by ascorbic acid under acidic conditions; the NO released is carried to the detector by inert gas where it reacts with ozone to produce a chemiluminescence signal. The signal intensity obtained is proportional to the concentration of NO generated (or NO_2^- present in the original sample). This method is used extensively in the author's lab for the determination of plasma NO_2^-.

$$NO_2^- + H^+ \leftrightarrow HNO_2 \quad [1]$$

$$HNO_2 + H^+ \leftrightarrow H_2NO_2^+ \leftrightarrow NO^+ + H_2O \quad [2]$$

$$H_2NO_2^+ + ASC \leftrightarrow NO + H_2O + ASC^\bullet + H^+ \quad [3]$$

$$NO^+ + ASC \leftrightarrow NO + ASC^\bullet + H^+ \quad [4]$$

2. Materials

2.1. Equipment

1. Nitric Oxide Analyzer: Sievers model 280 obtained from GE Analytical Instruments (Boulder, CO, USA) (For detailed information, *see* ref. 19).
2. Liquid program (NO Analysis version 3.21).
3. Origin program (version 6.1).
4. Oxygen and inert gas (argon or nitrogen).
5. Hamilton syringes: capacity 50, 100, and 250 μL with 5-inch long needles
6. Portable centrifuge (Labnet type Z 150).

2.2. Reagents

1. Water: Ultra pure water with resistivity ≥18.2 MΩ-cm. Prepared using PureLab Plus or Millipore water-Milli Q-Gard 2 purification pack.
2. Sodium hydroxide, 1 mM: prepared by 1,000-fold dilution of 1 M NaOH (Fisher Scientific) using water.
3. Preparation of NO_2^- standards:
   A. Stock solution 1 (100 mM): dissolve 0.690 g of sodium nitrite (Sigma-Aldrich) in 100 mL of 1 mM NaOH. The solution is stable at 4°C for 1 week.

   B. Stock solution 2 (1 mM): prepared by 100-fold dilution of stock solution 1 using 1 mM NaOH; stable for 24 h.

C. Working solution (1 μM): prepared by 1,000-fold dilution of stock solution 2 with water. Typically, 10 μL of stock solution 2 is diluted to 10 mL with water. Use within 2 to 3 h.

4. Dilute the working standard solution with water serially (1:2, 1:4,) down to a concentration of 7.812 nM (**Table 3.1**).

Table 3.1
Preparation of nitrite standards

Ratio	Volume of 1 μM nitrite std (mL)	Volume of water (mL)	Total volume (mL)	Final nitrite concentration (nM)
1:2	5	5	10	500
1:4	2.5	7.5	10	250
1:8	1.25	8.75	10	125
1:16	0.625	9.375	10	62.5
1:32	0.3125	9.6875	10	31.25
1:64	0.156	9.844	10	15.625
1:128	0.078	9.922	10	7.812

5. Diethylenetriaminepentaacetic acid (DTPA; 10 mM): suspend 39.3 mg DTPA in 5 mL of water. With the contents being constantly stirred, add 1 N NaOH drop wise until the DTPA is completely dissolved. Make up the final volume to 10 mL with water. This solution is stable at room temperature.

6. *N*-Ethylmaleimide (NEM; 100 mM): dissolve 125.1 mg NEM in 100 mL of water; fresh reagent should be prepared every day.

7. Ascorbic acid (0.5 mM): dissolve 0.881 g of ascorbic acid in 10 mL of water. Prepare fresh reagent every day.

3. Methods

3.1. Blood Collection

Collect venous blood samples (3 mL each) from subjects who fasted overnight into heparinized vacutainer tubes (*see* **Notes 2** and **3**). Nitrite is unstable in the whole blood. Therefore, it is advisable to use a bench-top centrifuge near the bedside of the subject in order to rapidly separate the plasma from red cells. The blood samples are centrifuged immediately (i.e., within 30 s following the collection) for 3 min at 5,000 rpm. Plasma (1 mL) free of hemolysis is transferred into dark colored microtubes that also contain 6.5 mM NEM (100 mM; 6.5 μL) and 0.1 mM DTPA (10 mM; 10 μL). NEM and DTPA prevent the destruction of plasma *S*-nitrosothiols to NO_2^-. The samples are immediately frozen on dry ice and stored at –85 or –150°C for future analysis.

3.2. Calibration Curve for Nitrite

1. Add 7 mL of glacial acetic acid and 1.0 mL of 0.5 M ascorbic acid into the purge vessel.
2. Antifoam reagent is generally not needed or added to the glacial acetic acid-ascorbic acid reagent.
3. A gas bubbler/sodium hydroxide trap is connected between the purge vessel and the detector.
4. The temperature of the water bath used for circulating water around the purge vessel is set at 37°C. Cold water circulation for condenser is not required.
5. Liquid program version 3.21 is used to acquire data. The reaction mixture is continuously purged with inert gas at constant cell pressure by adjusting the gas flow (*see* **Note 6**). The data are acquired every $^1/_4$ s. Once the baseline is stabilized, a 100 μL aliquot of the working standard of NO_2^- (concentration range: 7.812–500 nM) is injected into the purge vessel in duplicates and the chemiluminescence signals are recorded.
6. The data from the liquid program are imported into the Origin 6.1 program (*see* **Note 7**) and chemiluminescence signals are regenerated. The area under the curve (AUC) is determined using the method provided with the software (**Table 3.2**).

Table 3.2
Average area under the curve (AUC) of each signal

Nitrite concentration (nM)	Signal (AUC) (AU)
7.812	0.116
15.625	0.221
31.25	0.4569
62.5	0.897
125	1.912
250	3.76
500	7.59

The area of the blank value (water) 0.011 is subtracted from the area of all the nitrite standards. AU, arbitrary units.

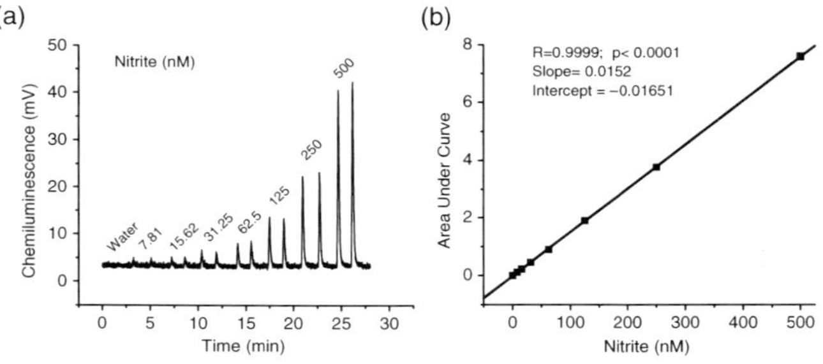

Fig. 3.1. (**a**). Generation of chemiluminescence signals following the addition of nitrite standard. Nitrite (7.812–500 nM; 100 μL) is injected into the purge vessel that contained 7 mL glacial acetic acid and 1 mL of 0.5 M ascorbic acid. (**b**) Calibration curve for nitrite is obtained by integrating the area under the curve.

7. Standard Curve: The calibration curve is constructed by plotting the peak area versus concentration of the nitrite (**Fig. 3.1a**). The slope and intercept of the calibration curve is determined by linear regression technique (**Fig. 3.1b**).

3.3. Measurement of Plasma Nitrite

1. The plasma is thawed under subdued light just prior to analysis and kept on ice until injected (*see* **Note 8**)
2. Using a Hamilton syringe, inject 100 μL of either fresh or stored plasma into the bottom of the purge vessel that contains 7 mL of glacial acetic acid and 1 mL of 0.5 M ascorbic acid (*see* **Notes 6** and **9**)
3. Antifoam reagent is not required even for plasma samples.
4. Each sample is injected in triplicate or quadruplicate. The reagents of the purge vessel are changed after four injections.

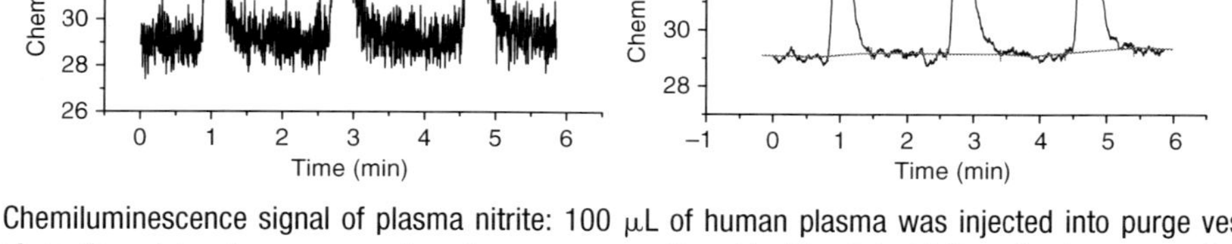

Fig. 3.2. Chemiluminescence signal of plasma nitrite: 100 μL of human plasma was injected into purge vessel. The contents of reaction mixture in purge vessel are the same as mentioned for **Fig. 3.1** . (**a**) Chemiluminescence signals from raw data. (**b**) Chemiluminescence signals from smoothened data used for calculating the area. Area of peak 1.2.33534: Peak 2, 2.50232; Peak 3, 2.64189.

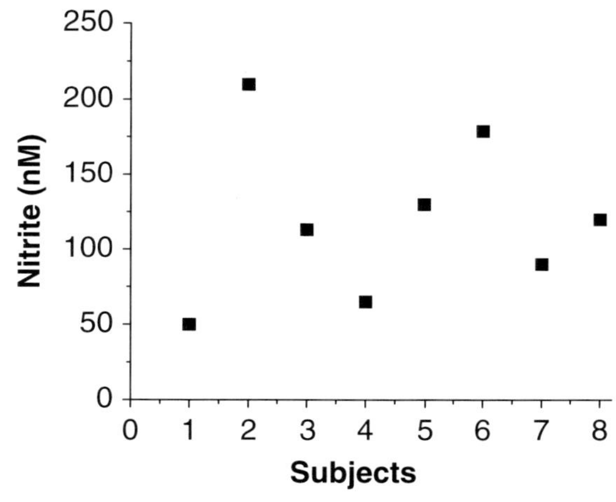

Fig. 3.3. Plasma nitrite values of eight human subjects.

5. Data are transferred to the Origin 6.1 program and the chemiluminescence signals are regenerated as mentioned for the NO_2^- standard curve (**Fig. 3.2a**).
6. To eliminate the noise and for baseline correction, the signals are smoothened using adjacent averaging of 20 points. The chemiluminescence signals are regenerated using the smoothened data. The area under the curve is determined as described for the standard curve (**Fig. 3.2b**).
7. The plasma nitrite level obtained for eight human subjects under fasting conditions are illustrated in **Fig. 3.3**. The following calculation is used.

   Calculations:

$$\begin{aligned}\text{Nitric (nM)} &= \frac{\text{Area} - \text{Y-intercept}}{\text{Slope}} \\ &\times \frac{\text{Volume of stds injected(ml)}}{\text{Volume of plasma injected(ml)}} \\ &\times \text{Dilution factor (NEM and DTPA)} \\ &= \frac{2.49 - (-0.01651)}{0.0152} \times \frac{0.1}{0.1} \times 1.0165 \\ &= 167.62\text{nM}\end{aligned}$$

4. Notes

1. Nitrite contamination is very common in solvents, reagents, glassware, and especially blood collecting tubes. It is important to check this nitrite contamination by injecting the test solutions into glacial acetic acid/ascorbic acid

reagent system (above) before using them. Appearances of any signals indicate the nitrite contamination.

2. Nitrite contamination in vacutainer tubes should be checked before use. The tubes are mixed well with ultra pure water and then the nitrite in water is determined. We found that heparin tubes are almost free of nitrite while EDTA tubes contain high levels of nitrite. One needs to check the nitrite contamination for each supplier of vacutainers.
3. Nitrite is rapidly oxidized to nitrate by oxyhemoglobin. Therefore it is necessary to take all precautions to avoid hemolysis while collecting and processing blood samples. Plasma should be free of hemolysis.
4. Always use fresh high quality water (Ultra pure water with resistivity $\geq$ 18.2 MΩ-cm) to prepare the solutions and to rinse the syringes and NOA purge system.
5. Always cap the reagent bottles tightly; otherwise atmospheric nitric oxide species dissolves in liquids over time.
6. Make sure that there is no gas leak in the purge system. Maintain constant instrument's cell pressure by adjusting the inert gas flow. Run the system at least 30–60 min for stabilization prior to initiating the sample analysis.
7. We strongly suggest the use of the Origin 6.1 program for data analysis. It can be purchased from Origin Lab Corporation, One Roundhouse Plaza, Suite 303, Northampton, MA 01060, USA, www.originlab.com.
8. Thaw the plasma samples just prior to use while protecting from light. Complete the analysis within 1 or 2 h.
9. Inject the samples at the bottom of the purge vessel using a syringe with a 5-inch long needle.
10. When high protein samples (plasma) are injected, leave a sufficient gap between injections until the gas bubbles are regenerated in the purge vessel.

References

1. Liu, X., Miller, M.J., Joshi, M.S., Sadowska-Krowicka, H., Clark, D.A., and Lancaster, J.R., Jr.. (1998) Diffusion-limited reaction of free nitric oxide with erythrocytes. *J. Biol. Chem.* **273**, 18709–18713.
2. Marin, J. and Rodriguez-Martinez, M.A. (1997) Role of vascular nitric oxide in physiological and pathological conditions. *Pharmacol. Ther.* **75**, 111–134.
3. Vanhoutte, P.M. (1997) Endothelial dysfunction and atherosclerosis. *Eur. Heart J.* **18**(Suppl E), E19–E29.
4. Tsikas, D. (2005) Methods of quantitative analysis of the nitric oxide metabolites nitrite and nitrate in human biological fluids. *Free. Radic. Res.* **39**, 797–815.
5. Lauer, T., Preik, M., Rassaf, T., Strauer, B.E., Deussen, A., Feelisch, M., and Kelm, M. (2001) Plasma nitrite rather than nitrate reflects regional endothelial nitric oxide synthase activity but lacks intrinsic vasodilator action. *Proc. Natl. Acad. Sci. U S A* **98**, 12814–12819.

6. Kleinbongard, P., Dejam, A., Lauer, T., Rassaf, T., Schindler, A., Picker, O., Scheeren, T., Godecke, A., Schrader, J., Schulz, R., Heusch, G., Schaub, G.A., Bryan, N.S., Feelisch, M., and Kelm, M. (2003) Plasma nitrite reflects constitutive nitric oxide synthase activity in mammals. *Free. Radic. Biol. Med.* **35**, 790–796.
7. Allen, J.D., Cobb, F.R., and Gow, A.J. (2005) Regional and whole-body markers of nitric oxide production following hyperemic stimuli. *Free Radic. Biol. Med.* **38**, 1164–1169.
8. Nagababu, E., Ramasamy, S., Abernethy, D.R., and Rifkind, J.M. (2003) Active nitric oxide produced in the red cell under hypoxic conditions by deoxyhemoglobin-mediated nitrite reduction. *J. Biol. Chem.* **278**, 46349–46356.
9. Cosby, K., Partovi, K.S., Crawford, J.H., Patel, R.P., Reiter, C.D., Martyr, S., Yang, B.K., Waclawiw, M.A., Zalos, G., Xu, X., Huang, K.T., Shields, H., Kim-Shapiro, D.B., Schechter, A.N., Cannon, R.O., 3rd, and Gladwin, M.T. (2003) Nitrite reduction to nitric oxide by deoxyhemoglobin vasodilates the human circulation. *Nat. Med.* **9**, 1498–1505.
10. Nagababu, E., Ramasamy, S., and Rifkind, J.M. (2006) S-nitrosohemoglobin: A mechanism for its formation in conjunction with nitrite reduction by deoxyhemoglobin. *Nitric Oxide* **15**, 20–29.
11. Li, H., Samouilov, A., Liu, X., and Zweier, J.L. (2001) Characterization of the magnitude and kinetics of xanthine oxidase-catalyzed nitrite reduction. Evaluation of its role in nitric oxide generation in anoxic tissues. *J. Biol. Chem.* **276**, 24482–24489.
12. Schmidt, H.H.H.W. and Kelm, M. (1996) Determination of nitrite and nitrate by the Griess Reaction. In: *Methods in Nitric Oxide Research* (Feelisch, M., Stamler, J.S., Eds.), John Wiley, New York, pp. 491–497.
13. Cox, R.D. (1980) Determination of nitrate and nitrite at the parts per billion level by chemi-luminescence. *Anal. Chem.* **52**, 322–330.
14. Feelisch, M., Rassaf, T., Mnaimneh, S., Singh, N., Bryan, N.S., Jourd'Heuil, D., and Kelm, M. (2002) Concomitant S-, N-, and heme-nitros(yl)ation in biological tissues and fluids: Implications for the fate of NO in vivo. *FASEB J.* **16**, 1775–1785.
15. Yang, B.K., Vivas, E.X., Reiter, C.D., and Gladwin, M.T. (2003) Methodologies for the sensitive and specific measurement of S-nitrosothiols, iron-nitrosyls, and nitrite in biological samples. *Free Radic. Res.* **37**, 1–10.
16. Rassaf, T., Bryan, N.S., Kelm, M., and Feelisch, M. (2002) Concomitant presence of N-nitroso and S-nitroso proteins in human plasma. *Free Radic. Biol. Med.* **33**, 1590–1596.
17. MacArthur, P.H., Shiva, S., and Gladwin, M.T. (2007) Measurement of circulating nitrite and S-nitrosothiols by reductive chemiluminescence. *J Chromatogr B Analyt Technol Biomed Life Sci.* **851**, 93–105.
18. Nagababu, E. and Rifkind, J.M. (2007) Measurement of plasma nitrite by chemiluminescence without interference of S-, N-nitroso and nitrated species. *Free Radic. Biol. Med.* **42**, 1146–1154.
19. Hampl, V., Waters, C.L., and Archer, S.L. (1996) Determination of nitric oxide by the chemiluminescence reaction with ozone. *Methods in Nitric Oxide Research*, 309–318.

Chapter 4

Determination of Glutathione, Mitochondrial Transmembrane Potential, and Cytotoxicity in H9c2 Cardiomyoblasts Exposed to Reactive Oxygen and Nitrogen Species

K. Sathishkumar, Xueli Gao, Achuthan C. Raghavamenon, Subramanyam N. Murthy, Philip J. Kadowitz, and Rao M. Uppu

Abstract

Quantitative measurement of cellular oxidative stress (COS) and cytotoxicity are important to establish their significance in pathophysiologic conditions and disease states. So far, ample methods have been described to determine these processes based on spectrophotometric analysis. The application of simple, rapid, and sensitive fluorescence methods to determine the cytotoxicity and COS is described in the present chapter. Murine H9c2 cells were exposed to various free radical and non-free radical oxidants through use of diethylamine NONOate, 3-morpholinosydnonimine (SIN-1), and a synthetic preparation of peroxynitrite (PN). The viability of control and the treated H9c2 cells was measured based on the reduction of resazurin to resorufin which generates a fluorescent signal. The mitochondrial membrane potential was quantified by determining the cellular uptake of a fluorescent dye, (5,5′,6,6′-tetrachloro-1,1′-3,3′-tetraethylbenzimidazolcarbocyanine iodide (JC-1)) and its segregation in the mitochondrial fraction. The intracellular GSH was determined by assaying the glutathione-*S*-transferase (GST)-catalyzed conjugation of GSH to monochlorobimane. This chapter describes the feasibility and potential of the above-described fluorescence approach as simple alternative methods to determine reactive oxygen and nitrogen species-induced cytotoxicity and oxidative stress using H9c2 cardiomyoblasts as a model system.

Key words: Reactive oxygen and nitrogen species, mitochondrial membrane potential, glutathione, cardiomyoblasts, nitric oxide, peroxynitrite, cellular oxidative stress.

1. Introduction

Reactive oxygen and nitrogen species (RONs) are a family of molecules that include molecular oxygen and its derivatives produced in all aerobic cells (1). The RONs include both radical

R.M. Uppu et al. (eds.), *Free Radicals and Antioxidant Protocols*, Methods in Molecular Biology 610,
DOI 10.1007/978-1-60327-029-8_4, © Humana Press, a part of Springer Science+Business Media, LLC 1998, 2010

and non-radical species such as superoxide ($O_2^{\bullet -}$), nitric oxide ($^{\bullet}NO$), nitrogen dioxide ($^{\bullet}NO_2$), carbonate radical ($CO_3^{\bullet -}$), singlet oxygen ($^1O_2$), hydrogen peroxide (H_2O_2), peroxynitrite (PN; $ONOO^-$) and peroxynitrous acid (ONOOH; pK_a 6.2), and hypohalous acids (HOX) (2, 3). RONs, in particular those derived from $^{\bullet}NO$ and $O_2^{\bullet -}$, have been shown to cause cellular oxidative damage and trigger specific signaling events (3–7) that culminate in altered cellular physiology. RONs have also been shown to cause the oxidative degradation/modification of DNA, proteins, carbohydrates, and polyunsaturated lipids resulting in oxidative stress (8, 9). Oxidative stress is implicated in the pathogenesis and/or progression of several life-threatening diseases including cancer, diabetes mellitus, arthritis, atherosclerosis, neurodegenerative diseases, and ischemia/reperfusion injury (1, 10–14). Oxidative stress has been shown to cause disruption of cellular redox homeostasis particularly the cellular thiol-redox, mostly regulated by glutathione (GSH). This is critical for cell survival under normal physiological conditions and, in disease states with or without toxicant-induced stress.

Analysis of RONs in intracellular compartments is complicated by their high reactivities with the surrounding biological macromolecules and relatively short duration of existence (15). Hence, the application of reliable and sensitive methods to assess the footprints of RONs is the most critical and sought after component by researchers in this area of investigation. The purpose of this chapter is to describe specific fluorescence methods that are simple, rapid, reliable, reproducible, and sensitive to determine cytotoxicity and cellular oxidative stress (COS). Along these lines, the methodologies for the determination of cell viability, measurement of intracellular RONs and GSH, and quantification of mitochondrial membrane potential are described. The emphasis of this chapter is on exposure of H9c2 cardiomyoblasts to $^{\bullet}NO$, $O_2^{\bullet -}$, and the products and radicals resulting from their combination (PN, $^{\bullet}NO_2$, and $CO_3^{\bullet -}$) through use of diethylamine (DEA) NONOate (generates $^{\bullet}NO$), 3-morpholinosydnonimine (SIN–1) (generates $^{\bullet}NO$ and $O_2^{\bullet -}$ simultaneously), and a synthetic preparation of PN (generates $^{\bullet}NO_2$ and $CO_3^{\bullet -}$ in combination with CO_2) to assess the methodology and the determination of cytotoxicity and cellular oxidative stress resulting from exposure to RONs.

2. Materials

2.1. Cell Culture and Treatments

1. Water (*see* **Note 1**)
2. Murine H9c2 cardiomyoblasts (American Type Culture Collection, Rockville, MD) (*see* **Note 2**).

3. Dulbecco's modified Eagle's medium (DMEM; Sigma, St. Louis, MO) containing glutamine (2 mM), glucose (4.5 g/L), sodium pyruvate (1.0 mM), penicillin (100 units/mL), streptomycin (100 μg/mL), and $NaHCO_3$ (1.5 g/L). Supplement DMEM with fetal bovine serum, FBS (Atlanta Biologicals, Lawrenceville, GA) at a final concentration of 10% (v/v).
4. Trypsin (0.25%, w/v)-EDTA (1 mM) solution (Sigma).
5. *Peroxynitrite*: synthesized in the laboratory based on nitrosation of the hydroperoxide anion by isoamyl nitrite in mixed solvents of isopropyl alcohol (IPA) water (16). A typical synthesis of 200 mL of 0.2 M PN in the homogeneous solvent system consisting of IPA/water (50/50, v/v) can be prepared as follows:

   (a) In a round bottom flask (capacity: 500 mL), dissolve 1.92 g NaOH (48 mmol) in 89.5 mL of water.
   (b) Add 100 mL of IPA. Stir the mixture for 1 min.
   (c) Add 4.9 mL of 9 M H_2O_2 (44 mmol). Stir for 1 min (*see* **Note 3**).
   (d) Add 5.6 mL of 7.14 M isoamyl nitrite (40 mmol). Stir the contents continuously during the addition and subsequently for another 10 min at room temperature (*see* **Note 4–6**).
   (e) Transfer the contents into a separatory funnel and extract with 3 × 1 vol of dichloromethane followed by 2 × 1 vol of hexane.
   (f) Pass the aqueous layer over granular MnO_2 (~25 g) packed in the form of column (*see* **Note 7**).
   (g) Purge the PN solution with Ar for 10 min (*see* **Note 8**).
   (h) Measure the concentration of PN in the final preparation based on absorbance at 302 nm ($\varepsilon = 1{,}670\ M^{-1}\ cm^{-1}$). Store in small aliquots (1 mL each) at –70°C until use (17). It is necessary to determine the concentration of PN each time a vial is thawed and used.
6. DEA NONOate and SIN-1 (Sigma): dissolve in water at a final concentration of 0.5 mM each and store in small aliquots at –70°C until use.
7. Phosphate buffered saline-glucose (PBSG) buffer: 50 mM Na_2HPO_4, 90 mM NaCl, 5 mM KCl, 0.8 mM $MgCl_2$, 0.2 mM $CaCl_2$, and 5 mM glucose; pH adjusted to 7.4.

2.2. Cell Viability

1. Resazurin reagent: dissolve resazurin sodium salt (Sigma) in DMEM at a concentration of 500 μg/mL. It is important that DMEM does not contain a pH indicator (phenol red). Following sterile filtration, store the resazurin reagent in the dark at 4°C and use within one week. There are ready to use resazurin reagents marketed by Promega and Invitrogen and these can also be used.
2. Krebs-Ringer-HEPES (KRH) buffer: 131 mM NaCl, 5 mM KCl, 1.3 mM $MgSO_4$, 1.3 mM $CaCl_2$, 0.4 mM KH_2PO_4, 6 mM glucose, and 20 mM HEPES; pH adjusted to 7.4.

2.3. Determination of Intracellular GSH and Mitochondrial Transmembrane Potential

1. ApoGSH GSH fluorometric assay kit (BioVision, Mountain View, CA) containing monochlorobimane and glutathione-*S*-transferase (GST). Alternatively, GST and monochlorobimane can be purchased from Sigma and Invitrogen, respectively. In such a case, GST reagent is prepared in PBS at a concentration of 50 units/mL, whereas monochlorobimane is dissolved in DMSO at 25 mM.
2. JC-1 (5,5′,6, 6′-tetrachloro-1,1′,3,3′-tetraethylbenzimidazolylcarbocyanine iodide) (Invitrogen): dissolve this fluorescent dye in DMSO at a concentration of 10 mM.
3. Triton X-100 (Sigma).

3. Methods

3.1. Cell Culture and Exposure to RONs and Antioxidants

1. Culture H9c2 cells in DMEM medium supplemented with 10% FBS in 95% humidified air and 5% CO_2 at 37°C.
2. Detach the cells using trypsin–EDTA. Wash the cells with DMEM (or PBS) and then seed them in a 24-well plate at a density of 30,000 per well in DMEM containing 10% FBS. Allow the cells to grow overnight. Make sure that the cells grow to subconfluence (~70%) and not to 100% confluence.
3. *Exposure to PN* – Wash the H9c2 cells once with PBSG. Replace the medium with 500 μL of PBSG/well. Add PN at a final concentration of 10–50 μM (bolus addition; *see* **Note 9**). After 20 min of incubation, replace the medium with DMEM containing 10% FBS (500 μL) and culture the cells for 24 h as described above. Make sure that control cells go through the entire protocol but using decomposed PN (*see* **Note 10**).
4. *Exposure to SIN-1 and DEA NONate* – Expose the H9c2 cardiomyoblasts to SIN-1 (0.05–5 mM) or DEA NONOate (0.05–5 mM) in serum-free DMEM for 60 min

(*see* **Note 11**). Following this, replace the medium with DMEM cantaining 10% FBS (500 μL) and culture the cells for 24 h as described above.

3.2. Determination of Cell Viability (CellTiter-Blue assay)

The resazurin assay is a fluorometric method for determining the number of viable cells in a cell population based on metabolic activity. Cells that are viable convert the dark blue oxidized form of the dye (resazurin) to a red reduced form (resorufin) that generates the fluorescence. Non-viable cells which rapidly lose metabolic activity do not reduce resazurin. The actual reduction takes place within the cell and the resorufin diffuses into the surrounding culture medium causing an increase in fluorescence, which is directly proportional to the number of viable cells. Results can be recorded by a fluorometer or spectrophotometer. Although tetrazolium dyes are routinely used to determine cell viability, the method described here has the advantage of simplicity and greater sensitivity. Tetrazolium reagents have a sensitivity to detect ~1,000 cells, whereas resazurin can detect ~50–400 cells.

1. To the control and treated H9c2 cells in 24-well plates (*see* **Section 3.1**, Steps 3 and 4), add 10 μL of the resazurin reagent per well.
2. Incubate for 2 h at 37°C in 95% humidified air and 5% CO_2 (*see* **Note 12**).
3. Measure the fluorescence at excitation and emission wavelengths of 560 and 590 nm (respectively) using a Spectramax Gemini EM spectrofluorometer (Molecular Devices, sunnyvale, CA).
4. The percent change in cell viability is expressed relative to untreated controls.

A representative determination of cytotoxicity using the resazurin reagent in cardiomyoblast exposed to PN, SIN-1, and DEA NONOate is presented in the **Fig. 4.1**. Exposure to PN, SIN-1, and DEA NONOate, but not PN-decomposed, caused a decrease in cell viability. Values of IC_{50} (concentration required to produce half-maximal cell death) were calculated by taking the log concentration–response curves by standard nonlinear regression analysis using GraphPad Prism 4.0 Software (San Diego, CA). Data are presented as mean ± SEM of at least five separate experiments. Under the present experimental conditions the IC_{50} for PN, SIN-1, and DEA NONOate was found to be 50 ± 4.2, 1,500 ± 270 and 1,320 ± 210 μM, respectively.

3.3. Determination of Intracellular GSH Content

Traditionally the intracellular redox status of cells is determined by measuring the GSH/GSSG ratio using the enzymatic recycling method. This method has been described extensively in the literature (18–20). In the present work, the monochlorobimane

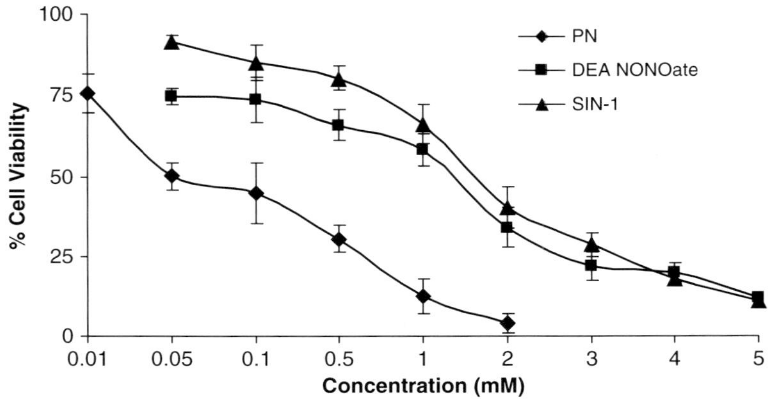

Fig. 4.1. Effect of PN, SIN-1, and DEA NONOate on the viability of H9c2 cardiomyoblasts. H9c2 cells were treated with PN (initial concentration: 0.01–2 mM) for 20 min or SIN-1 (0.05–5 mM) and DEA NONOate (0.05–5 mM) for 1 h. The cell viability/proliferation was measured 24 h later using the resazurin reagent. Data represent the mean ± SEM for five independent experiments.

fluorescence assay was chosen to measure intracellular GSH. This is based on the GST-catalyzed conjugation of GSH to monochlorobimane. Since the conjugation reaction is catalyzed by the enzyme and the product formed is fluorescent (excitation: 398 nm and emission: 488 nm), the assay has the advantage of both specificity and sensitivity.

1. Expose cells to PN (10–50 μM), SIN-1 (1–2 mM), or DEA NONOate (1–2 mM) as mentioned in **Section 3.1**, Steps 3 and 4.
2. Remove the culture medium after 8 h and wash the cells twice with PBSG.
3. Lyse the cells in 100 μL of PBSG containing 0.2% Triton X-100. Centrifuge the lysate at 13,000 × *g* for 5 min to remove debris if any.
4. Mix 2 μL of the monochlorobimane reagent (final concentration 50 μM) and 2 μL of GST (final concentration 0.1 unit/mL) with the supernatant (100 μL); incubate for 30 min at 37°C (*see* **Note 12**).
5. Using Spectramax Gemini EM spectrofluorometer, measure the fluorescence at excitation and emission wavelengths set at 398 and 488 nm, respectively.
6. After normalizing for cell numbers, the results are expressed as percent change in fluorescence relative to untreated controls.

Figure 4.2 presents the data on the quantitative measurement of GSH in the control and treated H9c2 cardiomyoblasts. There was a decrease in GSH levels measured after 8 h of

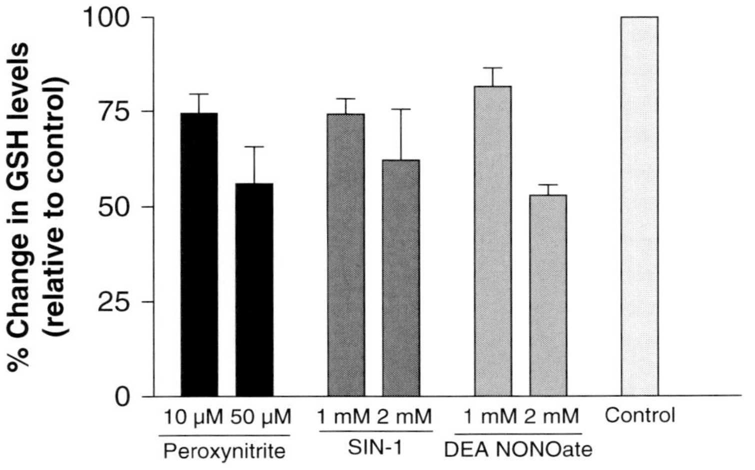

Fig. 4.2. Dose- and time-dependent decreases in the levels of GSH in H9c2 cardiomyoblasts measured after 8 h of exposure to PN (10–50 μM), SIN-1 (1–2 mM), or DEA NONOate (1–2 mM). Percentage change in GSH levels was expressed relative to untreated controls (100%). The concentration of GSH was measured based on GST-catalyzed conjugation of GSH to monochlorobimane resulting in a fluorescent product (for details, *see* **Section 3.3**). Data represent the mean ± SEM for three independent experiments.

exposure to PN, SIN-1, and DEA NONOate. Data are presented as mean ± SEM of three separate experiments.

3.4. Determination of Mitochondrial Transmembrane Potential

Alterations in mitochondrial transmembrane potential ($\Delta\Psi_m$) in response to stressful stimuli (i.e., oxidative bursts) in cultured cells, can be determined using a JC-1, a cationic dye. The assay is based on accumulation of JC-1 in mitochondria which is dependent on $\Delta\Psi_m$. The dye remains in the monomeric form in the cytoplasm and exhibits green fluorescence when the value of $\Delta\Psi_m$ is small. On the other hand, at high $\Delta\Psi_m$, there is greater accumulation of JC-1 in the mitochondria forming J-aggregates, and the fluorescence shifts to red (E_{max} 590 nm). Thus, the changes in the emission patterns of JC-1 of this cyanine dye can be used as a sensitive measure of mitochondrial trans membrane potential.

1. Expose H9c2 cells to PN (10–20 μM), SIN-1 (1–2 mM), or DEA NONOate (1–2 mM)) as described in **Section 3.1,** Steps 3 and 4.
2. Remove the culture medium after 8 h and wash the H9c2 cells twice with PBSG.
3. Add PBSG containing JC-1 at a final concentration of 10 μM (*see* **Notes 12–13**) and incubate for 30 min at 37°C with 95% humidity and 5% CO_2.
4. Remove the unbound JC-1 dye by washing the cells twice with PBSG and add fresh buffer (PBSG: 500 μL).

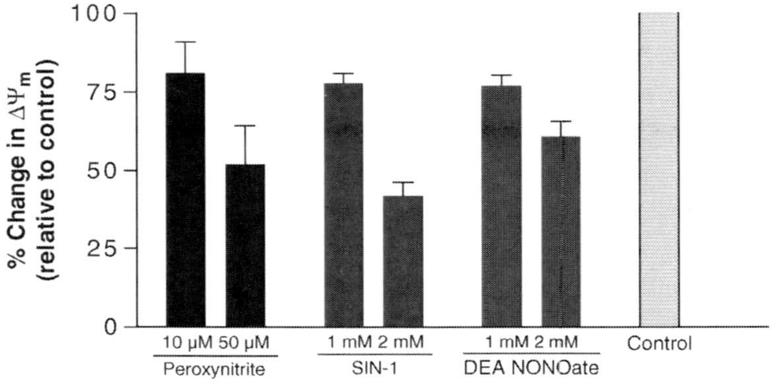

Fig. 4.3. Changes in mitochondrial transmembrane potential in H9c2 cardiomyoblasts exposed to PN (10–50 μM), SIN-1 (1–2 mM), or DEA NONOate (1–2 mM), Dose- and time-dependent changes in the mitochondrial transmembrane potential were determined 8 h following the exposure to PN, SIN-1, and DEA NONOate using JC-1 (*see* **Section 3.4**).

5. Using a Spectramax EM Gemeni spectrofluorometer, measure the fluorescence at excitation and emission wavelengths of 488 and 590 nm, respectively.
6. The results are expressed as percentage decrease in $\Delta\Psi_m$ compared to the untreated controls.

We measured changes in $\Delta\Psi_m$ after 8 h of exposure to PN, SIN-1, and DEA NONOate in the H9c2 cardiomyoblasts. The cationic dye JC-1 accumulates in the mitochondria of normal cells giving a maximal fluorescence at 590 nm (100%). In cells undergoing oxidative stress/cytotoxicity, the mitochondrial accumulation of the dye would be lower, provided there is a decrease in transmembrane potential. As shown in **Fig. 4.3**, we find evidence for lowering of $\Delta\Psi_m$ in H9c2 cells exposed to PN, SIN-1, and DEA NONOate. Data are presented as mean ± SEM of three separate experiments.

4. Notes

1. Unless specified otherwise, throughout this procedure "water" refers to ultrapure water with resistance $\geq$18.2 MΩ.
2. These are immature embryonic cardiomyocytes with electrical and hormonal signaling pathways preserved as in the adult cardiac cells. About 95% of the cells are mononucleated and express L-type calcium channels typical of cardiomyocytes. Also, they express N-cadherin and other specific cardiac markers and maintain the phenotypic characteristics for several passages. Sipido and Marban (21)

argue that the H9c2 cells had been neglected as potential surrogates for primary cardiac cells.

3. Commercial preparations of H_2O_2 (30%) can be used as such provided the concentration based on UV absorbance at 240 nm ($\varepsilon = 41\ M^{-1}\ cm^{-1}$) (16, 17) is around 9 M. Where necessary, the preparation needs to be diluted with water.
4. Isoamyl nitrite is a hypotensive agent. No direct exposure or contact must be allowed. Gloves (double) and lab coat must be worn at all times. All experiments involving isoamyl nitrite must be performed behind a safety shield in a fume hood.
5. The concentration of isoamyl nitrite in the neat solution (96% w/v; density: 0.872 g/mL at 25°C) is approximately 7.14 M.
6. Certain preparations of isoamyl nitrite contain Na_2CO_3 up to 2% (w/v) as a stabilizing agent. As a precaution, wash the preparation with 3 × 2 vol of water before using in the synthesis of PN (17).
7. Wash the MnO_2 column with 20 mL each of water and 0.1 N NaOH before passing the solution of PN. The first few milliliters of the PN solution eluting from the MnO_2 column should be discarded to minimize the dilution.
8. This helps to remove dissolved gases (O_2 and N_2O) and traces of organic solvent(s) used in the extraction process.
9. Peroxynitrite readily decomposes in aqueous media ($t_{1/2} \approx$ 1.5 s at pH 7.4). Therefore, the alkaline solutions of PN (pH $\geq$ 12.0) must be introduced as bolus additions of 2–10 μL and mixed in the shortest time possible. Also, cell cultures treated with decomposed PN (*see* **Note 10**) should be used as control and included in every experiment.
10. The decomposed solutions of PN (1–2 mM) can be prepared by incubating small volumes (10–20 μL) of PN stock solutions (~100 mM) with 2 mL of PBS, pH 7.4, for 5 min at room temperature.
11. This helps avoiding interaction of $O_2^{\bullet-}$, $^{\bullet}NO$, and the products of $^{\bullet}NO$–$O_2^{\bullet-}$ reaction (formed from the decomposition of DEA NONOate and SN-1) with serum components.
12. During incubations, care should be taken to protect the plate from external light. This helps to minimize the loss of fluorescence, if any.
13. As it is difficult to completely solublize the JC-1 dye in the buffer, try to dissolve to the maximum possible by vigorous vortexing. Centrifuge briefly and remove the undissolved dye before addition to the culture.

Acknowledgments

This publication was made possible by National Institutes of Health (NIH) Grants P20 RR16456 (from the BRIN Program of the National Center for Research Resources) and ES10018 (from the ARCH Program of the National Institute of Environmental Health Sciences), and US Department of Education (Title III, Part B – Strengthening Historically Black Graduate Institutions, HBGI; grant number: PO31B040030). Its contents are solely the responsibility of authors and do not necessarily represent the official views of the NIH, NSF, or US Department of Education.

References

1. Darley-Usmar, V., Wiseman, H., and Halliwell, B. (1995) Nitric oxide and oxygen radicals: A question of balance. *FEBS Lett.* **369**, 131–135.
2. Stanczyk, M., Gromadzinska, J., and Wasowicz, W. (2005) Roles of reactive oxygen species and selected antioxidants in regulation of cellular metabolism. *Int. J. Occup. Med. Environ. Health* **18**, 15–26.
3. Uppu, R.M., Squadrito, G.L., and Pryor, W.A. (2000) Nitration and nitrosation by peroxynitrite: Role of CO2 and evidence for common intermediates. *J. Am. Chem. Soc.* **122**, 6911–6916.
4. Fu, Y., Sies, H., and Lei, X.G. (2001) opposite roles of selenium-dependent glutathione peroxidase-1 in superoxide generator diquat- and peroxynitrite-induced apoptosis and signaling. *J. Biol. Chem.* **276**, 43004–43009.
5. Bauer, G. (2002) Signaling and proapoptotic functions of transformed cell-derived reactive oxygen species. *Prostaglandins Leukot. Essent. Fatty Acids* **66**, 41–56.
6. Forman, H.J. and Torres, M. (2001) Redox signaling in macrophages. *Mol. Aspects Med.* **22**, 189–216.
7. Uppu, R.M., Nossaman, B.D., Greco, A.J., Fokin, A., Murthy, S.N., Fonseka, V., and Kadowitz, P.J. (2007) Cardiovascular effects of peroxynitrite. *Clin. Exp. Pharmacol. Physiol.* **34**, 933–937.
8. O'Donnell, V.B. and Freeman, B.A. (2001) Interactions between nitric oxide and lipid oxidation pathways: Implications for vascular disease. *Circ. Res.* **88**, 12–21.
9. Uppu, R.M., Squadrito, G.L., and Pryor, W.A. (1996) Acceleration of peroxynitrite oxidations by carbon dioxide. *Arch. Biochem. Biophys.* **327**, 335–343.
10. Hadjigogos, K. (2003) The role of free radicals in the pathogenesis of rheumatoid arthritis. *Panminerva Med.* **45**, 7–13.
11. Sun, A.Y. and Chen, Y.M. (1998) Oxidative stress and neurodegenerative disorders. *J. Biomed. Sci.* **5**, 401–414.
12. Wolin, M.S., Davidson, C.A., Kaminski, P.M., Fayngersh, R.P., and Mohazzab, H.K. (1998) Oxidant – nitric oxide signaling mechanisms in vascular tissue. *Biochemistry (Mosc)* **63**, 810–816.
13. Moncada, S. and Bolanos, J.P. (2006) Nitric oxide, cell bioenergetics and neurodegeneration. *J. Neurochem.* **97**, 1676–1689.
14. Kamat, J.P. (2006) Peroxynitrite: A potent oxidizing and nitrating agent. *Indian J Exp. Biol.* **44**, 436–447.
15. Batandier, C., Fontaine, E., Keriel, C., and Leverve, X.M. (2002) Determination of mitochondrial reactive oxygen species: Methodological aspects. *J. Cell. Mol. Med.* **6**, 175–187.
16. Uppu, R.M. (2006) Synthesis of peroxynitrite using isoamyl nitrite and hydrogen peroxide in a homogeneous solvent system. *Anal. Biochem.* **354**, 165–168.
17. Uppu, R.M. and Pryor, W.A. (1995) Synthesis of peroxynitrite in a two-phase system using isoamyl nitrite and hydrogen peroxide. *Anal. Biochem.* **236**, 242–249.
18. Redegeld, F.A., van Opstal, M.A., Houdkamp, E., and van Bennekom, W.P. (1988) Determination of glutathione in biological material by flow-injection analysis using an enzymatic recycling reaction. *Anal. Biochem.* **174**, 489–495.
19. Baker, M.A., Cerniglia, G.J., and Zaman, A. (1990) Microtiter plate assay for the measurement of glutathione and glu-

tathione disulfide in large numbers of biological samples. *Anal. Biochem.* **190**, 360–365.

20. Teare, J.P., Punchard, N.A., Powell, J.J., Lumb, P.J., Mitchell, W.D., and Thompson, R.P. (1993) Automated spectrophotometric method for determining oxidized and reduced glutathione in liver. *Clin. Chem.* **39**, 686–689.
21. Sipido, K.R. and Marban, E. (1991) L-type calcium channels, potassium channels, and novel nonspecific cation channels in a clonal muscle cell line derived from embryonic rat ventricle. *Circ. Res.* **69**, 1487–1499.

Section II

Natural and Synthetic Antioxidants

Chapter 5

Phenolic Acids and Flavonoids: Occurrence and Analytical Methods

Constantine D. Stalikas

Abstract

Phenolics are structurally assorted and are generally part of a complex mixture isolated from plant and biological origin matrices. A wide gamut of natural products have been the focus of main study for phenolic compounds while urine and blood are the two main biological fluids that have been analyzed for metabolism studies. Traditional and more advanced techniques have come to prominence for sample preparation, detection, and identification. This review is devoted to a short discussion of the occurrence of phenolic acids and flavonoids, their role in human health, and focuses on a detailed presentation of the analytical methods, concluding with the advantages of analytical methods employed so far and prospects. Strategies and practical aspects for the determination of phenolic acids and flavonoids in biological fluids, beverages, plants, and food are reported. Novel and past applications are provided with significant treatment and detection-related developments on the basis of the employment of separation and non-separation analytical techniques.

Key words: Phenolic acids, flavonoids, health effects, analytical methods.

1. Introduction

Vascular plants synthesize a diversified range of organic molecules – phytochemicals, referred to as "secondary metabolites." Phenolic compounds are considered as secondary metabolites that are synthesized by plants during normal development (1) and under stress conditions such as infection, wounding, and UV radiation, among others (2). The term "phenolic compounds" or "phenolics" encompasses approximately 8,000 naturally occurring compounds, which share in common a phenol, i.e., an aromatic ring bearing at least one hydroxyl substituent.

R.M. Uppu et al. (eds.), *Free Radicals and Antioxidant Protocols*, Methods in Molecular Biology 610,
DOI 10.1007/978-1-60327-029-8_5, © Humana Press, a part of Springer Science+Business Media, LLC 1998, 2010

A major class of phenolic compounds is phenolic acids. The name "phenolic acids," in general, describes phenols that possess at least one carboxylic-acid functionality. However, when describing plant metabolites, it refers to a distinct group of organic acids which contain two distinguishing carbon frameworks: the hydroxybenzoic and the hydroxycinnamic structures (**Fig. 5.1**). Noteworthy is also the case of chlorogenic acid (5-*O*-caffeoylquinic acid) which is the caffeic acid esterified with quinic acid (**Fig. 5.2**).

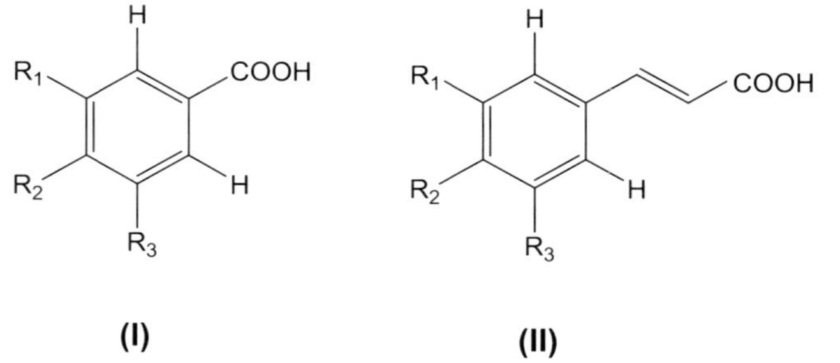

Fig. 5.1. General structures of naturally occurring phenolic acids. (**I**) Hydroxybenzoic and (**II**) hydroxycinnamic structures. R_1, R_2, and R_3 denote one of the following atoms and chemical groups: H, OH, OCH_3.

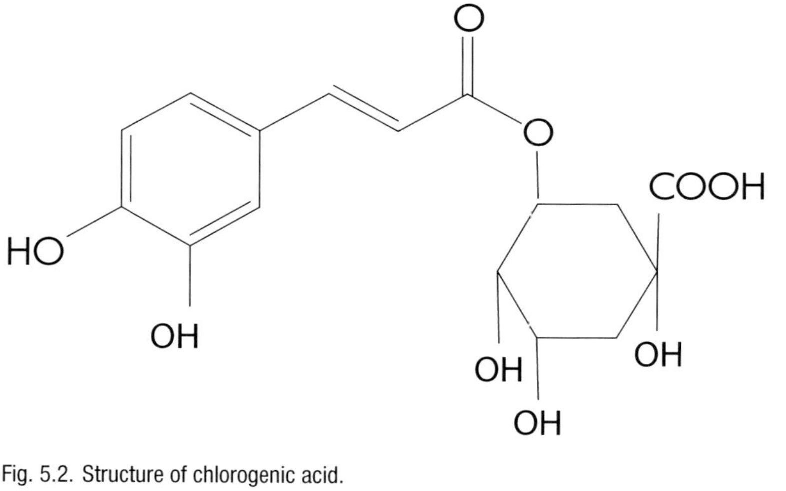

Fig. 5.2. Structure of chlorogenic acid.

Flavonoids belong to polyphenols and are usually glycosylated. They are planar molecules and their structural variation comes, in part, from the pattern of substitution: hydroxylation, methoxylation, prenylation, or glycosylation. Flavonoids have two aromatic rings enclosing a heterocyclic six-membered ring with oxygen (**Fig. 5.3**). They can be classified as anthocyanins, flavanols (catechins), flavones, flavanones, and flavonols. Isoflavones are structurally isomeric to flavonoids and along with

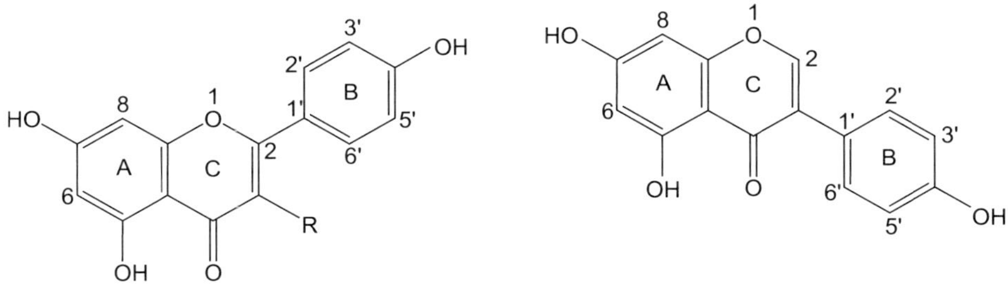

(I) (II)

Fig. 5.3. Basic structure of (**I**) flavonoids (R = OH: flavonols, R = H: flavones) and (**II**) isoflavones.

their metabolites they have structures similar to the mammalian estradiol (3).

The rapid and systematic measurement of phenolic acids and flavonoids is a serious challenge for analytical and food chemists, phytochemists, and biochemists because of the inherent structural diversity of the compounds, the dietary and health impact (4). This chapter focuses on the occurrence of phenolic acids and flavonoids in nature and their role in human health but most emphasis is placed on the different methodologies and techniques used in their analysis in a diversity of matrices. Steps in the sample treatment process are examined and chromatographic and less common spectrophotometric techniques are provided for the quantification of phenolics.

2. Sources and Their Role in Human Health

Phenolics are omnipresent in the plant kingdom. Insoluble phenolics are the components of cell walls, while soluble phenolics are compartmentalized within the plant cell vacuoles (5, 6). Phenolic acids are widely distributed in the diet, mostly in fruits, vegetables, coffee, wine, beer, and olive oil, mainly in esterified forms with organic acids, sugars, and lipids (7, 8). Although much is still unknown about the role of phenolic acids in plants, they have been associated with diverse functions, including nutrient uptake, protein synthesis, enzyme activity, photosynthesis, allelopathy, etc. Only a minor fraction exists in the free acid form; the major one is linked through ester, ether, or acetal bonds to larger polyphenols or smaller organic molecules (e.g., glucose, quinic, maleic, or tartaric acids) or other naturally occurring compounds (e.g., terpenes) (9, 10). This feature raises difficulties in the analysis of phenolic acids.

Yet, when discussing about phenolics in plant foods, flavonoids are the predominant class described, because they account for approximately two-thirds of the dietary phenols (11, 12). Although they are sometimes found as their aglycones, flavonoids most commonly exist in plant materials as flavonoid *O*-glycosides, in which one or more hydroxyl groups of the aglycones are bound to a sugar, forming an acid-labile glycosidic O–C bond (13). They contribute to the brilliant shades of blue, scarlet, and orange in leaves, flowers, and fruits. Apart from various vegetables and fruits, flavonoids are found in seeds, nuts, grains, spices, and different medicinal plants as well as in beverages, such as wine (particularly red wine), tea, and (at lower levels) beer. More specifically, the flavones apigenin and luteolin are common in cereal grains and aromatic herbs while their hydrogenated analogues hesperetin and naringin are almost exclusively present in citrus fruits. The flavonols quercetin and kaempferol are predominant in vegetables and fruits, isoflavones are found most often in legumes, including soybeans, black beans, green beans, and chick peas. The flavan-3-ols (+)-catechin, (−)-epicatechin, (−)-epigallocatechin and their gallate esters are widely distributed in plants and particularly in tea leaves. Anthocyanidins and their glycosides (anthocyanins) are natural pigments and abound in berries and red grape (14).

Phenolics are antioxidants with redox properties, which allow them to act as reducing agents, hydrogen donators, and singlet oxygen quenchers (15). Among dietary antioxidants, phenolics are by far the most abundant in most of the diets. Epidemiological studies have suggested associations between the consumption of phenolic-rich food and the prevention against oxidative stress-related diseases (11, 16, 17). In particular, the antioxidant properties of these compounds are often claimed for the protective effects of plant-based beverages against cardiovascular disease, certain forms of cancer, and photosensitivity reactions (18). A great deal of research has been performed on the in vitro and chemical antioxidant behavior of phenolics (19, 20). Rice-Evans et al. (21) reviewed the structure–antioxidant activity relationships, and Formica and Regelson (22) detailed the biology of flavonoids. Moreover, an increasing awareness and interest in the antioxidant behavior and potential health benefits associated with these simple phenolic acids has been noticed. Their role as dietary antioxidants has received the most attention in recent literature (23–25).

The estimated average consumption of phenolic acid has been reported to be in the order of 200 mg/day within a wide range (25 mg^{-1} g), depending on diet (fruit, vegetables, grains, teas, coffees, and spices) (7, 8). For individuals regularly consuming wine, coffee, beer, and tea, these beverages are likely the major sources of phenolics. However, the significance of this intake for metabolic effects and antioxidant status, in vivo unequivocally is affected by the bioavailability of these compounds.

They are important components in the human diet, although they are widely accepted as non-nutrients. The intake of flavonoids from diet is considerably high ranging between 50 and 800 mg/day, depending on the consumption of vegetables and fruit and of specific beverages such as red wine, tea, and unfiltered beer (26). Red wine and tea contain high levels (~200 mg per glass of red wine or cup of tea) of total phenols. Thus, variations in consumption of these beverages are mainly responsible for the overall flavonoid intake in different national diets. Another significant source of flavonoids is the different medicinal plants and related phytomedicines (27).

A small fraction of the ingested dietary flavonoids is absorbed in either the aglycon or glycoside form, while the major part is extensively degraded to different phenolic acids. Both the absorbed flavonoids and their metabolites may display an in vivo antioxidant activity, which seems to involve differently the physiological antioxidants (15).

Pharmacokinetic and pharmacodynamic studies are needed to investigate the fate of phenolics in the body after their ingestion with the diet. This will largely allow unraveling their potential role in the prevention of certain diseases and the entire human health.

3. Analytical Methods

3.1. Sample Treatment

The quality of a chemical analysis is as trustworthy as the step that is of the least accuracy. No doubt, sample treatment procedure is a step that significantly impacts data quality but also assay throughput. Over the years, many sample treatment methods have been developed to determine polyphenolic and simple phenolic compounds in various sample types. This step varies a great deal from protocol to protocol and among techniques owing to the diversity in matrices. The basic goals of the sample preparation process in the analysis of phenolic compounds are as follows:

I. Analytes isolation from the primary matrix and the associated interfering compounds present in the sample.

II. Phase switching to one suitable for the chosen analytical technique.

III. Enrichment of analytes to allow their determination by the analytical method chosen.

A parameter which should be taken into consideration when making the decision of the sample treatment technique consists of such information as the number of samples to be analyzed. The question is whether the planned procedure is going to be used in carrying out single, multiple, or even routine analysis. In the latter case, the techniques facilitating automation and low cost per analysis are favorable. However, one should bear in mind that the absence of the sample preparation stage in an analytical procedure is the most convenient and preferable way of sample preparation in order to minimize changes in the sample composition. Hence, reducing the number of operations performed on the primary sample should be aimed by the analyst.

It is significant to choose the optimal treatment of matrix, based on the chemical structures and properties of analyzed compounds. There are three main types of phenolic-containing matrices with respect to the treatment needs: plants, foods and liquid samples, the latter including biological fluids and beverages.

Sample preparation for the analysis of phenolics can range from the simple "filter-and-inject" procedures to the more elaborated hydrolysis/extraction/cleanup. Because of the assorted types of phenolics and the different matrices with many interfering components, the choice of the technique for the treatment differs from one another. In some cases, only a one-step extraction and/or simple cleanup procedure are sufficient before analysis, but the described assays almost invariably include two or more steps of preparation. Obviously, each step contributes to higher sensitivity and selectivity but, as mentioned above, it could increase the number of errors through introducing interferents and artifacts and decrease the recovery.

The solid samples are usually subject combinations of sieving, milling, or grinding and homogenization before further treatment for the determination of the extractable analytes. Liquid samples are filtered and/or centrifuged before isolation or separation and detection. In the case of wines, the alcohol is usually removed from the sample via rotary evaporation, and the residue is taken up in a small volume of the solvent for subsequent analysis. Air-drying or even more freeze-drying are universal steps for drying or condensation of primary samples.

An important aspect of phenolic analysis is whether to determine the target analytes in their various conjugated or free forms (e.g., aglycones). In neutraceuticals and food products, researchers are usually interested in the intact conjugates and, for the classification of plant species, intact flavonoid profiles in plants are determined. In many other instances, the knowledge of total aglycone content is required. In biological fluids, flavonoids exist as glucuronide and sulfate conjugates. Therefore, a hydrolysis-digestion step is used to disrupt glycoside or sulfur linkages. When the glycosylated form of the flavonoids is the case, digestion is omitted.

3.1.1. Hydrolysis of Phenolics

Acid hydrolysis and saponification are the most common means of releasing the phenolic acids, even though it is still unclear how much of the acids decompose under these conditions. Usually, acid ranges from 1 to 2 N aqueous HCl and the reaction times from 30 min to 1 h. Krygier et al. reported that loss under acidic conditions varies with the phenolic acid fluctuating from 15 to 95% for *o*-coumaric and sinapic acids, respectively (28). Saponification entails treating the sample with a solution of NaOH with reported concentrations ranging from 1 to 4 M. Some investigations recommend that such reactions be carried out in the dark, as well as under an inert atmosphere such as argon or nitrogen gases (29). Andreasen et al. discuss and compare several different enzyme preparations for the release of phenolic acids from the cell wall of rye grains (30). Yu et al. reported that a sequential acid, α-amylase, and cellulose hydrolysis might be applicable to the release of phenolic acids from barley (31).

In flavonoids, the hydrolyses frequently used to remove the sugar moieties from glycosides are acidic, basic or enzymatic. The hydrolysis process should be a compromise to minimize degradation reactions of glycosides and to achieve complete release of aglycones. Phenolic extract of sunflower honey was hydrolyzed in 2 N NaOH (32) while the glycosides of flavones and flavonols were hydrolyzed in refluxing 1.2 N HCl in 50% $MeOH/H_2O$ (v/v) (33). For plasma, serum, and urine, flavonoids may be first hydrolyzed with β-glucuronidase, sulfatase, or a mixture containing both enzymes (34). To ensure that these enzymes are active in the incubates, in certain cases, ^{13}C-labeled flavonoid conjugates have been available (35).

3.1.2. Extraction – Cleanup

Analytes have to be isolated, completely or to a certain degree, from the matrices in which they exist naturally, before analysis. Extraction serves this task along with the additional one to isolate analytes from potentially interfering sample components while bringing these analytes into a form suitable for analysis. Extraction of phenolic compounds is influenced, apart from the storage time and conditions, by their chemical nature, the extraction method employed, sample particle size as well as the presence of interfering substances.

Liquid–liquid and solid–liquid extractions are the most commonly used procedures prior to analysis of polyphenolics and simple phenolics. Widely accepted extraction solvents are alcohols (methanol, ethanol), acetone, diethyl-ether, and ethyl acetate. However, very polar phenolic acids (e.g., benzoic, cinnamic acids) cannot be extracted completely with pure organic solvents, and mixtures of alcohol–water or acetone–water are recommended. Process conditions such as pH, temperature, sample-to-solvent volume ratio, and the number and time intervals of individual extraction steps also play an important role in the extraction procedure. Extractions are repeated 2–3 times and extracts are combined.

Extraction of flavonoids from biological matrices is one of the fastest and less time consuming tasks. In a recent attempt, rat plasma was acidified with 0.25 N HCl, mixed with ethyl acetate, vortexed, and centrifuged (36). The upper organic phase was evaporated to dryness and the residue was reconstituted in the mobile phase for HPLC analysis. To quote another example, for quercetin and kaempferol in urine, 1 mL of 25% HCl was added to 4.0 mL of human urine and mixed well (37). After the urine sample had been hydrolyzed for 30 min at 80°C, 5.0 mL of ether was added. The hydrolyzed sample was extracted for 5 min and the 4.0 mL ethereal phase was evaporated to dryness under nitrogen stream. The residue was reconstituted with 100 μL of mobile phase and an aliquot of 20 μL of the resulting solution was injected into the HPLC system.

Soxhlet extraction is used less frequently to isolate flavonoids from solid samples. Various flavonoids or phenolic acids were extracted from *Tilia europea, Urtica dioica, Mentha spicata, Hypericum perforatum*, and *Echinacea purpurea* after 12 h Soxhlet extraction with methanol (38, 39).

A feature that exists with the above-mentioned conventional extractions is that they influence the integrity of flavonoid glycoside during the prolonged extraction, thus affecting the reproducibility and reliability (40). Supercritical fluid extraction has increasingly gained momentum in both food and pharmaceutical industries. The intrinsic low viscosity and high diffusivity of supercritical CO_2 has granted higher separation speed and efficiency to this mode of extraction, providing relatively clean extracts. The solvating power of a supercritical fluid is varied by controlling the pressure or by adding organic modifiers such as methanol. Supercritical fluid extraction was compared to Soxhlet extraction, steam distillation and maceration for the isolation of the active components present in chamomile flower heads (41). The recovery of the flavonoid apigenin obtained by supercritical CO_2 after a 30-min extraction at 200 atm and 40°C was 71.4% compared to Soxhlet extraction performed for 6 h and 124.6% compared to maceration performed for 3 days. However, the highly polar flavonoid apigenin-7-glucoside was not extracted by 100% CO_2 and the addition of the polar modifier methanol (5%, v/v) to the CO_2 fluid was indispensable. This technique is applicable to medicinal and other plant samples and it can be combined also with other sample preparation techniques.

Finally, microwave irradiation and sonication have been successfully used to enhance extraction of phenolic acids from *Echinacea purpurea* (42) while a newly designed dynamic microwave-assisted extraction system has been developed for the continuous and rapid extraction of flavonoids from *Saussurea medusa* Maxim dried cell cultures (43). By comparing the dynamic microwave-assisted extraction with dynamic solvent extraction without microwave assistance, the former showed obvious advantages related to the short extraction time and high efficiency.

Despite the merit of extraction in sample treatment, extra steps may be required for the removal of non-phenolic substances. Although the sample preparation for contemporary analytical techniques (e.g., liquid chromatography–(tandem) mass spectrometry) does not need to be as elaborate as others (e.g., liquid chromatography–ultraviolet), it remains pivotal to remove matrix components that might contaminate the system, when high sensitivity is needed, or give rise to unreliable results (44). In many cases, biological samples cannot be assayed directly but require a treatment to free from endogenous proteins, carbohydrates, salts, lipids, etc.

Solid-phase extraction is a reasonable choice for the preconcentration and cleanup procedure for crude plant extracts and biological matrices. Protein precipitation is the simplest means of sample treatment in order to analyze natural products in biofluids and should normally precede this kind of extraction. There is a consistence to the choice of sorbents for isolating the phenolic acids and flavonoids. The C_{18} bonded silica is the sorbent of choice; sample solution and solvents are usually slightly acidified to prevent ionization of the flavonoids, which would reduce their retention. In a recent study, different sample preparation methods for human plasma phenolic compounds (six phenolic acids, five flavonoids, *trans*-resveratrol, and tyrosol) were compared (45). These treatments included solid-phase extraction, extraction with methanol, removal of plasma proteins with different deproteinization agents, and inhibition of enzymatic plasma activity. Aiming to quantitate the whole set of compounds, in this case, the most suitable approach was to inhibit enzymatic activity and then deproteinize with acidified ethanol.

Hydrolyzed and non-hydrolyzed acidified urine were analyzed, elsewhere, by passing it through Amberlite XAD-2 particles and stirred to retain the phenolic compounds on the surface of the nonionic Amberlite particles (46). Two hundred and fifty microliters of plasma were diluted and acidified with 0.5% formic acid before application to a C_{18} sorbent. Urine was passed through Amberlite XAD-2 particles and stirred to retain the phenolic compounds on the surface of the nonionic Amberlite particles for hydrolyzed and non-hydrolyzed urine.

The prominent solid-phase microextraction mode was also employed to extract genistein and daidzein from human urine in combination with liquid chromatographic analysis. A Carbowax-templated poly(divinylbenzene) resin proved to be the best fiber, with a 5-min extraction at pH 4 and a temperature of 35°C, which accommodates the need for absence of organic solvents (47).

A relatively new solid-phase extraction method used a molecularly imprinted polymer as the sorbent to determine quercetin in red wine (48). The recovery was over 98% when using methanol containing 15% acetic acid or acetonitrile containing 10% aqueous triethylamine, as eluent. The molecularly imprinted polymer was proved to be highly selective for the target analyte enhancing, at the same time, the intensity of the quercetin and reducing the complexity of the chromatographic trace. Another molecularly imprinted polymer was evaluated toward six phenolic acids extracting selectively the analytes from *Melissa officinalis* (49).

Matrix solid-phase dispersion is an alternative for sample preparation workable for liquid and semi-liquid samples. Sample extraction and cleanup are carried out simultaneously with, generally, good recoveries and precision. Matrix solid-phase dispersion is frequently used to determine pesticides in, e.g., foods, but application to flavonoid analysis was reported only recently. For the determination of isoflavone aglycones and glycosides in *Radix astragali,* this extractive cleanup step was compared to Soxhlet and ultrasonic extraction with respect to the extraction capacity (50). For the aglycones, matrix solid-phase dispersion yielded the best extraction efficiency but for the glycosides Soxhlet proved to be more efficient.

4. Separation-Detection

The critical decision for the analyst to make is which analytical technique to use, depending on the expected composition of the sample, the designation of analytical expectations, and certainly on the instrument availability. In the main, separation techniques are particularly attractive, but also less common techniques such as spectrophotometric ones are applied.

4.1. Thin-Layer Chromatography (TLC)

Thin-Layer Chromatography (TLC) is especially useful for the rapid screening of plant or medicinal extracts for pharmacologically active substances, especially, because of the high sample throughput, prior to detailed analysis by other more modern instrumental techniques.

TLC plays a distinct role in the determination of phenolic acids in natural product analysis (51, 52). In most cases, silica is used as stationary phase, and plates are developed with either a combination of 2-(diphenyl boryoxo)ethylamine and polyethylene glycol or with $AlCl_3$. Detection is mainly performed using

UV light at 350–365 nm or 250–260 nm or with densitometry, at the same wavelengths.

Quantification, generally, is not the main goal of TLC studies. However, densitometry is used in several studies in this direction. Kaempferol and quercetin were determined in an extract of *Ginko biloba* leaves by scanning the HPTLC silica plates in the reflectance mode at 254 nm (53).

Two-dimensional TLC on cyanopropyl-bonded silica was used to separate eight flavonoids and three phenolic acids in *Flos sambuci* L. (54). The first dimension relied on a normal-phase separation for which seven binary eluents were tested, and the second one provided a reversed-phase separation, studied by using three binary eluents. From amongst the 21 combinations, the three best ones all contained *n*-hexane in the first, and water in the second dimension. More than 12 spots were discerned and 9 flavonoids and 3 phenolic acids were identified in the *Flos sambuci* L. extract.

Special attention is to be drawn on the native fluorescence of flavonoids. Their fluorescence properties were examined applying TLC separation with fluorodensitometric detection (55). The native fluorescence of 14 flavone and 26 flavonol type compounds was enhanced by their in situ reaction on the plate with 2-aminoethyl ester of diphenylboric acid.

4.2. Gas Chromatography

The high separation capacity and compatibility, with various kinds of detectors, renders gas chromatography (GC) a technique to be taken into consideration seriously, even if the additional stage of analytes derivatization has to be included into the analytical procedure. Gas chromatography is an important chromatographic technique employed for the analysis of phenolic compounds. In particular, when combined with mass spectrometry it offers high sensitivity and selectivity.

Preparation of samples for GC may include the removal of lipids or proteins, depending on the nature of sample, liberation of phenolics from ester and glycosidic bonds by alkali, acid, and enzymatic hydrolysis.

The significant concern with phenolic compounds is that they are not directly amenable to analysis by gas chromatography. In addition to sample extraction, isolation, and cleanup, analysis requires a chemical modification step, often referred to as derivatization. Nonetheless, Christov et al. described flame ionization and electron capture as detection methods in the analysis of underivatized acids (56). Earlier work with derivatized phenolics was typically performed with flame ionization detection but mass spectrometry (MS) has become widespread. Most of the GC–MS work is performed in the electron ionization mode, with the ionization voltage set to a standard 70 eV. The spectra are collected up to m/z 650 in scanning mode.

There are a variety of reagents used to modify and generate volatile derivatives via converting hydroxyl groups to ethers or esters. Prior to chromatography, phenolics are usually transformed to more volatile derivatives by methylation, conversion to trimethylsilyl derivatives, or derivatization with *N*- (*tert*-butyldimethylsilyl)-*N*-methyltrifluoroacetamide.Typically, in GC, flavonoids are hydrolyzed and converted into their trimethylsilyl derivatives, injected onto non-polar columns in the split or splitless mode and separated with a linear 30–90 min temperature program up to 300°C. The reaction involves dissolving the dried sample in a base (e.g., pyridine or ethylamine), addition of the trimethylsilyl reagent, and then heating the reaction vial up to 70°C, for 20–120 min. In an endeavor to speed up the silylation procedure, Chu et al. reasoned that the heat transfer was a slow process and devised a microwave derivatization procedure cutting the time to 30 s (57).

Hušek made use of ethyl and methyl chloroformate for the formation of ethyl and methyl esters, respectively (58). Dimethyl sulfoxide with methyl iodide in an alkaline medium is another alternative to methylation. However, methyl esters can lead to some confusion, as they are naturally occurring in some plant-based material. There are many advantages to generating the silylated derivatives instead of other derivatizating agents. Phenols and carboxylic acids are relatively reactive and are functional groups susceptible to silylation. Both functional groups (acids and phenols) are derivatized in the same step. Moreover, many of the minor products or artifacts have been well described and documented, are extremely volatile, and do not interfere with the analysis (59).

Plasma levels of catechin and its metabolite 3′-*O*-methylcatechin have been determined by GC–MS of the trimethylsilyl derivatives, after consumption of red wine (60). Glucuronide and sulfate conjugates were determined after enzymatic hydrolysis. In conventional GC, it is very difficult to analyze flavonoid glycosides even after derivatization. Pereira et al. used high-temperature–high-resolution GC–MS for the glucoside hesperidin, with columns that can withstand temperatures up to 400°C (61).

Phenolics of propolis were identified and quantitated subsequent to their derivatization with *N*,*O*-bis-(trimethylsilyl)-trifluoroacematide by flame ionization and MS detection (62, 63). Flavonoid content originating from the leaf extract of *Ginkgo biloba* in human urine (64) and in pharmaceutical preparations (65) has been determined in extracts immediately and subsequent to hydrolysis. One of the methods utilized gas chromatography – negative ion chemical ionization mass spectrometry of the trimethysilyl derivatives of the flavonoids. Hydrolysis of urine samples resulted in markedly higher quercetin and kaempferol content due to the fact that these two aglycones are present in the urine as glucuronides.

An improved derivatization procedure, by Stalikas et al., proposed an in-vial derivatization–extraction method for the GC–MS analysis of methylated flavonoids and phenolic acids in various herbs (38). Derivatization takes place under basic conditions so that the hydroxyl groups of the analytes are deprotonated. The anionic nucleophiles are transferred to the organic phase as ion-pairs using a phase-transfer catalyst and are next subjected to reaction with methyl iodide (66). Polymer-bound tri-*n*-butylmethylphosphonium chloride proved to be the best catalyst. In the selective-ion monitoring mode, good separation was attained (**Fig. 5.4**) and the limits of detection of the phenolics in the extracts were 4–40 ng/mL.

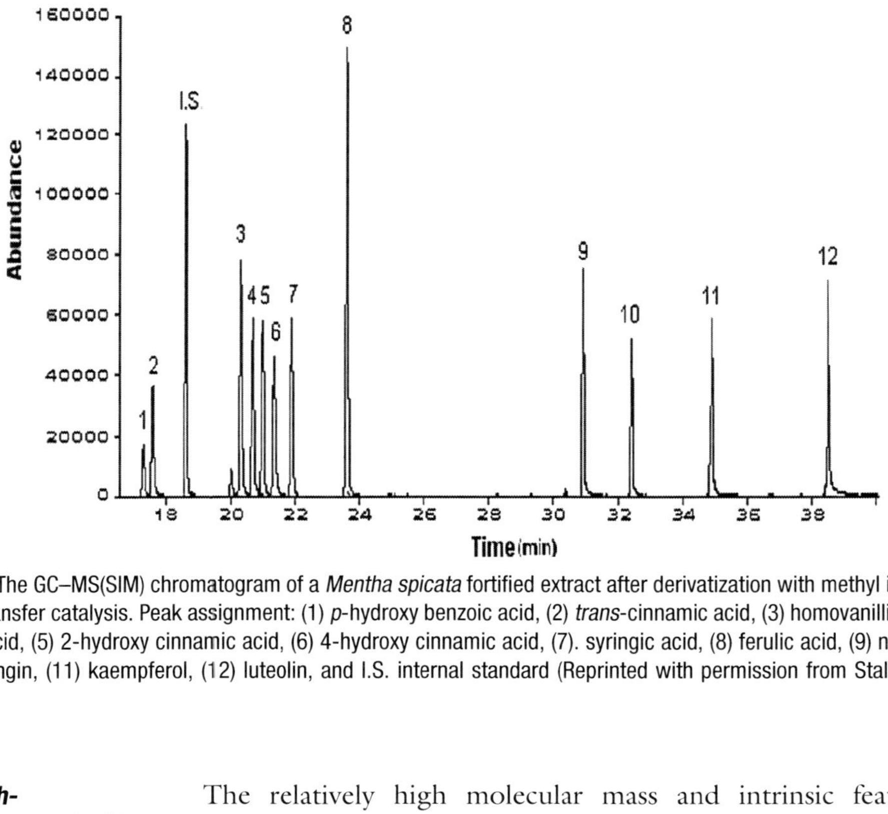

Fig. 5.4. The GC–MS(SIM) chromatogram of a *Mentha spicata* fortified extract after derivatization with methyl iodide and phase-transfer catalysis. Peak assignment: (1) *p*-hydroxy benzoic acid, (2) *trans*-cinnamic acid, (3) homovanillic acid, (4) vanillic acid, (5) 2-hydroxy cinnamic acid, (6) 4-hydroxy cinnamic acid, (7). syringic acid, (8) ferulic acid, (9) naringenin, (10) galangin, (11) kaempferol, (12) luteolin, and I.S. internal standard (Reprinted with permission from Stalikas et al. (38)).

4.3. High-Performance Liquid Chromatography

The relatively high molecular mass and intrinsic feature of hydrophobic flavonoid aglycones and hydrophilic flavonoid glycosides and phenolic acids prompt researchers to choose high-performance liquid chromatography (HPLC) for their assay. The introduction of reversed-phase (RP) columns has considerably enhanced HPLC separation of different classes of phenolic compounds so that, nowadays, most of the analytical work on phenolic compounds is performed in a RP mode. In particular, almost exclusively, RP C_{18} columns ranging from 100 to 250 mm in length and usually with an internal diameter of 3.9–4.6 mm are employed. Particle sizes are in the range of 3–10 μm. Narrow-bore columns (internal diameter 2 mm) are also recommended for special HPLC–MS applications (67).

The quality of sorbent (purity, end-capping) and the particle size do not have a very significant influence on the peak symmetry of the analyzed phenolic acids but the chromatographic resolution and the efficiency of the column are better for columns with very good free silanol group covering, end-capping, or embedding. For the analysis of phenolic compounds in beer with LC, separation conditions were optimized as regards 11 different stationary phases (all C_{18} bonded silicas) for a standard mixture of several flavone aglycones and glycosides (68). Four columns qualified as the most appropriate. In the same context, different RP columns of conventional dimensions were applied for the analysis of flavonoid glycosides. According to other authors, a column which is well end-capped should be preferred to counterparts lacking end-capping because residual silanol groups appear to impair the separation of flavonoid glycosides (69).

A fast RP-HPLC method for the simultaneous separation of 11 flavonoid aglycones was developed using a monolithic continuous-bed column (Chromolith Performance C_{18} end-capped, from Merck). The method was successfully applied to the analysis of these compounds in complex natural samples such as propolis and Ginkgo biloba (70).

Recently, improved chromatographic performance was achieved by the introduction of ultra-performance liquid chromatography (UPLC) capitalizing on basic chromatographic principles to perform separations using columns packed with smaller particles and/or at higher flow rates. Four flavonoids were determined in the flower of *Trollius ledibouri* from different sources (71). The analysis was performed on an AcQuity UPLC BEH C_{18} column using gradient elution with a mobile phase of 0.1% acetic acid and acetonitrile, over 20 min.

Stereochemistry in the studied field is rarely an issue in the recent literature. A systematic work on the enantiomeric separation of flavanones and the diastereomeric separation of flavanone glycosides has been reported by Ficarra et al., who utilized four chiral liquid chromatographic columns based on polysaccharide derivatives (72).

As regards the elution conditions, both isocratic and gradient elution are routinely applied for the analysis of phenolic compounds. The choice depends on the number and type of analytes and the nature of matrix of concern. Methanol, acetonitrile, and tetrahydrofuran are the most commonly used organic modifiers, as well as acidified aqueous solvents such as aqueous acetic acid, formic acid, phosphoric acid, or trifluoroacetic acid. In some cases, acetonitrile leads to better resolution in a shorter analysis time than methanol and, generally, acetonitrile gives sharper peak shapes, resulting in a higher plate number. Occasionally, tetrahydrofuran and 2-propanol as less polar solvents with their high elution strength have been used.

The greatest alteration observed in the mobile phases was the type of acid used as modifier to minimize peak tailing. Phenolic acids have pK_a values of about 4 while flavonoids presenting several ionizable hydroxyl groups have pK_a values relatively close to each other but certainly greater than 4. Therefore, the recommended pH range for the HPLC assay is 2–4. Dalluge et al. found that a deactivated C_{18} column in combination with trifluoroacetic acid as the acidic modifier of the mobile phase, greatly improves peak shape and reproducibility of retention times of catechins in tea (73). Besides, the use of aqueous buffers (citrate, phosphate, acetate) instead of the addition of acid is recommended at concentrations from 5 to 50 mM.

Free phenolic acids (chlorogenic, protocatechuic, *p*-hydroxybenzoic, caffeic, vanillic, syringic, *p*-coumaric, and ferulic) could be separated in medicinal plants or pharmaceutical preparations using a simple isocratic mobile phase (methanol–water–acetic acid) (74). When phenolic acids of different chemical structures and different polarities have to be analyzed simultaneously, gradient elution is indispensable. As a general observation, phenolic acids are eluted from RP columns according to decreasing polarities. The loss of polar hydroxy groups and the presence of the methoxy groups or ethylenic side chains could decrease the polarity and increase the retention time.

Some phenolic acids could be present in natural plants as geometric isomers. The greatest number of phenolic acids occurs in nature as *trans*-isomers, but on exposure to UV radiation or daylight they are gradually transformed to *cis*-isomers, which elute usually, before *trans*-isomers. Their simultaneous separation is usually possible using RP stationary phases including an optically active molecule in the mobile phase (75).

Finally, as far as detection means is concerned, phenolics are commonly detected using ultraviolet-visible (UV-VIS), photodiode array (DAD), and fluorescence detectors. Because every phenol exhibits a higher or lower absorption of UV or UV-VIS light, given the intrinsic existence of conjugated double and aromatic bonds, UV detection is the ideal method to localize a phenol in the effluent of a column. Most of the benzoic acid derivatives have absorption maxima at 246–262 nm, with a shoulder at 290–315 nm, except for gallic and syringic acids with absorption maxima at 271 and 275 nm, respectively. The cinnamic acids absorb in two ranges: 225–235 and 290–330 nm. Detection at 280 nm is the most generally used wavelength for the simultaneous separation of mixtures of phenolic acids.

All flavonoid aglycones contain at least one aromatic ring and, consequently, efficiently absorb UV light. The first maximum, which is localized in the 240–285 nm range, is due to the A-ring and the second maximum in the 300–550 nm range is attributed to the substitution pattern and conjugation of the C-ring. Reasonably, UV detection became the preferred tool in LC-based analyses and, even today, LC with multiple-wavelength or DAD is the prevalent tool in studies dealing with, e.g., screening, quantification, and provisional sub-group classification.

For a comprehensive HPLC-UV method of flavonol determination in human urine and plasma, readers can refer to the publication of Hollman (76).

Fluorescence detection in phenolic acid analysis is scant. The nature of the functional groups and their substitution pattern determine whether a particular flavonoid is fluorescent or not. In case that fluorescence detection is feasibile in combination with UV, it offers the possibility to discriminate between fluorescent and non-fluorescent co-eluting compounds (77). To extend the application range of fluorescence detection, derivatization has been used. For example, quercetin, kaempferol, and morin, with their 3-OH, 4-keto substituents, can form complexes with metal cations, some of which are highly fluorescent (78, 79).

Other systems used for the detection of phenolics encompass electrochemical, mass spectrometric, and nuclear magnetic resonance (NMR) detectors.

Electrochemical detection is very sensitive for the compounds that can be oxidized or reduced at low-voltage potentials. Phenolic acids in food and human plasma extracts are routinely detected by HPLC-electrochemical coulometric detection +600 mV (80). A multi-channel coulometric detection system, being compatible with gradient elution, may serve as a highly sensitive way for detecting phenolic acids and flavonoids in a wide range of samples well as it can be applicable to the overall characterization of antioxidants (81).

For many years, liquid chromatography–mass spectrometry (MS) systems have been applied for the detection and identification of flavonoid glycosides in plants extracts and various biological fluids. In some cases, HPLC with different sensitivity detectors (UV, electrochemical, fluorescence) and HPLC–MS are simultaneously used for the identification and quantification of phenolics in natural plants and related food products.

Electrospray ionization MS has been employed for structural confirmation of phenolics in plums, peaches, grapeseeds, soyfood, cocoa, olive oil, and walnut leaves (82) in human urine and plasma (83) with a polymeric chromatographic column. Conventional RP columns were coupled to DAD detector and a magnetic sector-type MS equipped with an electrospray ionization source was applied to the analysis of flavonoid glycosides in Crataegus extract (69). It was demonstrated that UV spectra and first-order electrospray ionization mass spectra allowed a fast characterization of flavonoids even if reference compounds are not at hand or available.

Electrospray ionization and atmospheric pressure chemical ionization interfaces were both used combined with HPLC for the quantitative analysis of flavonoids and their metabolites in biofluids (84). The detection was carried out in the positive or negative ion mode; data could be collected in multiple-reaction monitoring, selective-reaction monitoring, and selective-ion monitoring mode.

Detailed investigations on the identification of flavonoid metabolites after the consumption of onions were published by Mullen et al. (85). The analysis was performed by LC-ion-trap MS and 23 metabolites of quercetin were identified, which are illustrated in HPLC traces of **Fig. 5.5**. The different classes of metabolites that were detected can be summarized as follows: quercetin monoglucuronides, quercetin diglucuronides, methylquercetin monoglucuronides, methylquercetin diglucuronides, quercetin, quercetin sulfates, quercetin glucuronide sulfates, quercetin glucosides, quercetin glucoside glucuronides, and quercetin glucoside sulfates.

Unquestionably, NMR is the technique that generates more information for unambiguous identification of a molecule. In a

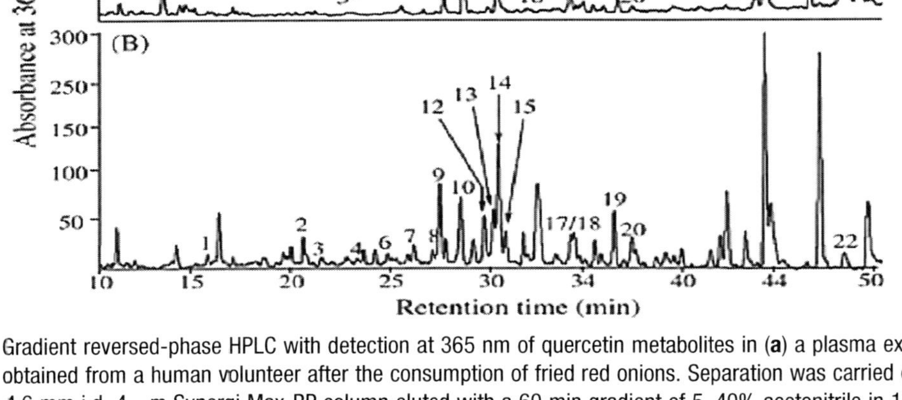

Fig. 5.5. Gradient reversed-phase HPLC with detection at 365 nm of quercetin metabolites in (**a**) a plasma extract and (**b**) urine obtained from a human volunteer after the consumption of fried red onions. Separation was carried out using a 250 × 4.6 mm i.d. 4 μm Synergi Max-RP column eluted with a 60 min gradient of 5–40% acetonitrile in 1% formic acid, at a flow rate of 1 mL/min and maintained at 40°C. Peaks 1–23 belong to the compound classes reported in the text (Reprinted with permission from Mullen et al. (85)).

recent study on the flavonoid constituents of a red clover extract, stopped-flow LC-NMR and stand-alone NMR were used to identify structural isomers that could not be distinguished on the basis of MS/MS information (86). A recently developed cryoflow NMR probe exhibits detectability about fourfold better than with conventional probes or, alternatively, the scan time is 16-fold shorter for the same amount of sample. The probe has been applied for the analysis of an oregano extract where five flavonoids were identified using an LC–UV–solid-phase extraction–NMR–MS setup (87).

Other less common means of detection, coupled to LC, have been through refractive index and evaporative light scattering detection. The latter offers freedom from some of the limitations of spectroscopic detection because it is not limited to compounds that contain UV-absorbing chromophores and it is immune to mobile-phase variations and gradient baseline shifting. Although being less sensitive than those previously described, both of the detectors have been successfully used: the HPLC-refractive index system in the quantification of (3,3′,4,4′,5,7-hexahydroxyflavan) in unripe banana pulp (88) and the HPLC-evaporative light scattering detection in the determination of *Radix Astragali* flavonoids (89).

4.4. Capillary Electrophoretic and Capillary Electrochromatographic Methods

Capillary electrophoretic and electrochromatographic modes gained intrance as routine methods ranging from environmental to pharmaceutical and biochemical analysis. Likewise, they can be useful additions to the arsenal of chromatographers for phenolics. Typically, phosphate or borate buffers, capillaries of 50–100 μm I.D., voltages of 10–30 kV, and 10–50 nL injection volumes have been used. Detection is usually performed with UV but also electrochemical and MS detectors are not neglected. In a number of papers, the merits and drawbacks of HPLC and electromigration methods, as utilized in routine analysis of phenolics were discussed (90, 91).

All background electrolytes are alkaline to maintain a degree of ionization and sufficient separation of polyphenols thanks to the complex-formation ability of borate (92). It was also found that the migration times of the analytes increased with the increase in borate concentration due to the stronger interaction of borate and flavonoids at high borate concentration.

Separation of flavonoid-3-*O*-glycosides differing in their sugar moiety and flavonoid-7-*O*-glycosides varying in their aglycones were separated as borate complexes by capillary electrophoresis (CE) (93) and micellar electrokinetic chromatography (MEKC) (94), applying UV detection. In both cases, authors decided that CE and MEKC gave higher efficiency, selectivity, and speed compared to HPLC. In CE, 0.2 M borate buffer (pH 10.5) was employed, while for MEKC, the separation of flavonoid-7-*O*-glycosides in the presence of sodium dodecyl sulfate under neutral conditions (pH 7.1) proved to be the optimum.

Huang et al. compared the microemulsion electrokinetic chromatography with MEKC methods for the analysis of phenolic compounds (95). A higher voltage and a higher column temperature improved the separation efficiency, without any noticeable reduction in resolution, for microemulsion electrokinetic chromatography, although they caused a poor resolution for the MEKC system.

Other authors used an amperometric system based on the end-column wall-jet configuration, in which the working electrode is placed at the outlet of the separation capillary (96). This configuration allows the use of normal size (i.e., >100-μm diameter) working electrodes without introducing significant post-capillary zone broadening which could jeopardize separation efficiency.

A handful of papers discuss the use of CE–MS for the determination of flavonoids and phenolic acids (97). In the CE–ESI–MS study by Lafont et al., a standard mixture of eight phenolic compounds was analyzed (98). With selective-ion monitoring–MS the authors were able to identify all eight compounds based on their retention times and characteristic fragment ions ($[M-H]^-$, loss of CO, CO_2, and CH_3) and obtained LODs of 0.1–40 μg/L. Huck et al. developed a CE method for the analysis of a flavonoid mixture consisting of 5-methoxyflavone, biochanin A, hesperetin, and naringenin obtained from plant extracts (99). carrier electrolyte based on ammonium acetate and a bare fused silica capillary suitable for coupling to a quadrupole MS were selected. Optimization of parameteric values for both CE and MS provided detection limits comparable to UV detection and conclusive proof-of-application was produced by determining naringenin in a phytomedicine prepared from five different herbal drugs (**Fig. 5.6**). The authors concluded that the selectivity of UV detection was much too low to enable reliable quantitative analysis in this complex sample. Although CE–MS is not a technique that will replace others, it can be made into a complementary analytical tool for the analysis of phenolics.

4.5. Spectrophotometric Methods

The development of a satisfactory ultraviolet spectrophotometric assay is a rather cumbersome and difficult task and highly dependent on the material to be analyzed. In addition, the characteristics of absorption spectra of phenolics are affected by the nature of the solvent and the pH of the solution (100). Visible spectroscopic techniques are often used for the identification of isolated phenolic compounds, particularly flavonoids to identify the presence of groups of predominant phenolic compounds (101, 102). The Folin–Denis assay is the first and most widely used rapid reaction procedure for the quantification of total

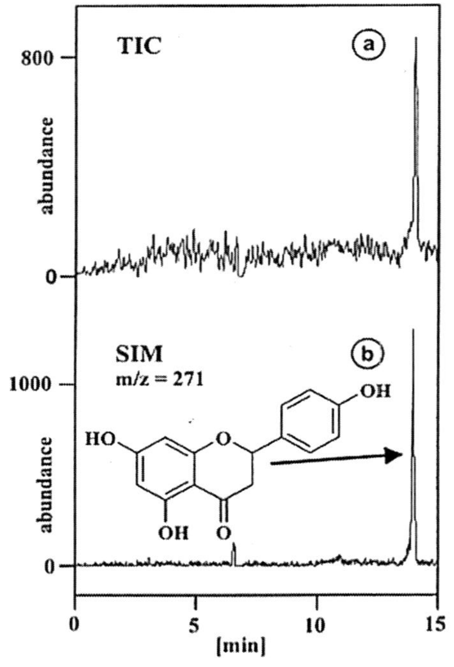

Fig. 5.6. CE–ESI–MS of naringenin in a phytomedicine prepared from five different herbal drugs: (**a**) total ion current (TIC) and (**b**) single-ion monitoring trace (SIM). Capillary, fused silica, 60 cm, 50 μm I.D.; buffer, 40 mM ammonium acetate (pH 9.5), 15% acetonitrile (v/v); detection, negative mode (Reprinted with permission from Huck et al. (99)).

phenolics in plant materials (103). The method relies on the reduction of phosphomolybdic–phosphotungstic acid (Folin–Denis) reagent to a blue colored complex in an alkaline solution. The generated phosphomolybdic–phosphotungstic–phenol complex gives an easily detected absorbance at 760 nm. The Folin–Ciocalteu assay is also used for the determination of the total content of plant food phenolics (104, 105). Neither of the above-mentioned is specific as they detect all phenolic groups; besides, they suffer from interference from reducing agents such as ascorbic acid.

The vanillin method is specific for flavan-3-ols, dihydrochalcones, and proanthocyanidins which have a single bond at the 2,3-position and possess free *meta*-hydroxy groups on the B-ring (106). This is the reason that this assay is mainly recognized as a useful method for the detection and quantification of proanthocyanidins in plant materials.

The complexation of phenolics with Al(III) has been reported for the development of spectrophotometric methods to determine of total caffeic acid, total flavonoids, and total tannins at 425 nm (107, 108). The method is based on the formation of a complex between the aluminum ion, Al(III), and the carbonyl and hydroxyl groups of the flavonoid.

For the quantification of flavanones and dihydroflavonols, the spectrophotometric method based on the interaction of these compounds with 2,4-dinitrophenylhydrazine in acidic media (sulfuric acid) to form colored phenylhydrazones has been reported (109). The absorbance was measured at 486 nm.

Quantification of anthocyanins takes advantage of their characteristic behavior in acidic media. The analytical procedure for quantification of anthocyanins was first developed by Sondheimer and Kertesz (110) and modified by Swain and Hillis (111) who suggested expressing the concentration of pigments in terms of the change in the absorbance at λ_{max} between pH 3.5 and pH $\leq$ 1.0.

Finally, Schulz et al. (112) used a near-infrared reflectance spectroscopic method for prediction of polyphenols in the leaves of green tea (*Camelia sinensis* L) while Edelmann et al. (113) developed a rapid method of discrimination of Austrian red wines based on mid-infrared spectroscopy of phenolic extracts of wine.

5. Conclusions

Research on phenolic acids and flavonoids is of current interest due to their important biological and pharmacological properties which have shown in vitro benefits to human health, although more in vivo studies are needed to ascertain the propitious effects. Several hundred papers on the analysis of phenolic compounds have been published in the past 20 years or so. Because of the great assortment of matrices where phenolics exist, the impetus for developing analytical methods for phenolics has been proved to be multifaceted. Sample preparation is a critical step of the entire assay. Special attention should be paid to the need for detecting phenolic compounds in their conjugated or free forms, before extracting-isolating the analytes. For the separation of phenolic compounds, although all the gamut of separation techniques can be of great use, liquid chromatography has become a rather indispensable integral part of most of the analytical methods. However, as with any analytical technique, it has its own unique set of potential caveats mostly centered on the incomplete separation and sensitivity to unwanted sample components of the several matrices. But the possibility of coupling it to several detection devices has turned it into an even more valuable tool. In addition, the emergence of UPLC coupled to MS appeared as an alternative to traditional HPLC techniques. The strengths of UPLC technology promote the ability to separate and identify drug compounds with significant gains in resolution and sensitivity and marked reductions in the overall analysis time.

References

1. Harborne, J.B. (1982) *Introduction to Ecological Biochemistry*. 2nd edn., Academic Press, New York, NY.
2. Beckman, C.H. (2000) Phenolic-storing cells: Keys to programmed cell death and periderm formation in wilt disease resistance and in general defence responses in plants? *Physiol. Mol. Plant Pathol.* **57**, 101–110.
3. Bingham, S.A., Atkinson, C., Liggins, J., Bluck, L., and Coward, A. (1998) Phytooestrogens: Where are we now? *Br. J. Nutr.* **79**, 393–406.
4. Stalikas, C.D. (2007) Extraction, separation, and detection methods for phenolic acids and flavonoids. *J. Sep. Sci.* **30**, 3268–3295.
5. Wink, M. (1997) Compartmentation of secondary metabolites and xenobiotics in plant vacuoles. *Adv. Bot. Res.* **25**, 141–169.
6. Bengoechea, M.L., Sancho, A.I., Bartolomé, B., Estrella, I., Gómez-Cordovés, C., and Hernández, T. (1997) Phenolic Composition of industrially manufactured purées and concentrates from peach and apple fruits. *J. Agric. Food Chem.* **45**, 4071–4075.
7. Clifford, M.N. (1999) Chlorogenic acids and other cinnamates-nature, occurrence and dietary burden. *J. Sci. Food Agric.* **79**, 362–372.
8. Herrmann, K. (1989) Occurrence and content of hydroxycinnamic and hydroxybenzoic acid compounds in foods. *Crit. Rev. Food Sci. Nutr.* **28**, 315–347.
9. Hartley, R.D. and Jones, E.C. (1976) Carbohydrates and carbohydrate esters of ferulic acid released from cell walls of Lolium multiflorum by treatment with cellulolytic enzymes. *Phytochemistry* **15**, 305–307.
10. Winter, M. and Herrmann, K. (1986) Esters and glucosides of hydroxycinnamic acids in vegetables. *J. Agric. Food Chem.* **34**, 616–620.
11. Scalbert, A. and Williamson, G. (2000) Dietary intake and bioavailability of polyphenols. *J. Nutr.* **130**, 2073S–2085S.
12. Bors, W., Heller, W., Michel, C., and Stettmaier, K. (1996) Flavanoids and polyphenols: Chemistry and biology. In *Handbook of Antioxidants* (Packer, L., and Cadenas, E., Eds.), Marcel Dekker, New York, NY, pp. 409–466.
13. Croteau, R., Kutchan, T.M., and Lewis, N.G. (2000) Natural products (Secondary metabolites). In *Biochemistry & Molecular Biology of Plants (*Buchanan, B., Gruissem, W., Jones, R., Eds.), American Society of Plant Physiologists, Rockville, IN, pp. 1250–1318.
14. Haslam, E. (1989) *Plant Polyphenols*, Cambridge University Press, Cambridge, UK.
15. Pietta, P.-G. (2000) Flavonoids as antioxidants. *J. Nat. Prod.* **63**, 1035–1042.
16. Block, G., Patterson, B., and Subar, A. (1992) Fruit, vegetables and cancer prevention: A review of the epidemiological evidence. *Nutr. Cancer* **18**, 1–30.
17. Fuhrman, B., Lavy, M., and Aviram, M. (1995) Consumption of red wine with meals reduces the susceptibility of human plasma and low-density lipoprotein to lipid peroxidation. *Am. J. Clin. Nutr.* **61**, 549–554.
18. Bravo, L. (1998) Polyphenols: Chemistry, dietary sources, metabolism, and nutritional significance. *Nutr. Rev.* **56**, 317–333.
19. Laranjinha, J.A.N., Almeida, L.M., and Madeira, V.M.C. (1994) Reactivity of dietary phenolic acids with peroxyl radicals: Antioxidant activity upon low-density lipoprotein peroxidation. *Biochem. Pharmacol.* **48**, 487–494.
20. Morton, L.W., Croft, K.D., Puddey, I.B., and Byrne, L. (2000) Phenolic acids protect low-density lipoproteins from peroxynitrite-mediated modification in vitro. *Redox Rep.* **5**, 124–125.
21. Rice-Evans, C.A., Miller, N.J., and Paganga, G. (1996) Structure-antioxidant activity relationships of flavonoids and phenolic acids. *Free Radic. Biol. Med.* **20**, 933–956.
22. Formica, J.V. and Regelson, W. (1995) Review of quercetin and related bioflavonoids. *Food Chem. Toxicol.* **33**, 1061–1080.
23. Rice-Evans, C.A., Miller, N.J., and Paganga, G. (1997) Antioxidant properties of phenolic compounds. *Trends Plant Sci.* **2**, 152–159.
24. Laranjinha, J. (2002) Caffeic acid and related antioxidant compounds: Biochemical and cellular effects. In *Handbook of Antioxidants* (Packer, L., and Cadenas, E., Eds.), Marcel Dekker, New York, NY, pp. 279–302.
25. Lodovici, M., Guglielmi, F., Meoni, M., and Dolara, P. (2001) Effect of natural phenolic acids on DNA oxidation in vitro. *Food Chem. Toxicol.* **39**, 1205–1210.
26. Larson, R.A. (1988) The antioxidants of higher plants. *Phytochemistry* **27**, 969–978.
27. Pietta, P.G. (1998) Flavonoids in medicinal plants. In *Flavonoids in Health and Disease*

(Rice-Evans, C.A., and Packer, L., Eds.), Marcel Dekker, New York, NY, pp. 61–110.

28. Krygier, K., Sosulski, F., and Hogge, L. (1982) Free, esterified, and insoluble-bound phenolic acids. 1. Extraction and purification procedure. *J. Agric. Food Chem.* **30**, 330–334.
29. Rommel, A. and Wrolstad, R.E. (1993) Influence of acid and base hydrolysis on the phenolic composition of red raspberry juice. *J. Agric. Food Chem.* **41**, 1237–1241.
30. Andreasen, M.F., Christensen, L.P., Meyer, A.S., and Hansen, Å. (1999) Release of hydroxycinnamic and hydroxybenzoic acids in rye by commercial plant cell wall degrading enzyme preparations. *J. Sci. Food Agric.* **79**, 411–413.
31. Yu, J., Vasanthan, T., and Temelli, F. (2001) Analysis of phenolic acids in barley by high-performance liquid chromatography. *J. Agric. Food Chem.* **49**, 4352–4358.
32. Sabatier, S., Amiot, M.J., Tacchini, M., and Aubert, S. (1992) Identification of flavonoids in sunflower honey. *J. Food Sci.* **57**, 773–777.
33. McDonald, M.S., Hughes, M., Burns, J., Lean, M.E.J., Matthews, D., and Crozier, A. (1998) Survey of the free and conjugated myricetin and quercetin content of red wines of different geographical origins. *J. Agric. Food Chem.* **46**, 368–375.
34. Prasain, J.K., Wang, C.-C., and Barnes, S. (2004) Mass spectrometric methods for the determination of flavonoids in biological samples. *Free Radic. Biol. Med.* **37**, 1324–1350.
35. Setchell, K.D., Faughnan, M.S., Avades, T., Zimmer-Nechemias, M.F., Brown, N.M., Wolfe, B.E., Brashear, W.T., Desai, P., Oldfield, M.F., Botting, N.P., and Cassidy, A. (2003) Comparing the pharmacokinetics of daidzein and genistein with the use of ^{13}C-labeled tracers in premenopausal women. *Am. J. Clin. Nutr.* **77**, 411–419.
36. Li, L., Liang, S., Du, F., and Li, C. (2007) Simultaneous quantification of multiple licorice flavonoids in rat plasma. *J. Am. Soc. Mass Spectrom.* **18**, 778–782.
37. Wang, F.M., Yao, T.W., and Zeng, S. (2003) Determination of quercetin and kaempferol in human urine after orally administrated tablet of ginkgo biloba extract by HPLC. *J. Pharm. Biomed. Anal.* **33**, 317–321.
38. Fiamegos, Y.C., Nanos, C.G., Vervoort, J., and Stalikas, C.D. (2004) Analytical procedure for the in-vial derivatization-extraction of phenolic acids and flavonoids in methanolic and aqueous plant extracts followed by gas chromatography with mass-selective detection. *J. Chromatogr. A* **1041**, 11–18.
39. Glowniak, K., Zgorka, G., and Kozyra, M. (1996) Solid-phase extraction and reversed-phase high-performance liquid chromatography of free phenolic acids in some Echinacea species. *J. Chromatogr. A* **730**, 25–29.
40. Song, J.-Z., Mo, S.-F., Yip, Y.-K., Qiao, C.-F., Han, Q.-B., and Xu, H.-X. (2007) Development of microwave assisted extraction for the simultaneous determination of isoflavonoids and saponins in *Radix astragali* by high-performance liquid chromatography. *J. Sep. Sci.* **30**, 819–824.
41. Scalia, S., Giuffreda, L., and Pallado, P. (1999) Analytical and preparative supercritical fluid extraction of chamomile flowers and its comparison with conventional methods. *J. Pharm. Biomed. Anal.* **21**, 549–558.
42. Pomponio, R., Gotti, R., Hudaib, M., and Cavrini, V. (2002) Analysis of phenolic acids by micellar electrokinetic chromatography: Application to *Echinacea purpurea* plant extracts. *J. Chromatogr. A* **945**, 239–247.
43. Gao, M., Song, B.-Z., and Liu, C.-Z. (2006) Dynamic microwave-assisted extraction of flavonoids from Saussurea medusa Maxim cultured cells. *Biochem. Engin. J.* **32**, 79–83.
44. Guo, J.F., Zhao, Y.M., Zhao, L., Zhang, W.Q., Zhang, A.J., and Xu, B. (2006) Simultaneous quantification of CTN986 and its deglycosylation products in rat serum using liquid chromatography/tandem mass spectrometry. *Rapid Commun. Mass Spectrom.* **20**, 1701–1708.
45. Martinez-Ortega, M.V., Garcia-Parilla, M.C., and Troncoso, A.M. (2004) Comparison of different sample preparation treatments for the analysis of wine phenolic compounds in human plasma by reversed phase high-performance liquid chromatography. *Anal. Chim. Acta* **502**, 49–55.
46. Watson, D.G. and Pitt, A.R. (1998) Analysis of flavonoids in tablets and urine by gas chromatography/mass spectrometry and liquid chromatography/mass spectrometry. *Rapid Commun. Mass Spectrom.* **12**, 153–156.
47. Satterfield, M. and Brodbelt, J.S. (2001) Structural characterization of flavonoid glycosides by collisionally activated dissociation of metal complexes. *J. Am. Soc. Mass Spectrom.* **12**, 537–549.
48. Molinelli, A., Weiss, R., and Mizaikoff, B. (2002) Advanced solid phase extraction

using molecularly imprinted polymers for the determination of quercetin in red wine. *J. Agric. Food Chem.* **50**, 1804–1808.
49. Karasová, G., Lehotay, J., Sádecká, J., Skačáni, I., and Lachová, M. (2005) Selective extraction of derivates of *p*-hydroxybenzoic acid from plant material by using a molecularly imprinted polymer. *J. Sep. Sci.* **28**, 2468–2476.
50. Xiao, H.B., Krucker, M., Albert, K., and Liang, X.M. (2004) Determination and identification of isoflavonoids in *Radix astragali* by matrix solid-phase dispersion extraction and high-performance liquid chromatography with photodiode array and mass spectrometric detection. *J. Chromatogr. A* **1032**, 117–124.
51. Ragazzi, E. and Veronese, G. (1973) Quantitative analysis of phenolic compounds after thin-layer chromatographic separation. *J. Chromatogr.* **77**, 369–375.
52. Schulz, J.M. and Herrmann, K. (1980) Analysis of hydroxybenzoic and hydroxycinnamic acids in plant material. I. Sample preparation and thin-layer chromatography. *J. Chromatogr.* **195**, 85–94.
53. Jamshidi, A., Adjvadi, M.,, and Husain, S.W. (2000) Determination of kaempferol and quercetin in an extract of *Ginkgo biloba* leaves by high-performance thin-layer chromatography (HPTLC). *J. Planar Chromatogr.* **13**, 57–59.
54. Hawryl, M.A., Hawryl, A., and Soczewinski, E. (2002) Application of normal- and reversed-phase 2D TLC on a cyanopropyl-bonded polar stationary phase for separation of phenolic compounds from the flowers of *Sambucus nigra* L. *J. Planar Chromatogr.* **15**, 4–10.
55. Kartnig, T. and Göbel, I. (1996) Effect of fluorescence intensifiers on the fluorodensitometric determination of flavones and flavonols after detection with diphenylboric acid 2-aminoethyl ester. *J. Chromatogr. A* **740**, 99–107.
56. Christov, R. and Bankova, V. (1992) Gas chromatographic analysis of underivatized phenolic constituents from propolis using an electron-capture detector. *J. Chromatogr.* **623**, 182–185.
57. Chu, T.-Y., Chang, C.-H., Liao, Y.-C., and Chen, Y. (2001) Microwave-accelerated derivatization processes for the determination of phenolic acids by gas chromatography-mass spectrometry. *Talanta* **54**, 1163–1171.
58. Hušek, P. (1992) Fast derivatization and GC analysis of phenolic acids. *Chromatographia* **34**, 621–626.
59. Little, J.L. (1999) Artifacts in trimethylsilyl derivatization reactions and ways to avoid them. *J. Chromatogr. A* **844**, 1–22.
60. Donovan, J.L., Bell, J.R., Kasim-Karakas, S., German, J.B., Walzem, R.L., Hansen, R.J., and Waterhouse, A.L. (1999) Catechin is present as metabolites in human plasma after consumption of red wine. *J. Nutr.* **129**, 1662–1668.
61. Pereira, A.D.S., Padilha, M.C., and Neto, F.R.D.A. (2004) Two decades of high temperature gas chromatography (1983–2003). What's Next? *Microchem. J.* **77**, 141–149.
62. Bankova, V., Christov, R., Stoev, G., and Popov, S. (1992) Determination of phenolics from propolis by capillary gas chromatography. *J. Chromatogr.* **607**, 150–153.
63. Prytzyk, E., Dantas, A.P., Salamo, K., Pereira, A.S., Bankova, V.S., DeCastro, S.L., and Neto, F.R.A. (2003) Flavonoids and trypanocidal activity of Bulgarian propolis. *J. Ethnopharmacol.* **88**, 189–193.
64. Watson, D.G. and Oliveira, E.J. (1999) Solid-phase extraction and gas chromatography-mass spectrometry determination of kaempferol and quercetin in human urine after consumption of Ginkgo biloba tablets. *J. Chromatogr. B* **723**, 203–210.
65. Deng, F. and Zito, S.W. (2003) Development and validation of a gas chromatographic–mass spectrometric method for simultaneous identification and quantification of marker compounds including bilobalide, ginkgolides and flavonoids in *Ginkgo biloba* L. extract and pharmaceutical preparations. *J. Chromatogr. A* **986**, 121–127.
66. Fiamegos, Y.C. and Stalikas, C.D. (2005) Phase-transfer catalysis in analytical chemistry. *Anal. Chim. Acta* **550**, 1–12.
67. Ryan, D., Robards, K., and Lavee, S. (1999) Determination of phenolic compounds in olives by reversed-phase chromatography and mass spectrometry. *J. Chromatogr. A* **832**, 87–96.
68. Řehová, L., Škeříková, V., and Jandera, P. (2004) Optimisation of gradient HPLC analysis of phenolic compounds and flavonoids in beer using a CoulArray detector. *J. Sep. Sci.* **27**, 1345–1359.
69. Cuyckens, F. and Claeys, M. (2002) Optimization of a liquid chromatography method based on simultaneous electrospray ionization mass spectrometric and ultraviolet photodiode array detection for analysis of flavonoid glycosides. *Rapid Commun. Mass Spectrom.* **16**, 2341–2348.

70. Repollés, C., Herrero-Martínez, J.M., and Ràfols, C. (2006) Analysis of prominent flavonoid aglycones by high-performance liquid chromatography using a monolithic type column. *J. Chromatogr. A* **1131**, 51–57.
71. Li, X., Xiong, Z., Ying, X., Cui, L., Zhu, W., and Li, F. (2006) A rapid ultra-performance liquid chromatography-electrospray ionization tandem mass spectrometric method for the qualitative and quantitative analysis of the constituents of the flower of *Trollius ledibouri* Reichb. *Anal. Chim. Acta* **580**, 170–180.
72. Ficarra, P., Ficarra, R., Bertucci, C., Tommasini, S., Calabro, M., Costantino, D., and Carulli, M. (1995) Direct enantiomeric separation of flavanones by high performance liquid chromatography using various chiral stationary phases. *Planta Med.* **61**, 171–176.
73. Dalluge, J.J., Nelson, B.C., Thomas, J.B., and Sander, L.C. (1998) Selection of column and gradient elution system for the separation of catechins in green tea using high-performance liquid chromatography. *J. Chromatogr. A* **793**, 265–274.
74. Zgorka, G. and Dawka, S. (2001) Application of conventional UV, photodiode array (PDA) and fluorescence (FL) detection to analysis of phenolic acids in plant material and pharmaceutical preparations. *J. Pharm. Biomed. Anal.* **24**, 1065–1072.
75. Yáñez, J.A., Andrews, P.K., and Davies, N.M. (2007) Methods of analysis and separation of chiral flavonoids. *J. Chromatogr. B* **848**, 159–181.
76. Hollman, P.C.H. (2001) Determination of flavonols in body fluids. *Methods Enzymol.* **335**, 97–103.
77. Rodríguez-Delgado, M.A., Malovaná, S., Perez, J.P., Borges, T., and García Montelongo, F.J. (2001) Separation of phenolic compounds by high-performance liquid chromatography with absorbance and fluorimetric detection. *J. Chromatogr. A* **912**, 249–257.
78. Hollman, P.C.H., van Trijp, J.M.P., and Buysman, M.N.C.P. (1996) Fluorescence detection of flavonols in HPLC by postcolumn chelation with aluminum. *Anal. Chem.* **68**, 3511–3515.
79. Gutierrez, A.C. and Gehlen, M.H. (2002) Time resolved fluorescence spectroscopy of quercetin and morin complexes with Al^{3+}. *Spectrochim. Acta A* **58**, 83–89.
80. Nardini, M. and Ghiselli, A. (2004) Determination of free and bound phenolic acids in beer. *Food Chem.* **84**, 137–143.
81. Jandera, P., Škeříková, V., Řehová, L., Hájek, T., Baldriánová, L., Škopová, G., Kellner, V., and Horna, A. (2005) RP-HPLC analysis of phenolic compounds and flavonoids in beverages and plant extracts using a CoulArray detector. *J. Sep. Sci.* **28**, 1005–1022.
82. Satterfield, M. and Brodbelt, J.S. (2000) Enhanced detection of flavonoids by metal complexation and electrospray ionization mass spectrometry. *Anal. Chem.* **72**, 5898–5906.
83. Cremin, P., Kasim-Karakas, S., and Waterhouse, A.L. (2001) LC/ES-MS Detection of hydroxycinnamates in human plasma and urine. *J. Agric. Food Chem.* **49**, 1747–1750.
84. de Rijke, E., Out, P., Niessen, W.M.A., Ariese, F., Gooijer, C., and Brinkman, U.A.Th. (2006) Analytical separation and detection methods for flavonoids. *J. Chromatogr. A* **1112**, 31–63.
85. Mullen, W., Boitier, A., Stewart, A.J., and Crozier, A. (2004) Flavonoid metabolites in human plasma and urine after the consumption of red onions: Analysis by liquid chromatography with photodiode array and full scan tandem mass spectrometric detection. *J. Chromatogr. A* **1058**, 163–168.
86. de Rijke, E., de Kanter, F., Ariese, F., Gooijer, C., and Brinkman, U.A.Th. (2004) Liquid chromatography coupled to nuclear magnetic resonance spectroscopy for the identification of isoflavone glucoside malonates in *T. pratense* L. leaves. *J. Sep. Sci.* **27**, 1061–1070.
87. Exarchou, V., Godejohann, M., van Beek, T.A., Gerothanassis, I.P., and Vervoort, J. (2003) LC-UV-solid-phase extraction-NMR-MS combined with a cryogenic flow probe and its application to the identification of compounds present in Greek oregano. *Anal. Chem.* **75**, 6288–6294.
88. Lewis, D.A., Fields, W.N., and Shaw, G.P. (1999) A natural flavonoid present in unripe plantain banana pulp (*Musa sapientum* L. var. paradisiaca) protects the gastric mucosa from aspirin-induced erosions. *J. Ethnopharmacol.* **65**, 283–288.
89. Yu, Q.-T., Qi, L.-W., Li, P., Yi, L., Zhao, J., and Bi, Z. (2007) Determination of seventeen main flavonoids and saponins in the medicinal plant Huang-qi (*Radix astragali*) by HPLC-DAD-ELSD. *J. Sep. Sci.* **30**, 1292–1299.
90. Molnár-Perl, I. and Füzfai, Zs. (2005) Chromatographic, capillary electrophoretic and capillary electrochromatographic techniques in the analysis of flavonoids. *J. Chromatogr. A* **1073**, 201–227.

91. Herrero, M., Ibáñez, E., and Cifuentes, A. (2005) Analysis of natural antioxidants by capillary electromigration methods. *J. Sep. Sci.* **28**, 883–897.
92. Jáč, P., Polášek, M., and Pospišilová, M. (2006) Recent trends in the determination of polyphenols by electromigration methods. *J. Pharm. Biomed. Anal.* **40**, 805–814.
93. Morin, Ph., Villard, F., Dreux, M., and André, P. (1993) Borate complexation of flavonoid-*O*-glycosides in capillary electrophoresis: II. Separation of flavonoid-3-*O*-glycosides differing in their sugar moiety. *J. Chromatogr.* **628**, 161–169.
94. Morin, Ph., Villard, F., and Dreux, M. (1993) Borate complexation of flavonoid-*O*-glycosides in capillary electrophoresis: I. Separation of flavonoid-7-*O*-glycosides differing in their flavonoid aglycone. *J. Chromatogr.* **628**, 153–160.
95. Huang, H.-Y., Lien, W.-C., and Chiu, C.-W. (2005) Comparison of microemulsion electrokinetic chromatography and micellar electrokinetic chromatography methods for the analysis of phenolic compounds. *J. Sep. Sci.* **28**, 973–981.
96. Ye, J. and Baldwin, R.P. (1993) Amperometric detection in capillary electrophoresis with normal size electrodes. *Anal. Chem.* **65**, 3525–3527.
97. Huck, C.W., Stecher, G., Scherz, H., and Bonn, G. (2005) Analysis of drugs, natural and bioactive compounds containing phenolic groups by capillary electrophoresis coupled to mass spectrometry. *Electrophoresis* **26**, 1319–1333.
98. Lafont, F., Aramendia, M.A., Garcia, I., Borau, V., Jiménez, C., Marinas, J.M., and Urbano, F.J. (1999) Analyses of phenolic compounds by capillary electrophoresis electrospray mass spectrometry. *Rapid Commun. Mass Spectrom.* **13**, 562–567.
99. Huck, C.W., Stecher, G., Ahrer, W., Stöggl, W.M., Buchberger, W., and Bonn, G.K. (2002) Analysis of three flavonoids by CE-UV and CE-ESI-MS. Determination of naringenin from a phytomedicine. *J. Sep. Sci.* **25**, 904–908.
100. Owades, J.L., Rubin, G., and Brenner, M.W. (1958) Determination of food tannins by ultraviolet spectrophotometry. *J. Agric. Food Chem.* **6**, 44–46.
101. Mabry, T.J., Markham, K.R., and Thomas, M.B. (1970) *The Systematic Identification of Flavonoids*, Springer-Verlag, New York, NY.
102. Macheix, J.-J., Fleuriet, A., and Billot, J. (1990) *Fruit Phenolics*, CRC Press, Bota Raton, FL.
103. Folin, O. and Denis, W. (1912) Phosphotungstic-phosphomolybdic compounds as color reagents. *J. Biol. Chem.* **12**, 239–243.
104. Slinkard, K. and Singleton, V.L. (1977) Total phenol analysis-automation and comparison with manual methods. *Am. J. Enol. Vitic.* **28**, 49–55.
105. Brune, M., Hallberg, L., and Skanberg, A.B. (1991) Determination of iron-binding phenolic groups in foods. *J. Food Sci.* **56**, 128–131.
106. Goldstein, J.L. and Swain, T. (1963) Changes in tannins in ripening fruits. *Phytochemistry* **2**, 371–378.
107. Lamaison, J.L.C. and Carnat, A. (1990) Levels of principal flavonoids in flowers and leaves of Crataegus-Monogyna Jacq and Crataegus-Laevigata (*Poiret rosaceae*). *Pharm. Acta Helv.* **65**, 315–320.
108. Zaporozhets, O.A., Krushynska, O.A., Lipkovska, N.A., and Barvinchenko, V.N. (2004) A new test method for the evaluation of total antioxidant activity of herbal products. *J. Agric. Food Chem.* **52**, 21–25.
109. Nagy, M. and Grancai, D. (1996) Colorimetric determination of flavanones in propolis. *Pharmazie* **51**, 100–101.
110. Sondheimer, E. and Kertesz, Z.I. (1948) Anthocyanin pigments. Colorimetric determination in strawberries and strawberry preserves. *Anal. Chem.* **20**, 245–248.
111. Swain, T. and Hillis, W.E. (1959) The phenolic constituents of *Prunus domestica*. I. – The quantitative analysis of phenolic constituents. *J. Sci. Food Agric.* **10**, 63–68.
112. Schulz, H., Engelhardt, U.H., Wegent, A., Drews, H.-H., and Lapczynski, S. (1999) Application of near-infrared reflectance spectroscopy to the simultaneous prediction of alkaloids and phenolic substances in green tea leaves. *J. Agric. Food Chem.* **47**, 5064–5067.
113. Edelmann, A., Diewok, J., Schuster, K.C., and Lendl, B. (2001) Rapid method for the discrimination of red wine cultivars based on mid-infrared spectroscopy of phenolic wine extracts. *J. Agric. Food Chem.* **49**, 1139–1145.

Chapter 6

Design, Synthesis, and Action of Antiatherogenic Antioxidants

Osamu Cynshi, Kunio Tamura, and Etsuo Niki

Abstract

Ample evidence supports the critical role of oxidized low-density lipoprotein (ox-LDL) in initiation and progression of atherosclerosis. Oxidation of LDL is a complex process involving several steps (processes) of reactions such as initiation and propagation. Both proteins and lipids in LDL undergo free radical-mediated oxidations leading to the formation of ox-LDL that plays a pivotal role in atherosclerosis. Antioxidants of various types (both aqueous and lipophilic) either arrest or retard the oxidation of LDL at various steps of the oxidation process (e.g., initiation or propagation). Certain lipophilic antioxidants act as the chain-terminating antioxidants leading to the inhibition of LDL oxidation. The current chapter describes the designing and efficacy of two novel lipophilic antioxidants (benzofuranol, BO-653 and aniline, BO-313) in inhibiting the LDL oxidation and atherogenesis in experimental animal model. Furthermore, the characteristics of an effective antioxidant to inhibit LDL oxidation and atherogenesis which dictates the designing of the antioxidant drug and its mechanism(s) of antiatherogenic action are discussed.

Key words: Antioxidant, low-density lipoprotein (LDL), LDL oxidation, thiobarbituric acid reactive substances (TBARS), gel permeation chromatography (GPC), Watanabe heritable hyperlipidemic (WHHL) rabbit, atherosclerosis, 2,2'–azobis 2-amidinopropane (AAPH), lipoxygenase, *N*-(3,7,11,15-tetramethylhexadecyl)-3,4,5-trimethoxyaniline (BO-313), 4,6-Di-*tert*-butyl-2,3-dihydro-2,2-dipentyl-5-benzofuranol (BO-653).

1. Introduction

Many lines of evidence suggest that oxidative modification of the low-density lipoprotein (LDL) has a critical role in the initiation and progression of atherosclerosis (1, 2). Lipophilic antioxidants are believed to reduce the LDL oxidation, thus exerting

R.M. Uppu et al. (eds.), *Free Radicals and Antioxidant Protocols*, Methods in Molecular Biology 610,
DOI 10.1007/978-1-60327-029-8_6, © Humana Press, a part of Springer Science+Business Media, LLC 1998, 2010

their antiatherogenic activities. For the last two decades, various lipophilic antioxidants have been exhaustively investigated in animal models with the goal of reducing the progression of atherosclerosis (3–7). However, some animal models have failed to show any reduction in atherosclerosis (4, 8, 9). A potential difficulty in developing lipophilic antioxidants as antiatherogenic agents may be the identification of the specific step(s) in LDL oxidation which needs to be suppressed. The LDL oxidation process contains various sequences of chemical reactions and identifying a "key" process (step) is essential for the design of an effective antiatherogenic molecule. Here, we hypothesize that a systematic design of antioxidants would clarify and precisely identify the activity desired to effectively suppress the key step in LDL oxidation and such a design might indicate the suitability of the antioxidant as an antiatherogenic agent. The synthesis of a lipophilic benzofuranol (BO-653) and a lipophilic aniline (BO-313) illustrates two types of lipophilic antioxidants exhibiting distinctly different antiatherogenic activities but with similar abilities to attenuate LDL oxidation.

2. Materials

2.1. Synthesis of Antioxidants

1. Column chromatography: Wako-gel C-200, 70–230 mesh (Wako Pure Chemical Industries, Osaka, Japan).
2. Thin-layer chromatography (TLC): Kieselgel F^{254} (E. Merck and Co.).
3. High performance liquid chromatography (HPLC) analysis: Hitachi L-4000 UV detector with a Hitachi L-6200 pump using a YMC-Pack C8 A-212 column eluted with CH_3CN-*i*-PrOH-water (8:1:1) at a flow rate of 1.0 mL/min.
4. Melting point (mp): Yanagimoto micro melting point apparatus or Mettler FP62 melting point apparatus.
5. ^1H NMR spectra: Hitachi R-24B spectrometer (60 MHz) or JEOL EX-270 spectrometer (270 MHz) with tetramethylsilane as an internal standard.
6. Mass spectra (MS): Shimadzu GCMS-QP1000 spectrometer.
7. Infrared spectra (IR): Hitachi 270-30 infrared spectrophotometer.
8. High resonance mass spectra (HRMS): Measured at the Toray Research Center (Tokyo, Japan).

9. BO-653: 4,6-Di-*tert*-butyl-2,3-dihydro-2,2-dipentyl-5-benzofuranol.
10. BO-313: *N*-(3,7,11,15-Tetramethylhexadecyl)-3,4,5-trimethoxyaniline.

2.2. Evaluation of LDL Oxidation and Atherosclerosis in Rabbits

1. Animals: Male Watanabe heritable hyperlipidemic (WHHL) rabbits, 2–3 months old (Kitayama Labs, Ina, Japan) as the atherosclerosis model and female Japanese white (JW) rabbits, 3 months old (CSK Research Park, Tokyo, Japan) for preparation of rabbit LDL
2. Ultracentrifugation system with a 70-Ti rotor (Beckman)
3. High performance liquid chromatography with electrochemical detector (ECD-HPLC) using a column SUPELCOIL LC-318 (Supelco)
4. ECD-HPLC elution buffer (methanol/*tert*-butyl alcohol [90/10, v/v] containing 50 mM $NaClO_4$)
5. Gel permeation chromatography (GPC) using a G5000PW column (Tosoh)
6. GPC elution buffer (0.1 M NaCl + 0.05 M Tris buffer)
7. Thiobarbituric acid reactive substances (TBARS) assay kit (Lipid Peroxide Test Wako, Wako Pure Chemical)
8. Phosphate-buffered saline (PBS)
9. Dulbecco's Modified Eagle Medium (DMEM)
10. Ethylenediamine-N,N,N',N'-tetraacetic acid (EDTA)
11. Copper sulfate ($CuSO_4$)
12. 2,2'–azobis 2-amidinopropane dihydrochloride (AAPH)
13. Soybean 15-lipoxygenase (LO)
14. SigmaScan Pro software (Jandel Scientific Software)

3. Methods

3.1. Drug Design

Probucol, a lipophilic antioxidant, is reported to inhibit LDL oxidation leading to a suppression of atherosclerosis development in animal models (3). However, there are two serious problems: it induces cardiac toxicity (QT_C prolongation) (10) and decreases high-density lipoproteins (HDL) (11). In order to find a drug without encountering those problems caused by probucol, we examined: (i) the effects of lipophilic antioxidants on the inhibition of Cu^{2+}-catalyzed oxidation of rabbit LDL, (ii) the pharmacokinetic (PK) profile of the lipophilic antioxidants in a rabbit model atherosclerosis for affinity to LDL, and (iii) the efficacy of lipophilic antioxidants on atherosclerosis in a rabbit model atherosclerosis (WHHL rabbits). Probucol was used as the reference antiatherosclerotic drug in these experiments. We found that one of the tested lipophilic antioxidants (BO-313, an aniline-type antioxidant which bears a phytyl group as a lipophilic moiety) inhibited LDL oxidation with an affinity to LDL but exacerbated the atherosclerotic condition. It was further characterized as an LDL antioxidant which does not inhibit either AAPH- or lipoxygenase-catalyzed LDL oxidation. This suggested that BO-313 does not inhibit the propagation steps (reactions) common in LDL oxidation, but inhibits the Cu^{2+}-catalyzed oxidation of LDL. This finding also suggests that the design of an antioxidant with a desired mode of action and confirmation of its action are important in identifying novel LDL antioxidants.

Inhibitors against the propagation of LDL oxidation are called the chain-breaking antioxidants and the well-known natural chain-breaking antioxidant is α-tocopherol. α-Tocopherol exists not in the core but in the surface layer of LDL due to its polarity and exhibits both its antioxidant and prooxidant activities (12). We concluded that a more efficient antioxidant could be designed as an LDL antioxidant, which consists of the following characteristics (13): (i) its high hydrogen-donating activity and low prooxidant activity as a result of the radical intermediate, (ii) its localization inside the core of the LDL particle, and (iii) its efficient delivery to LDL.

The first characteristic was rationally achieved by lowering the dissociation energy of the phenolic O-H bond, which is determined in part by the resonance stabilization of the resulting phenoxyl radical. Ingold and his coworkers reported that the p-type lone-pair orbital of the oxygen atom located in the *para* position of the phenolic hydroxyl group can optimally stabilize the phenoxyl radical when the oxygen atom belongs to a five-member ring, although vitamin E has a six-member ring (14). Thus, we designed a 2,3-dihydro-5-benzofuranol structure to achieve high hydrogen-donating activity and low prooxidant activity (**Fig. 6.1**).

To achieve the second characteristic, we utilized the two *tert*-butyl groups on the *ortho* position of the hydroxyl group instead

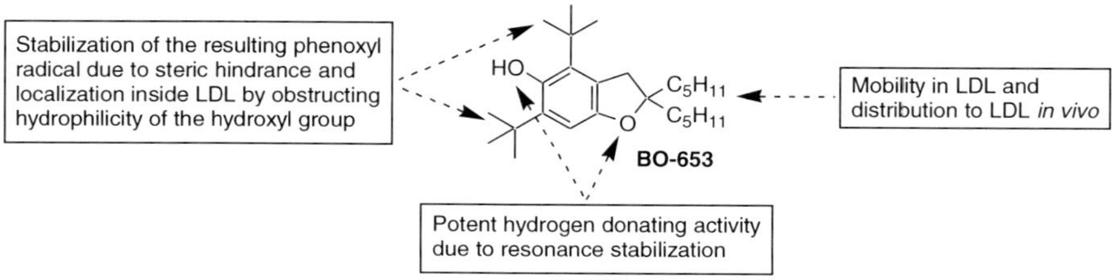

Fig. 6.1. Characteristics required for potent LDL antioxidants.

of the methyl groups of α-tocopherol. The two *tert*-butyl groups were designed to obstruct the hydrophilicity of the hydroxyl group and result in the preferential localization to the hydrophobic region. Moreover, they result in the kinetically persistent phenoxyl intermediate due to their steric hindrance, which leads to a reduction in the prooxidant action of the phenoxyl intermediate. Furthermore, in addition to high reactivity, the relative stability of the intermediate also leads to a reduction in the coexisting prooxidant action of α-tocopherol (15).

The third characteristic of efficient delivery is essential for LDL antioxidants, and we examined their distribution to lipoproteins using a series of compounds. We synthesized 2,3-dihydro-5-benzofuranol derivatives bearing different alkyl substituents in position 2 and evaluated their distribution using Watanabe heritable hyperlipidemic (WHHL) rabbits selective for the 2,2-dipentyl derivative, BO-653.

3.2. Synthesis

3.2.1. BO-653 (see Fig. 6.2)

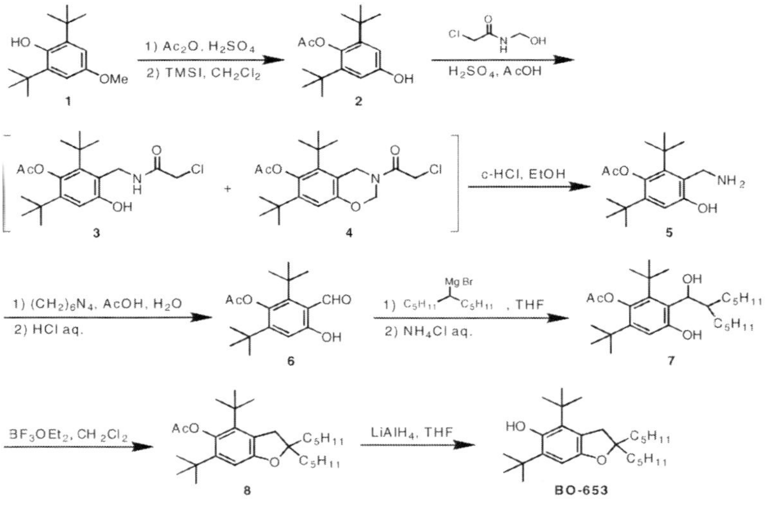

Fig. 6.2. Synthetic scheme of BO-653.

3.2.1.1. 4-Acetoxy-3,5-di-*tert*-butylanisole

1. Stir a mixture of 3,5-di-*tert*-butyl-4-hydroxyanisole 1 (23.6 g, 0.1 mol) and conc. H_2SO_4 (0.5 mL) in acetic anhydride (150 mL) at 70°C for 2 h.
2. Concentrate the reaction mixture under reduced pressure, and add saturated aqueous $NaHCO_3$ to the concentrate.
3. Extract the mixture with EtOAc, dry over anhydrous $MgSO_4$, and concentrate.

4. Recrystallize the residue from MeOH–water (2:1) to give the title compound (24.5 g, 88%) as a white solid, mp 97°C: ^{1}H NMR (60 MHz, $CDCl_3$) δ1.06 (s, 18H), 2.02 (s, 3H), 3.47 (s, 3H), 6.53 (s, 2H), MS *m/e* 278 (M^+).

3.2.1.2. 4-Acetoxy-3,5-di-*tert*-butylphenol (2)

1. Add iodotrimethylsilane (0.31 mL, 2.2 mmol) drop wise at 0°C, to a solution of 4-acetoxy-3,5-di-*tert*-butylanisole (0.50 g, 1.8 mmol) in CH_2Cl_2 (2 mL).
2. Allow the mixture to warm to room temperature and stir for 2 days.
3. Add a saturated aqueous solution of $NaHCO_3$ to the reaction mixture, and extract the mixture with Et_2O.
4. Wash the extract with saturated brine, dry over anhydrous $MgSO_4$, and concentrate.
5. Purify the residue by silica gel column chromatography (15% EtOAc in *n*-hexane) to give **2** (0.38 g, 80%) as a white solid, mp 157°C: ^{1}H NMR (60 MHz, $CDCl_3$) δ 1.27 (s, 18H), 2.27 (s, 3H), 5.22 (br, 1H), 6.67 (s, 2H), MS *m/e* 264 (M^+).

3.2.1.3. 4-Acetoxy-3,5-di-*tert*-butyl-2-(chloroacetylaminomethyl) phenol (3) and 6-Acetoxy-5,7-di-*tert*-butyl-3-chloroacetyl-2,3-dihydro-4*H*-1,3-benzoxazine (4)

1. Dissolve phenol **2** (29.0 g, 0.11 mol) in a 9:1 mixed solution (200 mL) of acetic acid and H_2SO_4 and add 2-chloro-*N*-(hydroxymethyl)acetamide (34.0 g, 0.28 mol).
2. Stir the mixture at room temperature for 48 h. Subsequently, pour the reaction mixture into water, neutralize with 1 N NaOH, and extract with EtOAc.
3. Dry the organic layer over anhydrous $MgSO_4$ and concentrate.
4. The concentrate can be used in the subsequent reaction without further purification. When a portion of the concentrate was purified by silica gel column chromatography (20% EtOAc in *n*-hexane), the title compounds are obtained: **3** as a colorless oil, ^{1}H NMR (60 MHz, $CDCl_3$) δ 1.30 (s, 9H), 1.43 (s, 9H), 2.28 (s, 3H), 4.00 (s, 2H), 4.73 (d, $J =$ 6.0 Hz, 2H), 6.88 (s, 1H), 7.54 (t, $J = 6.0$ Hz, 1H), MS *m/e* 369 (M^+), 327, 234, 57, and **4** as a colorless oil, ^{1}H NMR (60 MHz, $CDCl_3$) δ 1.30 (s, 9H), 1.47 (s, 9H), 2.30 (s, 3H), 4.17 (s, 2H), 5.00 (s, 2H), 5.33 (s, 2H), 6.83 (s, 1H), MS *m/e* 381 (M^+), 339, 304, 57.

3.2.1.4. 4-Acetoxy-2-aminomethyl-3,5-di-*tert*-butylphenol (5)

1. Dissolve the concentrate obtained in the above reaction in a 10:3 mixed solution (550 mL) of EtOH and concentrated HCl.
2. Heat the reaction solution to reflux for 2 h.

3. After being cooled, pour the solution into water and the mixture, neutralize with 1 N NaOH, followed by extraction with EtOAc.
4. Dry the organic layer over anhydrous $MgSO_4$ and concentrate. The concentrate is used in the subsequent reaction without further purification. When a portion of the concentrate was purified by silica gel column chromatography (20% MeOH in $CHCl_3$), phenol **5** is obtained: ^{1}H NMR (60 MHz, $CDCl_3$) δ 1.27 (s, 9H), 1.37 (s, 9H), 2.25 (s, 3H), 4.22 (s, 2H), 5.18 (bs, 3H), 6.85 (s, 1H), MS *m/e* 293 (M^+), 234, 191, 57.

3.2.1.5. 5-Acetoxy-4,6-di-*tert*-butyl-2-hydroxybenzaldehyde (6)

1. Dissolve the concentrate (obtained in the above reaction) in an 11:3 mixed solution (636 mL) of acetic acid and water.
2. After addition of hexamethylenetetramine (19.3 g, 0.14 mol), heat the mixture to reflux for 4 h.
3. Subsequently, add 4.5 N HCl (85 mL) to the mixture and heat the resulting mixture to reflux for additional 20 min.
4. After being cooled, pour the reaction mixture into water, neutralize with 1 N NaOH, and extract with EtOAc.
5. Dry the organic layer over anhydrous $MgSO_4$ and concentrate.
6. Purify the concentrate by silica gel column chromatography ($CHCl_3$) to afford **6** (19.0 g, 59% for three steps) as a fine-grained pale yellow crystal, mp 79°C: IR (KBr) 2,976, 1,758 cm^{-1}, ^{1}H NMR (60 MHz, $CDCl_3$) δ 1.35 (s, 9H), 1.54 (s, 9H), 2.35 (s, 3H), 6.92 (s, 1H), 10.67 (s, 1H), 12.32 (s, 1H), MS *m/e* 292 (M^+), 250, 235, 217, 57.

3.2.1.6. 4-Acetoxy-3,5-di-*tert*-butyl-2-(1-hydroxy-2-pentylheptyl)phenol (7)

1. To a Grignard reagent prepared with Mg (30 g, 1.23 mol), 6-bromoundecane (290 g, 1.23 mol) in the usual manner and dry THF (1.2 L) under a N_2 atmosphere, add drop wise solution 6 (120 g, 0.41 mol) in dry THF (400 mL).
2. After the reaction mixture stirred at room temperature for 2 h, add saturated aqueous NH_4Cl solution to the mixture, followed by extraction with EtOAc.
3. Dry and concentrate the organic layer over anhydrous $MgSO_4$.
4. Purify the concentrate by silica gel column chromatography (5% EtOAc in *n*-hexane) to give 7 (74.3 g, 40%) as a white solid, mp 126°C: IR (KBr) 3,493, 1,761, 1,369, 1,190, 908 cm^{-1}, ^{1}H NMR (270 MHz, $CDCl_3$) δ 0.74 (t, $J = 6.8$ Hz, 3H), 0.91 (t, $J = 6.6$ Hz, 3H), 0.95–1.63 (m, 16H), 1.29 (s, 9H), 1.40 (s, 9H), 2.12 (m, 1H), 2.28 (s, 3H), 2.50 (d, $J = 2.6$ Hz, 1H), 5.22 (dd, $J = 2.6$, 9.9 Hz, 1H), 6.77 (s, 1H), 7.89 (s, 1H), MS *m/e* 448 (M^+).

3.2.1.7. 5-Acetoxy-4,6-di-*tert*-butyl-2,3-dihydro-2,2-dipentylbenzofuran (8).

1. To **7** (74.3 g, 0.166 mol) in CH_2Cl_2 (500 mL) add drop wise under a N_2 atmosphere BF_3 diethyl etherate (20 mL, 0.16 mol).
2. After being stirred at room temperature for 4.5 h, add the reaction mixture to ice cold water, neutralize with 2 N KOH, and extract with CH_2Cl_2.
3. Wash the extract with saturated aqueous NaCl, dried over anhydrous $MgSO_4$, and concentrate to give solution **8** as pale yellow oil. The concentrate is used in the subsequent reaction without further purification: IR (film) 1,760, 1,567, 1,365, 1,214, 1,172, 943 cm^{-1}, ^1H NMR (270 MHz, $CDCl_3$) δ 0.88 (t, J = 6.6 Hz, 6H), 1.22–1.39 (m, 12H), 1.30 (s, 9H), 1.37 (s, 9H), 1.55–1.75 (m, 4H), 2.29 (s, 3H), 3.10 (d, J = 15.3 Hz, 1H), 3.21 (d, J = 15.3 Hz, 1H), 6.72 (s, 1H), MS m/e 430 (M^+).

3.2.1.8. 4,6-Di-*tert*-butyl-2,3-dihydro-2,2-dipentyl-5-benzofuranol (BO-653)

1. Suspend LiAlH4 (6.3 g, 0.166 mol) in dry THF (200 mL) under a N_2 atmosphere.
2. Add solution **8** in dry THF (400 mL) drop wise to the suspension at 0°C.
3. After being heated to reflux for 3 h, allow the reaction mixture to cool to room temperature, and add EtOAc drop wise to quench the excess $LiAlH_4$.
4. After the addition of 10% HCl, extract the mixture with EtOAc.
5. Wash the extract with saturated brine, dry over anhydrous $MgSO_4$, and concentrate. The residue is purified by silica gel column chromatography (*n*-hexane) to afford **BO-653** (55.0 g, 85% for two steps) as a colorless oil: IR (film) 3,652, 2,956 cm^{-1}, ^1H NMR (270 MHz, $CDCl_3$) δ 0.88 (t, J = 6.9 Hz, 6H), 1.30 (br, 12H), 1.40 (s, 9H), 1.49 (s, 9H), 1.58–1.70 (m, 4H), 3.18 (s, 2H), 4.66 (s, 1H), 6.62 (s, 1H), MS m/e 388 (M^+), HPLC analysis (CH_3CN–*i*-PrOH–water (8:1:1)) t_R = 15.7 min (95.1%). HRMS calculated for $C_{26}H_{44}O_2$: 388.3341. Found: 388.3339.

3.2.2. BO-313 (see **Fig. 6.3***)*

3.2.2.1. Phytyl bromide

1. Dissolve phytol (350 g, 1.18 mol) in Et_2O (500 mL).
2. To the solution add drop wise a solution of PBr_3 (61 mL, 0.64 mol) in Et_2O (50 mL) under ice cooling.
3. After stirring the mixture overnight, pour the reaction mixture into ice water.

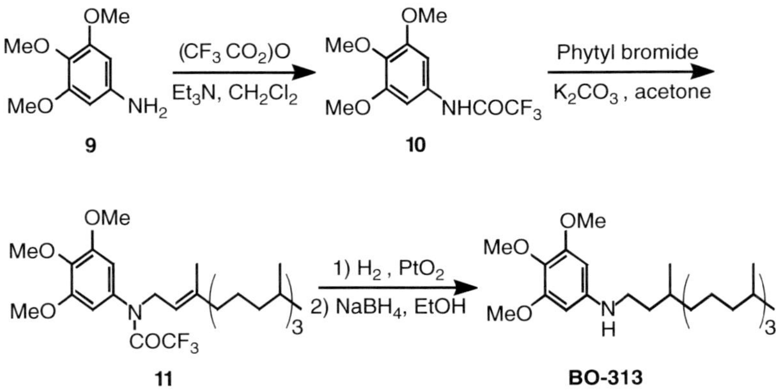

Fig. 6.3. Synthetic scheme of BO-313.

4. Separate the resulting mixture, and wash the obtained organic layer with water, 1 N aqueous KOH solution, and saturated brine.
5. Dry the organic layer over anhydrous $MgSO_4$, filtrate, and concentrate to give oily phytyl bromide (404 g, 95%): ^{1}H NMR (60 MHz, $CDCl_3$) δ 0.85 (d, J = 6.0 Hz, 12H), 1.00–2.30 (m, 21H), 1.70 (s, 3H), 3.94 (d, J = 8.5 Hz, 2H), 5.49 (t, J = 8.5 Hz, 1H).

3.2.2.2. 3',4',5'-Trimethoxytrifluoroacetoanilide (10)

1. Dissolve 3,4,5-trimethoxyaniline **9** (183 g, 1.0 mol) and triethylamine (145 mL, 1.05 mol) in CH_2Cl_2 (500 mL).
2. To the solution add drop wise trifluoroacetic anhydride (150 mL, 1.05 mol) under ice cooling.
3. Stir the mixture was overnight, concentrate under reduced pressure.
4. Add water to the concentrate and stir the resulting mixture vigorously to give a crude solid.
5. Filter, the crude solid, dry, and recrystallize from benzene to afford **10** (266 g, 95%): ^{1}H NMR (60 MHz, $CDCl_3$) δ 3.76 (s, 9H), 6.76 (s, 2H), 8.00 (bs, 1H), mp 140–141°C.

3.2.2.3. *N*-(3,7,11,15-Tetramethylhexadecyl)-3,4,5-trimethoxyaniline (**BO-313**)

1. Dissolve anilide **10** (5.58 g, 20 mmol) and phytyl bromide (7.19 g, 20 mmol) in acetone (100 mL).
2. To the solution add anhydrous K_2CO_3 (2.76 g, 20 mmol). And heat the reaction mixture to reflux overnight,
3. Add water to the mixture and extract with EtOAc.
4. Wash the extract with saturated brine, dry over anhydrous $MgSO_4$, filtrate, and concentrate to give *N*-(3,7,11,15-tetramethyl-2-hexadecenyl)-3',4',5'-trimethoxytrifluoroacetoanilide **11**.

5. Dissolve the residue in EtOH (50 mL) and hydrogenate under 1 atm H_2 gas in the presence of a catalytic amount of PtO_2 for 4 d.
5. After filtration of the PtO_2 catalyst, add $NaBH_4$ (0.76 g, 20 mmol) to the filtrate under ice cooling, and stir the mixture overnight.
6. Evaporate the reaction mixture, mix with water and EtOAc, and separate the organic layer. Wash the organic layer with saturated brine, dry over anhydrous $MgSO_4$, filtrate, and concentrate.
7. Dissolve the residue, purify by silica gel column chromatography (20% EtOAc in *n*-hexane) to give oily **BO-313** (7.53 g, 93%): ^1H NMR (60 MHz, $CDCl_3$) δ 0.85 (d, $J =$ 6.0 Hz, 15H), 1.05–1.80 (m, 24H), 3.06 (t, $J = 7.0$ Hz, 2H), 3.71 (s, 3H), 3.78 (s, 6H), 5.79 (s, 2H).

3.3. Evaluation of Antioxidant Activities

3.3.1. Rabbit LDL Preparation

1. LDL is separated from freshly isolated plasma. Establishing a reproducible quality of LDL preparation is useful for the systematic screening of antioxidants and so LDL is prepared from specific pathogen-free (SPF) laboratory rabbits.
2. Female JW rabbits are housed singly in cages and given free access to food (RC-4 commercial pellet diet) until 12 days before bleeding in order to enrich the LDL content.
3. Collect blood from the carotid using EDTA as an anticoagulant after fasting. (Use ketamine to anesthetize.)
4. Separate plasma by centrifugation and isolate LDL fraction ($1.019 < d < 1.063$) from the plasma according to the ultracentrifugal method described by Havel et al. (16).
5. Dialyze isolated LDL against a solution containing 0.15 mol/L NaCl + 0.01% NaN_3 + 0.01% EDTA, and then against a solution containing 0.15 mol/L NaCl + 2 mmol/L disodium hydrogen phosphate without NaN_3 and EDTA.
6. Analyze the protein content in the LDL solution by the Lowry method (17) and prepare the LDL solution based on the protein concentration.

3.3.2. LDL Oxidation

1. Examine a test compound in three types of LDL oxidation. Subject LDL to oxidative modification by incubating with (i) copper ions (18), (ii) AAPH (19), or (iii) LO (20) under the following conditions.
2. Dissolve the study compounds in DMSO and then add to the LDL solution to achieve final concentration of 1% DMSO before starting oxidation.

3. Incubate LDL at 200 μg/mL concentration with (i) 5 μmol/L $CuSO_4$ in DME without EDTA, (ii) 2 mmol/L AAPH in PBS, and (iii) 40 μg/mL LO in LO buffer (0.45% NaCl + 1 mmol/L $CaCl_2$ + 0.1 mmol/L EDTA + 50 mmol/L borate buffer at pH 9.0) at 37°C for 24 h.

3.3.3. Analysis of Oxidized LDL

Oxidation of LDL is evaluated as lipid peroxidation or protein oxidation. Lipid peroxidation products generated in the LDL oxidation are usually measured as the thiobarbituric acid reactive substances (TBARS) using a commercial TBARS determination kit (Lipid Peroxide Test Wako, Wako Pure Chemical) and according to the instructions provided by the manufacturer. The kit is designed on the basis of Yagi's method (21). The results for BO-313 and BO-653 are shown in **Table 6.1**. Both antioxidants suppress the TBARS generated in the Cu^{2+}-catalyzed LDL oxidation more potently than probucol, indicating that both compounds have the potential to act as LDL antioxidants more potent than probucol.

Table 6.1
Inhibitory effects of BO-653 and BO-313 on Cu^{2+}-catalyzed LDL oxidation[a]

	μM	TBARS nmol/mg		μM	TBARS nmol/mg
Control	0	24.2 ± 0.8	Control	0	25.4 ± 0.4
BO-313	1	15.7 ± 0.1	BO-653	1	23.4 ± 0.7
	10	5.9 ± 0.0		10	3.3 ± 0.4
	100	3.1 ± 0.2		100	3.1 ± 0.2
Probucol	1	19.8 ± 0.1	Probucol	1	21.1 ± 0.6
	10	5.9 ± 0.6		10	7.6 ± 0.3
	100	4.9 ± 0.3		100	6.1 ± 0.1

[a] Rabbit LDL at a concentration of 200 μg/mL was incubated with Cu^{2+} at 37°C for 24 h in the presence or absence of the test compound. Control contained 1% DMSO as vehicle. Each determination was performed in duplicate and results are shown as mean ± range.

Protein oxidation during the LDL oxidation was measured by evaluating fluorescence in the protein (22). Change in the fluorescence intensity of LDL was analyzed by gel permeation chromatography (GPC) using a gel filtration column (TOSOH G5000PW). The LDL solution (100 μl) was applied to GPC and eluted with 0.1 mol/L NaCl + 0.05 mol/L Tris buffer (pH 7.4). The elution buffer was monitored using a fluorometric detector (Ex: 360 nm, Em: 430 nm, JASCO FP-210) and ultraviolet photometric detector (280 nm, JASCO 875-UV). Fluorescence intensity was measured by integrating the fluorescence intensity of substances with molecular weight greater than 150 kDa (18).

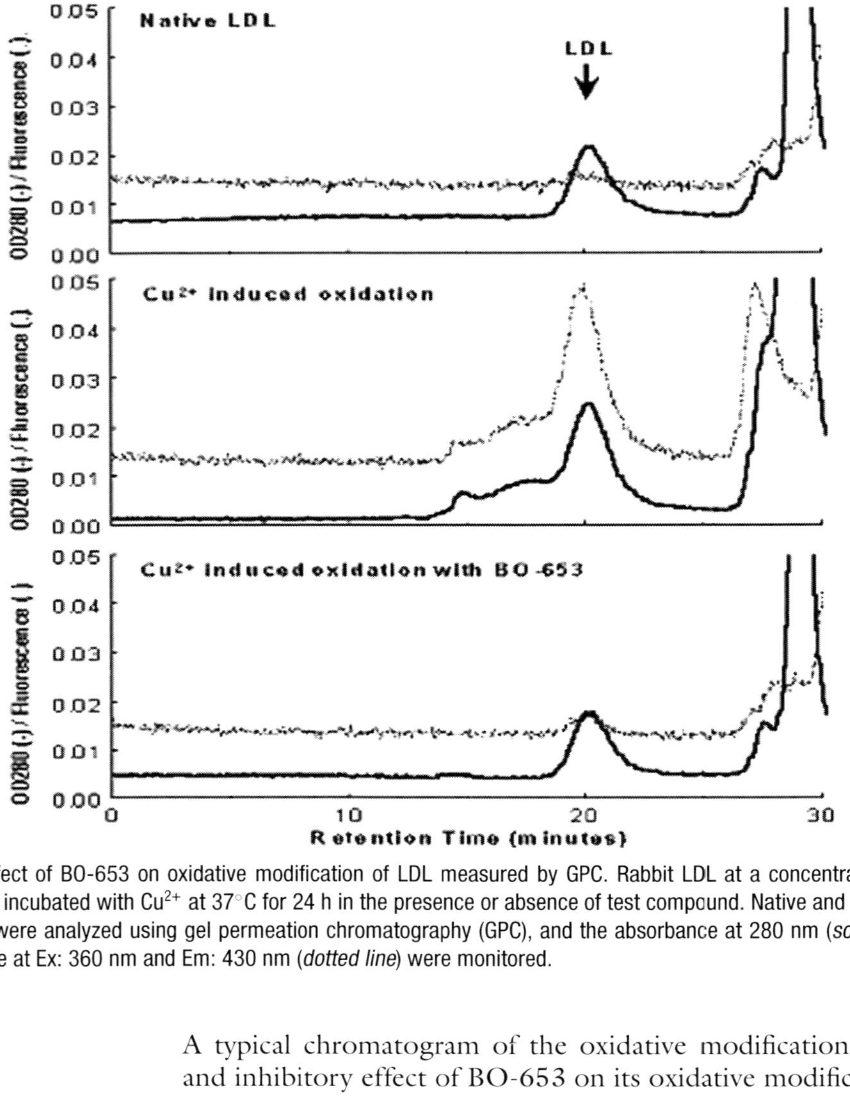

Fig. 6.4. Effect of BO-653 on oxidative modification of LDL measured by GPC. Rabbit LDL at a concentration of 200 μg/mL was incubated with Cu^{2+} at 37°C for 24 h in the presence or absence of test compound. Native and the oxidized LDL levels were analyzed using gel permeation chromatography (GPC), and the absorbance at 280 nm (*solid line*) and fluorescence at Ex: 360 nm and Em: 430 nm (*dotted line*) were monitored.

A typical chromatogram of the oxidative modification of LDL and inhibitory effect of BO-653 on its oxidative modification are shown in **Fig. 6.4**. After the incubation of LDL with copper, various changes such as TBARS were evaluating and the generation of the aggregate substances as well as fluorescence was measured by GPC. After incubation of the LDL-copper solution in the presence of BO-653 (10 μmol/L), the LDL showed a similar chromatogram to that of native LDL, indicating that the generation of both aggregates and fluorescent substances was suppressed.

Quantitative comparison of changes in the chromatogram was shown in **Table 6.2**. Notably, BO-313 showed inhibition of the LDL oxidation catalyzed by copper; however it did not show inhibition in that catalyzed by AAPH or LO. On the other hand, BO-653 inhibited not only Cu^{2+}-catalyzed oxidation but also both AAPH- and LO-catalyzed oxidation. This discrepancy between BO-313 and BO-653 indicates that BO-653 inhibits all the three

Table 6.2
Inhibitory effects of BO-653 and BO-313 on AAPH- and lipoxygenase-catalyzed LDL oxidation[a]

	Conc.	Fluorescence (% inhibition) AAPH[b]	LO		Conc.	Fluorescence (% inhibition) AAPH	LO
Control		892 ± 38	308 ± 5	Control		221 ± 15	245 ± 2
BO-313	Low	972 (+9)	324 (+5)	BO-653	Low	239 (+8)	258 (+6)
	High	700 (−22)	356 (+16)		High	148 (−33)	34 (−91)
Probucol	Low	798 (−11)	306 (−1)	Probucol	Low	179 (−19)	248 (+1)
	High	371 (−88)	63 (−84)		High	111 (−50)	61 (−79)

[a] Rabbit LDL at a concentration of 200 μg/mL was incubated with AAPH or SLO at 37°C for 24 h in the presence or absence of test compound. Fluorescence (Ex: 360 nm, Em: 430 nm) was integrated in whole LDL fraction determined by gel permeation chromatography (GPC). Low and high concentrations of test compound were 10 and 100 μM for AAPH-catalyzed oxidation, respectively, and 0.5 and 5 μM for LO-catalyzed oxidation, respectively. The control contained 1% DMSO as vehicle.

[b] The fluorophotometric detector in the GPC system was replaced with another one in this assay.

types of LDL oxidation due to the design of the drug that is responsible for the inhibition of the propagation process (steps) of LDL oxidation.

3.4. Evaluation of PK Profiles and Antiatherogenic Activity

3.4.1. Plasma Concentration of the Antioxidants Administered to WHHL Rabbits

The PK profile of antioxidants is important to clarify in vivo antiatherogenic activity. Our design for LDL antioxidants leads to highly lipophilic compounds and the realization that low solubility in water might cause low absorption. Furthermore, experimental animals show low plasma cholesterol levels in general and most of the plasma cholesterol in rodents comes from the HDL. Normal animals fed a normal diet show low absorption of lipophilic compounds. In order to overcome this low absorption in normal animals, we used WHHL rabbits with a high LDL level thus allowing the assessment of absorption of lipophilic compounds. WHHL rabbits were orally administered BO-653 or BO-313 (250 mg/kg) once a day for 9 days and the concentrations in plasma were measured.

Plasma concentration of the test compound was measured by HPLC equipped with an electrochemical detector (ECD) (23). Plasma samples (100 μL) from WHHL rabbit were extracted to 2 mL of acetone and then separated by centrifugation. The acetone extract was then evaporated under nitrogen on a rotary evaporator. The residue was then resolved with 100 μL of methanol and the solution was then analyzed on an ECD-HPLC system with methanol/*tert*-butyl alcohol (90/10, v/v) containing 50 mM $NaClO_4$ as the eluents at a flow rate of 0.8 mL/min.

Figure 6.5 shows that both compounds were absorbed and emerged in plasma more easily than probucol. Following fractionating the plasma into very-low-density lipoprotein (VLDL), LDL, and HDL fractions at 24 h after the final administration of the test drug, we confirmed the affinity of the test compound to LDL. The results indicate that the oral administration of both compounds distributed in the VLDL and LDL fractions rather than in the HDL fraction (data not shown) (13),

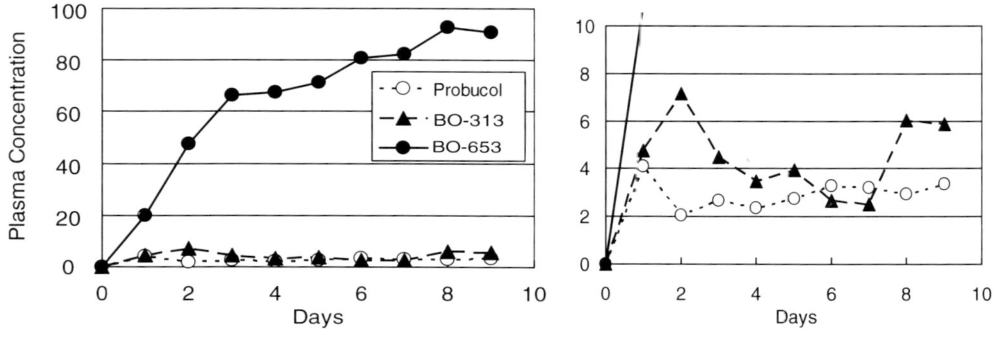

Fig. 6.5. Plasma concentration of BO-653 and BO-313 in WHHL rabbits. Rabbits (n=2) were orally administered once a day with the test compound (250 mg/kg) or probucol (500 mg/kg) for 9 days and were evaluated at 24 h after each daily administration. The graph on the *right* shows a magnification of the plasma concentration from 0 to 10 (μg/mL).

3.4.2. Evaluation of Atherosclerotic Lesions of the Antioxidant-Treated WHHL Rabbits

Male WHHL rabbits were maintained for 6 or 8 months on a normal diet (CR3) without additional antioxidants or with 0.5% BO-653 or BO-313. A diet with 1% probucol was used as the reference for an antiatherogenic compound. The atherosclerotic lesion area of the WHHL rabbits was determined as the area of fatty region on the aortic inner surface using SigmaScan Pro software (Jandel Scientific Software) with the photos of the aortic inner surface (24). To evaluate the inhibitory potency against atherosclerosis, we conducted an experiment for each compound; that is, experiment 1 is with BO-313 and experiment 2 is with BO-653.

Although both compounds inhibited the Cu^{2+}-catalyzed LDL oxidation, BO-313 and BO-653 also showed contrary results as shown in **Fig. 6.6**. BO-653 ameliorated atherosclerosis in WHHL rabbits, whereas BO-313 deteriorated it. This result suggests that the mechanism of an antioxidant is essential for exerting its antiatherogenic effect in vivo and that the inhibition of a common process in various types of LDL oxidation, e.g., the propagation process in LDL oxidation, is more critical for designing an effective antiatherogenic antioxidant.

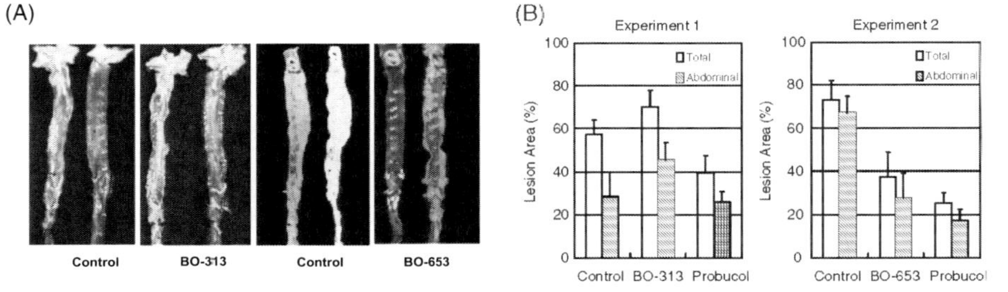

Fig. 6.6. Inhibitory effect of BO-653 and BO-313 on atherosclerosis in WHHL rabbits. Male rabbits (n=5) were maintained for 6 or 8 months on a diet containing 0.5, 0.5, or 1% concentration of BO-313, BO-653, or probucol, respectively. (**a**) Representative photographs of the aortic inner surface of WHHL rabbits. (**b**) Effects of BO-313 or BO-653 on intimal surface area with atherosclerotic lesions of the whole aorta and abdominal aorta in two experiments. All results are shown as mean ± SD.

4. Notes

1. In GPC analysis, the LDL solution, in particular its oxidation form, easily adheres to the gel permeation column. Therefore, a column prepared specifically for lipoprotein analysis should be used. An eluent specially used for lipoprotein analysis is currently available and provided by TOSOH Corporation which offers a stable GPC analysis.
2. For PK analysis, the test compound was administered orally as a suspension in 3% gum Arabic solution. If the gum Arabic content were lower, a uniform suspension would be difficult to prepare due to separation of the test compound from the gum Arabic in solution. As most of the test compounds in this study were oils, a uniform suspension was prepared with an electric homogenizer and had to be used within 1 h of its preparation. If a stock solution of the suspension is used, the solution would need to be resuspended with the electric homogenizer.

Acknowledgments

The authors thank Drs. T. Kodama at Tokyo University and T. Kita at Kyoto University for providing excellent advice on atherosclerosis in animal models. We also thank Drs. Y. Kawabe, Y. Kato, T. Suzuki, Y. Takashima, M. Takeda, H. Kaise, M. Kim, J. Aono, and Y. Ohba for their collaboration on this project in Chugai Pharmaceutical Co., Ltd. and Ms Ford Frances for her editorial assistance.

References

1. Steinberg, D., Parthasarathy, S., Carew, T.E., Khoo, J.C., and Witztum, J.L. (1989) Beyond cholesterol. Modifications of low-density lipoprotein that increase its atherogenicity. *N Engl. J. Med.* **320**, 915–924.
2. Steinberg, D. and Witztum, J.L. (2002) Is the oxidative modification hypothesis relevant to human atherosclerosis? Do the antioxidant trials conducted to date refute the hypothesis? *Circulation* **105**, 2107–2111.
3. Kita, T., Nagano, Y., Yokode, M., Ishii, K., Kume, N., Ooshima, A., Yoshida, H., and Kawai, C. (1987) Probucol prevents the progression of atherosclerosis in Watanabe heritable hyperlipidemic rabbit, an animal model for familial hypercholesterolemia. *Proc. Natl. Acad. Sci. USA* **84**, 5928–5931.
4. Mao, S.J., Yates, M.T., Rechtin, A.E., Jackson, R.L., and Van Sickle, W.A. (1991) Antioxidant activity of probucol and its analogues in hypercholesterolemic Watanabe rabbits. *J. Med. Chem.* **34**, 298–302.
5. Bjorkhem, I., Henriksson-Freyschuss, A., Breuer, O., Diczfalusy, U., Berglund, L., and Henriksson, P. (1991) The antioxidant butyrate hydroxytoluene protects against atherosclerosis. *Arterioscler. Thromb. Vasc. Biol.* **11**, 15–22.
6. Sparrow, C.P., Doebber, T.W., Olszewski, J., Wu, M.S., Ventre, J., Stevens, K.A., and Chao, Y.S. (1992) Low density lipoprotein is protected from oxidation and the progression of atherosclerosis is slowed in cholesterol-fed rabbits by the antioxidant N, N′-diphenyl-phenylenediamine. *J. Clin. Invest.* **89**, 1885–1891.
7. Tangirala, R.K., Casanada, F., Miller, E., Witztum, J.L., Steinberg, D., and Palinski, W. (1995) Effect of the antioxidant N, N′-diphenyl 1, 4-phenylenediamine (DPPD) on atherosclerosis in apoE-deficient mice. *Arterioscler. Thromb. Vasc. Biol.* **15**, 1625–1630.
8. Fruebis, J., Steinberg, D., Dresel, H.A., and Carew, T.E. (1994) A comparison of the antiatherogenic effects of probucol and of a structural analogue of probucol in low density lipoprotein receptor-deficient rabbits. *J. Clin. Invest.* **94**, 392–398.
9. Fruebis, J., Bird, D.A., Pattison, J., and Palinski, W. (1997) Extent of antioxidant protection of plasma LDL is not a predictor of the antiatherogenic effect of antioxidants. *J. Lipid Res.* **38**, 2455–2464.
10. Reinoehl, J., Frankovich, D., Machado, C., Kawasaki, R., Baga, J.J., Pires, L.A., Steinman, R.T., Fromm, B.S., and Lehmann, M.H. (1996) Probucol-associated tachyarrhythmic events and QT prolongation: Importance of gender. *Am. Heart J.* **131**, 1184–1191.
11. Johansson, J., Olsson, A.G., Bergstrand, L., Elinder, L.S., Nilsson, S., Erikson, U., Molgaard, J., Holme, I., and Walldius, G. (1995) Lowering of HDL2b by probucol partly explains the failure of the drug to affect femoral atherosclerosis in subjects with hypercholesterolemia. A Probucol Quantitative Regression Swedish Trial (PQRST) Report. *Arterioscler. Thromb. Vasc. Biol.* **15**, 1049–1056.
12. Upston, J.M., Terentis, A.C., and Stocker, R. (1999) Tocopherol-mediated peroxidation of lipoproteins: Implications for vitamin E as a potential antiatherogenic supplement. *FASEB J.* **13**, 977–994.
13. Tamura, K., Kato, Y., Ishikawa, A., Himori, M., Yoshida, M., Takashima, Y., Suzuki, T., Kawabe, Y., Cynshi, O., Kodama, T., Niki, E., and Shimizu, M. (2003) Design and synthesis of 4, 6-di-tert-butyl-2, 3-dihydro-5-benzofuranols as a novel series of antiatherogenic antioxidants. *J. Med. Chem.* **46**, 3083–3093.
14. Ingold, K.U., Burton, G.W., Foster, D.O., Zuker, M., Hughes, L., Lacelle, S., Lusztyk, E., and Slaby, M. (1986) A new vitamin E analogue more active than alpha-tocopherol in the rat curative myopathy bioassay. *FEBS Lett.* **205**, 117–120.
15. Cynshi, O., and Stocker, R. (2005) Inhibition of lipoprotein lipid oxidation. *Handb. Exp. Pharmacol.* **170**, 563–590.
16. Havel, R.J., Eder, H.A., and Bragdon, J.H. (1955) The distribution and chemical composition of ultracentrifugally separated lipoproteins in human serum. *J. Clin. Invest.* **34**, 1345–1353.
17. Lowry, O.H., Rosebrough, N.J., Farr, A.L., and Randall, R.J. (1951) Protein measurement with the Folin phenol reagent. *J. Biol. Chem.* **193**, 265–275.
18. Cynshi, O., Takashima, Y., Suzuki, T., Kawabe, Y., Ohba, Y., and Kodama, T. (1994) Characterization of aggregated low density lipoproteins induced by copper-catalyzed oxidation. *J. Atheroscler. Thromb.* **1**, 87–97.
19. Kawabe, Y., Cynshi, O., Takashima, Y., Suzuki, T., Ohba, Y., and Kodama, T. (1994) Oxidation-induced aggregation of rabbit low-density lipoprotein by azo initiator. *Arch. Biochem. Biophys.* **310**, 489–496.

20. Cathcart, M.K., McNally, A.K., and Chisolm, G.M. (1991) Lipoxygenase-mediated transformation of human low density lipoprotein to an oxidized and cytotoxic complex. *J. Lipid. Res.* **32**, 63–70.
21. Yagi, K. (1998) Simple assay for the level of total lipid peroxides in serum or plasma. *Methods Mol. Biol.* **108**, 101–106.
22. Koller, E., Quehenberger, O., Jurgens, G., Wolfbeis, O.S., and Esterbauer, H. (1986) Investigation of human plasma low density lipoprotein by three-dimensional fluorescence spectroscopy. *FEBS Lett.* **198**, 229–234.
23. Noguchi, N., Okimoto, Y., Tsuchiya, J., Cynshi, O., Kodama, T., and Niki, E. (1997) Inhibition of oxidation of low-density lipoprotein by a novel antioxidant, BO-653, prepared by theoretical design. *Arch. Biochem. Biophys.* **347**, 141–147.
24. Cynshi, O., Kawabe, Y., Suzuki, T., Takashima, Y., Kaise, H., Nakamura, M., Ohba, Y., Kato, Y., Tamura, K., Hayasaka, A., Higashida, A., Sakaguchi, H., Takeya, M., Takahashi, K., Inoue, K., Noguchi, N., Niki, E., and Kodama, T. (1998) Antiatherogenic effects of the antioxidant BO-653 in three different animal models. *Proc. Natl. Acad. Sci. USA* **95**, 10123–10128.

Chapter 7

Preparation of Drug-Loaded Polymeric Nanoparticles and Evaluation of the Antioxidant Activity Against Lipid Peroxidation

Adriana R. Pohlmann, Scheila Rezende Schaffazick, Tânia B. Creczynski-Pasa, and Sílvia S. Guterres

Abstract

Antioxidants have been found to be effective as prophylatic and therapeutic agents for different diseases such as diabetes, cancer, and neurodegenerative disorders. However, antioxidant substances can present poor solubility in water, inefficient permeability, gastrointestinal degradation, first-pass effect, and/or instability during storage. These drawbacks can be potentially circumvented by encapsulating the susceptible antioxidants. Polymeric nanoparticles (nanocapsules or nanospheres) have been used to improve the drug efficacy and release. Our group has shown that the in vitro antioxidant effect of melatonin against lipid peroxidation in microsomes and liposomes can be improved by encapsulation of the antioxidant drug in polymeric nanoparticles.

Key words: Polymeric nanoparticles, nanospheres, nanocapsules, antioxidant drug, melatonin, lipid peroxidation, microsomes, liposomes.

1. Introduction

Nanoencapsulation of drugs is based on the preparation of drug-loaded particles with diameters less than 1 μm (1–3). Among the different nanocarriers, the polymeric nanoparticles are employed as drug delivery systems. The term polymeric nanoparticle is widely used for either nanocapsules or nanospheres (3). Nanocapsules are carriers generally composed of an oil core surrounded by a polymeric wall (vesicular systems), whereas nanospheres constitute a polymeric matrix-type structure (3, 4). The nanoparticles

R.M. Uppu et al. (eds.), *Free Radicals and Antioxidant Protocols*, Methods in Molecular Biology 610,
DOI 10.1007/978-1-60327-029-8_7, © Humana Press, a part of Springer Science+Business Media, LLC 1998, 2010

are stabilized by different surfactant systems (5). A variety of therapeutic agents can be encapsulated in the polymeric nanoparticles depending on their lipophilicities. These agents exhibit different mechanisms of association with the nanoparticles including their dissolution in the oil and/or the polymer, or adsorption on the particles, as well as they can be attached to or dispersed in the nanoparticles (2, 6, 7).

Depending on their size and the chemical nature of their surfaces the nanoparticles can cross certain biological barriers and penetrate the tissues through small capillaries to be subsequently taken up by cells (2, 8). In this way, an efficient uptake of the drug by target sites and sustained drug release can be achieved (2). Amidst the versatility of the methods available to prepare drug-loaded nanoparticles, the nanoparticle size, the size distribution, the chemical nature of the surface, and the drug release mechanism can be controlled. The surface functionalities can be used to prolong the circulation time of the nanoparticles in the blood stream or target to specific cells or tissue by using specific ligands (2, 8). The potential applications of nanoparticles include the controlled release of drugs (7), drug targeting to specific sites (as tumor tissues) improving the therapeutic index or efficacy (1, 4), drug delivery to the brain (6), protection of drugs against inactivation in the gastrointestinal tract, the improvement of bioavailability of drugs (9), protection of gastrointestinal mucosa from toxicity of drugs (10, 11), delivery of poorly water-soluble compounds (5), coating of microparticles (12), and the entrapment of agents for diagnosis (4).

The polymeric nanoparticles have been recently studied in order to increase the stability of antioxidants such as the highly lipophilic coenzyme Q_{10} (13). Polymeric nanoparticles are able to protect catalase, an antioxidant enzyme, from external proteolysis (14).

The ability of nanoparticles to passively target the tumor tissue can be explored to deliver antioxidants directly in tumors cells (15). The encapsulation of idebenone (16) and vitamin E (17) in nanocapsules or nanospheres improved their antioxidant actions. A higher intestinal permeation of poorly water-soluble antioxidants and their sustained release can be achieved by drug encapsulation in nanoparticles, as reported for ellagic acid (18). These data suggest that nanoparticles can be effectively used for oral delivery in order to improve the bioavailability of antioxidants.

We have developed melatonin-loaded polymeric nanoparticles (19, 20) which increased its antioxidant activity against lipid peroxidation (19). In these studies, the nanoparticles were prepared using the methods of nanoprecipitation (nanospheres) or interfacial deposition (nanocapsules). Besides, the particles were characterized by dynamic light scattering to determine the particle size and the polydispersion index by electrophoretic mobility to measure the zeta potential and by electron microscopy to observe their shape and size is exploited. Further, the application of liquid chromatography to determine the total drug content and the drug encapsulation efficiency of the formulations provide relevant information. Using phosphatidylcholine liposomes or liver microsomes as oxidizable substrates, the melatonin-loaded Eudragit S100®-nanocapsules or nanospheres have been tested against lipid peroxidation and were observed to have enhanced drug effect than the drug in solution. The lipid peroxidation was induced by iron–ascorbate system and was evaluated by the thiobarbituric method (TBARS) (19).

2. Materials

2.1. Polymeric Nanoparticles

1. Polymer: poly(ε-caprolactone) (MW= 65,000 g/mol; Aldrich, Strasbourg, France), Eudragit S100®, or Eudragit RS100® (Röhm GmbH, Darmstadt, Germany).
2. Oil (in the case of nanocapsules): caprylic/capric triglyceride (Brasquim, Porto Alegre, Brazil).
3. Polysorbate 80 (Tween 80®, Delaware, Porto Alegre, Brazil).
4. Sorbitan monooleate (Span 80®, Sigma, St. Louis, USA) or sorbitan monostearate (Span 60®, Delaware, Porto Alegre, Brazil).
5. Antioxidant drug: melatonin (Acros Organics, Belgic).

2.2. Particle Size

1. Brookhaven Instruments standard setup (BI-200 M goniometer, BI-9000AT digital correlator and a BI9863 detection system) using a Spectra Physics He-Ne laser (model 127, λ_0 = 632,8 nm) as light source (21) or Zetasizer® Nano Series (ZEN 3600 model) from Malvern Instruments.
2. To dilute the samples: water (MilliQ®).

2.3. Zeta Potential

1. Zetasizer® Nano Series (ZEN 3600 model) from Malvern Instruments.
2. Solution to prepare the samples: 1 mM NaCl (Merck AG, Darmstadt, Germany) aqueous solution (MilliQ® water) filtered across membrane (0.22 μm).

2.4. Transmission Electron Microscopy (TEM)

1. Transmission electron microscope model JEM 1200 Exll, from Jeol.
2. Formvar–carbon support film on specimen grid (Electron Microscopy Sciences).

3. Negative stain solution: 2% (w/v) uranyl acetate in water (MilliQ®). Store at 4°C protected from the light (*see* **Note 1**).

2.5. Liquid Chromatography

1. High-performance liquid chromatography (HPLC): pump 600 controller, 2487 Dual λ absorbance detector, 717 plus auto-sampler; Waters, USA.
2. Reversed phase column: LiChrospher® RP-18 (5 μm; 250 × 4 mm: Merck, Germany).
3. Mobile phase: acetonitrile/MilliQ® water (55:45; v/v).
4. To separate nanoparticles and free drug: device Ultrafree-MC® (MW-cut off 10,000 Da; Millipore).

2.6. Liposomes

1. Buffer solution at pH 8.0 to dissolve the phospholipids: 10 mM tricine (Sigma), 20 g/L sodium cholate (Aldrich), and 10 g/L sodium deoxycholate (Sigma).
2. Phospholipids (phosphatidylcholine, Sigma): Dissolve phosphatidyl choline 50 mg/mL in the buffer solution (pH 8.0).
3. Buffer solution at pH 8.0 for dialysis: 10 mM tricine (Sigma) and 2.5 mM $MgCl_2$ (Merck).
4. Membrane of dialysis (MW-cut off 5,000 Da; 63 mm of diameter; Diachema).

2.7. Microsomes

1. Liver of Wistar rats are homogenized (1:5 w/v) in 10 mM Tris–sucrose pH 7.4 (Merck).
2. Aqueous solution to precipitate microsomal fraction: 8 mM $CaCl_2$ (Merck).
3. Wash buffer: 15 mM KCl (Merck) and 10 mM Tris, pH 7.4 (Merck).
4. Stocking solution: 20% (w/v) glycerol (Sigma) in 0.1 M Na_2HPO_4 (Merck).

2.8. Lipid Peroxidation

1. Buffer solution: 1 M Tris–HCl, pH 7.4 (Merck).
2. $FeSO_4$ (Merck) at 500 μM is prepared using distilled water (*see* **Note 2**).
3. Ascorbic acid (Sigma): prepare 10 mM solution using distilled water (*see* **Note 2**).
4. Thiobarbituric acid (Sigma): dissolve at 0.73% (w/v) in distilled water under heating. Stored up to 1 week.
5. Trichloroacetic acid (Merck): dissolve at 12% (w/v) in distilled water.

3. Methods

Nanoparticles can be prepared by methods based on the polymerization of dispersed monomers or the precipitation of preformed synthetic or natural polymers (3). The methods based on preformed polymers avoid the potential toxicity caused by the residual monomers or oligomers and/or drug-to-monomer cross-reaction related to the polymerization methods (3, 13, 22). The classical methods based on spontaneous emulsification-solvent diffusion, generally referred as interfacial deposition or nanoprecipitation method, have been widely employed to produce nanoparticles from preformed polymers (3, 13).

The characterization of the nanoparticle suspensions involves the determination of the mean particle size, the polydispersion index, the zeta potential, the electronic microscopy image analysis, the drug total content, and the drug encapsulation efficiency (4).

The control of the particle size, the size distribution, as well as the carrier surface characteristics is the key to predict the biological effect of the encapsulated drug. After intravenous administration, the biological fate of nanoparticles is influenced by those parameters, since they either define the uptake by the mononuclear phagocytes system or direct the nanoparticles to target pathological sites, like tumors or regions with inflammation (1). In general, by the oral route the intracellular uptake of nanoparticles can be enhanced by the small particle sizes and by the high hydrophobicity of their surfaces (18). The positively charged polymers may interact with the negatively charged mucus and open up the tight junctions of epithelial cells representing a tool to improve the drug bioavailability (23). In the skin, the nanoparticles have high accumulation in the follicular regions (24).

3.1. Preparation of Polymeric Nanocapsules

1. Nanocapsule suspension containing the antioxidant drug is prepared by interfacial deposition of preformed polymers, as previously described by Fessi et al. (25).
2. The organic solution consists of polymer (250 mg), oil (0.8 mL), and sorbitan monooleate or monostearate (192 mg) dissolved in 67 mL of acetone. This organic phase is maintained under magnetic stirring at a temperature between 25 and 40°C.
3. After the total dissolution of all components in the organic phase, the melatonin (12.5 mg) is added to the organic solution (*see* **Note 3**).
4. The polysorbate 80 (192 mg) is dissolved in 133 mL of water (MilliQ®).

5. The organic phase is poured into the aqueous solution, using a funnel, under moderate magnetic stirring (between 500 and 800 rpm). The stirring is maintained for 10 min at room temperature.
6. After this period, the acetone is removed and the solution is concentrated by evaporation under reduced pressure (below 40°C) and the final volume of the suspension is adjusted to 25 mL in a volumetric flask (*see* **Note 4**).

3.2. Preparation of Polymeric Nanospheres

1. Nanosphere suspension containing the antioxidant drug is prepared by nanoprecipitation of preformed polymers, as previously described by Fessi et al. (26).
2. Nanosphere suspension is prepared as described above for nanocapsules (*see* **Section 3.1**) without oil in the organic phase.

3.3. The Particle Size and Polydispersion Index

1. A sample of the final suspension is diluted with water (MilliQ®) (*see* **Note 5**).
2. The particle size and the polydispersion of the nanocapsules are determined by photon correlation spectroscopy (PCS) using Zetasizer® Nano Series or Brookhaven Instruments standard setup (*see* **Note 6**). An example of the results is shown in **Fig. 7.1**.

3.4. Zeta Potential

1. A sample of the final nanoparticle suspension is diluted with 1 mM NaCl aqueous solution (*see* **Note 5**).

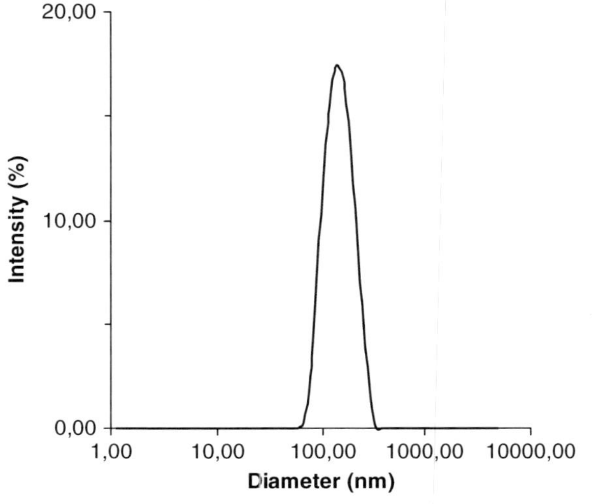

Fig. 7.1. Graphic representation of the size distribution of melatonin-loaded Eudragit RS100® nanocapsules obtained by Zetasizer® Nano Series (polydispersion indexes = 0.065; mean diameter = 142 nm).

2. The zeta potential is determined using Zetasizer® Nano Series (*see* **Note 7**).

3.5. Transmission Electron Microscopy

1. The nanoparticle suspension is diluted with water (MilliQ®) (*see* **Note 8**).
2. The diluted sample (around 20 μL) is deposited on the formvar–carbon-coated grid for 2–3 min. The grid is handled with a fine point tweezers. The sample preparation is carried out on a piece of parafilm.
3. The excess of water is removed by the edge of the grid using a piece of filter paper.
4. After 3–5 min, the deposited sample is negatively stained by adding approximately 20 μL of uranyl acetate solution (20).
5. After 3–5 min, the excess of water is removed by the edge of the grid using a piece of filter paper. The samples are air-dried before measurement.
6. The transmission electron microscopy (TEM) is carried out using a voltage of 80 kV at the magnifications of 50,000-, 120,000-, and 250,000-folds. An example of the photomicrographs is shown in **Fig. 7.2**.

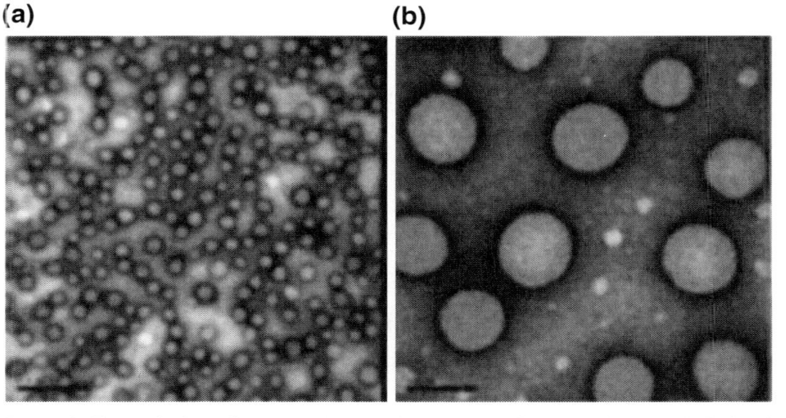

Fig. 7.2. Transmission electron microscopic images of melatonin-loaded Eudragit RS100® nanocapsules: (**a**) bar = 500 nm and (**b**) bar = 100 nm.

3.6. Encapsulation Efficiency

1. The total content of melatonin in the nanoparticle suspension is determined using HPLC after the dissolution of the nanoparticles by acetonitrile. The solution is filtered (0.22 or 0.45 μm). The analytical method to quantify the melatonin (*see* **Section 2.5**) is previously validated. The injected volume was 20 μL and the drug was detected at 229 nm (20) (*see* **Note 9**).
2. The nanoparticle suspension (approximately 400 μL) is added in the Ultrafree-MC® unit (MW-cutoff 10,000 Da) and centrifuged at 15,300×*g* for 10 min.

3. Free drug (melatonin non-associated with the nanoparticles) is determined in the ultrafiltrate by HPLC as described above (19, 20).
4. The associated melatonin with the nanoparticles (encapsulation efficiency, *EE*) is calculated from the ratio between the difference of the total and the free drug concentrations (C_T and C_F, respectively) divided by the total concentration, multiplied by 100 (**Eq. [1]**) (19, 20). **Table 7.1** shows examples of the physicochemical characteristics of unloaded and drug-loaded nanoparticles.

$$EE = \frac{(C_T - C_F)}{C_T} \times 100 \qquad [1]$$

Table 7.1
Physicochemical properties of empty and melatonin-loaded nanoparticle suspensions[a]

Nanoparticle suspensions	Diameter (nm)	Zeta potential (mV)	Encapsulation efficiency (%)
Melatonin-loaded nanospheres	126 ± 12	–36.6 ± 1.9	54
Unloaded nanospheres	192 ± 46	–35.4 ± 2.3	–
Melatonin-loaded nanocapsules	236 ± 20	–43.5 ± 0.8	56
Unloaded nanocapsules	254 ± 52	–45.0 ± 1.5	–

[a]Adapted from (19).

3.7. Preparation of Liposomes as Substrate for the Antioxidant Activity Assay

1. Bilayer liposomes are prepared by dialysis (27).
2. Dissolution of the phospholipids in the buffer solution at pH 8.0 (tricine/sodium cholate/sodium deoxycholate).
3. The dialysis procedure is performed at 30°C for 5 h against 2 L buffer solution at pH 8.0 (tricine/$MgCl_2$) using a membrane with 2 nm pores.

3.8. Preparation of Microsomes as Subtract for the Antioxidant Activity Assay

1. Microsomes are obtained by differential centrifugation with calcium aggregation (28, 29).
2. After the homogenization of liver in Tris–sucrose, pH 7.4, the mixture is centrifuged at 600×*g* for 5 min at 0°C.
3. The supernatant is centrifuged at 12,000×*g* for 10 min.
4. The microsomal fraction is resulted from the precipitation of the supernatant with $CaCl_2$ and centrifuged at 25,000×*g* for 15 min.

5. The pellet resulted is washed with KCl/Tris, pH 7.4, and centrifuged at 25,000×*g* for 15 min.
6. After washing, microsomes are re-suspended with glycerol in Na_2HPO_4.
7. The fraction is immediately placed in liquid nitrogen at –170°C up to the antioxidant activity evaluation.

3.9. *In vitro* Lipid Peroxidation

1. The lipid peroxidation is induced by ascorbyl radical (iron–ascorbate) (30).
2. Buffer is placed in test tubes: 20 μL (v_1) of 1 M Tris–HCl, pH 7.4, is added to reach 0.1 M Tris–HCl, pH 7.4, in the final volume of 200 μL (v_{total} of the reactional medium).
3. Distilled water (v_2) is added to the test tubes after its calculation (*see* **Note 10**).
4. The volume (v_3) of drug solution (*see* **Note 11**) or melatonin-loaded formulations (nanocapsule or nanosphere suspensions) corresponding to the antioxidant dose of 0.2 mM (18.6 μL) or 0.4 mM (37.2 μL) is added in the test tubes. The controls for the experiments correspond to the respective formulations prepared in the absence of the drug (unloaded formulations) (*see* **Note 12**).
5. In the test tubes, 10 μL (v_4) of 500 μM $FeSO_4$ solution is added to reach 25 μM $FeSO_4$.
6. In the test tubes, 10 μL (v_5) of 10 mM ascorbic acid is added to reach 500 μM ascorbic acid.
7. The membrane substrates, microsomes or liposomes, are added (v_6) to the reactional medium in the test tubes at 1 mg/mL microsomal protein/mL (25 μL) or lipids (liposomes) at 12.5 mg/mL (53.6 μL) (*see* **Note 13**).
8. The samples are then incubated for 30 min at 37°C.
9. After the incubation period, 200 μL of 12% trichloroacetic acid and 200 μL of 0.73% thiobarbituric acid are added to stop the reaction.
10. The mixture is then maintained at 100°C for 30 min (*see* **Note 14**).
11. The mixture is cooled and centrifuged (10,000×*g* for 5 min) (*see* **Note 15**).
12. The supernatants are monitored at 535 nm.
13. The amount of lipid peroxidation is determined using the molar extinction coefficient of 1.56×10^5/M/cm and expressed as thiobarbituric acid reactive substances

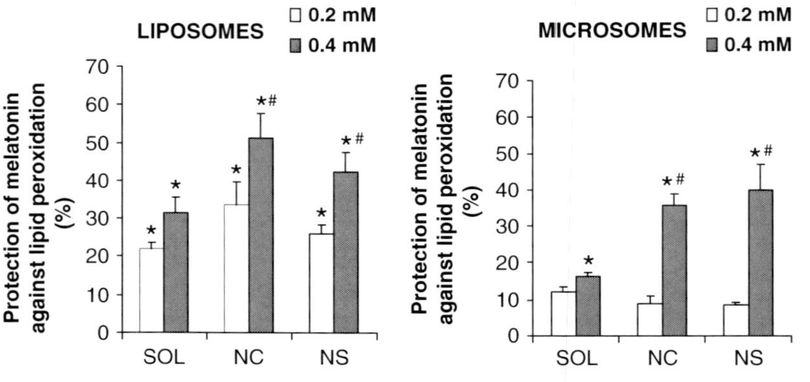

Fig. 7.3. Effect of melatonin encapsulation on lipid peroxidation induced by the ascorbyl-free radical (SOL = melatonin solution containing 1% ethanol; NC = Eudragit S100® nanocapsule suspension; NS= Eudragit S100® nanosphere suspension). This protection is relative to the formulations prepared without melatonin (controls). *Statistical differences between each formulation and its respective control established using *t* test ($p < 0.05$); #statistical differences between the solution (0.4 m*M* of melatonin) and the formulations established using *t* test ($p < 0.05$). (Data published in 19).

(TBARS). The calculated protection against lipid peroxidation is relative to the formulations prepared without melatonin (controls). An example of the results is shown in **Fig. 7.3**.

4. Notes

1. The solution is prepared under exhaustion (fumed hood) and it is filtered (pore size: 0.22 or 0.45 μm) before use.
2. The solution is prepared just before the experiment.
3. In the case of the antioxidant drug insolubility in water-miscible solvents (acetone or ethanol), this method cannot be employed. Other methods can be used, as the emulsification–diffusion, which employs ethyl acetate, benzyl alcohol, or propylene carbonate as solvents (3, 22).
4. The melatonin-loaded nanoparticle suspensions must be stored to protect from the light. In this way, evaporation is carried out in an amber flask and the suspension is stored in an amber recipient.
5. The usual dilution is 500-fold, but changing the qualitative composition of the suspension a study of dilution is necessary. In general, the dilution ranges between 500- and 2,000-folds.

6. The diameters usually observed for nanoparticles prepared using preformed polymers range between 100 and 500 nm (3, 4). Polydispersion indexes lower than 0.2 indicate homogeneous systems presenting a satisfactory narrow particle distribution.
7. Generally, zeta potential values higher than 30 mV (positive or negative) indicate relatively stable nanoparticle suspensions due to the repulsion between particles reducing their aggregation (4).
8. Generally, the suspension is diluted 10-fold in water, but depending on the composition a study should be carried out to optimize the dilution.
9. The melatonin calibration curve is prepared between 2.5 and 17.5 μg/mL in acetonitrile. The drug analytical method is validated following the ICH or the USP 26 (31, 32).
10. Volume v_2 is calculated by the following: $v_2 = 200 - v_1 - v_3 - v_4 - v_5 - v_6$, where v_1, v_3, v_4, v_5, and v_6 correspond to the volumes of the solution: Tris–HCl, antioxidant, $FeSO_4$, ascorbic acid, and substrates (liposomes or microsomes), respectively.
11. The 0.5 mg/mL melatonin solution is prepared using distilled water containing 1% ethanol (v/v).
12. Each test tube must have a respective control prepared in the same condition omitting the drug.
13. Volume v_6 can change depending on the total volume of the resulting liposome suspension or microsomal protein content.
14. Using a hot syringe needle, a small orifice in the test tubes must be performed for pressure compensation.
15. The supernatants must be transparent. The use of an ultracentrifuge or a solvent can be necessary.

Acknowledgments

The authors acknowledge the financial support from Conselho Nacional de Desenvolvimento Científico e Tecnológico (CNPq), Fundação de Amparo à Pesquisa do Estado do Rio Grande do Sul (FAPERGS), Fundação de Amparo à Pesquisa do Estado de Santa Catarina (FAPESC), Coordenação de Aperfeiçoamento de Pessoal de Nível Superior (CAPES), Rede Nanocosméticos CNPq/MCT-Brazil.

References

1. Brigger, I., Dubernet, C., and Couvreur, P. (2002) Nanoparticles in cancer therapy and diagnosis. *Adv. Drug Deliv. Rev.* **54**, 631–651.
2. Sahoo, S.K. and Labhasetwar, V. (2003) Nanotech approaches to drug delivery and imaging. *Drug Discov. Today* **8**, 1112–1120.
3. Pinto Reis, C., Neufeld, R.J., Ribeiro, A.J., and Veiga, F. (2006) Nanoencapsulation I. Methods for preparation of drug-loaded polymeric nanoparticles. *Nanomed.* **2**, 8–21.
4. Couvreur, P., Barrat, G., Fattal, E., Legrand, P., and Vauthier, C. (2002) Nanocapsule technology: A review. *Crit. Rev.Ther. Drug Carrier Syst.* **19**, 99–134.
5. Legrand, P., Barratt, G., Mosqueira, V., Fessi, H., and Devissaguet, J.-P. (1999) Polymeric nanocapsules as drug delivery systems: A review. *STP Pharma Sci.* **9**, 411–418.
6. Garcia-Garcia, E., Andrieux, K., Gil, S., and Couvreur, P. (2005) Colloidal carriers and blood-brain barrier (BBB) translocation: A way to deliver drugs to the brain? *Int. J. Pharm.* **298**, 274–292.
7. Soppimath, K.S., Aminabhavi, T.M., Kulkarni, A.R., and Rudzinski, W.E. (2001) Biodegradable polymeric nanoparticles as drug delivery devices. *J. Control. Release* **70**, 1–20.
8. Fahmy, T.M., Fong, P.M., Goyal, A., and Saltzman, W.M. (2005) Targeted for drug delivery. *Nanotoday*, 18–26.
9. Allémann, E., Leroux, J.-C., and Gurny, R. (1998) Polymeric nano- and microparticles for the oral delivery of peptides and peptidomimetics. *Adv. Drug Deliv. Rev.* **34**, 171–189.
10. Guterres, S.S., Fessi, H., Barratt, G., Puisieux, F., and Devissaguet, J.-P. (1995) Poly(D,L-lactide) nanocapsules containing non-steroidal anti-inflammatory drugs: Gastrointestinal tolerance following intravenous and oral administration. *Pharm. Res.* **12**, 1–3.
11. Schaffazick, S.R., Pohlmann, A.R., Dalla-Costa, T., and Guterres, S.S. (2003) Freeze-drying polymeric colloidal suspensions: Nanocapsules, nanospheres and nanodispersion. A comparative study. *Eur. J. Pharm. Biopharm.* **56**, 501–505.
12. Beck, R.C.R., Pohlmann, A.R., and Guterres, S.S. (2004) Nanoparticle-coated microparticles: Preparation and characterization. *J. Microencapsul.* **21**, 499–512.
13. Kwon, S.S., Nam, Y.S., Lee, J.S., Ku, B.S., Han, S.H., Lee, J.Y., and Chang, I.S. (2002) Preparation and characterization of coenzyme Q_{10}-loaded PMMA nanoparticles by a new emulsification process based on microfluidization. *Colloids Surfaces A: Physicochem. Eng. Asp.* **210**, 95–104.
14. Dziubla, T.D., Karim, A., Vladimir, R., and Muzykantov, V.R. (2005) Polymer nanocarriers protecting active enzyme cargo against proteolysis. *J. Control. Release* **102**, 427–439.
15. Ratnam, D.V., Ankola, D.D., Bhardwaj, V., Sahana, D.K., and Ravi Kumar, M.N.V. (2006) Role of antioxidants in prophylaxis and therapy: A pharmaceutical perspective. *J. Control. Rel.* **113**, 189–207.
16. Palumbo, M., Russo, A., Cardile, V., Renis, M., Paolino, D., Puglisi, G., and Fresta, M. (2002) Improved antioxidant effect of idebenone-loaded polyethyl-2-cyanoacrylate nanocapsules tested on human fibroblast. *Pharm. Res.* **19**, 71–78.
17. Shea, T.B., Ortiz, D., Nicolosi, R.J., Kumar, R., and Watterson, A.C. (2005) Nanosphere-mediated delivery of vitamin E increases its efficacy against oxidative stress resulting from exposure to amyloid beta. *J. Alzheimer's Dis.* **7**, 297–301.
18. Bala, I., Bhardwaj, V., Hariharan, S., Kharade, S.V., Roy, N., and Kumar, M.N.V.R. (2006) Sustained release nanoparticulate formulation containing antioxidant-ellagic acid as potential prophylaxis system for oral administration. *J. Drug. Target.* **14**,27–34.
19. Schaffazick, S.R., Pohlmann, A.R., de Cordova, C.A.S., Creczynski-Pasa, T.B., and Guterres, S.S. (2005) Protective properties of melatonin-loaded nanoparticles against lipid peroxidation. *Int. J. Pharm.* **289**, 209–213.
20. Schaffazick, S.R., Pohlmann, A.R., Mezzalira, G., and Guterres, S.S. (2006) Development of nanocapsule suspensions and nanocapsule spray-dried powders containing melatonin. *J. Braz. Chem. Soc.* **17**, 562–569.
21. Pohlmann, A.R., Weiss, V., Mertins, O., Pesce da Silveira, N., and Guterres, S.S. (2002) Spray-dried indometacin-loaded polyester nanocapsules and nanospheres: Development, stability evaluation and nanostructure models. *Eur. J. Pharm. Sci.* **16**, 305–312.
22. Quintanar-Guerrero, D., Allémann, E., Fessi, H., and Doelker, E. (1998) Preparation techniques and mechanisms of formation of biodegradable nanoparticles from preformed polymers. *Drug Dev. Ind. Pharm.* **24**, 1113–1128.

23. Ubrich, N., Schmidt, C., Bodmeier, R., Hoffman, M., and Maincent, P. (2005) Oral evaluation in rabbits of cyclosporin-loaded Eudragit RS or RL nanoparticles. *Int. J. Pharm.* **288**, 169–175.
24. Alvarez-Román, R., Naik, A., Kalia, Y.N., Guy, R.H., and Fessi, H. (2004) Skin penetration and distribution of polymeric nanoparticles. *J. Control. Rel.* **99**, 53–62.
25. Fessi, H., Puisieux, F., Devissaguet, J.-P., Ammoury, N., and Benita, S. (1989) Nanocapsule Formation by interfacial polymer deposition following solvent displacement. *Int. J. Pharm.* **55**, r1–r4.
26. Fessi, H., Devissaguet, J.-P., Puisieux, F., and Thies, C. (1986) Procedé de préparation des systèmes colloidaux dispersibles d'une substance sous forme de nanoparticules. Fr. Patent Application No. 8618446.
27. Creczynski-Pasa, T.B. and Gräber, P. (1994) ADP binding and ATP synthesis by reconstituted H^1 –ATPase from chloroplasts. *FEBS Lett.* **350**, 195–198.
28. Schenkman, J.B. and Cinti, D.L. (1978) Preparation of microsomes with calcium. *Methods Enzymol.* **52**, 83–89.
29. Cordova, C.A.S., Siqueira, I.R., Netto, C.A., Yunes, R.A., Volpato, A.M., Filho, V.C., Curi-Pedrosa, R., and Creczynski-Pasa, T.B. (2002) Protective properties of butanolic extract of the *Calendula officinalis* L.(marigold) against lipid peroxidation of rat liver microsomes and action as free radical scavenger. *Redox Report* 7, 95–102.
30. Teixeira, A., Morfim, M.P., Cordova, C.A.S., Charão, C.C.T., Lima, V.R., and Creczynski-Pasa, T.B. (2003) Melatonin protects against pro-oxidant enzymes and reduces lipid peroxidation in distinct membranes induced by the hydroxyl and ascorbyl radicals and by peroxinitrite. *J. Pineal Res.* **35**, 262–268.
31. Validation of Analytical Procedures: Methodology, ICH-Harmonised Tripartity Guideline, IFPMA, Geneva, Switzerland, 1996.
32. The United States Pharmacopoeia. 27th Edn., The United State Phamacopoeial Convention, Rockville, USA, 2003.

Chapter 8

Nanoparticle and Iron Chelators as a Potential Novel Alzheimer Therapy

Gang Liu, Ping Men, George Perry, and Mark A. Smith

Abstract

Current therapies for Alzheimer disease (AD) such as the acetylcholinesterase inhibitors and the latest NMDA receptor inhibitor, Namenda, provide moderate symptomatic delay at various stages of the disease, but do not arrest the disease progression or bring in meaningful remission. New approaches to the disease management are urgently needed. Although the etiology of AD is largely unknown, oxidative damage mediated by metals is likely a significant contributor since metals such as iron, aluminum, zinc, and copper are dysregulated and/or increased in AD brain tissue and create a pro-oxidative environment. This role of metal ion-induced free radical formation in AD makes chelation therapy an attractive means of dampening the oxidative stress burden in neurons. The chelator desferrioxamine, FDA approved for iron overload, has shown some benefit in AD, but like many chelators, it has a host of adverse effects and substantial obstacles for tissue-specific targeting. Other chelators are under development and have shown various strengths and weaknesses. Here, we propose a novel system of chelation therapy through the use of nanoparticles. Nanoparticles conjugated to chelators show unique ability to cross the blood–brain barrier (BBB), chelate metals, and exit through the BBB with their corresponding complexed metal ions. This method may provide a safer and more effective means of reducing the metal load in neural tissue, thus attenuating the harmful effects of oxidative damage and its sequelae. Experimental procedures are presented in this chapter.

Key words: Alzheimer disease, chelation therapy, metal dysregulation, nanoparticles.

1. Introduction

Alzheimer disease (AD) is a devastating neurodegenerative disease with progressive and irreversible damage to thought, memory, and language. AD is the most common form of dementia among people aged 65 and older, progressing slowly from mild

R.M. Uppu et al. (eds.), *Free Radicals and Antioxidant Protocols*, Methods in Molecular Biology 610,
DOI 10.1007/978-1-60327-029-8_8, © Humana Press, a part of Springer Science+Business Media, LLC 1998, 2010

forgetfulness to the need for total care (reviewed in (1)). Unfortunately, an explicative etiology or a viable cure is not available.

Compared with other tissues, the central nervous system may be particularly susceptible to oxidative damage (2, 3). Accumulating evidence supports the hypothesis that oxidative stress generated by various mechanisms may be among the major intermediary risk factors that initiate and promote neurodegeneration, leading to AD (4–12). Oxidation reactions can be catalyzed by transition metals such as iron and copper (13) and, as such, the likelihood that an oxidation reaction will take place is probably increased by the regional concentrations of transition metals (14). Substantial studies show that the metabolism of iron is involved in AD and that the concentration of iron in the brain of AD patients is elevated (8, 15). Aluminum has also received attention in AD, although a role has never been convincingly demonstrated. Nonetheless, aluminum has been found in high concentrations in both senile plaques and intraneuronal neurofibrillary tangles in the brains of subjects with AD, which suggests that this metal may be involved in the etiopathology of AD (8, 15–19). Aluminum, unlike transition metal ions, is unable to participate in redox cycles of electron transfer reactions because of a fixed oxidation state of 3+ in biological systems, but evidence suggests that it can act synergistically with iron to increase free radical damage (2, 20). Strong evidence also shows that other metals implicated in the development of AD include copper (9, 10, 21–25) and zinc (8, 21, 26–28). In the AD brain, the concentration of zinc is significantly elevated in senile plaques and the concentration of copper is elevated in the rim of senile plaques. Overall, these studies indicate that the environmental conditions in AD, due to imbalances of several metals, have the potential for catalyzing and stimulating free radical formation and enhancing neuron degeneration. Moreover, growing studies reveal that all of the aforementioned metals that accumulated in the central nervous system modulate amyloid-β formation and deposition (29). The metals and amyloid-β can form complexes tightly, which also cause neurooxidative damage (11, 20, 24, 30–32).

Simultaneously elevated concentrations of various metals promoting oxidative damage, and hence promoting neurodegeneration, present a complex system of pathophysiology not yet fully understood. Despite this complexity, metal dysregulation may in fact be the Achilles' heel of AD, opening a door for chelation therapy. An iron chelator, regardless of synthetic or natural origin, can have high affinity for iron, but it may also undesirably chelate other metals in various tissues leading to serious side effects. Affinity for multiple metals such as aluminum, copper, and zinc may pose useful rather than detrimental since various metals are implicated as oxidative instigators. Perhaps this may be the reason why desferrioxamine (DFO), a specific iron chelator with high affinities for aluminum, copper, and zinc has demonstrated some therapeutic benefits for patients with AD.

1.1. Iron Chelators in the Treatment of AD

DFO is a hexadentate iron chelator and has been found to significantly slow the progression of AD in one clinical trial (33). In this study, the chelation of aluminum was examined, but it is possible that the therapeutic effect may also have been due to removal of iron since DFO preferably chelates iron (34, 35). DFO also has an appreciable affinity for copper and zinc (34, 36). The affinity constants of DFO for Fe(III), Al(III), Cu(II), and Zn(II) are 30.6, 22.0, 14.1, and 11.1 (log K), respectively (37). In this clinical study, copper and zinc were not monitored. Interestingly, 2 years after the initial publication, a verbal report at the International Conference on Alzheimer's Disease (Padua, Italy, 1992) provides evidence that iron and zinc concentrations are decreased in a postmortem analysis for DFO-treated patients (34, 38).

DFO is one of the only two iron chelators approved by the FDA for iron overload disease. DFO is a hexadentate iron chelator (**Fig. 8.1a**). Its therapy promotes iron excretion and has led to great improvements in the quality and duration of life of patients who suffer from β-thalassemia and other refractory anemias. In addition, DFO also inhibits nigrostriatal degeneration induced by 6-hydroxydopamine (39). Unfortunately, DFO has serious side effects including neurotoxicity and neurological changes (13, 33, 40–45). Furthermore, DFO is poorly absorbed by the

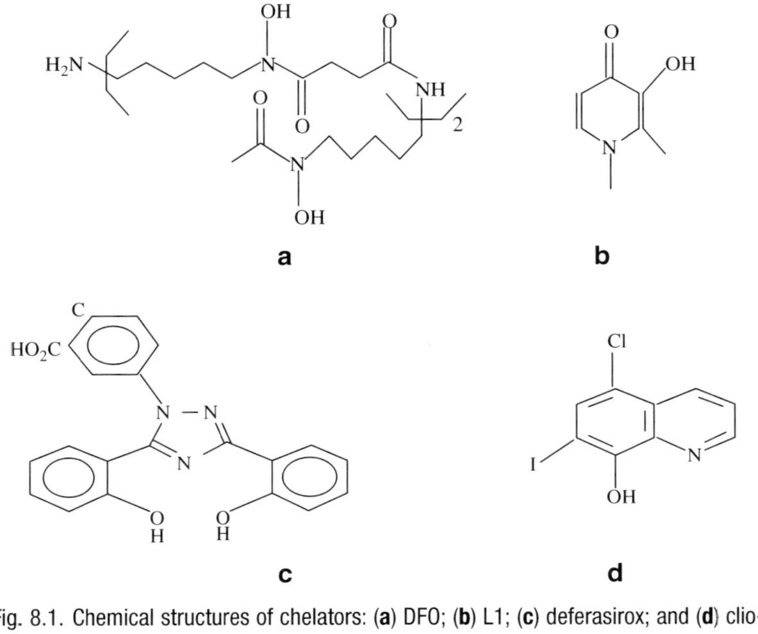

Fig. 8.1. Chemical structures of chelators: (**a**) DFO; (**b**) L1; (**c**) deferasirox; and (**d**) clioquinol.

gastrointestinal tract and rapidly degrades after administration (46). Therefore, it requires long subcutaneous administration to yield significant iron excretion (35, 47). Moreover, some studies show DFO does not easily penetrate the blood–brain barrier (BBB) due to its hydrophilic nature (48), although this point remains open to debate (49). Some penetration may occur due to a compromised BBB via lesion sites (48). But the neurotoxicity and difficulty of administration and delivery present serious hindrances to the use of DFO for AD treatment.

Deferiprone or L1 (1,2-dimethyl-3-hydroxyl-4-pyridinone) is a bidentate iron and aluminum chelator (**Fig. 8.1b**) approved in Europe, but not in the United States (50). Although L1 has high oral activity and BBB penetration ability due to its lipophilicity, its use is limited because of serious side effects (51, 52). In addition, studies have shown that L1 lacks the ability to remove iron from the brain (49) probably due to strong hydrophilicity of the iron–L1 complex. Additionally, there is no carrier-mediated transport system available to remove the complex from the brain. Other L1 derivatives with higher lipophilicity also have the ability to cross the BBB and complex brain iron, but they also possess considerable neurotoxicity (40, 43, 51).

Currently, FDA has approved another iron chelator (Deferasirox) for treatment of transfusional iron overload in thalassemia. Deferasirox is a tridentate iron chelator (**Fig. 8.1c**) with oral bioavailability (53). However, its long-term profiles are not yet available (54, 55). Furthermore, its lipophilic nature like L1 may raise questions concerning potential toxicity in AD treatment.

Thus, the use of the currently available iron chelators to simultaneously remove several excess metals in the brain of AD is limited because of their toxicity and/or poor transference across the BBB. Most bi- or tridentate iron chelators with small molecular weight and high lipophilicity have the ability to penetrate the BBB, but show toxicity (56). On the other hand, hexadentate iron chelators are considered better candidates for chelation therapy than bi- and tridentate ones because of their lower toxicity before and after chelation (56), but they have difficulty penetrating the BBB (49, 56, 57) due to their hydrophilicity and relative high molecular weight. One strategy to increase the BBB penetration is by enhancing the lipophilicity of the iron chelators (58); however, this is believed to increase toxicity (59). In addition, the increase in lipophilicity of iron chelators will decrease the solubility in aqueous solution with probably a decrease in bioavailability (52). Also, it is possible that some lipophilic chelators, which normally should cross the brain endothelial cells, are rapidly pumped back into the bloodstream by extremely effective efflux pumps. These include multiple organic anion transporter and P-glycoprotein (multidrug resistance protein) (60). Many promising attempts have been made to develop iron chelators with abilities to penetrate BBB and prevent oxidative damage (61–67). However, there is a great need to develop safer and more effective iron chelators for the treatment of AD and other neurodegeneration diseases.

The role of metals in the AD development and the usefulness of chelators for AD treatment have also been demonstrated in the studies with Iodochlorhydroxyquin (clioquinol) (34, 68–70). Clioquinol is a copper and zinc-specific chelator (**Fig. 8.1d**) with BBB penetrable and is able to dissolve amyloid-β plaques. With clioquinol therapy, the clinical rate of cognitive decline is slowed in a subset of AD patients compared with that in controls (71). However, clioquinol is reported being associated with subacute myelo-optic neuropathy and withdrawn from the market as an antibiotic. In this regard, the second generation of clioquinol has been developed and is under clinical investigations (72).

1.2. Nanoparticle Systems with Iron Chelators: Increased BBB Permeability and Lower Toxicity

Nanoparticles made of natural or artificial polymers ranging in size from about 10–1,000 nm (60, 73) present a possible tool to transport drugs across the BBB (60) and nanoparticles of a size around or less than 300 nm coated with surfactants such as polysorbate 80 have been demonstrated to possess this ability (74–76). Recent studies have also shown the possibility of nanoparticulated drugs for the treatment of AD (77–80). The advantages of nanoparticles include reduced drug toxicity, improved biodistribution, and therapeutic efficacy (81). The mechanism by which the nanoparticles deliver drugs into brain may be involved in preferential absorption of apolipoprotein E (ApoE) and/or B. The particles also appear to mimic LDL and interact with the LDL receptor, resulting in their uptake by brain endothelial cells (60, 74, 82–84). The transferrin transcytosis systems may be also employed by the particulated drug delivery systems to deliver drugs into the brain (60, 82, 85). If an iron chelator can be covalently bonded to a nanoparticle, the particle may serve as a targeting vehicle to deliver the chelator to the brain and cross the BBB. There are three advantages to this approach. First, the chelators need not be lipophilic to cross the BBB. Second, the lipophilic character of the chelator no longer contributes to potential toxicity. And third, hydrophilic hexadentate iron chelators with large molecular weights may be used, as previously demonstrated with nanoparticle technology (75, 76).

For iron chelation to be effective, the chelators must be capable of leaving the brain with the corresponding complexed metal ions. If the nanoparticles are not or controlled biodegradation and can mimic lipoprotein particles by preferentially absorbing Apo-AI, known to facilitate the removal of particles from the brain (60, 86), the same carrier-mediated transport systems will be able to carry the iron complex nanoparticles out of the brain.

This is in contrast to lipophilic chelators that can enter the brain, but when complexed, they are unable to cross the BBB due to a change in their lipophilicity. For example, the distribution coefficient (DC) of free L1 determined in *n*-octanol/Tris–HCl buffer system is 0.24, but when complexed is down to 0.0009 (87). Although L1 can reportedly penetrate the BBB, it fails to remove iron from the brain (49).

Our studies show that nanoparticles have the potential to transfer chelators in and out brain as well, thus effectively preventing metal-associated oxidative damage (80, 88). This novel approach of chelation will provide not only a useful means of AD treatment, but also insights into the mechanisms of AD pathophysiology. It may also show utility in other iron-mediated neurodegenerative diseases such as Friedreich's ataxia, Parkinson's disease, and Hallervorden-Spatz Syndrome. More studies are warranted to demonstrate the protective efficacy of the chelator–nanoparticle systems, to evaluate their toxicity and to optimize their capability to cross the BBB.

2. Materials

All chemicals and biochemicals are purchased from Aldrich-Sigma (St. Louis, MO), unless specifically mentioned. The materials obtained are used without further purification. Solutions are prepared following standard protocols.

3. Methods

In order to conjugate nanoparticles covalently, iron chelators must have a functional group to react with an active moiety on the particle surface. The functional group introduced into the iron chelators should not possess adverse effects on the chelator-metal binding. Synthetic methods to produce a series of iron chelators with such functional side chains have been developed (89–91). The metal binding properties of these chelators and some biological properties such as the in vitro ability to remove iron from tissue sections of AD brains and from ferritin (an important protein for iron storage) have been examined (89, 90). Methods for conjugation of various iron chelators to nanoparticles have also been developed. After conjugation, the amounts of chelator that conjugate to the particles and the ability of the chelator–particle systems to bind iron are determined. The human plasma protein absorption patterns on iron chelator particle systems and their iron complexes are examined using 2-D PAGE technology to evaluate the ApoE and Apo-AI absorptions (60, 92). These studies indicate that iron chelator–nanoparticle systems have the potential to enter the brain and bring excess metals out of the brain, thus effectively preventing metal-associated oxidative damage. As prototypes, the syntheses of two kinds of iron chelators containing active functional groups have been described (*see* **Note 1**). Key experiments are briefly described as follows.

3.1. Synthesis of 2-Methyl-N-(2′-aminoethyl or 3′-aminopropyl)-3-hydroxyl-4-pyridinone (MAEHP and MAPHP) (see Note 2), An Iron Chelator with Functional Groups for Nanoparticle Conjugation

1. Mix 3-hydroxyl-2-methyl-4-pyranone with benzyl chloride in a molar ration of 1:1.1 in aqueous methanol solution containing NaOH. Reflux for 6 h with the contents being constantly stirred on magnetic stirrer.
2. Remove methanol under vacuum and add water. Extract the product 3-benzyloxy-2-methyl-4-pyranone into methylene chloride.
3. Wash the organic (methylene chloride) layer with 5% (w/v) NaOH followed by water and dried it over anhydrous $MgSO_4$.
4. Evaorate the solvent under vacuum. Add 1,2-diaminoethane or 1,3-diaminopropane in aqueous ethanol solution to the residue containing 3-benzyloxy-2-methyl-4-pyranone reacted and allow the reaction to proceed at the ambient temperature for about 1 week.
5. Evaporate the solvents and residual diamines under vacuum. Dissolve the residue in chloroform. Wash the chloroform solution with water and dry it over anhydrous Na_2SO_4.
6. Remove the solvent under vacuum and dissolve the residue in methanol. Adjust the pH to approximately 1.0 with HCl. The product 1-(2′-aminoethyl)-3-benzyloxy-2-methyl-4-pyridinone or 1-(3′-aminopropyl)-3-benzyloxy-2-methyl-4-pyridinone separates from methanolic solution as dihydrochloride salt. Collect the dihydrochloride salts by filtration and recrystallize them from a solution of methanol and ether to obtain the pure product(s).
7. Mix the products with BBr_3 (1.0 M CH_2Cl_2 solution) in CH_2Cl_2 and stir overnight at room temperature under a nitrogen atmosphere.
8. Add water and stirring for an additional 4 h at room temperature. The aqueous phase containing MAEHP or MAPHP is separated and evaporated under vacuum.
9. The MAEHP and MAPHP are purified further through recrystallization from an ethanol/ether solution.

3.2. Synthesis of 2-Methyl (or Ethyl)-N-(2′-hydroxyethoxy)methyl-3-hydroxyl-4-pyridinone (MHEMHP or EHEMHP) (see Note 3), An Iron Chelator with Functional Groups for Nanoparticle Conjucation

1. Synthesize 3-benzyloxyl-2-alkyl-4-pyridinone as described in **Section 3.1, Step 1.**
2. Replace the ring oxygen of 3-benzyloxyl-2-alkyl-4-pyranone by a nitrogen atom via a substitution reaction with aqueous ammonia for 48 h at room temperature.
3. Silylate the 3-benzyloxyl-2-alkyl-4-pyridinone using hexamethyldisilazane under refluxing and nitrogen gas for 2 h.
4. Remove the solvent under vacuum. Dissolve the residue in 1,2-dichloroethane and then add benzyloxyethoxymethylchloride (*see* **Note 4**) in the presence of a catalytic amount of trimethylsilyl trifluoromethanesulfonate (*see* **Note 5**).
5. Stir the mixture at room temperature for 4 h and then treat with an aqueous solution saturated with sodium bicarbonate.
6. Discard the aqueous phase. Dry the organic phase over anhydrous Na_2SO_4 and then evaporate the solvent under vacuum.
7. Remove the two protection groups simultaneously by hydrogenation with H_2/Pt on active carbon in acidic aqueous ethanol at room temperature for 24 h (*see* **Note 6**).
8. Finally, recrystallize the chelators from a 1:1 solution of CH_3Cl/MeOH (*see* **Note 7**).

3.3. Titration of Chelators with Iron Ions in Buffer Solution

1. To 2.3 mL of 25 mM Tris–HCl buffer, pH 7.5, containing chelators (0.474 mM), add freshly prepared $Fe(NO_3)_3$ solution (15.1 mM) in Tris buffer gradually in small aliquots of 5 μL each.
2. Monitor the change in absorbance due to the formation of chelator–iron complexes photometrically at 450 nm or higher. The chelators and iron form purple complexes with typical absorption in the visible range over 450 nm, whereas free chelators absorb maximally 280 nm (*see* **Note 8**).

3.4. Iron Removal by Chelators from Ferritin

1. To study the mobilization of iron from ferritin, incubate horse spleen ferritin (9.2 μL of 100 mg/mL stock solution) with chelators (0.474 mM) in 2.3 mL of Tris buffer solution (25 mM, pH 7.5) at 37°C for 72 h.
2. Monitor the changes in absorbance due to the formation of iron–chelator complex spectrophotometrically at different time intervals. The kinetics of the iron release was investigated for periods up to 72 h (*see* **Note 9**).

3.5. Iron Removal by Chelators from AD Brain Sections In Vitro

The ability of chelators to mobilize iron from brain sections can be examined via histochemical method (80, 93).

1. Fix the tissue specimens, collected from the hippocampal region of AD patients (*see* **Note 10**), in methacarn (*see* **Note 11**) overnight at 4°C.
2. Place the brain tissues in 50% ethanol, dehydrate in ascending concentrations of ethanol and finally embed them in paraffin.
3. Thin sections (thickness: 6 μm) of the tissue and mounted on silane-coated slides.
4. Deparaffinize the tissue sections with two changes of xylene (10 min each) and then re-hydrate through graded ethanol/TBS mixture.
5. Apply 40 μL of PBS containing various concentrations chelator(s) to each section and incubate overnight at 37°C.
6. At the end of incubation period, rinse the tissue sections thoroughly with TBS.
7. Incubate at 37°C for 2 h in 7% (w/v) potassium ferrocyanide in 3% (w/v) HCl in water.
8. Rinse the sections Tris–HCl buffer and subsequently incubate in 0.75 mg/mL 3,3′-diaminobenzidine and 0.015% H_2O_2 for 5–10 min.
9. Finally, dehydrate tissue sections through graded ethanol, put coverslip, and examine using differential interference microscopy (*see* **Note 12**).

3.6. Conjugation of Iron Chelators with Nano or Microsphere Particles

As a prototypic procedure (*see* **Note 13**), a simple method of particle–chelator conjugation by forming an amido bond is presented. Monodispersed polystyrene particles with carboxyl groups on the surface (Bangs Laboratories, Fishers, IN) were used to conjugate MAEHP or MAPHP as prototypic chelators, each of which contained a free primary amino group available for the conjugation (*see* **Note 14**).

1. Prepare a suspension of carboxylated particles by pipetting and vortexing and immediately transfer into a microcentrifuge tube.
2. Remove the supernatant by centrifugation.
3. Re-suspend the particles in 0.01 N NaOH solution, mix well, and repeat the process.
4. Wash the particles twice with 0.1 M of MES (2-[*N*-morpholino]ethane sulfonic acid) buffer (pH 5.0) and once with cold Milli-Q water.
5. Carboxyl groups on the particles are to be activated by adding cold Milli-Q water containing *N*-cyclohexyl-*N′*-(2-morpholinoethyl)carbodiimide methyl-*p*-toluensulfonate (CMC, 0.01 M) and incubating for 10 min at 4°C with slow tilt rotation.

6. After removing the supernatant, add CMC solution again along with MES buffer (0.3 M, pH 5.0).
7. Vortex the mixture and incubate as described above for 30 min.
8. Wash the activated carboxyl groups containing particles twice with cold 0.1 M MES as quickly as possible and resuspend in MES buffer (0.1 M, pH 5.0) containing excess MAEHP or MAPHP (0.01 M) (*see* **Note 15**).
9. Vortex the mixture followed by incubation for 30 min at room temperature with tilt rotation.
10. Wash twice the chelator–particle systems with 0.1 M MES buffer and PBS, store in PBS at 4°C.
11. Determine the yields of chelator conjugation by measurements of the free chelator concentrations in the solutions before and after conjugation using UV-visible spectrometer (or HPLC) at the wavelength of maximum absorption (94).
12. Concentrations and size distributions of the chelator–particle systems could be determined using a Beckman Coulter Multisizer II in a counting cuvette containing Isoton II diluent or using a Coulter N4 Plus Sub-micron Particle Sizer.

3.7. Reaction of Chelator–Particle Systems with Ferric Iron

1. Add an aliquot of freshly prepared ferric iron solution ($Fe(NO_3)_3$, 0.002 M in MES buffer 0.01 M, pH 5.0) to MES (0.01 M, pH 5.0) solution containing suspended MAPHP–particle systems as prototype, or plain particles as a control.
2. Allow the mixture to rotate at room temperature for 4 h. The iron–chelator–particle systems and supernatant are separated by centrifugation.
3. Wash the systems thoroughly with MES buffer 5 times to remove non-complexed iron ions.
4. After combination of the supernatants, add excess MAPHP in MES buffer (0.01 M) to complex the iron ions that does not react with the chelator–particle systems. The visible absorbance of the iron–MAPHP complex is measured using UV-visible spectrophotometry at a maximum wavelength of 455 nm (ε 3.02×10^3) after the chelating reaction reached equilibrium.
5. Obtain a standard curve for iron concentration by measuring several solutions of iron–MAPHP complex with known iron concentrations to estimate the amount of non–complexed iron with the chelator–particle systems (*see* **Note 16**).

3.8. Protein Absorptions of Chelator–Particle Systems and Chelator–Particle Systems with Iron

The absorbed proteins on chelator–particle systems and chelator–particle systems with iron, which are obtained by reaction of ferric iron with chelator–particle systems, were evaluated using 2-D PAGE analyses.

1. Incubate separately the chelator–particle systems that are overcoated with polysorbate 80 at room temperature and the chelator–particle systems with iron (100 μL of each system, 2.5% w/v in PBS buffer) in 1 mL of citrated human plasma for 5 min at 37°C (92).
2. After separating by centrifugation, wash the systems four times with Milli-Q water.
3. Elute the adsorbed proteins from the particle surface with a protein solubilizing solution (5% SDS), 5% dithioerythritol, 10% glycerol, and 60 mM Tris, pH 6.8) (92).
4. In the first dimension of the 2D-PAGE analysis, isoelectric focusing (IEF), the proteins are separated according to their isoelectric points (pI). Carry out the IEF in glass tubes of inner diameter 2.0 mm using 2.0% pH 3.5–10 ampholines for 9,600 V-h.
5. In the second dimension of SDS-PAGE, the separation is based on molecular weight (MW). Equilibrate each tube for 10 min in 62.5 mM Tris, pH 6.8, buffer containing 2.3% SDS, 50 mM dithioerythritol, and 10% glycerol.
6. Seal to the top of a stacking gel that is on the top of a 10% acrylamide slab gel (145 × 145 × 0.75 mm).
7. Perform SDS slab gel electrophoresis for about 4 h at 12.5 mA/gel.
8. After SDS-PAGE, dry the gels between sheets of cellophane and silver-stained (92) (*see* **Note 17**).

4. Notes

1. The synthetic procedures are straightforward and product yields are high. The chelators have been characterized using standard methods such as ^{1}H-NMR, MS, UV-vis, and elemental analysis.
2. The chelators are prepared using a modified procedure as described in **Scheme 8.1** (91, 95).
3. These chelators are synthesized using established methods (**Scheme 8.2**) (90, 96).
4. (2-Acetoxyethoxy)methyl bromide can be used to replace benzyloxyethoxymethylchloride (97).

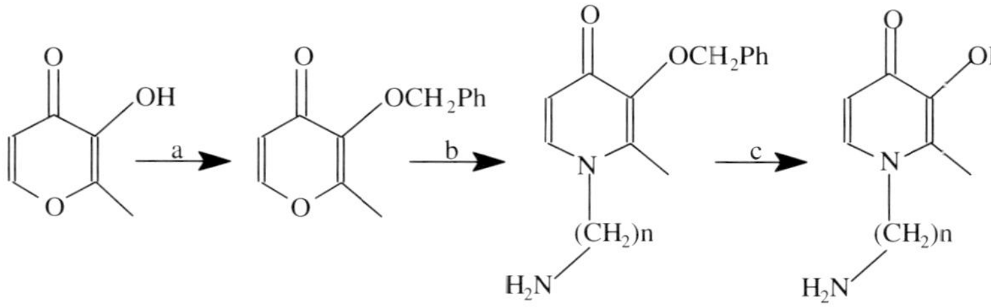

Scheme 8.1. Synthesis of 2-methyl-N-(2′-aminoethyl ($n = 2$) or 3′-aminopropyl ($n = 3$))-3-hydroxy-4-pyridinone: (**a**). benzylchloride/NaOH; (**b**) $NH_2(CH_2)_nNH_2$, $n = 2, 3$; and (**c**). BBr_3 in CH_2Cl_2 at 4°C or hydrogenation with H_2/Pt on active carbon.

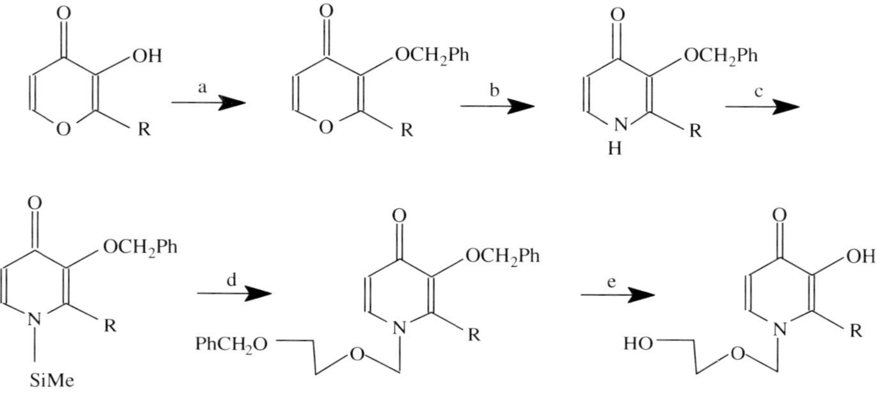

Scheme 8.2. Synthesis of 2-methyl (or ethyl)-*N*-(2′-hydroxyethoxy)methyl-3-hydroxy-4-pyridinone: (**a**). $PhCH_2Cl$/NaOH/refluxing/6 h; (**b**). NH_4OH/rt./48 h; (**c**). hexamethyldisilazane/chlorotrimethylsilane; (**d**). benzyloxyethoxymethylchloride, trimethylsilyl trifluoromethanesulfonate in 1,2-dichloroethane; and (**e**) H_2, Pd/C, AcOH in 95% EtOH. R = Me or Et.

5. $SnCl_4$ could also be used as catalyst in the alkylation reaction but might result in separation difficulties and low yields (98).
6. The de-protection procedure can be achieved by using BBr_3 in CH_2Cl_2 at 4°C (99–101).
7. To evaluate whether the linked (2′-hydroxyethoxy)methyl moiety affected the geometry of the iron binding site in the chelators, molecular and crystal structures of EHEMHP were determined by X-ray crystallographic analysis. A piece of colorless crystal (0.33 × 0.33 × 0.11 mm) formed in methanol-ethyl acetate solution was used for X-ray measurement with an Enraf-Nonius CAD-4 diffractometer equipped with a graphite monochromator of Mo Kα (0.71073 Å) (90). The results indicate that there is no significant change in the geometry of iron binding site. An ORTEP stereo-view of the EHEMHP molecular structure was depicted in **Fig. 8.2**.

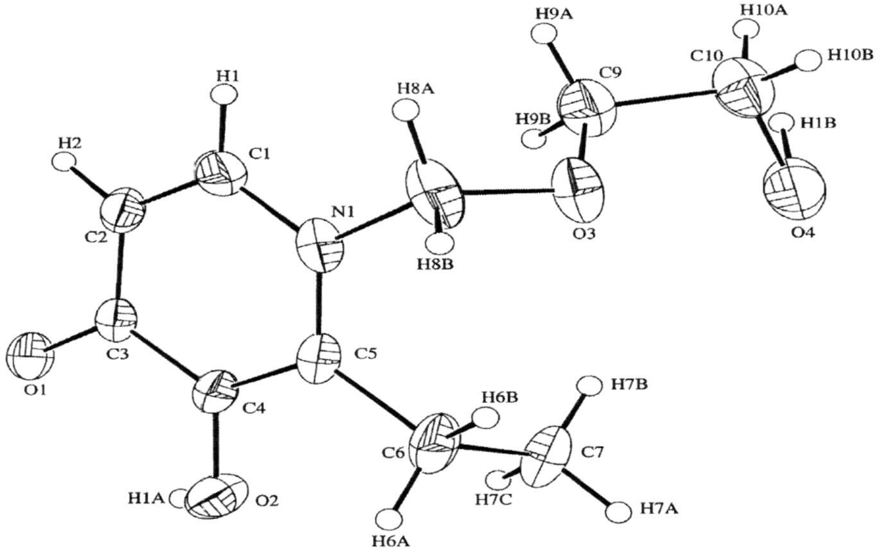

Fig. 8.2. ORTEP stereoview of chelator EHEMHP.

8. Typical titration curves using MHEMHP and EHEMHP as prototype are presented in **Fig. 8.3**. The endpoints of the titration indicate the formation of chelator/iron (3:1) complexes (89, 90).

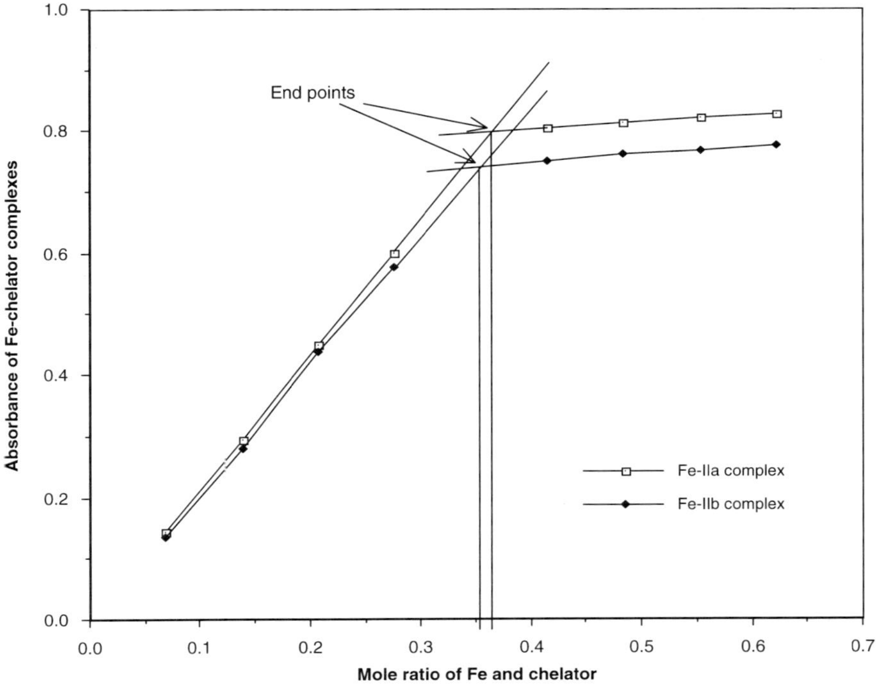

Fig. 8.3. Titration of MHEMHP (IIa) and EHEMHP (IIb) with iron.

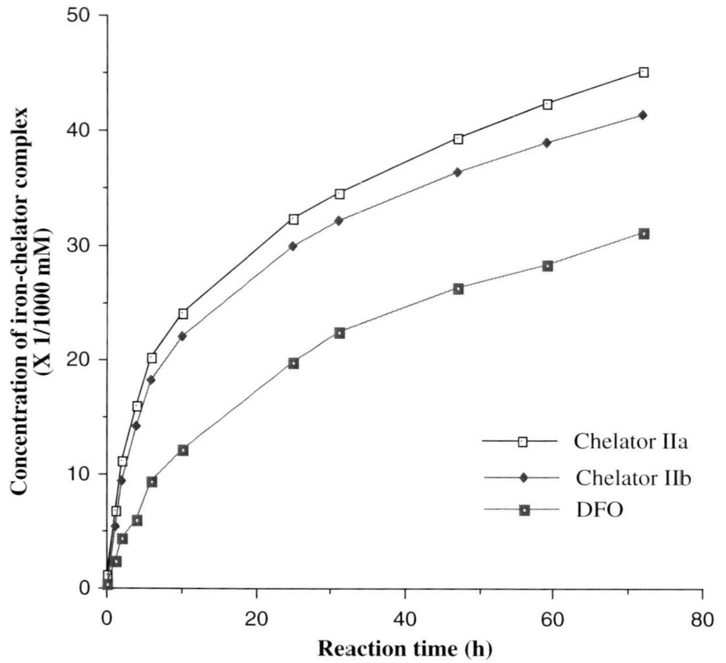

Fig. 8.4. Removal of iron from ferritin by the chelators of MHEMHP (IIa), EHEMHP (IIb), and DFO.

9. The concentrations of iron–chelator complexes are estimated from ε_{max} values at the wavelength of λ_{max} of the complexes (89, 90). **Fig. 8.4** shows the iron removal from ferritin by MHEMHP and EHEMHP as a prototype compared with DFO. It also shows that the chelators are more effective to remove iron from ferritin than DFO.
10. Brain tissue from transgenic mouse models can also be used in this kind of studies (93, 102).
11. The use of methacarn instead of formalin for fixation can provide more accurate results (93, 102).
12. The results show that chelators are capable of depleting iron from the AD brain tissue sections (**Fig. 8.5**), which depends on the chelator chemical structures and concentrations used (80, 103). This method also provides a useful tool to screen potential chelators for mobilization of iron from the AD brain.
13. A variety of covalent bonds including amido, amino, ether, and thioether can be easily formed for linking chelators and particles, which are dependent on the existing functional groups located on chelator side chains and on the surface of particles (94, 104, 105).

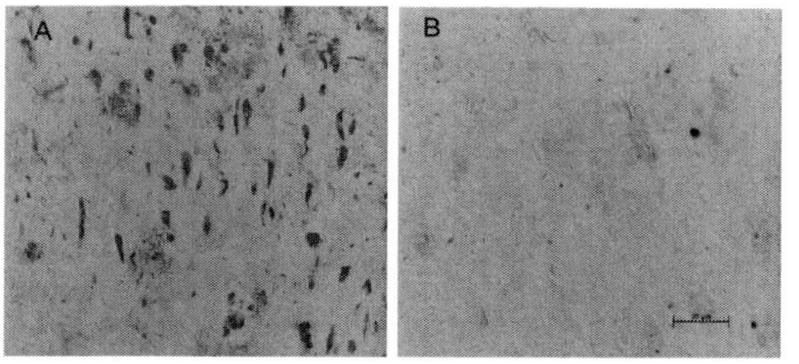

Fig. 8.5. Lesion-associated chelatable iron in AD brain sections was depleted with iron chelator (MAEHP as a prototypal chelator), which was detected histochemically with a modified Perl Stain. Saline- (**a**) and MAEHP-treated (**b**) sections.

14. The preparation of the chelator–particle conjugates is presented in **Scheme 8.3** (103).

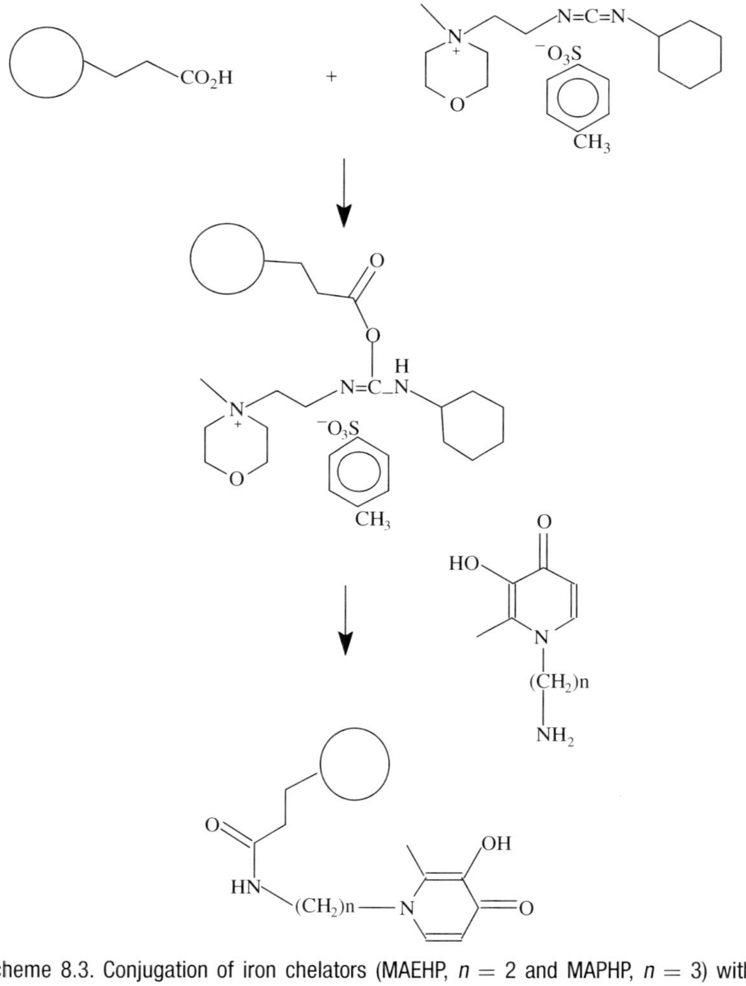

Scheme 8.3. Conjugation of iron chelators (MAEHP, $n = 2$ and MAPHP, $n = 3$) with particles.

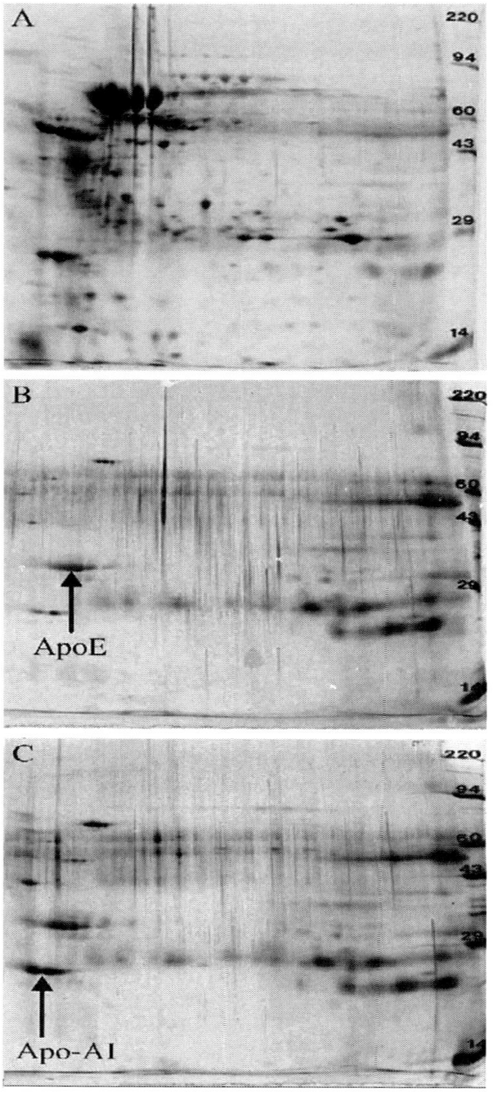

Fig. 8.6. Images of plasma protein patterns examined by 2D PAGE. (**a**) Plasma; (**b**) CNPS (MAPHP conjugated) coated with polysorbate 80; and (**c**) ICNPS.

15. The particles are rapidly washed with cool Mill-Q water since the active intermediate ester is unstable and undergoes hydrolysis. Alternatively, a water-soluble *N*-hydroxyl compound like sulfo-*N*-hydroxysuccinimide (NHS) could be added to increase the coupling yield. This is because NHS is known to form a more stable intermediate ester by replacing the oacylisourea intermediate formed by carbodiimide. The NHS-formed ester is less susceptible to hydrolysis but still highly reactive toward amino groups (106, 107).

16. Interestingly, this bi-dentate iron chelator converts to hexandentate chelators after conjugation to particles because the particles provided backbone linkages. This phenomenon greatly improved the metal binding stability and lowered toxicity associated with metal–chelator complexes. DFO still retains its hexadentate iron binding property after conjugation to particles (103).
17. These studies show that the protein absorption pattern on the iron chelator particle systems is totally different from that of the human plasma proteins (**Fig. 8.6a**). Through changing the system-surface properties, such as chelators and surfactants, the chelator–particle systems can preferentially absorb ApoE (**Fig. 8.6b**). With the same kind of changes, it is also found that the chelator–particle systems after binding metals can preferentially absorb Apo A-I (**Fig. 8.6c**). Such preferential absorptions allow the systems to mimic the ApoE or Apo A-I nanoparticles and to cross the BBB through LDL transport mechanisms (76, 86). Uniform coating of the systems with ApoE, B, or A-I can also be achieved by overcoating of these apolipoproteins, which may enable the systems to cross the BBB with high efficiency (74). Studies indicate the potential to obtain chelator–nanoparticle systems with optimal surface properties via changing chelators, linkages, coating materials, and nanoparticles with different surfaces. The particles can be made of biocompatible synthetic or natural macromolecules (60, 73, 108) with functional groups on their surface for covalent bonding with chelators (94, 104).

Acknowledgments

The work in the authors' laboratories is supported by the National Institutes of Health, the Alzheimer's Association and Philip Morris USA Inc. and Philip Morris International.

References

1. Smith, M.A. (1998) Alzheimer disease. *Int. Rev. Neurobiol.* **42**, 1–54.
2. Gutteridge, J.M. (1994) Hydroxyl radicals, iron, oxidative stress, and neurodegeneration. *Ann. NY Acad. Sci.* **738**, 201–213.
3. Evans, P.H. (1993) Free radicals in brain metabolism and pathology. *Br. Med. Bull.* **49**, 577–587.
4. Perry, G., Castellani, R.J., Hirai, K., and Smith, M.A. (1998) Reactive oxygen species mediate cellular damage in Alzheimer disease. *J. Alzheimer's Dis.* **1**, 45–55.
5. Smith, M.A., Sayre, L.M., Monnier, V.M., and Perry, G. (1995) Radical AGEing in Alzheimer's disease. *Trends Neurosci.* **18**, 172–176.
6. Prasad, K.N., Hovland, A.R., Cole, W.C., Prasad, K.C., Nahreini, P., Edwards-Prasad, J., and Andreatta, C.P. (2000) Multiple antioxidants in the prevention and treat-

ment of Alzheimer disease: Analysis of biologic rationale. *Clin. Neuropharmacol.* **23**, 2–13.
7. Pitchumoni, S.S. and Doraiswamy, P.M. (1998) Current status of antioxidant therapy for Alzheimer's disease. *J. Am. Geriatr. Soc.* **46**, 1566–1572.
8. Christen, Y. (2000) Oxidative stress and Alzheimer disease. *Am. J. Clin. Nutr.* **71**, 621S–629S.
9. Kennard, M.L., Feldman, H., Yamada, T., and Jefferies, W.A. (1996) Serum levels of the iron binding protein p97 are elevated in Alzheimer's disease. *Nat. Med.* **2**, 1230–1235.
10. Jefferies, W.A., Food, M.R., Gabathuler, R., Rothenberger, S., Yamada, T., Yasuhara, O., and McGeer, P.L. (1996) Reactive microglia specifically associated with amyloid plaques in Alzheimer's disease brain tissue express melanotransferrin. *Brain Res.* **712**, 122–126.
11. Harman, D. (1995) Free radical theory of aging: Alzheimer's disease pathogensis. *Age*, **18**, 97–119.
12. Casadesus, G., Smith, M.A., Zhu, X., Aliev, G., Cash, A.D., Honda, K., Petersen, R.B., and Perry, G. (2004) Alzheimer disease: Evidence for a central pathogenic role of iron-mediated reactive oxygen species. *J. Alzheimer's Dis.* **6**, 165–169.
13. Olanow, C.W. (1992) An introduction to the free radical hypothesis in Parkinson's disease. *Ann. Neurol.* **32**(Suppl), S2–S9.
14. Halliwell, B. and Gutteridge, J.M.C. (Ed.) (1999) *Free Radicals in Biology and Medicine*, Oxford University, New York.
15. Markesbery, W.R. and Ehmann, W.D. (1999) Oxidative stress in Alzheimer disease. In: *Alzheimer Disease* (Terry, R.D., Katzman, R., Bick, K.L., and Sisodia, S.S., Eds.), Lippincott Williams & Wilkins, Philadelphia, pp. 401–414.
16. Ohtawa, M., Seko, M., and Takayama, F. (1983) Effect of aluminum ingestion on lipid peroxidation in rats. *Chem. Pharm. Bull.* (Tokyo), **31**, 1415–1418.
17. Evans, P.H., Klinowski, J., Yano, E., and Urano, N. (1989) Alzheimer's disease: A pathogenic role for aluminosilicate-induced phagocytic free radicals. *Free Radic. Res. Commun.* **6**, 317–321.
18. Garrel, C., Lafond, J.L., Guiraud, P., Faure, P., and Favier, A. (1994) Induction of production of nitric oxide in microglial cells by insoluble form of aluminium. *Ann. NY Acad. Sci.* **738**, 455–461.
19. Kong, S., Liochev, S., and Fridovich, I. (1992) Aluminum(III) facilitates the oxidation of NADH by the superoxide anion. *Free Radic. Biol. Med.* **13**, 79–81.
20. Bondy, S.C., Guo-Ross, S.X., and Truong, A.T. (1998) Promotion of transition metal-induced reactive oxygen species formation by beta-amyloid. *Brain Res.* **799**, 91–96.
21. Lovell, M.A., Robertson, J.D., Teesdale, W.J., Campbell, J.L., and Markesbery, W.R. (1998) Copper, iron and zinc in Alzheimer's disease senile plaques. *J. Neurol. Sci.* **158**, 47–52.
22. Markesbery, W.R. and Carney, J.M. (1999) Oxidative alterations in Alzheimer's disease. *Brain Pathol.* **9**, 133–146.
23. Multhaup, G., Schlicksupp, A., Hesse, L., Beher, D., Ruppert, T., Masters, C.L., and Beyreuther, K. (1996) The amyloid precursor protein of Alzheimer's disease in the reduction of copper(II) to copper(I). *Science*, **271**, 1406–1409.
24. Sayre, L.M., Perry, G., and Smith, M.A. (1999) Redox metals and neurodegenerative disease. *Curr. Opin. Chem. Biol.* **3**, 220–225.
25. Linder, M.C. and Hazegh-Azam, M. (1996) Copper biochemistry and molecular biology. *Am. J. Clin. Nutr.* **63**, 797S–811S.
26. Bush, A.I., Pettingell, W.H., Multhaup, G., d Paradis, M., Vonsattel, J.P., Gusella, J.F., Beyreuther, K., Masters, C.L., and Tanzi, R.E. (1994) Rapid induction of Alzheimer A beta amyloid formation by zinc. *Science*, **265**, 1464–1467.
27. Hensley, K., Carney, J.M., Mattson, M.P., Aksenova, M., Harris, M., Wu, J.F., Floyd, R.A., and Butterfield, D.A. (1994) A model for beta-amyloid aggregation and neurotoxicity based on free radical generation by the peptide: Relevance to Alzheimer disease. *Proc. Natl. Acad. Sci. USA*, **91**, 3270–3274.
28. Butterfield, D.A. (1997) beta-Amyloid-associated free radical oxidative stress and neurotoxicity: Implications for Alzheimer's disease. *Chem. Res. Toxicol.* **10**, 495–506.
29. Pratico, D., Clark, C.M., Liun, F., Rokach, J., Lee, V.Y., and Trojanowski, J.Q. (2002) Increase of brain oxidative stress in mild cognitive impairment: A possible predictor of Alzheimer disease. *Arch. Neurol.* **59**, 972–976.
30. Atwood, C.S., Scarpa, R.C., Huang, X., Moir, R.D., Jones, W.D., Fairlie, D.P., Tanzi, R.E., and Bush, A.I. (2000) Characterization of copper interactions with alzheimer amyloid beta peptides: Identification of an attomolar-affinity copper binding site on amyloid beta1-42. *J. Neurochem.* **75**, 1219–1233.

31. Pratico, D., Uryu, K., Sung, S., Tang, S., Trojanowski, J.Q., and Lee, V.M. (2002) Aluminum modulates brain amyloidosis through oxidative stress in APP transgenic mice. *FASEB J.* **16**, 1138–1140.
32. House, E., Collingwood, J., Khan, A., Korchazkina, O., Berthon, G., and Exley, C. (2004) Aluminium, iron, zinc and copper influence the in vitro formation of amyloid fibrils of Abeta42 in a manner which may have consequences for metal chelation therapy in Alzheimer's disease. *J. Alzheimer's Dis.* **6**, 291–301.
33. McLachlan, D.R., Kruck, T.P., Lukiw, W.J., and Krishnan, S.S. (1991) Would decreased aluminum ingestion reduce the incidence of Alzheimer's disease? *CMAJ*, **145**, 793–804.
34. Cuajungco, M.P., Faget, K.Y., Huang, X., Tanzi, R.E., and Bush, A.I. (2000) Metal chelation as a potential therapy for Alzheimer's disease. *Ann. NY Acad. Sci.* **920**, 292–304.
35. Richardson, D.R. and Ponka, P. (1998) Development of iron chelators to treat iron overload disease and their use as experimental tools to probe intracellular iron metabolism. *Am. J. Hematol.* **58**, 299–305.
36. Keberle, H. (1964) The biochemistry of desferrioxamine and its relation to iron metabolism. *Ann. NY Acad. Sci.* **119**, 758–768.
37. Hider, R.C. and Hall, A.D. (1991) Clinically useful chelators of tripositive elements. *Prog. Med. Chem.* **28**, 41–173.
38. Finefrock, A.E., Bush, A.I., and Doraiswamy, P.M. (2003) Current status of metals as therapeutic targets in Alzheimer's disease. *J. Am. Geriatr. Soc.* **51**, 1143–1148.
39. Ben-Shachar, D., Riederer, P., and Youdim, M.B. (1991) Iron-melanin interaction and lipid peroxidation: Implications for Parkinson's disease. *J. Neurochem.* **57**, 1609–1614.
40. Floor, E. (2000) Iron as a vulnerability factor in nigrostriatal degeneration in aging and Parkinson's disease. *Cell Mol. Biol.* (Noisy-le-grand), **46**, 709–720.
41. Blake, D.R., Winyard, P., Lunec, J., Williams, A., Good, P.A., Crewes, S.J., Gutteridge, J.M., Rowley, D., Halliwell, B., Cornish, A. et al. (1985) Cerebral and ocular toxicity induced by desferrioxamine. *Q. J. Med.* **56**, 345–355.
42. Kruck, T.P., Fisher, E.A., and McLachlan, D.R. (1993) A predictor for side effects in patients with Alzheimer's disease treated with deferoxamine mesylate. *Clin. Pharmacol. Ther.* **53**, 30–37.
43. Struck, M., Waldmeier, P., and Berdoukas, V. (1993) The treatment of iron overload-psychiatric implication. In: *Iron in Central Nervous System Disorders* (Riederer, P. and Youdim, M.B.H., Eds.) Springer Verlag, Wien, pp. 189–196.
44. Klaasen, C.D. (1996) Heavy metals and heavy-metalantagonists. In: *Goodman and Gilman's The Pharmacological Basis of Therapeutics* (Hardman, J.G., Limbird, L.E., Molinoff, P.B., Ruddon, R.W., and Gilman, A.G., Eds.), McGraw Hill, New York, pp. 1649–1671.
45. Jenner, P. and Olanow, C.W. (1998) Understanding cell death in Parkinson's disease. *Ann. Neurol.* **44**, S72–S84.
46. May, P.M. and Bulman, R.A. (1983) The present status of chelating agents in medicine. *Prog. Med. Chem.* **20**, 225–336.
47. Olivieri, N.F. and Brittenham, G.M. (1997) Iron-chelating therapy and the treatment of thalassemia. *Blood*, **89**, 739–761.
48. Lynch, S.G., Fonseca, T., and Levine, S.M. (2000) A multiple course trial of desferrioxamine in chronic progressive multiple sclerosis. *Cell Mol. Biol.* (Noisy-le-grand), **46**, 865–869.
49. Crowe, A. and Morgan, E.H. (1994) Effects of chelators on iron uptake and release by the brain in the rat. *Neurochem. Res.* **19**, 71–76.
50. Kontoghiorghes, G.J. (1995) New concepts of iron and aluminium chelation therapy with oral L1 (deferiprone) and other chelators. A review. *Analyst*, **120**, 845–851.
51. Ward, R.J., Dexter, D., Florence, A., Aouad, F., Hider, R., Jenner, P., and Crichton, R.R. (1995) Brain iron in the ferrocene-loaded rat: Its chelation and influence on dopamine metabolism. *Biochem. Pharmacol.* **49**, 1821–1826.
52. Richardson, D.R. (1999) The therapeutic potential of iron chelators. *Expert Opin. Investig. Drugs*, **8**, 2141–2158.
53. Medical News Today (2005) FDA grants priority review for Exjade(R) for the treatment of chronic iron overload due to blood transfusions, http://www.medicalnewstoday.com/medicalnews.php?newsid=26610.
54. Neufeld, E.J. (2006) Oral chelators deferasirox and deferiprone for transfusional iron overload in thalassemia major: New data, new questions. *Blood*, **107**, 3436–3441.
55. Piga, A., Galanello, R., Forni, G.L., Cappellini, M.D., Origa, R., Zappu, A., Donato, G., Bordone, E., Lavagetto, A., Zanaboni, L., Sechaud, R., Hewson, N., Ford, J.M.,

Opitz, H., and Alberti, D. (2006) Randomized phase II trial of deferasirox (Exjade, ICL670), a once-daily, orally-administered iron chelator, in comparison to deferoxamine in thalassemia patients with transfusional iron overload. *Haematologica*, **91**, 873–880.

56. Hider, R.C., Porter, J.B., and Singh, S. (1994) The design of therapeutically useful iron chelators. In: *The Development of Iron Chelators for Clinical Use* (Bergeron, R.J. and Brittenham, G.M., Eds.), CRC, Boca Raton, pp. 353–371.
57. Gassen, M. and Youdim, M.B. (1997) The potential role of iron chelators in the treatment of Parkinson's disease and related neurological disorders. *Pharmacol. Toxicol.* **80**, 159–166.
58. Lee, J.Y., Friedman, J.E., Angel, I., Kozak, A., and Koh, J.Y. (2004) The lipophilic metal chelator DP-109 reduces amyloid pathology in brains of human beta-amyloid precursor protein transgenic mice. *Neurobiol. Aging*, **25**, 1315–1321.
59. Porter, J.B., Morgan, J., Hoyes, K.P., Burke, L.C., Huehns, E.R., and Hider, R.C. (1990) Relative oral efficacy and acute toxicity of hydroxypyridin-4-one iron chelators in mice. *Blood*, **76**, 2389–2396.
60. Kreuter, J. (2001) Nanoparticulate systems for brain delivery of drugs. *Adv. Drug Deliv. Rev.* **47**, 65–81.
61. Raymond, K.N. and Xu, J. (1994) Siderophore-based hydroxypyridonate sequestering agents. In: *The Development of Iron Chelators for Clinical Use* (Bergeron, R.J. and Brittenham, G.M., Eds.), CRC Press, Boca Raton, pp. 354–371.
62. Hider, R.C., Choudhury, R., Rai, B.L., Dehkordi, L.S., and Singh, S. (1996) Design of orally active iron chelators. *Acta Haematol.* **95**, 6–12.
63. Martell, A.E., Motekaitis, R.J., Sun, Y., Ma, R., Welch, M.J., and Pajeau, T. (1999) New chelating-agents suitable for the treatment of iron overload. *Inorg. Chim. Acta*, **291**, 238–246.
64. Caravan, P. and Orvig, C. (1997) Tripodal aminophenolate ligand complexes of aluminum(III), gallium(III), and indium(III) in water. *Inorg. Chem.* **36**, 237–248.
65. Faller, B., Spanka, C., Sergejew, T., and Tschinke, V. (2000) Improving the oral bioavailability of the iron chelator HBED by breaking the symmetry of the intramolecular H-bond network. *J. Med. Chem.* **43**, 1467–1475.
66. Bergeron, R.J. and McManis, J.S. (1994) Synthesis and biological activity of hydroxamate-based iron chelators. In: *The Development of Iron Chelators for Clinical Use* (Bergeron, R.J. and Brittenham, G.M., Eds.), CRC, Boca Raton, pp. 237–273.
67. Richardson, D.R. and Ponka, P. (1998) Pyridoxal isonicotinoyl hydrazone and its analogs: Potential orally effective iron-chelating agents for the treatment of iron overload disease. *J. Lab. Clin. Med.* **131**, 306–315.
68. Cherny, R.A., Atwood, C.S., Xilinas, M.E., Gray, D.N., Jones, W.D., McLean, C.A., Barnham, K.J., Volitakis, I., Fraser, F.W., Kim, Y., Huang, X., Goldstein, L.E., Moir, R.D., Lim, J.T., Beyreuther, K., Zheng, H., Tanzi, R.E., Masters, C.L., and Bush, A.I. (2001) Treatment with a copper-zinc chelator markedly and rapidly inhibits beta-amyloid accumulation in Alzheimer's disease transgenic mice. *Neuron*, **30**, 665–676.
69. Cherny, R.A., Legg, J.T., McLean, C.A., Fairlie, D.P., Huang, X., Atwood, C.S., Beyreuther, K., Tanzi, R.E., Masters, C.L., and Bush, A.I. (1999) Aqueous dissolution of Alzheimer's disease Abeta amyloid deposits by biometal depletion. *J. Biol. Chem.* **274**, 23223–23228.
70. Loske, C., Gerdemann, A., Schepl, W., Wycislo, M., Schinzel, R., Palm, D., Riederer, P., and Munch, G. (2000) Transition metal-mediated glycoxidation accelerates cross-linking of beta-amyloid peptide. *Eur. J. Biochem.* **267**, 4171–4178.
71. Ritchie, C.W., Bush, A.I., Mackinnon, A., Macfarlane, S., Mastwyk, M., MacGregor, L., Kiers, L., Cherny, R., Li, Q.X., Tammer, A., Carrington, D., Mavros, C., Volitakis, I., Xilinas, M., Ames, D., Davis, S., Beyreuther, K., Tanzi, R.E., and Masters, C.L. (2003) Metal-protein attenuation with iodochlorhydroxyquin (clioquinol) targeting Abeta amyloid deposition and toxicity in Alzheimer disease: A pilot phase 2 clinical trial. *Arch. Neurol.* **60**, 1685–1691.
72. Doraiswamy, P.M. and Xiong, G.L. (2006) Pharmacological strategies for the prevention of Alzheimer's disease. *Expert Opin. Pharmacother.* 7, 1–10.
73. Brem, H., Walter, K.A., Tamargo, R.J., Olivi, A., and Langer, R. (1994) Drug delivery to the brain. In: *Polymeric Site-Specific Pharmacotherapy* (Domb, A.J., Ed.), John Wiley & Sons, New York, pp. 117–139.
74. Kreuter, J., Shamenkov, D., Petrov, V., Ramge, P., Cychutek, K., Koch-Brandt, C., and Alyautdin, R. (2002) Apolipoprotein-mediated transport of nanoparticle-bound drugs across the blood-brain barrier. *J. Drug Target.* **10**, 317–325.

75. Schroeder, U., Sommerfeld, P., Ulrich, S., and Sabel, B.A. (1998) Nanoparticle technology for delivery of drugs across the blood-brain barrier. *J. Pharm. Sci.* **87**, 1305–1307.
76. Alyautdin, R.N., Tezikov, E.B., Ramge, P., Kharkevich, D.A., Begley, D.J., and Kreuter, J. (1998) Significant entry of tubocurarine into the brain of rats by adsorption to polysorbate 80-coated polybutylcyanoacrylate nanoparticles: An in situ brain perfusion study. *J. Microencapsul.* **15**, 67–74.
77. Siegemund, T., Paulke, B.R., Schmiedel, H., Bordag, N., Hoffmann, A., Harkany, T., Tanila, H., Kacza, J., and Hartig, W. (2006) Thioflavins released from nanoparticles target fibrillar amyloid beta in the hippocampus of APP/PS1 transgenic mice. *Int. J. Dev. Neurosci.* **24**, 195–201.
78. Cui, Z., Lockman, P.R., Atwood, C.S., Hsu, C.H., Gupte, A., Allen, D.D., and Mumper, R.J. (2005) Novel D-penicillamine carrying nanoparticles for metal chelation therapy in Alzheimer's and other CNS diseases. *Eur. J. Pharm. Biopharm.* **59**, 263–272.
79. Shea, T.B., Ortiz, D., Nicolosi, R.J., Kumar, R., and Watterson, A.C. (2005) Nanosphere-mediated delivery of vitamin E increases its efficacy against oxidative stress resulting from exposure to amyloid beta. *J. Alzheimer's Dis.* **7**, 297–301.
80. Liu, G., Men, P., Harris, P.L., Rolston, R.K., Perry, G., and Smith, M.A. (2006) Nanoparticle iron chelators: A new therapeutic approach in Alzheimer disease and other neurologic disorders associated with trace metal imbalance. *Neurosci. Lett.* **406**, 189–193.
81. Ravi Kumar, M.N. (2000) Nano and microparticles as controlled drug delivery devices. *J. Pharm. Sci.* **3**, 234–258.
82. Dehouck, B., Fenart, L., Dehouck, M.P., Pierce, A., Torpier, G., and Cecchelli, R. (1997) A new function for the LDL receptor: Transcytosis of LDL across the blood-brain barrier. *J. Cell Biol.* **138**, 877–889.
83. Muller, R.H., Jacobs, C., and Kayser, O. (2001) Nanosuspensions as particulate drug formulations in therapy. Rationale for development and what we can expect for the future. *Adv. Drug Deliv. Rev.* **47**, 3–19.
84. Ramge, P. (Ed.) (1998) *Untersuchungen zur Ueberwindung der Blut-Hirn-Schranke mit Hilfe von Nanopartikeln*, Shaker Verlag, Aachen.
85. Fenart, L., Casanova, A., Dehouck, B., Duhem, C., Slupek, S., Cecchelli, R., and Betbeder, D. (1999) Evaluation of effect of charge and lipid coating on ability of 60-nm nanoparticles to cross an in vitro model of the blood-brain barrier. *J. Pharmacol. Exp. Ther.* **291**, 1017–1022.
86. Davson, H. and Segal, M.B. (Ed.) (1996) *Physiology of the CSF and Blood-Brain Barriers*, CRC Press, Boca Raton, FL.
87. Porter, J.B., Gyparaki, M., Burke, L.C., Huehns, E.R., Sarpong, P., Saez, V., and Hider, R.C. (1988) Iron mobilization from hepatocyte monolayer cultures by chelators: The importance of membrane permeability and the iron-binding constant. *Blood*, **72**, 1497–1503.
88. Liu, G., Garrett, M.R., Men, P., Zhu, X., Perry, G., and Smith, M.A. (2005) Nanoparticle and other metal chelation therapeutics in Alzheimer disease. *Biochim. Biophys. Acta*, **1741**, 246–252.
89. Liu, G., Bruenger, F.W., Miller, S.C., and Arif, A.M. (1998) Molecular structure and biological and pharmacological properties of 3-hydroxy-2-methyl-1-(beta-D-ribofuranosyl or pyranosyl)-4-pyridinone: Potential iron overload drugs for oral administration. *Bioorg. Med. Chem. Lett.* **8**, 3077–3080.
90. Liu, G., Men, P., Kenner, G.H., Miller, S.C., and Bruenger, F.W. (2004) Acyclonucleoside iron chelators of 1-(2-hydroxyethoxy)methyl-2-alkyl-3-hydroxy-4-pyridinones: Potential oral iron chelation therapeutics. *Nucleosides Nucleotides Nucleic Acids*, **23**, 599–611.
91. Liu, G., Miller, S.C., and Bruenger, F.W. (1995) Synthesis of lipophilic 3-hydroxy-2-methyl-4-pyridinone derivatives. *Syn. Commun.* **25**, 3247–3253.
92. Blunk, T., Hochstrasser, D.F., Sanchez, J.C., Muller, B.W., and Muller, R.H. (1993) Colloidal carriers for intravenous drug targeting: Plasma protein adsorption patterns on surface-modified latex particles evaluated by two-dimensional polyacrylamide gel electrophoresis. *Electrophoresis*, **14**, 1382–1387.
93. Smith, M.A., Harris, P.L., Sayre, L.M., and Perry, G. (1997) Iron accumulation in Alzheimer disease is a source of redox-generated free radicals. *Proc. Natl. Acad. Sci. USA*, **94**, 9866–9868.
94. Bangs Laboratories, Inc. (1999) TechNote 201, Working with microspheres.
95. Dobbin, P.S., Hider, R.C., Hall, A.D., Taylor, P.D., Sarpong, P., Porter, J.B., Xiao, G., and van der Helm, D. (1993) Synthesis, physicochemical properties, and biological evaluation of N-substituted 2-alkyl-3-hydroxy-4(1H)-pyridinones: Orally active iron chelators with clinical potential. *J. Med. Chem.* **36**, 2448–2458.

96. Liu, G., Miller, S.C., and Bruenger, F.W. (1996) Efficient synthesis of N-[2-hydroxyethoxy)methyl]-2-alkyl-3-hydroxy-4-pyridinone by a modified Hilbert-Johnson reaction. *Syn. Commun.* **26**, 2681–2686.
97. Robins, M.J. and Hatfield, P.W. (1982) Nucleic acid related compounds. 37. Convenient and high-yield synthesis of N-[(2-hydroxyethoxy)methyl] heterocycles as "acyclic nucleoside" analogues. *Can. J. Chem.* **60**, 547–553.
98. Schaeffer, H.J., Gurwara, S., Vince, R., and Bittner, S. (1971) Novel substrate of adenosine deaminase. *J. Med. Chem.* **14**, 367–369.
99. Streater, M., Taylor, P.D., Hider, R.C., and Porter, J. (1990) Novel 3-hydroxy-2(1H)-pyridinones. Synthesis, iron(III)-chelating properties, and biological activity. *J. Med. Chem.* **33**, 1749–1755.
100. Nelson, W.O., Timothy, B., Karpishin, T.B., Retting, S.J., and Orvig, C. (1988) Aluminum and gallium compounds of 3-hydroxy-4-pyridinones: Synthesis, characterization, and crystallography of biologically active complexes with unusual hydrogen bonding. *Inorg. Chem.* **27**, 1045–1051.
101. Harris, R.L.N. (1976) Potential wool growth inhibitors. Improved synthesis of mimosine and related 4(1H)-pyridones. *Australian J. Chem.* **29**, 1329–1334.
102. Sayre, L.M., Perry, G., Harris, P.L., Liu, Y., Schubert, K.A., and Smith, M.A. (2000) In situ oxidative catalysis by neurofibrillary tangles and senile plaques in Alzheimer's disease: A central role for bound transition metals. *J. Neurochem.* **74**, 270–279.
103. Liu, G., Men, P., Perry, G., and Smith, M.A. (2007) Nanoparticles for the treatment of Alzheimer's disease: Theoretical rationale, present status and future perspectives. In: *Nanomaterials for Medical Diagnosis and Therapy* (Kumar, C.S.S.R., Ed.), Wiley-VCH Verlag GmbH & Co. KGaA. (Nanotechnologies for the Life Sciences) Weinheim, pp. 644–706.
104. Wong, S.S. (Ed.) (1991) *Chemistry of Protein Conjugation and Cross-Linking*, CRC Press, Boca Raton, Fl.
105. Arano, Y., Matsushima, H., Tagawa, M., Koizumi, M., Endo, K., Konishi, J., and Yokoyama, A. (1991) A novel bifunctional metabolizable linker for the conjugation of antibodies with radionuclides. *Bioconjug. Chem.* **2**, 71–76.
106. Anjaneyulu, P.S. and Staros, J.V. (1987) Reactions of N-hydroxysulfosuccinimide active esters. *Int. J. Pept. Protein Res.* **30**, 117–124.
107. Staros, J.V., Wright, R.W., and Swingle, D.M. (1986) Enhancement by N-hydroxysulfosuccinimide of water-soluble carbodiimide-mediated coupling reactions. *Anal. Biochem.* **156**, 220–222.
108. Kreuter, J. (2004) Influence of the surface properties on nanoparticle-mediated transport of drugs to the brain. *J. Nanosci. Nanotechnol.* **4**, 484–488.

Chapter 9

Synthesis and Characterization of Polymer Nanocarriers for the Targeted Delivery of Therapeutic Enzymes

Eric Simone, Thomas Dziubla, Vladimir Shuvaev, and Vladimir R. Muzykantov

Abstract

Protein drugs, such as recombinant enzymes useful for detoxification and replacement therapies, have extraordinary specificity and potency. However, inherently inadequate delivery to target sites and rapid inactivation limit their medical utility. Using chaperone polymeric particles designed within an injectible size range (sub-micron) may help solve these shortcomings. Such nanocarriers would (i) prevent premature inactivation of encapsulated therapeutic protein cargoes, (ii) provide a carrier that can be surface decorated by targeting ligands, and (iii) optimize sub-cellular localization of the drug. This chapter describes the techniques successfully employed for the preparation of polymer nanocarriers (PNC) loaded with the antioxidant enzyme, catalase, and targeted to endothelial cells. Methods of PNC synthesis, loading with catalase, characterization, coupling of a targeting moiety, and in vitro testing of the enzymatic and targeting activities are provided here. Advantages and disadvantages of specific designs are discussed. Due to the modular nature of the targeting methodology employed, it is believed that these protocols will provide a solid foundation for the formulation of a wide variety of enzymatic drug targeting strategies.

Key words: Polymer, biomaterials, nanocarriers, drug delivery, enzyme therapy, catalase, antioxidants, oxidative stress, vascular endothelium.

1. Introduction

Due to high potency and specificity of their activities, protein drugs are rapidly gaining clinical use. Recent advances in the biotechnology of recombinant human proteins can provide large amounts of homogenous, GMP-standard protein therapeutics. Among current clinically used protein therapies fibrinolytic plasminogen activators (1, 2), interferons and cytokines (3–5) growth factors and hormones (6–9), proteases and their inhibitors such as serpins (10–12), and diverse enzymes (e.g., lysosomal

R.M. Uppu et al. (eds.), *Free Radicals and Antioxidant Protocols*, Methods in Molecular Biology 610,
DOI 10.1007/978-1-60327-029-8_9, © Humana Press, a part of Springer Science+Business Media, LLC 1998, 2010

hydrolases for replacement therapies) (13–15) are making substantial progress. Further examples include an extensive and rapidly growing list of antibodies, antibody fragments, and their conjugates with toxins and other biologically active agents (16–18). Interestingly enough, although slightly longer clinical testing and approval phases seem to be the trend (19), the number of FDA approvals of proteins for therapeutic applications increased dramatically in the last decade (20). This reflects an unprecedented translation of protein drugs from the realm of small biotech start-ups into specialized resourceful departments of big pharmaceutical companies.

However, protein drugs are still in need of further improvement, particularly in the area of their delivery. They are typically expensive (e.g., cost of a recombinant plasminogen activator for fibrinolytic therapy is ~$1,000). Therefore, both economic and biological reasons (e.g., immune response) restrict the practicality of multiple injections for chronic interventions. Even at a single injection, the higher the dose of protein therapeutics, the higher the cost and probability of side effects. Unfortunately, high doses are required when pharmacokinetics is unfavorable due to factors such as rapid elimination via renal clearance and hepatic uptake, which turns out to be the case for most protein therapeutics.

In addition, due to the specific nature of their effects (e.g., enzymatic conversion of selected substrates or binding to selected ligands) many, if not all, protein drugs require localization in specific areas of the body – an organ, tissue, cell, or even sub-cellular compartment. However, only a few protein drugs have a natural affinity to their desired site of therapeutic action and the major fraction of injected dose represents a potentially harmful waste. Even when these drugs reach the site of action, factors such as inactivation by inhibitors and proteolytic degradation greatly limit activity and duration of therapeutic effect.

Therefore, the optimization of protein delivery remains an extremely important, yet challenging and not fully achieved, goal. Through the addition of a carrier or vehicle, therapeutic protein delivery can be optimized in several key ways. First, loading into and, in some cases, even coupling to the surface of the carriers can protect proteins from inactivation by proteases, inhibitors, and other aggressive components of blood and/or other compartments in the body (21). Second, by creating a barrier between a drug on one side and blood and non-target tissues, cells, or organs on another side, a carrier can minimize potentially harmful side effects and avoid eliciting an immune response, both of which are commonly inherent of protein therapeutics. Third, loading into long-circulating carriers decelerate blood clearance of drugs that otherwise are rapidly excreted and also protect the cargo against premature inactivation en route to the target (22). Fourth, by varying size, mechanical properties, and surface decoration, targeted delivery of carrier cargoes to therapeutic sites can be achieved by either passive mechanisms (e.g., utilizing mechanical or electrostatic retention in capillaries or effect of enhanced permeability and retention in areas of inflammation and tumors) (23, 24) or active mechanisms (e.g., utilizing affinity moieties on the surface of carriers that bind to specific determinants in targets) (18, 25, 26). This feature can even extend to achieving more adequate sub-cellular addressing of cargoes, utilizing specific endocytic pathways and intracellular sorting and traffic machinery (27), depicted in **Fig. 9.1** for nanocarriers coated with targeting antibodies. Finally, slowly degrading carriers produce a prolonged and more stable profile of drug activity in the target site (28).

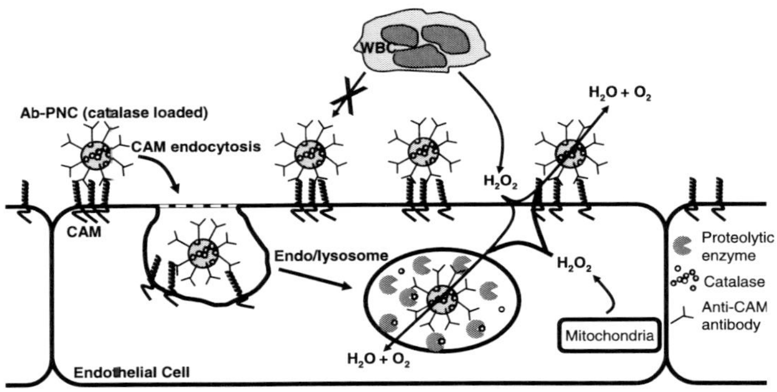

Fig. 9.1. Concept of cell adhesion molecule-targeted PNC delivery to endothelial cells (EC). Antibody (Ab)-coated PNC targeted to cell adhesion molecules (CAM) are internalized via CAM-mediated endocytosis, before trafficking to endo/lysosomes (33). PNC-encapsulated catalase is protected from proteolytic digestion during residence within these recycling compartments. Both internalized and surface-bound PNC can detoxify the highly diffusive reactive oxygen species, hydrogen peroxide (produced under pathological conditions by either adherent white blood cells (WBC) or by endothelial cells (e.g., in mitochondria). Masking of endothelial CAM by the targeted PNC may also attenuate pro-inflammatory vascular adhesion of white blood cells.

This chapter describes the methodology and protocols to formulate polymer nanocarriers (PNC) targeted to endothelial cell adhesion molecules (CAM) designed for enzyme encapsulation and thus protection from lysosomal degradation. While the immobilization of *active* enzymes within PNC has been elusive, the freeze–thaw double emulsion solvent evaporation technique described in this chapter has demonstrated ample loading of catalytically active enzyme (catalase) and protection of this cargo from proteolytic degradation (28). To conjugate targeting CAM antibodies to the PNC formed in this method, a biotin–streptavidin conjugation scheme is used. This protocol attaches antibody–streptavidin conjugates to PNC-containing biotinylated polymer through a single-step addition process, avoiding undesirable aggregation of PNC. These techniques taken together are believed to provide a solid basis for the design of any targeted PNC enzyme delivery strategy.

2. Materials

2.1. Equipment

1. Rotovap (Safety Vap 205; Buchi, Switzerland).
2. Freeze dryer (RCT 60; Jouan, Winchester, VA).
3. Homogenizer (Kinemetica Polytron 3100 with a 7 mm PDTA3007/2 generator; Brinkmann Instruments, Westbury, NY).
4. Dynamic light scattering 90 Plus Particle Sizer (Brookhaven Instruments Corp., Holtsville, NY).
5. UV-VIS spectrophotometer (Cary 50 UV-Vis; Varian, Palo Alto, CA).
6. Microplate reader (Model 2550-UV; Bio-Rad Labs, Hercules, CA).
7. Gamma counter (1470 Wallac Wizard; PerkinElmer, Wellesley, MA).
8. Fluorescence microscope.
9. Desalting Quick Spin Protein Columns (Roche Applied Science, Indianapolis, IN).
10. Laboratory low-speed centrifuge.
11. General cell culture equipment: CO_2 incubators, sterile hood, etc.

2.2. Reagents

All reagents obtained from Sigma-Aldrich (St. Louis, MI), unless otherwise stated.

1. Diamine polyethylene glycol (PEG) (Nektar, Huntsville, AL).
2. 5,000 MW mono-methoxyPEG (mPEG) (Polysciences, Warrington, PA).
3. Lactide (3,6-dimethyl-1,4-dioxane-2,5-dione).
4. Stannous octoate.
5. Poly(lactic co glycolic acid) free acid ended (PLGA – 5050 DL 4A) (Alkermes, Cambridge, MA).
6. *N,N'*-Dicyclohexylcarbodiimide (DCC).
7. Dichloromethane (DCM).
8. Polyvinyl alcohol (PVA) or Pluronic® F68.
9. Catalase from bovine liver (Calbiochem, San Diego, CA).
10. Hydrogen peroxide (H_2O_2).
11. Sodium hydroxide (NaOH).
12. Hydrochloric acid (HCl).
13. Lactate oxidase.

14. 10-Acetyl-3, 7-dihydroxyphenoxazine(Amplex red; Molecular Probes, Eugene, OR).
15. Horseradish peroxidase (HRP) (Calbiochem).
16. Barium chloride ($BaCl_2$).
17. Potassium iodide (KI).
18. Iodine (I_2).
19. Succinimidyl 4-(*N*-maleimidomethyl) cyclohexane-1-carboxylate (SMCC) (Pierce, Rockford, IL).
20. *N*-Succinimidyl-*S*-acetylthioacetate (SATA) (Pierce).
21. Streptavidin from *Streptomyces avidinii* (Calbiochem).
22. Biotin *N*-hydroxysuccinimide ester (Biotin-NHS) (Pierce).
23. Murine monoclonal anti-PECAM antibody (clone 62 was generously provided by Dr. Nakada; Centocor, Malvern, PA).
24. Non-immune mouse IgG (Jackson ImmunoResearch Labs, West Grove, PA).
25. *N*, *N*-Dimethylformamide (DMF).
26. Dimethyl sulfoxide (DMSO).
27. Hydroxylamine-HCl.
28. Pronase (Protease type XIV – bacterial from *Streptomyces griseus*).

2.3. Buffers, Media, and Cells

1. Phosphate-buffered saline (PBS): 0.1 M sodium phosphate, 0.15 M sodium chloride, pH 7.2.
2. Human umbilical vein endothelial cells (HUVEC; Lonza, Basel, Switzerland).
3. REN human mesothelioma cells and REN-hPECAM cells transfected with human PECAM (kindly provided by Dr. Steve Albelda, University of Pennsylvania).
4. HUVEC culture medium: M199 medium (GIBCO, Grand Island, NY), 10% FCS (GIBCO) supplemented with 100 μg/mL heparin (Sigma, MO), 2 mM L-glutamine (GIBCO), 15 μg/mL endothelial cell growth supplement (Upstate, Lake Placid, NY), 100 U/mL penicillin, and 100 μg/mL streptomycin.
5. REN cell culture medium: RPMI 1640 medium (GIBCO), 10% fetal bovine serum (Sigma), 2 mM L-glutamine (GIBCO), 100 U/mL penicillin, and 100 μg/mL streptomycin. For REN-hPECAM cells the medium was supplemented with 0.5 mg/mL G418-sulfate.
6. Lysis buffer: 1% Triton X-100, 1.0 N NaOH.

3. Methods

3.1. Preparation of Biotinylated PEG-PLA/PLGA

1. To remove small (MW < 1 kDa) compounds, dissolve 5 g PLGA (38 kDa) into 100 mL dichloromethane and add solution drop wise to a 1 L bath of cold ether, precipitating the PLGA polymer. Perform the same step with the 20 g D, L-lactide in place of PLGA.
2. Filter out precipitates and dry under nitrogen in a rotovap followed by a freeze dryer.
3. Add the PEG diamine and PLGA in a 1:6 molar ratio (e.g., 2 g PLGA to ~100 mg of 3.4 kDa PEG) into a common vessel such as a 250 mL Erlenmeyer flask containing a magnetic stir bar.
4. Remove excess water from this powder mixture by freeze drying overnight.
5. Remove powder from freeze dryer and cap with a rubber septum. Purge vessel with nitrogen 3–5 times.
6. Using a needleline transfer technique, add 100 mL of anhydrous DCM to polymer powder and fully dissolve by mixing on a stir plate.
7. Prepare 5 mL of a 50 mM solution of DCC in anhydrous DCM. DCC is a highly reactive primary amine reagent. Use exclusively in a hood, or preferably using a glove bag or glove box. Since DCC is readily hydrolyzed, prepare solution immediately before use.
8. Next, add 1.16 mL of the DCC solution to the polymer solution to create a final 1:1.1 PLGA:DCC molar ratio.
9. React under constant stirring overnight.
10. Filter solution through a filter (pore size: 3 μm) followed by concentrating polymer solution on a rotovap to ~50 mL.
11. Add 500 mL of cold ether to precipitate polymer and dry sample on a rotovap.
12. Add 50 mL of DCM and repeat ether precipitation.
13. To remove unreacted PEG, dissolve polymer in acetone (50 mL). Acetone may need to be heated to fully dissolve polymer.
14. Add, drop wise, acetone solution to cold water.
15. Collect precipitate and freeze dry. The consistency of the polymer may prevent filtration. If this occurs, precipitate can be centrifuged instead (15,000 × g for 1 h at 4°C).

16. Dissolve 1 g of PEG-PLGA into a 50 mL 1:1 solution DCM:DMSO.
17. Prepare a 50 mM solution of biotin-NHS in DMSO.
18. Add biotin-NHS to the PEG-PLGA solution for a final 1.5:1 biotin-NHS:PEG-PLGA molar ratio.
19. Purify polymer by evaporating off DCM and then adding solution to cold DI water.
20. Collect precipitate, dissolve again into acetone and re-precipitateinto cold DI water.
21. Freeze dry polymer and store under desiccation until ready for use.
22. For synthesis of PEG-PLA, add 5 g of the purified D, L-lactide from step 1, with 1.25 g of mPEG to a round bottom flask with a Teflon stir bar. PEG and lactide are mixed in stoichiometric amounts to achieve the desired polymer molecular weights. For example, the above amounts result in a mPEG$_5$-PLA$_{20}$ (5 kDa PEG block and 20 kDa PLA block) diblock copolymer. To synthesize mPEG$_5$-PLA$_{30}$, simply mix 5 g of lactide with 0.8333 g of mPEG.
23. Seal the flask with a rubber septum and purge with dry nitrogen.
24. Heat the sample to 140°C for 2 h, sealed under dry nitrogen to bring the reagents to a melt and drive off residual water. A three neck, round bottom flask can be used instead, with a Drierite drying column if one is concerned about excessive moisture. Trace water severely hinders the polymerization.
25. Reduce the temperature to 130°C and add stannous octoate at 1% of the total reagent mass (for this example that amounts to about 46.6 μL, taking into account the density of stannous octoate). Let polymerization proceed for approximately 6 h.
26. Reduce temperature to 25°C and dissolve the polymer in approximately 30 mL of DCM.
27. Precipitate this DCM–polymer mixture into cold diethyl ether, adding drop wise. Repeat twice, collecting filtrate.
28. Freeze dry polymer and store under desiccation until ready for use.

3.2. Polymer Nanocarrier (PNC) Formulation and Catalase Encapsulation

1. Prepare a mPEG-PLA solution containing ~20 mol% biotin-PEG-PLA. To do this, dissolve 5 mg biotinylated mPEG-PLA/PLGA (prepared in **Section 3.1**) and 20 mg mPEG$_5$-PLA$_{20-30}$ diblock copolymer in 1 mL DCM for a total 25 mg/mL polymer solution, per prep. It should be

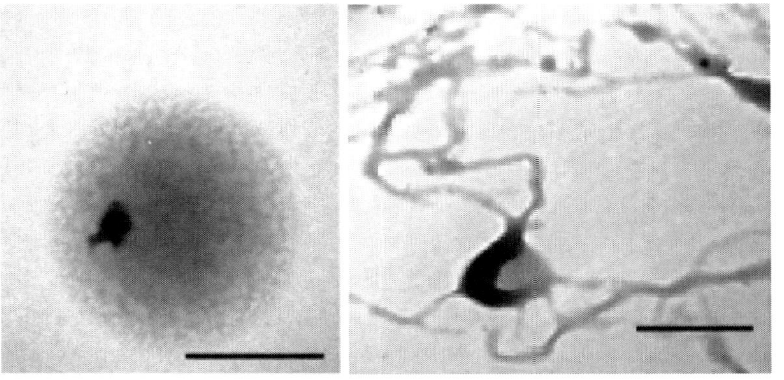

Fig. 9.2. Transmission electron microscopy (TEM) of mPEG-PLA PNC. Re-suspended PNC pellet was diluted 1:4 in DI water before application to TEM grid and subsequent staining with 2% (w/v) uranyl acetate. Scale bar is 500 nm. *Left*: sphere morphologies are produced when larger fractions of amphiphilic diblock copolymer consist of hydrophilic block mPEG. In the *left* figure, a 22:78 MW ratio of mPEG:PLA was used to form PNC. *Right*: Worm/branched morphology was achieved with a higher MW fraction of hydrophobic polymer. In this case, only 7% of the diblock polymer MW consisted of mPEG. In general, increasing the hydrophilic fraction of diblock copolymers can lead to a transition from worm micelle and branched morphologies to more spherical particles, if polymers with near monodisperse MW are utilized (34, 35). Due to the more disperse MWs of polymers synthesized with a ring-opening polymerization (degree of MW polydispersity increases further with higher MW), a slightly more heterogeneous PNC morphology can be observed at higher polymer MW with the system described in this chapter.

noted that a $mPEG_5$-PLA_{20} polymer yields spherical particles, while $mPEG_5$-PLA_{30} results in filamentous particles, as illustrated in **Fig. 9.2** (29).

2. Prepare 100 μL of a protein solution (1 mg/mL) solution of catalase in 20 mM PBS. Maximum protein concentration used is limited by the solubility of protein in solution. The key parameter is the volume of the aqueous phase, and should be maintained, for this particular formulation, at ~10% of the primary emulsion (e.g., 100 μL catalase/PBS added to 1 mL of polymer solution).
3. Prepare surfactant solution of PVA or F68 pluronic (2% wt) in 20 mM PBS. Higher % hydrolyzed PVA may not go entirely into solution. If this occurs, filter solution with a 0.2 μm filter.
4. Combine 100 μL catalase solution (aqueous phase) and 1 mL of the polymer–DCM solution (oil phase) in a 12 × 75 mm glass (borosilicate) test tube for the primary emulsion, and place on ice. Use glass test tubes to prevent contamination (e.g., DCM dissolves polystyrene plastic test tubes) and to minimize sample loss due to vessel wall adherence.

5. Add 5 mL of the surfactant solution to a 16 × 100 mm borosilicate test tube for the second emulsion, and place on ice.
6. Add 10 mL of the surfactant solution to a small vessel (a 20-mL scintillation vial will suffice) and place on a stir plate set at 1,400 rpm.
7. Prepare a dry-ice and acetone bath in a small vessel deep enough to submerge the entire liquid portion of the test tube containing the primary emulsion and allow it to equilibrate (–80°C) (*see* **Note 1**).
8. Set up the 7 mm generator on the homogenizer and wash first with water and then acetone, drying thoroughly. Preset the homogenizer at the desired mixing rate prior to emulsification. For this protocol, set speed to 15,000 rpm.
9. Homogenize the primary emulsion, at 15,000 rpm. Once the homogenizer has reached 15,000 rpm, add the dry ice bath, ensuring that the entire liquid portion of the emulsion is chilled. Homogenize the solution for 1 min. To minimize protein denaturing, minimize the time between starting the homogenizer and submersing into the dry ice bath. Note, however, that adding the dry ice bath before the homogenizer has come up to speed will prematurely freeze the aqueous phase before it can blend with the organic polymer-containing phase.
10. After the primary emulsion is complete, pipette the solution into the 5 mL surfactant containing test tube. Homogenize for an additional 1 min at 15,000 rpm, without the dry-ice acetone bath. The speed of this step will determine the final size of the nanocarriers. Important: minimize the transfer time to enhance final loading, carrier yield, and minimize final PNC size.
11. Add the second emulsion into the 10 mL surfactant solution. Allow to stir at ambient conditions for at least 3 h to allow for DCM evaporation and PNC hardening.
12. Centrifuge the 15 mL sample at 1,000 × *g* for 10 min to remove large aggregate/microparticle fraction (*see* **Note 2**).
13. Centrifuge the nanoparticle-containing supernatant at 20,000 × *g* for 30 min. Minimize excessive centrifugation to avoid irreversible PNC aggregation.
14. Remove the supernatant and resuspend the PNC pellet in 500 μL PBS.
15. Dilute sample again to 15 mL and repeat steps 13 and 14, in order to remove any residual surfactant.

3.3. PNC Characterization (*see* Note 3)

Protein loading can be determined either by mass (tracing of radioisotope-labeled protein) or catalytic activity.

3.3.1. Loading Determination

Loading determination by tracing catalase mass:

1. Radiolabel catalase with $Na^{125}I$ (PerkinElmer, Boston, MA) via the Iodogen method (Pierce Biotech., Rockford, IL). Unbound ^{125}I is removed from catalase by gel permeation chromatography (GPC, Biospin 6 columns, Bio-Rad Labs, Hercules, CA).
2. Form PNC, as described in **Section 3.2**, substituting catalase in the primary emulsion with a mixture of ^{125}I-catalase and non-iodinated catalase. Make sure a sufficient amount of ^{125}I-catalase is added, since reliable measurements are obtained only when the radioactivity of the resultant samples is above 100 cpm, which usually means that $\sim$ 50,000 cpm/prep should be added.
3. Measure total solution ^{125}I-catalase content in the sample before centrifugation. Centrifuge the samples as outlined above (**Section 3.2**) to separate super and submicron fractions of catalase-loaded PNC. Resuspend PNC pellet to determine radioactivity of the ^{125}I-catalase/PNC with a Wizard 1470 gamma counter (PerkinElmer, Boston, MA).

Loading determination by enzyme catalytic activity:

Similarly, to determine loading via enzymatic activity, a catalase activity assay (30) is used, both for the total sample before and resuspended pellet after centrifugation.

1. Add 900 μL of 5 mM H_2O_2 in PBS and 100 μL of enzyme-loaded PNC to a quartz cuvette and mix, avoiding vigorous shaking which results in foaming and errors in measurement. In general, aim for a total volume of 1 mL, but amount of PNC can vary from 10 to 100 μL. If PNC concentration is too high, diffraction and other light scattering phenomena can interfere with the spectrophotometer reading. This can increase background noise and make measurements of samples with low activity difficult.
2. Measure loss of H_2O_2 absorbance (242 nm) over time with a spectrophotometer to determine degradation by catalase associated with carriers. Calculate slopes (ΔA/min), which correlate with loaded enzymatic activity, of decreasing absorbance. Calculate the activity using following equation:

$$\text{Catalase activity, U/mg} = 23.0(\Delta A/\text{min})/\text{mg protein}$$

3.3.2. PNC Enzyme Protection Assay

PNC protection of encapsulated catalase can be shown via incubation with a proteolytic enzyme, such as pronase (28). Protection can be shown by either preservation of catalase activity, via the above described activity assay in **Section 3.3.1**, or by mass of loaded protein degraded.

1. If determining protection by mass, prepare PNC loaded with ^{125}I-catalase as described above in **Sections 3.2** and **3.3.1**.
2. Incubate final PNC prep with a 0.2% (w/v) pronase solution at 37 °C in a shaker bath set at 60 rpm for 1 h, or longer if desired. Dependent on how many time points one wishes to make, 0.5–1 mL total samples are typically enough.
3. If determining protection by mass, remove samples from shaker bath and centrifuge at 16,000 × *g* for 20 min for a 1 mL sample. Collect and place in the gamma counter the supernatant containing products of ^{125}I-catalase proteolytic degradation and pellet containing intact/protected protein, encapsulated within PNC.
4. If determining protection by activity, simply measure enzymatic activity before and after incubation with protease, via the activity assay in **Section 3.3.1.**

3.3.3. Morphology Characterization

Adequate characterization of sizing and geometry can be accomplished with dynamic light scattering (DLS) and transmission electron microscopy (TEM).

Dynamic light scattering:

From DLS, approximate values of diameter can be obtained. *Relative* concentration, proportional to the number of PNC within a specific volume, can be determined from the average scattering intensity measurements. Please refer to the method by Shuvaev et al. for practical details in experimental theory and data interpretation (30).

1. Add 20 μL of reconstituted PNC to a glass NMR tube and dilute with 200 μL of filtered DI water or buffer. Amount of PNC solution added should be adjusted to keep scattering within the optimum range of the instrument, typically for nanoparticles between 200 and 800 kcps.
2. Place in dynamic light scattering apparatus containing an NMR to Cuvette adapter insert (DLS, 90 Plus Particle Sizer; Brookhaven Instruments, Holtsville, NY). Record for 2 min at a 90° angle.

Transmission electron microscopy:

TEM provides a means of validating DLS sizing data and determining actual carrier morphology, (e.g., spherical, branched, or wormlike). TEM is not an effective means of determining concentration since drying phenomena can lead to heterogeneous PNC distribution.

1. Apply 5 μL of sample to a TEM mesh grid (Formvar Film 200 mesh; Electron Microscopy Sciences, Hatfield, PA). Cover only the dull (not polished) grid side, completely with sample.
2. Remove excess by touching edge of grid with filter paper, before drying at ambient conditions. Removal of excess sample promotes a monolayer of PNC on the grid surface, improving image quality. Also be sure to try multiple concentrations/dilutions of PNC samples. If samples are too concentrated, aggregation is common upon grid staining and further limited quality imaging results. Be careful not to touch top surface of grid and introduce any particulate contaminants.
3. Prepare 2% (w/v) uranyl acetate (UA) solution in DI water. Cover with aluminum foil and avoid exposure to light. UA solution stored at 4 °C typically yields decent sample staining if used within a week of preparation. Longer storage times often lead to excessive UA precipitation.
4. Filter UA through a 0.1 μm syringe filter, placing a drop on parafilm.
5. Touch the top of the sample – containing grid to UA drop. Turn grid sample-side up and cover to block out light. Do not allow any UA to stain the backside of the grid, as this will obscure imaging. Let sample stain for 5 min, before drop wise washing with filtered DI water.
6. Remove excess water as described in **step 2** (above) and allow grid to dry at ambient conditions for 1 h before imaging with TEM.

It is imperative that fluid UA exposure to light be minimized, as this will lead to contrast agent precipitation and imaging artifacts. When imaging, be sure to use the minimum necessary accelerating voltage to avoid PNC sample melting.

3.3.4. Yield Determination

A convenient method of PNC yield determination is a colorimetric PEG assay based on the PEG-barium iodide complex (31). Alternatively, one could perform a lactic acid enzymatic assay (*see* **Note 4**)

1. Rapidly hydrolyze a 50 μL aliquot of PNC sample by adding 200 μL of 5 N NaOH and reacting overnight at 80 °C, before neutralizing with 5 N HCl.
2. Prepare solution A, consisting of 2.4 g barium chloride, 8.0 mL 6 N HCl and 32 mL deionized (DI) water.
3. Prepare solution B, consisting of 800 mg of potassium iodide, 500 mg of iodine, and 40 mL of DI water. This solution is very concentrated and not all of the iodine may go into solution. Preparing a more diluted version is not advised since iodine is volatile and may not store well.

4. Prepare a 0.2 mg/mL calibration solution of PEG in DI water (preferably of the same MW used in the PNC, but this is not essential and can be accounted for in the analysis).
5. In a 96-well plate, prepare a calibration series of 50–0 μL of the stock PEG solution. Ideally, calibration should be in 10 μL increments for 50–10 μL, and 2 μL increments for 10–0 μL. This setup is good for the detection of 10–0.1 μg PEG.
6. Add PEG sample from **Step 1** to wells in appropriate volumes such that approximately 10–0.1 μg of PEG is present.
7. Dilute sample and calibration volumes to 170 μL total volume with DI water.
8. Add 40 μL of solution A to each well and mix thoroughly.
9. Make a 1/5 dilution of solution B and add 40 μL to each well.
10. Allow to develop for 10 min and measure absorbance at 535 nm on the microplate reader. A 550 nm filter on the reader is adequate.

3.4. Preparation of Antibody–Streptavidin Conjugates

Sulfhydryls are introduced onto the molecule of antibody or control IgG through primary amines using SATA. In parallel, heterobifunctional cross-linker, SMCC, is used to introduce stable maleimide reactive groups into streptavidin molecules. Conjugation takes place as a specific reaction between the SH group of the antibody and the maleimide group of streptavidin. Introducing one sulfhydryl per antibody molecule prevents multivalent cross-linking and aggregation whereas several reactive groups may be introduced into streptavidin to increase the yield of the conjugation. In addition, such a conjugation of antibody molecule to streptavidin limits the availability of biotin-binding sites on streptavidin and thus prevents possible cross-linking of the PNC, in contrast to unconjugated streptavidin that may function to cross-link PNC together. Further, streptavidin can be replaced by less immunogenic neutral avidin derivative, neutravidin.

1. Dissolve IgG (antibody or non-immune IgG) and streptavidin in 0.1 M PBS to a concentration of 5.0 and 6.0 mg/mL, respectively (33.3 and 100 μM, respectively). Make sure that there is no contamination with amine-containing components such as Tris buffer. Keep proteins on ice.
2. Prepare fresh 40 mM SATA in DMSO (9.25 mg/mL).
3. Add freshly prepared SATA to 200 μL of IgG in a 1:5 IgG:SATA molar ratio and incubate for 2 h at room temperature.

4. Separate modified protein from unreacted components using desalting Quick Spin Protein Columns (Roche Applied Science, Indianapolis, IN).
5. Add 20 μL of 0.5 M hydroxylamine-HCl (NH_2OH) to SATA-modified IgG and incubate for 2 h at room temperature before again passing IgG sample through desalting column.
6. Prepare fresh 20 mg/mL (45.8 mM) SMCC in DMF. Add SMCC solution to 100 μM streptavidin to achieve a final streptavidin:SMCC molar ratio of 1:20 (4.4 μL SMCC/100 μL streptavidin). The reaction takes 1 h at room temperature. This step may be done during previous step 5. Separate SMCC-modified streptavidin using desalting column.
7. Conjugation is performed by mixing modified proteins at IgG:streptavidin molar ratio 2:1 and incubation for 1 h on ice.
8. The confirmation of conjugation products may be performed by SDS polyacrylamide gel electrophoresis (4–15% gradient gel; Bio-Rad) and following Coomassie staining, as shown in **Fig. 9.3**.

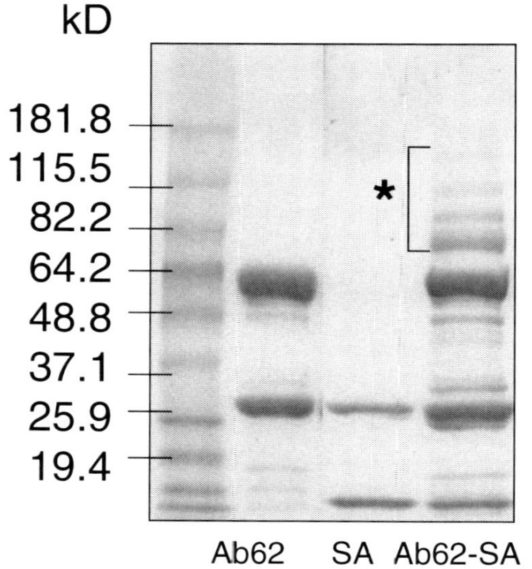

Fig. 9.3. Electrophoretic analysis of anti-PECAM/streptavidin conjugate. Anti-PECAM monoclonal antibody (Ab 62) was conjugated to streptavidin (SA) using heterobifunctional cross-linkers SATA and SMCC (*see* **Section 3.4**). Figure shows Coomassie blue staining of gel electrophoresis in reducing conditions (4–15% PAGE gel, Bio-Rad). A *star* indicates formed conjugated species. Due to denaturing and reducing conditions of the electrophoresis only heavy chains of antibody and only monomers of streptavidin that are directly cross-linked appear as conjugated species. Other heavy and light chains of antibody and non-cross-linked streptavidin monomers migrate as intact bands.

9. The conjugates may be frozen in appropriate aliquots (i.e., 50 μL/tube) and store at –80 °C before use.

3.5. Conjugation of PNC and Targeting Antibodies

1. Prepare antibody–streptavidin conjugates as described above in **Section 3.4**.
2. Mix 10 μL of 25 mg/mL PNC with 70 μg (total protein mass) of antibody–streptavidin conjugate solution (7 μL of a 50 μM conjugate stock solution by streptavidin basis). This results in a 1:10,000 Ratio of PNC:antibodies for 300 nm PNC.
3. Dilute solution to 100 μL and vortex for 1 min.
4. Incubate under gentle agitation for 1 h at ambient temperature.
5. Dilute sample with 1 mL of PBS and centrifuge (16,000 × *g* for 10 min).
6. Resuspend pellet and measure size and binding function based on previously reported methods (32).

3.6. Characterization of Targeted PNC In Vitro – PNC Binding to Cells

3.6.1. Fluorescence Detection of PNC Binding to Cells

To visualize binding and to show its specificity we used epifluorescence imaging.

1. Plate REN and REN-hPECAM cells on 24-well dishes with inserted 12-mm glass coverslips and cultivate to confluence.
2. Add anti-PECAM/PNC to confluent cells and incubate for 1 h in regular cell medium at 37°C.
3. Wash out unbound nanocarriers 5 times with fresh RPMI medium, fix cells, and stain with secondary FITC-labeled antibody against mouse IgG.
4. Visualize the bound nanocarriers in REN-hPECAM cells vs. REN cells as a negative control using fluorescence microscope.

3.6.2. Detection of PNC Binding to Cells by Radioisotope Tracing

Quantification of the binding to human endothelial cells was performed by radioactive tracing of nanocarriers.

1. Plate HUVEC on 24-well dishes and cultivate until culture is confluent.
2. Add anti-PECAM/PNC or control IgG/PNC labeled with ^{125}I-IgG-SA tracer to cells at serial dilutions, starting with ~25,000 cpm per well and incubate for 1 h in regular cell medium at 37°C.
3. Wash out unbound nanocarriers 5 times with fresh RPMI medium and lyse cells with 1% Triton X-100, 1.0 N NaOH.
4. Transfer cell lysate in glass tubes and measure radioactivity of bound material using Wallac 1470 Wizard™ gamma counter. Bound material is expressed as number of bound particles per cell.

4. Notes

1. *Formulation*. Do not allow second emulsion test tube containing surfactant to sit on ice longer than 15–30 min prior to the actual second emulsion. When the surfactant temperature of the second emulsion is at 4°C, the addition of the primary emulsion (–80°C) brings the entire solution below the gel point for the surfactant solution resulting in poor mixing.
2. *Serial centrifugation size fractionation*. Serial centrifugation to separate different size populations (28) is best used when synthesized particles possess two distinct population sizes, as is the case in this freeze–thaw solvent evaporation PNC formulation. Centrifugation time and speed can be determined by calculating the time needed to settle larger particles using Stokes' law:

   $$V = 2ga^2 \frac{\rho_1 - \rho_2}{9\eta}$$

   where V is the settling velocity, g the relative centrifugal force, a the particle radius, ρ_1 and ρ_2 are the particle density and the buffer density, respectively, and η is the buffer viscosity. After centrifugation, the supernatant can be collected and centrifuged at a greater speed to collect the nanoparticle fraction. This procedure is an efficient and rapid method of both size selection and removal of residual surfactant. An example of a centrifugation scheme is represented in **Fig. 9.4**. To remove any residual surfactant that may be bound to the PNC surface, resuspension of nanocarriers in buffer and an additional centrifugation purification step is recommended. As with any centrifugation fractionation technique, resuspension of PNC can be difficult. deally centrifugation rates will be selected such that PNC will be pelleted but not aggregated. However, if aggregation does occur, several laboratory techniques can be used to aid in resuspension. For instance, allow the sample to solvate for 5 min to 1 h with mild vortexing. Allowing the PNC to rehydrate slowly can aid in the hydration of the PNC's PEG shell, enhancing particle–particle repulsion. If, after this period, resuspension is still difficult, application of mild pulse sonication (5 W, 1 s intervals) can be used. However, such energy addition may result in disruption of PNC, releasing cargo and/or denaturing loaded protein.
3. *PNC size control*. The final size of PNC is dictated by a balance of both kinetic and thermodynamic forces.

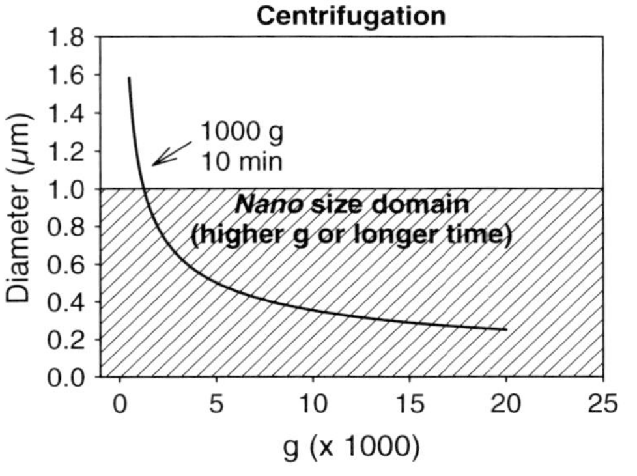

Fig. 9.4. Centrifugation and size fractionation. The scheme represents approximate centrifugation rates (g) necessary to sediment different size populations in a 10 min centrifugation of PLA-based particles. Higher rate, g, or longer spin time, can bring down smaller particles in the desired sub-500 nm range.

Mechanical disruption of the two-phase liquid system results in droplet formation. As more energy is added (either through faster shear rates, longer times, larger impeller diameters) the droplets are disrupted and form smaller droplets. However, as the interface between the two phases increases, the destabilizing energy also increases. Unless there is a surfactant that can function to stabilize these two phases, droplets will start to coalesce and form larger particles. Yet this coalescence is rate limiting and determined by factors including aqueous phase viscosity, polymer phase viscosity, surfactant concentration, and temperature. **Figure 9.5** demonstrates this balance between energy and size.

To summarize these effects, smaller PNC can be obtained by higher energy rates and increased surfactant content (either the amphiphilic PNC polymer or added solution surfactant). However, as PNC size decreases, so does the internal volume to surface area ratio. With smaller internal volumes, the likelihood of successfully encapsulating protein drops off significantly. Furthermore, longer duration and higher shear rate homogenizations can denature enzymes, dramatically decreasing recovered activity. As such, ultimate carrier size is selected by a compromise between cellular/vascular compatibility and protein loading efficiency.

4. *Particle yield and mass determination.* Double emulsion PNC synthesis typically results in a bimodal distribution, with a significant fraction of undesired large (>1 μm) particles. During process optimization, reduction in this loss fraction will be a significant technological hurdle. However, proper surfactant concentration may provide one avenue of achieving higher PNC yields. Increasing PVA surfactant concentration from 0.1 to 4 wt% increased the PNC fractional

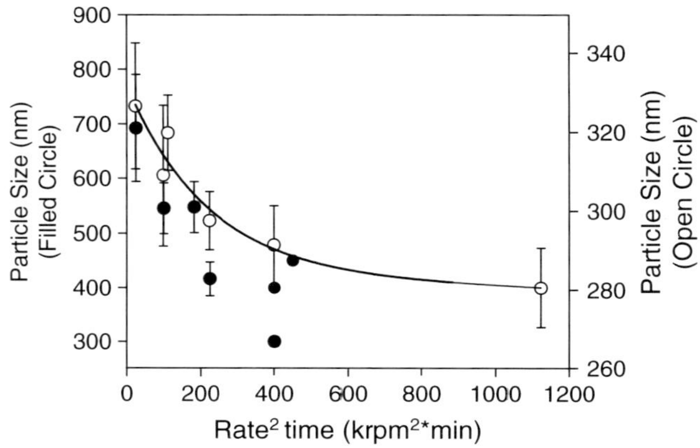

Fig. 9.5. PNC size decreases with increasing energy input. Mean diameter (measured by DLS) is a function of both energy input (which scales with homogenization rate^2 t) and polymer surfactant capacity (● 5% and ○ 11% PEG content). When surfactant capacity of system is saturated, further increases in energy input do not decrease PNC size. However, greater PEG content (greater intrinsic surfactant capacity of PNC polymer) results in formulation of even smaller PNC carriers.

yield, but also simultaneously increased the carrier diameter as well. This result is most likely due to two contributing factors. With greater PNC mass, the overall surface area between phases is increasing, resulting in a saturation of PVA's surfactant capacity at larger sizes. Also, the increased viscosity of the aqueous phase reduced the rate of droplet breakup under the mixing conditions, further adding to the larger carrier size.

Polymer concentration analysis provides the only reasonable method to determine PNC yield. Yet, the PEG assay approach possesses several potential complications. Polymeric surfactant used in PNC synthesis can participate in the barium iodide complex, resulting in overestimating the actual PNC concentration. This is of particular concern with Pluronic®, a PEG-containing surfactant. As such, careful particle purification in the form of extra centrifugation steps and/or dialysis may be necessary to determine the actual PNC concentration. PEG assays of the supernatants and dialysates provide an indicator of amount of surfactant released through each step. Once no measurable PEG is found, PNC can be assumed to be clear of surfactant. If dialyzing, keep in mind that Pluronic®, although less than 9 kDa, is a linear polymer and hence has a hydrodynamic radius on the order of a 500 kDa protein.

Erratic results with the PEG assay may also be obtained in the presence of high protein content (above 3 mg/mL) (31). As an alternative, an enzymatic assay may be utilized for the detection of lactic acid, the monomeric building block of PLA (*see* **Note 5**).

5. *Enzymatic lactic acid assay.* The working buffer used in this concentration assay is very sensitive to pH, and therefore 100 mM PBS is recommended for solutions and dilutions.
   1. Rapidly hydrolyze a 50 μL aliquot of PNC sample by adding 200 μL of 5 N NaOH and reacting overnight at 80°C, before neutralizing with 5 N HCl. This assay is sensitive only to the L-lactic acid enantiomer, and hence this step is necessary to convert all L-lactic acid in monomeric form.
   2. Prepare a 0.01 mg/mL calibration solution of L-lactic acid in DI water.
   3. In a 96-well microplate, prepare lactic acid calibration wells as described above in the PEG assay (**Section 3.3.4**).
   4. Add sample from step 1 to wells in appropriate volumes such that L-lactic acid is within the calibration range.
   5. Dilute sample and calibration volumes to 50 μL using 100 mM PBS.
   6. Prepare working buffer of 4.75 mL of 100 mM PBS, 50 μL of 10 mM Amplex red in DMSO, 100 μL of 10 U/mL HRP, and 100 μL of 50 U/mL lactate oxidase.
   7. Add 50 μL of working buffer to each well.
   8. Allow to develop for 10 min and measure absorbance at 530 nm on the microplate reader. A 550 nm filter on the reader is adequate.As mentioned above, this method is very pH sensitive, and 100 mM PBS may not be adequate. In such an event, careful titration of sample to neutral pH, after saponification, will be necessary. Also, the reagents of the working buffer, primarily lactate oxidase, are extremely sensitive to denaturation at room temperature and hence should not be prepared until ready to use. Make sure working buffer enzymes are stored at –20°C.

5. Conclusions

Several factors must be weighed when formulating the "optimal" PNC. Properties such as yield, enzyme loading, and PNC sizing can be tuned through process manipulation. Yet these process manipulations (e.g., energy input, surfactant concentration) can have complex and competing effects in terms of sizing and enzyme activity. Along with the notion that the definition of optimal is highly dependent upon the therapy being pursued, it is not possible to know a priori what the ideal compromise in PNC properties is. Coupled with the capacity to vary biodistribution through the non-aggregating coupling of antibodies to the carrier surface, in vivo studies must always be performed as a feedback mechanism to tune the synthesis of carrier for each specific therapeutic goal. Only through this continual interaction between the engineering and the biology can PNC delivery strategies result in clinically relevant solutions.

References

1. Harker, L.A., Hanson, S.R., and Kelly, A.B. (1997) *Thromb. Haemost.* **78**, 736–741.
2. Wieland, H.A., Laux, V., Kozian, D., and Lorenz, M. (2003) *Curr. Opin. Investig. Drugs* **4**, 264–271.
3. Barnes, P.J. (2003) *Cytokine Growth Factor Rev.* **14**, 511–522.
4. Burke, F. (1999) *Cytokines Cell Mol. Ther.* **5**, 51–61.
5. Younes, H.M. and Amsden, B.G. (2002) *J. Pharm. Sci.* **91**, 2–17.
6. Bremer, U., Horres, C.R., and Francoeur, M.L. (1997) *Pharm. Biotechnol.* **10**, 239–254.
7. Chen, R.R. and Mooney, D.J. (2003) *Pharm. Res.* **20**, 1103–1112.
8. Peppas, N.A., Wood, K.M., and Blanchette, J.O. (2004) *Expert. Opin. Biol. Ther.* **4**, 881–887.
9. Rosier, R.N., O'Keefe, R.J., and Hicks, D.G. (1998) *Clin. Orthop. Relat. Res.*, S294–S300.
10. Barnes, P.J. and Hansel, T.T. (2004) *Lancet* **364**, 985–996.
11. Rosenblum, J.S. and Kozarich, J.W. (2003) *Curr. Opin. Chem. Biol.* **7**, 496–504.
12. Schimmoller, F., Higaki, J.N., and Cordell, B. (2002) *Curr. Pharm. Des.* **8**, 2521–2531.
13. Layer, P., Keller, J., and Lankisch, P.G. (2001) *Curr. Gastroenterol. Rep.* **3**, 101–108.
14. Meikle, P.J. and Hopwood, J.J. (2003) *Eur. J. Pediatr.* **162**(Suppl 1), S34–S37.
15. Mignani, R. and Cagnoli, L. (2004) *J. Nephrol.* **17**, 354–363.
16. Thorpe, P.E. (2004) *Clin. Cancer. Res.* **10**, 415–427.
17. Foster, C.A. (1996) *J. Allergy. Clin. Immunol.* **98**, S270–S277.
18. Muzykantov, V.R. (2001) *J. Control. Release* **71**, 1–21.
19. Reichert, J.M. (2006) *Trends Biotechnol.* **24**, 293–298.
20. Reichert, J.M. (2004) *Regul. Aff. J* **15**, 491–497.
21. Rao, K.P. (1995) *J. Biomater. Sci. Polym. Ed.* **7**, 623–645.
22. Moghimi, S.M. and Szebeni, J. (2003) *Prog. Lipid. Res.* **42**, 463–478.
23. Maeda, H., Wu, J., Sawa, T., Matsumura, Y., and Hori, K. (2000) *J. Control Release* **65**, 271–284.
24. Torchilin, V.P. (2000) *Eur. J. Pharm. Sci.* **11** (Suppl **2**), S81–S91.
25. Gref, R., Minamitake, Y., Peracchia, M.T., Trubetskoy, V., Torchilin, V., and Langer, R. (1994) *Science* **263**, 1600–1603.
26. Moghimi, S.M., Hunter, A.C., and Murray, J.C. (2001) *Pharmacol. Rev.* **53**, 283–318.
27. Muro, S., Muzykantov, V.R., and Murciano, J.C. (2004) *Methods Mol. Biol.* **283**, 21–36.
28. Dziubla, T.D., Karim, A., and Muzykantov, V.R. (2005) *J. Control Release* **102**, 427–439.
29. Simone, E.A., Dziubla, T.D., Colon-Gonzalez, F., Discher, D.E., and Muzykantov, V.R. (2007) *Biomacromolecules* **8**, 3914–3921.
30. Shuvaev, V.V., Dziubla, T., Wiewrodt, R., and Muzykantov, V.R. (2004) *Methods Mol. Biol.* **283**, 3–19.
31. Sims, G.E. and Snape, T.J. (1980) *Anal. Biochem.* **107**, 60–63.
32. Muro, S., Dziubla, T., Qiu, W., Leferovich, J., Cui, X., Berk, E., and Muzykantov, V.R. (2006) *J. Pharmacol. Exp. Ther.* **317**, 1161–1169.
33. Muro, S., Cui, X., Gajewski, C., Murciano, J.C., Muzykantov, V.R., and Koval, M. (2003) *Am. J. Physiol. Cell. Physiol.* **285**, C1339–C1347.
34. Discher, D.E. and Eisenberg, A. (2002) *Science* **297**, 967–973.
35. Zhang, L. and Eisenberg, A. (1995) *Science* **268**, 1728–1731.

Chapter 10

Assessment of Antioxidant Activity of Eugenol In Vitro and In Vivo

Enika Nagababu, Joseph M. Rifkind, Sesikeran Boindala, and Lakshmaiah Nakka

Abstract

Reactive oxygen species are implicated in many human diseases and aging process. Much of the evidence is based on experimental data indicating increasing rates of lipid peroxidation in disease states and the ameliorating effects of antioxidants. It is becoming increasingly evident that the natural antioxidants, which have phenolic structure, play an important role in protecting the tissues against free radical damage. Eugenol (4-allyl-2 methoxyphenol) is one such naturally occurring phenolic compound. The antioxidant activity of eugenol was evaluated by the extent of protection offered against free radical-mediated lipid peroxidation using both in vitro and in vivo studies. The in vitro lipid peroxidation was induced in mitochondria by (Fe(II)-ascorbate) or (Fe(II) + H_2O_2). The lipid peroxidation was assessed colorimetrically by measuring the formation of thiobarbituric acid reactive substances (TBARS) following the reaction of oxidized lipids with TBA. Eugenol completely inhibited both iron and Fenton reagent-mediated lipid peroxidation. The inhibitory activity of eugenol was about fivefold higher than that observed for α-tocopherol and about tenfold less than that observed for BHT. The in vivo lipid peroxidation-mediated liver damage was induced by administration of CCl_4 to rats. Eugenol significantly inhibited the rise in SGOT activity and cell necrosis without protecting the endoplasmic reticulum (ER) damage as assessed by its failure to prevent a decrease in cytochrome p450 and G-6-phosphatase activity. The protective action of eugenol has been found to be due to interception of secondary radicals derived from ER lipids rather than interfering with primary radicals of CCl_4($^\bullet CCl_3/CCl_3OO^\bullet$).

Key words: Reactive oxygen species, antioxidants, eugenol, lipid peroxidation, TBARS, carbon tetrachloride hepatotoxicity.

1. Introduction

Reactive oxygen species (ROS) are continuously generated in aerobic cells as by-products of oxidative metabolism (1). ROS at low levels have been shown to play an important role in acting as

R.M. Uppu et al. (eds.), *Free Radicals and Antioxidant Protocols*, Methods in Molecular Biology 610,
DOI 10.1007/978-1-60327-029-8_10, © Humana Press, a part of Springer Science+Business Media, LLC 1998, 2010

secondary messengers for signal transduction pathways associated with cell growth and differentiation (2). Increased production of ROS above the physiological levels can damage lipids, proteins, and DNA leading to cellular injury (3). However, cells possess enzymatic and non-enzymatic antioxidant defense systems to protect against such damage. Oxidative stress is caused by excessive production of oxidants surpassing the antioxidant capacity of the cells (4). Oxidative stress has been shown to be involved in many pathological and age-associated diseases (3, 5). The toxicity of a wide range of chemicals and drugs is also associated with oxidative stress (6). The contentious issue of research in this area is to find whether exogenously given antioxidants elicit beneficial responses against ROS-mediated injury. Several studies have suggested that a diet rich in antioxidants either delays or protects against age-associated diseases (7–9).

Compounds, that have phenolic groups, have long been known to exhibit antioxidant properties, and by quenching free radical species through loss of a hydrogen atom. During this process the antioxidant phenolic group is converted into a radical species. Usually this phenoxyl radical species is more stable due to electron delocalization within adjacent benzene rings or dimerization. The phenoxyl radical is less harmful and does not react with many substrate molecules. The antioxidant property of the phenolic group is dependent on the ease with which it can lose electron(s) to scavenge a radical.

$$ROH + {}^{\bullet}OH \rightarrow RO^{\bullet} + H_2O$$

(${}^{\bullet}OH$ = hydroxyl radical, ROH = phenolic antioxidant). Several studies suggest that natural and synthetic antioxidants protect against ROS-mediated cellular damage (8). The synthetic antioxidants such as 3-*tert*-butyl-4-hydroxyanisole (BHA), 3,5-di-*tert*-butyl-4-hydroxytoluene (BHT), and propyl gallate (PG) are extensively used in food processing industries as preservatives (9). The toxicity concerns of some synthetic antioxidants have therefore shifted the focus on the identification of naturally occurring antioxidants (10). By and large, the natural antioxidants have phenolic structure, and are found in a variety of plants. Examples of common plant phenolic antioxidants include flavonoids, cinnamic acid derivatives, coumarins, tocopherols, and polyfunctional organic compounds (11).

Spices are used in food preparations not only as flavoring and aromatic agents but also for a multitude of the medicinal values. The active principles, which are mostly phenolic compounds, are responsible for the flavor, aroma, pungency, and the medicinal value of whole spices (12). Eugenol is one of the several phenolic compounds and is found in various plants. It is the major component of clove oil, known for its aroma and medicinal values. The structure of eugenol is depicted in **Fig. 10.1**.

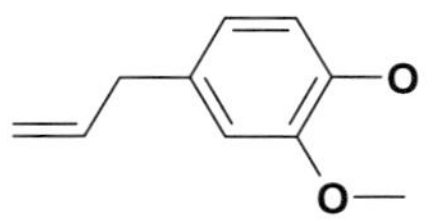

4-Allyl-2-methoxy-phenol

$\mathbf{C_{10}H_{12}O_2}$

Mr 64. 2 Datons

Fig. 10.1. Chemical structure of eugenol. Eugenol is a naturally occurring phenolic compound found in many plant products.

The antioxidant activity of eugenol is evaluated using both in vitro and in vivo methods. In view of the higher susceptibility of polyunsaturated fatty acids to free radical-mediated damage, mitochondrial preparations, which contain abundant amounts of polyunsaturated fatty acids and serve as source of ROS, are used as substrates for in vitro studies. The lipid peroxidation is induced in mitochondria by (Fe(II)-ascorbate) or (Fe(II) + H_2O_2) and the protective effect of eugenol against this lipid peroxidation is investigated.

The inhibition of lipid peroxidation in vitro by an antioxidant may not necessarily imply that the same antioxidant would protect against free radical-mediated injury in vivo as well. In order to elicit protective responses from free radical-mediated damage in vivo, the antioxidant has to gain access to be present at the loci in the cell where the free radicals react. One of the frequently used methods to induce oxidative stress in animals is to expose them to carbon tetrachloride (CCl_4) or acetaminophen. This model has been used for many years to investigate the effect of antioxidants on free radical-mediated hepatotoxicity (13, 14). Therefore, oxidative stress is induced in rats by administering CCl_4 in our studies to evaluate the antioxidant activity of eugenol in *vivo*.

2. Materials

2.1. In Vitro *Lipid Peroxidation*

2.1.1. Fe(II)-Ascorbate or Fe(II)-H_2O_2-Induced Peroxidation

1. *Isolation of Mitochondria.* Sacrifice adult SD rats, fasted overnight, by decapitation (*see* **Notes 1–3**). Excise the livers and place them immediately in ice-cold 0.15 M KCl solution. Homogenize 10 g of minced liver tissue in 30 mL of ice-cold KCl solution using a Potter–Elvehjem homogenizer (*see* **Notes 4** and **5**). Centrifuge the homogenate at 800×*g* for 10 min at 2–4°C to sediment cell debris and nuclei. Separate the supernatant and centrifuge again at 14,000×*g* for 10 min at 2–4°C. Resuspend the mitochondrial fraction (the pellet) in either 0.15 M KCl or 0.15 M

NaCl and store in refrigerator (*see* **Note 6**). Estimate the protein content of the mitochondrial preparation by the method of Lowry et al. (15) and dilute the preparation with ice-cold 0.15 M KCl or 0.15 M NaCl to a final protein concentration of 1 mg mL^{-1}.

2. Tris–HCl buffer, 0.025 M, pH 7.4, containing 0.125 M KCl (Tris–HCl–KCl buffer): dissolve 18.635 g of KCl in 1 L of deionized distilled (dd) water. Dissolve 3.03 g of Tris-base in 500 mL of KCl solution, adjust the pH to 7.4 with 1 M HCl, and make up the volume to 1 L with ddH_2O. This reagent is stable for 3 months at 4°C.
3. Ethanol (70%, v/v): mix 70 mL absolute ethanol with 30 mL water.
4. Eugenol stock solution (40 mM): dilute 154 μL (164.2 mg) of eugenol (Sigma, cat No. E5504) in 25 mL 70% (v/v) ethanol (*see* **Note 7**). Keep in a dark bottle at 4°C. The stock solution is prepared once a week.
5. Working standards of eugenol (2–10 mM): dilute 0.05, 0.1, 0.15, 0.2, or 0.25 mL of 40 mM stock eugenol solution to 1 mL final volume with 70% (v/v) alcohol to give concentrations of 2, 4, 6, 8, and 10 mM, respectively.
6. Stock solution of MDA (4.0 mM): mix 9.6 μL of 1,1,3, 3-tetraethoxypropane (Sigma, cat. No. T9889) with 9.990 mL of 0.1 N HCl and keep at 100°C for 5 min. The stock solution can be stored for a maximum of 1 week at 4°C.
7. Working MDA standard: dilute 100 μL of stock solution to 10 mL (100-fold) in Tris–HCl–KCl buffer, pH 7.4, to obtain 40 μM or 40 nmol mL^{-1} MDA. Dilute this working standard to 1:2, 1:4, 1:8, 1:16, and 1:32 with Tris–HCl–KCl buffer to give 20, 10, 5, 2.5, and 1.25 μM or 1.25 nmol mL^{-1}, respectively. Working standards should be prepared fresh every day.
8. α-Tocopherol (50 mM): dissolve small aliquots of α-tocopherol (Sigma, cat. No: 47783) in 10 mL of ethanol and measure the concentration based on absorbance at 292 nm ($\varepsilon = 3.26 \times 10^3\ M^{-1}\ cm^{-1}$) after appropriate dilutions (*see* **Note 7**). Store in a brown bottle in a refrigerator. The concentration of stock α-tocopherol is diluted to 50 mM with 70% (v/v) ethanol. Dilute 0.2, 0.4, 0.6, 0.8, and 1 mL of this solution to a final volume of 1 mL 70% alcohol to obtain 10, 20, 30, 40, and 50 mM working α-tocopherol solutions, respectively.
9. Butylated hydroxytoluene (BHT; 25 mM stock): dissolve 139 mg BHT (Aldrich, cat. No.W218405) in 25 mL of 70% (v/v) ethanol. This reagent can be stored in cold for up to 3 months. Dilute 8, 16, 24, 32, and 40 μL of stock to

1 mL 70% alcohol to obtain 0.2, 0.4, 0.6, 0.8, and 1 mM working BHT solutions, respectively (*see* **Note 7**).

10. Ferrous sulfate (1 mM): dissolve 13.90 mg in 50 mL of ddH_2O. Prepare this solution just before use (*see* **Note 8**).
11. Ascorbic acid (10 mM): dissolve 17.61 mg in 10 mL of ddH_2O. Prepare this solution on the day of experiment.
12. Trichloroacetic acid, 20% (w/v): dissolve 20 g of TCA to 100 mL of ddH_2O.
13. Thiobarbituric acid, 0.67% (w/v): dissolve 670 mg of TBA (Sigma, cat. No.5500) in 80 mL of ddH_2O and heat at 50°C for 1 h in a water bath. Make up the volume to 100 mL with ddH_2O.
14. NaCl (0.15 M): dissolve 8.766 g of NaCl in 1 L of ddH_2O.
15. Ferrous sulfate (8 mM): dissolve 22.24 mg in 10 mL ddH_2O (*see* **Note 8**).
16. *Hydrogen peroxide* (10 mM): dilute 56.6 μL of stock H_2O_2 (Sigma, cat. No. H1009) to 50 mL with ddH_2O. Calculate the concentration of H_2O_2 based on absorbance at 240 nm ($\varepsilon_{240\ nm} = 40\ M^{-1}\ cm^{-1}$).

2.2. *In Vivo* Lipid Peroxidation

1. Rats from Wistar/National Institute of Nutrition weighing 225–275 g.
2. KCl (0.15 M): dissolve 1.15 g of KCl in 100 mL of ddH_2O.
3. Starch [5% (w/v)]: mix 5 g of soluble starch in 80 mL of ddH_2O and warm until it forms slurry and make up the volume to 100 mL.
4. *Cytochrome P450 (CYP450)*
   (a). Sodium dithionite.
   (b). Carbon monoxide gas: CO gas from commercial sources.
   (c). Phosphate buffer (PB), 0.1 M, pH 7.4: dissolve 14.186 g of Na_2HPO_4 in 1 L of ddH_2O. Similarly, prepare 0.1 M NaH_2PO_4 by dissolving 12 g in 1 L of ddH_2O. Adjust the pH of Na_2HPO_4 to 7.4 with NaH_2PO_4.
5. *Serum glutamate–oxaloacetate transaminase (SGOT)*
   (a). L-Aspartate (0.2 mM): dissolve 3.424 g aspartate (Sigma, cat.No. A6558) in 80 mL of PB 0.1 M, pH 7.4, and adjust the pH with 1 M NaOH and make up the volume to 100 mL with 0.1 M PB, pH 7.4. Dispense the final solution in aliquots and store frozen; thaw an aliquot on the day of use.
   (b). α-Ketoglutarate (0.1 mM): dissolve 1.681 g of α-ketoglutarate sodium (Sigma, cat.No.K1875) in

80 mL of PB 0.1 M, pH 7.4. Adjust the pH with 1 M NaOH and make up the volume to 100 mL with 0.1 M PB, pH 7.4. Dispense in aliquots and store frozen; thaw an aliquot on the day of use for dilution and use.

(c). NADH (1 mM): dissolve 7.4 mg NADH dipotassium salt (Sigma, cat. No. N 4505) in 10 mL of 0.1 M PB, pH 7.4. Prepare fresh before use and keep in dark (photo sensitive).

(d). Malic dehydrogenase (MDH) (Sigma, cat. No. M 9004): keep refrigerated until use.

6. *Assay of glucose 6-phosphatase (G6Pase)*

   (a). Glucose-6-phosphate (G-6P): suspend 130 mg G-6P (Sigma, cat. No. G7879) and 130 mg $BaCl_2$ in 2 mL ddH_2O. Add a minimum amount of 1 N HCl to dissolve the suspension. Then add 72 mg of Na_2SO_4 (anhydrous), mix well, and remove $BaSO_4$ (precipitation) by centrifugation. Test the supernatant for the completion of precipitation by the addition of a very small amount of Na_2SO_4. Adjust the pH of the supernatant to 6.5 with 0.5 N NaOH and then make up the volume to 5 mL.

   (b). Maleate buffer (100 mM): dissolve 1.1607 g of maleic acid in 70 mL of ddH_2O. Adjust the pH to 6.5 with NaOH and then make up the volume to 100 mL.

   (c). TCA (10% (w/v)): dissolve 10 g in 100 mL of ddH_2O.

7. Assay of TBARS

   (a). TCA, 20% (w/v)

   (b). TBA, 0.67% (w/v)

   (c). *n*-Butanol

3. Methods

Thiobarbituric acid (TBA) test is a commonly used method to determine lipid peroxidation. This method is based on the reaction of TBA with MDA, one of the aldehyde products of lipid peroxidation. The sample is heated with TBA under acidic conditions; MDA forms an adduct with TBA and produces a pink-colored product, which is measured spectrophotometrically at 532 nm (16). There is considerable controversy regarding the specificity of TBA for MDA because several other substances in tissues and body fluids can also react nonspecifically with TBA producing a chromogen with absorbance between 530 and 535 nm. However, MDA has been shown to be a predominant product when cellular organelles are subjected to peroxidation in vitro. Hence, TBA assay has been used for studies of lipid peroxidation in vitro and is extensively used.

3.1. In Vitro *Lipid Peroxidation*

3.1.1. Fe(II)/Ascorbate-Induced Peroxidation

Lipid peroxidation is induced in mitochondria using ferrous sulfate and ascorbate (17). Ferrous iron is known to initiate lipid peroxidation, and ascorbate enhances the ferrous ion-mediated peroxidation by its ability to maintain the iron in the reduced state. The exact mechanism of iron-mediated lipid oxidation is not known. It is believed that iron perferryl ion (Fe(II)-$O^{\bullet}_2$), hydroxyl radical ($^{\bullet}OH$), maintenance of specific Fe(II):Fe(III) ratio, or combination of all these radical species initiates the peroxidation.

1. Incubate mitochondrial suspension (1 mL mg^{-1} protein) with 200 μM ascorbate (10 mM, 20 μL) and 10 μM ferrous sulfate (1 mM, 10 μL) in a total volume of 2 mL KCl–Tris–HCl buffer pH 7.4 at 37°C for 60 min.
2. Add test compounds, eugenol (2–10 mM, 10 μL) or α-tocopherol (10–50 mM, 10 μL) or BHT (0.2–1 mM, 10 μL) 2 min prior to addition of peroxidation inducers (Fe(II)-ascorbate) and mix well (*see* **Notes 7** and **9**).
3. Add only alcohol (70% w/v, 10 μL) to control (Fe(II)-ASC) and blank (mitochondria only) samples.
4. Terminate the peroxidation reaction by adding 1 mL of 20% TCA (w/v).
5. Add 50 μL BHT (0.2% w/v) and 2 mL thiobarbituric acid reagent (0.67% w/v).
6. Heat the tubes for 10 min in a boiling water bath and chill the tubes on ice (*see* **Notes 10** and **11**).
7. Centrifuge the tubes at 3,000 rpm for 15 min and separate the supernatant.
8. Measure the absorbance of supernatant at 532 nm using a spectrophotometer.
9. Subtract the blank value from control and sample values.
10. MDA calibration curve:
    (a). Mix MDA working standards (2 mL, 1.25–20 μM) with 1 mL of 20% TCA (w/v).
    (b). Add 2 mL thiobarbituric acid (0.67% w/v).
    (c). Heat the tubes for 10 min in a boiling water bath.
    (d). Measure the absorbance at 532 nm.
    (e). The calibration curve is constructed (*see* **Fig. 10.2** and **Note 13**).

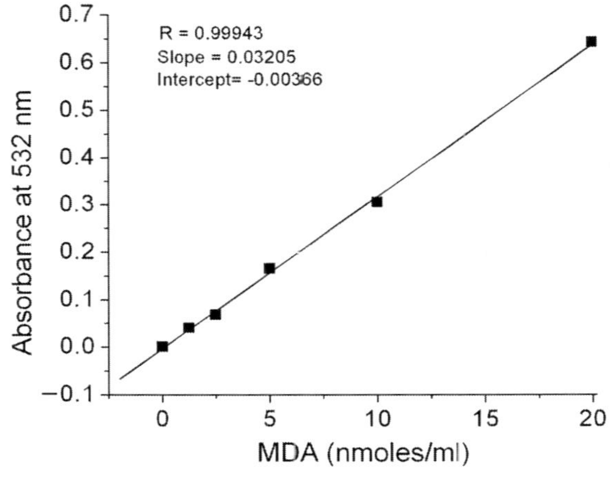

Fig. 10.2. Malondialdehyde calibration curve. The preparation of MDA, the contents of reaction mixture, and incubation conditions are described in Materials (**Section 2.1.1**, **Steps** 6 and 7) and Methods (**Section 3.1.1**).

Table 10.1
IC_{50} values for the free radical scavengers in lipid peroxidation system

	IC_{50} (μM)	
Inhibitor	**Fe(II)–ASC system**	**Fe(II)–H_2O_2 system**
Eugenol	10.0	14.0
α-Tocopherol	54.0	68.2
BHT	1.15	1.25

IC_{50}, the amount required for 50% inhibition of peroxidation; ASC, ascorbic acid.

(f). Calculate TBA reactive substances in samples using MDA calibration curve.

(g). MDA (nmol mg^{-1} protein) = (absorbance − intercept)/(slope)
Eugenol inhibited the Fe(II)-ascorbate-induced lipid peroxidation in a concentration-dependent manner (*see* **Fig. 10.3**). The amount of eugenol, tocopherol, and BHT required for inhibition of 50% lipid peroxidation (IC_{50}) is shown in **Table 10.1**.

3.1.2. Fenton Reagent (Fe(II)+H_2O_2)-Induced Lipid Peroxidation

Ferrous iron in presence of H_2O_2 is known to induce lipid peroxidation (18). This reaction is conducted in 0.15 M NaCl, pH 7.0, since some buffering agents are known to scavenge hydroxyl radicals.

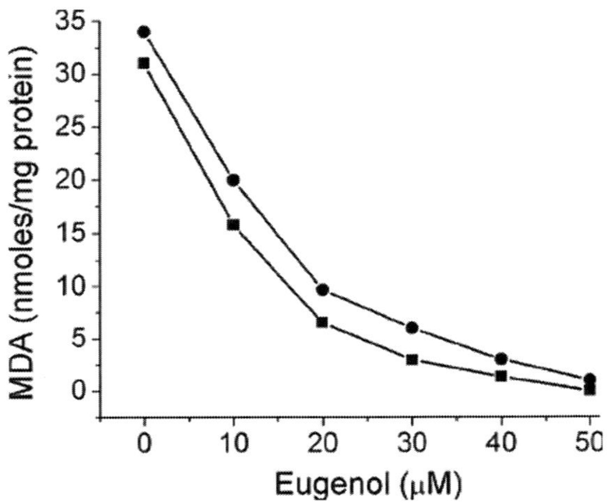

Fig. 10.3. Inhibitory effect of eugenol on (■) Fe(II)–ascorbate and (•) Fe(II)–H_2O_2-induced lipid peroxidation in liver mitochondria. The reaction mixture (2 mL) contained mitochondria (1 mg protein), 10 μM $FeSO_4$, 200 μM ascorbate and Tris–HCl buffer, pH 7.4, for the Fe(II)-ascorbate system, or 100 μM H_2O_2, 200 μM $FeSO_4$, and 0.15 M NaCl, pH 7.0, for the Fe(II)–H_2O_2 system. Indicated concentration of eugenol was added 2 min prior to addition of peroxidation inducers. The reaction mixture was incubated at 37°C for 60 min.

1. Incubate mitochondrial suspension (1 mL mg^{-1} protein) with 100 μM H_2O_2 (10 mM, 20 μL) and 200 μM ferrous sulfate (8 mM, 50 μL) in a total volume of 2 mL NaCl buffer, pH 7.4, at 37°C for 60 min.
2. Add eugenol (2–10 mM, 10 μL) or α-tocopherol (10–50 mM, 10 μL) or BHT (0.2–1 mM, 10 μL) to mitochondria 2 min prior to addition of peroxidation inducers (H_2O_2–Fe(II)) and mix well.
3. Follow samples treatment, measurement of absorbance, and calculation of MDA values as mentioned in **Section 3.1.1**.

Eugenol inhibited the Fe(II)–H_2O_2-induced lipid peroxidation in a concentration-dependent manner (*see* **Fig. 10.3**). The amount of eugenol, tocopherol, and BHT required for inhibition of 50% lipid peroxidation (IC_{50}) is shown in **Table 10.1**.

3.2. *In Vivo Lipid Peroxidation*

$$CCl_4 + CYP450 \longrightarrow {}^{\bullet}CCl_3 + Cl^-$$

$${}^{\bullet}CCl_3 + O_2 \longrightarrow CCl_3OO^{\bullet}$$

CCl_4 is metabolically activated to ${}^{\bullet}CCl_3$ radical by liver CYP450. These ${}^{\bullet}CCl_3$ and $CCl_3OO^{\bullet}$ radicals initiate lipid peroxidation in the liver cells leading to hepatotoxicity.

3.2.1. Experimental Animals and Treatment

1. Divide rats into five groups, each group containing a minimum of six rats.
2. Prepare varying doses (0.2, 1.0, 5.0, and 25 mg kg^{-1} body wt) of eugenol in 5% (w/v) soluble starch and administer to rats by oral gavage such that the administered volume would correspond to 2.5 mL kg^{-1} body wt.
3. Give each dose of eugenol thrice at different time intervals. These are prior to (–1 h), along with (0 h), and after (+3 h) of administration of CCl_4.
4. Suspend CCl_4 (0.4 mL kg^{-1} body wt) in peanut oil and administer by i.p. injection such that the administered volume would correspond to 2.5 mL kg^{-1} body wt.
5. Collect blood after 24 h from the orbital sinus.
6. Sacrifice rats by decapitation after 48 h of dosing with CCl_4.
7. Collect blood and liver tissue.
8. Separate the plasma and store at –80°C until use.
9. *Preparation of liver microsomes.* Centrifuge mitochondrial supernatant (*see* **Section 2.1.1**) that consists of microsomes at 105,000×*g* for 60 min at 2–4°C. Resuspend the pellet (microsomes) in 0.15 M KCl. Estimate the protein content of the microsomal fraction by the method of Lowry (mentioned above). Store microsomes at –85°C and complete all the analyses within a week.

3.2.2. Determination of TBARS in Liver (see **Note 5***)*

1. Place 2 g of liver in 5 mL of cold 20% (w/v) TCA.
2. Mince and homogenize using a Polytron homogenizer.
3. Centrifuge the homogenate at 3,000 rpm for 10 min.
4. Incubate 2 mL of the supernatant with 2 mL of 0.67% (w/v) TBA for 15 min at 100°C.
5. At the end of the incubation, chill the tubes on ice.
6. Extract the colored complex into 2 mL of *n*-butanol, read the absorbance of the organic layer immediately at 532 nm, and calculate the concentration of MDA ($\varepsilon = 1.56 \times 10^5$ M^{-1} cm^{-1}) (**Table 10.2**).

3.2.3. Assay of SGOT

1. Mix 20 μL clear non-hemolysed serum with 680 μL of 0.1 M PB, pH 7.4 (19).
2. Add 100 μL of 0.24 mM aspartate, 100 μL of 1 mM NADH, and 50 μL MDH. Mix the contents and leave the tubes at room temperature for 15 min.
3. Start the enzymatic reaction by addition of 50 μL of α-ketoglutarate and incubate for 5 min at 25°C. Read the

Table 10.2
Effect of eugenol treatment on serum GOT, microsomal CYP450, G6Pase, TBARS, and hepatic necrosis in rats administered CCl_4

Treatment	Serum GOT (units mL^{-1})[a]	CYP450 (nmol mg^{-1} protein)[b]	G6Pase (μmol Pi released per h mg^{-1} protein)[b]	TBARS (nmol g^{-1} per protein)	Hepatic necrosis (score)[b]
Control	18 ± 4	0.74 ± 0.10	10.5 ± 0.57	4.86 ± 0.25	0.0
CCl_4 (0.4 mL kg^{-1})	380 ± 45	0.38 ± 0.04	2.6 ± 0.41	9.2 ± 0.47	2.0 ± 0.0
CCl_4 (0.4 mL kg^{-1}) + eugenol (0.2 mg kg^{-1})	225 ± 35*[c]	0.36 ± 0.03	2.8 ± 0.24	8.4 ± 0.46	2.0 ± 0.0
CCl_4 (0.4 mL kg^{-1}) + eugenol (1.0 mg kg^{-1})	160 ± 20**[c]	0.41 ± 0.04	2.4 ± 0.11	7.3±0.36*[c]	1.5 ± 0.28*[c]
CCl_4 (0.4 mL kg^{-1}) + eugenol (5.0 mg kg^{-1})	150 ±21***[c]	0.34 ± 0.028	2.5 ± 0.22	6.6 ± 0.44*[c]	1.25 ± 0.25*[c]
CCl_4 (0.4 mL kg^{-1}) + eugenol (25 mg kg^{-1})	220 ± 27*[c]	0.45 ± 0.036	2.2 ± 0.10	6.8 ± 0.31*[c]	2.0 ± 0.25

[a]Serum collected after 24 h of administration of CCl_4 and assayed for GOT activity.
[b]Rats were killed after 48 h of administration of CCl_4. Microsomal CYP450, G6Pase, and TBARS were assayed in the liver samples as described in the text. Results represent mean ± SE for 4–6 rats.
[c]Significantly different from CCl_4 treatment at $^{*}p < 0.05$, $^{**}p < 0.01$, and $^{***}p < 0.001$.

change in absorbance at 340 nm against the blank that did not contain α-ketoglutarate.

4. Defining that a change in absorbance of 0.001 min^{-1} as one unit, calculate the SGOT activity as units per milliliter of serum (**Table 10.2**).

3.2.4. Determination of CYP450 Content

1. Take 1 mL of microsomal suspension containing 1 mg protein in 0.1 M PB, pH 7.0 (20).
2. Treat the sample with few grains of dithionite and take in the reference and sample cuvettes.
3. Correct the baseline between 400 and 500 nm using a spectrophotometer.
4. Transfer the contents of the cuvette into test tube and gently bubble CO gas for 1 min in a fume hood.
5. Transfer the contents back to sample cuvette and read the absorbance between 400 and 500 nm.

6. Calculate the CYP450 concentration based on the difference in absorbance at 450 and 490 nm of reduced P450 with CO (as CO complex) and without CO using extinction coefficient of 91 mM^{-1} cm^{-1} (**Table 10.2**).

3.2.5. G6Pase

G6Pase activity (21) can be determined based on the release of inorganic phosphate according to Chen et al. (22).

1. Mix 50 μL microsomes from the stock of 2 mg mL^{-1} with 850 μL 0.05 M maleic acid buffer, pH 6.5.
2. Add 100 μL substrate solution and mix well.
3. Incubate the mixture at 37°C for 15 min.
4. Terminate the reaction by adding 1 mL of 10% (w/v) TCA and chill on ice.
5. Centrifuge the sample at 3,000 rpm for 15 min.
6. An aliquot of the supernatant can be used for determination of inorganic phosphate (**Table 10.2**).

3.2.6. Histopathological Examination of Liver

Fix small portions of liver sample (ca. 20 mm length × 15 mm width × 2–3 mm thickness) isolated from the middle lobe in 10% neutral formalin. Process the tissue in an automated tissue processor in ascending grades of isopropanol (70, 80, and 100%) for a total duration of 6–7 h at the rate of 1 h in each station, followed by 1 h each of three changes of chloroform for clearing, and finally 3 h each of two changes of paraffin infiltration (melting point of paraffin 58–60°C), the second change being under vacuum. Prepare paraffin blocks of the tissue and make sections (thickness 6 μm) in a rotary manual or automated microtome using disposable knife blades. Stain the paraffin sections with Meyer's; hematoxylin–eosin (AFIP methods) and examine under ×10 objective with a final magnification of ×125. Grade hepatocellular necrosis, if any, as follows: no necrosis = 0; necrosis around centrilobular vein = 1.0; necrosis or fatty changes involving 1/3 of the lobule = 2.0; necrosis of more than 1/3 of lobule = 3.0 (*see* **Fig. 10.4**) and (**Table 10.2**).

3.3. Results and Conclusions

An antioxidant is defined as any substance that, when present at low concentrations compared with those of an oxdizable substrate, significantly delays or prevents oxidation of that substrate (23). The oxidation of substrates is prevented by either directly scavenging ROS or indirectly by inhibiting the production of ROS and promoting the antioxidant defense system. This chapter describes the evaluation of the antioxidant activity of eugenol by testing the free radical scavenging activity and/or inhibition of radical production. Since eugenol is a lipophilic compound, the scavenging ability of chain propagating radicals (alkoxyl and peroxyl), which form during lipid peroxidation, is mainly

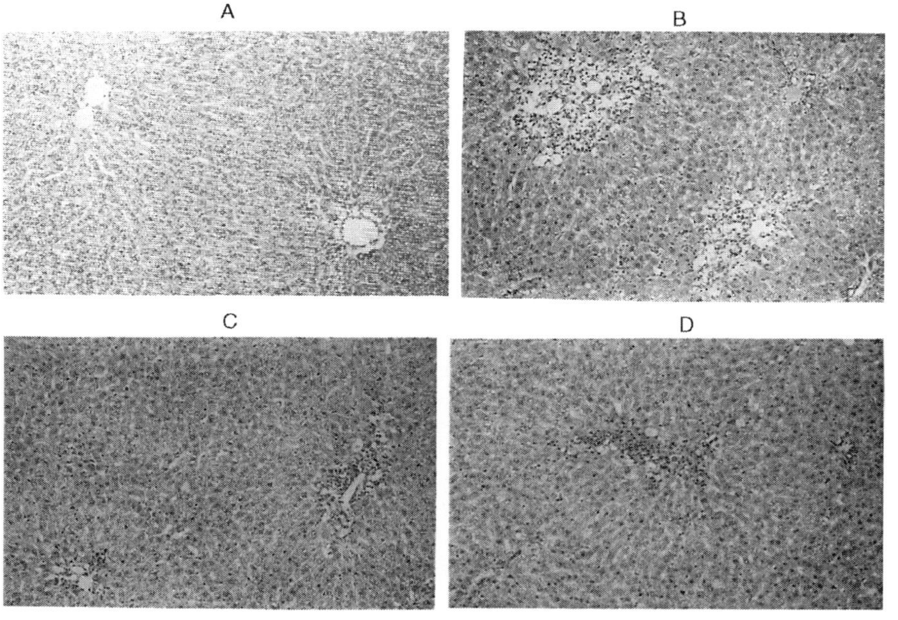

Fig. 10.4. Effect of eugenol administration on CCl_4-induced liver necrosis. (**a**) Photograph of liver from control rat, which received only vehicles (starch and peanut oil) exhibited normal liver architecture. **(b)** Histology of liver from CCl_4-treated rat indicating centrizonal necrosis (necrosis score 2.0.) (**c**) Liver necrosis of rat treated with eugenol (1 mg kg^{-1} body wt) + CCl_4 (necrosis score 1.5). (**d**) Liver necrosis showing moderate centrizonal necrosis from a rat, which received eugenol (5 mg kg^{-1} body wt) + CCl_4 (necrosis score 1.25).

investigated. Mitochondria are the major sources for free radical generation, rich in polyunsaturated fatty acids, and highly susceptible to lipid peroxidation. Therefore, liver mitochondria are used as substrate for lipid peroxidation studies.

The results show that eugenol inhibits iron and •OH radical initiated lipid peroxidation with IC_{50} values of 10 and 14 μM, respectively. The inhibitory activity of eugenol is five times more than α-tocopherol and 10 times less than BHT (**Table 10.1**). Eugenol incorporates into mitochondrial membrane and microsomal membrane inhibits lipid peroxidation by acting as chain-breaking agent (24, 25) (*see* **Fig. 10.3**).

Eugenol (0.2, 1.0, 5.0, or 25 mg kg^{-1} body wt) when given orally at three different times in relation to the time of CCl_4 dosing (i.p administration of 0.4 mg kg^{-1} body wt), i.e., prior to (−1 h), along with (0 h), or after (+ 3 h), prevented significantly the rise in SGOT activity, lipid peroxidation, as well as liver necrosis (*see* **Fig. 10.4**) (**Table 10.2**). The protective effect is more evident at 1 and 5 mg eugenol doses than that of 0.2 and 25 mg doses. However, the decrease in microsomal G6Pase activity and CYP450 content by CCl_4 treatment is not prevented by eugenol, suggesting that the damage to endoplasmic reticulum is not protected (26) (**Table 10.2**). The protective effect of eugenol against

CCl_4-induced hepatotoxicity is due to interception of secondary radicals derived from oxidized lipids of endoplasmic reticulum rather than interference with generation and reactions of primary radicals ($^{\bullet}CCl_3/CClOO^{\bullet}$).

4. Notes

1. Always use double-distilled water or ultra pure water with resistivity of $\geq$18.2 MΩ cm for preparation of all reagents.
2. Metal contamination is common in water and buffer solutions. Metals like iron and copper, in presence of reducing agents, activate oxygen which oxidizes lipids, proteins, and DNA. Therefore, it is very important to use metal-free water for the preparation of reagents. Metal-free buffers and water can be prepared by passing through Chelex 100 (200–400 mesh, sodium form) resin.
3. Metal contamination in solutions can be tested by adding ascorbic acid to test solutions and monitor the change in absorbance at 265 nm. The loss of absorbance is an indicator of metal contamination. Metals promote autoxidation of ascorbic acid (27).
4. Sucrose could interfere in the estimation of malondialdehyde. Avoid using sucrose media for preparation of mitochondria or microsomes.
5. When preparing mitochondria or microsomes, keep liver on ice and homogenize in ice-cold saline to minimize the oxidation of lipids. To determine TBARS in whole tissue, keep the tissue piece in ice-cold TCA solution and homogenize in same media.
6. Store microsomes and mitochondria in deaerated isotonic solutions in the refrigerator. Use within 72 h. Do not freeze the mitochondria or microsomes when using them as substrates for lipid peroxidation.
7. Phenolic antioxidants, vitamin E, and BHT are soluble in ethanol. Ethanol also scavenges radical species. Care should be taken to ensure that the total ethanol concentration in incubation mixture is less than 0.5%.
8. Prepare ferrous sulfate solution in water just before use. Do not prepare iron solution in buffers. Always use freshly prepared solutions.
9. Add test compounds (antioxidant) 2–5 min prior to addition of peroxidation inducers to substrates and mix it well.

10. Break-down products of lipid hydroperoxides formed during the heating process also produce color similar to MDA. Add 50 μL of 0.2% BHT to minimize these lipid hydroperoxides contribution to MDA color soon after the reaction is completed.
11. Keep marbles on test tubes to minimize the evaporation of water while heating the reaction mixtures for MDA determination. Replace if there is any water loss.
12. Stock hydrogen peroxide is standardized with potassium permanganate method or by molar extinction coefficient. Working standard solutions of lower concentrations should be prepared just before use.
13. In some instances, antioxidants can interfere with the TBA reaction and suppresses the color development. This possibility can be tested by adding a test compound to MDA standard or sample control soon after stopping the reaction with TCA.

Acknowledgments

This research was supported by National Institute of Nutrition, Indian Council of Medical Research, Hyderabad, India, and Intramural Research Program of the NIH, National Institute on Aging, USA.

References

1. Stogner, S.W. and Payne, D.K. (1992) Oxygen toxicity. *Ann. Pharmacother.* **26**, 1554–1562.
2. Hoidal, J.R. (2001) Reactive oxygen species and cell signaling. *Am. J. Respir. Cell Mol. Biol.* **25**, 661–663.
3. Beckman, K.B. and Ames, B.N. (1998) The free radical theory of aging matures. *Physiol. Rev.* **78**, 547–581.
4. McCord, J.M. (1993) Human disease, free radicals, and the oxidant/antioxidant balance. *Clin. Biochem.* **26**, 351–357.
5. Droge, W. (2002) Free radicals in the physiological control of cell function. *Physiol. Rev.* **82**, 47–95.
6. Valko, M., Morris, H., and Cronin, M.T. (2005) Metals, toxicity and oxidative stress. *Curr. Med. Chem.* **12**, 1161–1208.
7. Ames, B.N., Shigenaga, M.K., and Hagen, T.M. (1993) Oxidants, antioxidants, and the degenerative diseases of aging. *Proc. Natl. Acad. Sci. USA* **90**, 7915–7922.
8. Lopez-Velez, M., Martinez-Martinez, F., and Del Valle-Ribes, C. (2003) The study of phenolic compounds as natural antioxidants in wine. *Crit. Rev. Food Sci. Nutr.* **43**, 233–244.
9. Daniel, J.W. (1986) Metabolic aspects of antioxidants and preservatives. *Xenobiotica* **16**, 1073–1078.
10. Shahidi, F. (2000) Antioxidants in food and food antioxidants. *Nahrung* **44**, 158–163.
11. Scalbert, A. and Williamson, G. (2000) Dietary intake and bioavailability of polyphenols. *J. Nutr* **130**, 2073S–2085S.
12. Singh, U.P., Singh, D.P., Maurya, S., Maheshwari, R., Singh, M., Dubey, R.S., and Singh, R.B. (2004) Investigation on the phenolics of some spices having pharmacotherapeuthic properties. *J. Herb. Pharmacother.* **4**, 27–42.

13. Yasuda, H., Izumi, N., Shimada, O., Kobayakawa, T., and Nakanishi, M. (1980) The protective effect of tinoridine against carbon tetrachloride hepatotoxicity. *Toxicol. Appl. Pharmacol.* **52**, 407–413.
14. Gee, D.L., Bechtold, M.M., and Tappel, A.L. (1981) Carbon tetrachloride-induced lipid peroxidation: simultaneous in vivo measurements of pentane and chloroform exhaled by the rat. *Toxicol. Lett.* **8**, 299–306.
15. Lowry, O.H., Rosebrough, N.J., Farr, A.L., and Randall, R.J. (1951) Protein measurement with the Folin phenol reagent. *J. Biol. Chem.* **193**, 265–275.
16. Wilbur, K.M., Bernheim, F., and Shapiro, O.W. (1949) The thiobarbituric acid reagent as a test for the oxidation of unsaturated fatty acids by various agents. *Arch. Biochem.* **24**, 305–313.
17. Fortney, S.R. and Lynn, W.S., Jr.. (1964) Role of ascorbate and cysteine on swelling and lipid peroxidation in rat liver mitochondria. *Arch. Biochem. Biophys.* **104**, 241–247.
18. Minotti, G. and Aust, S.D. (1987) The requirement for iron (III) in the initiation of lipid peroxidation by iron (II) and hydrogen peroxide. *J. Biol. Chem.* **262**, 1098–1104.
19. Karmen, A. (1955) A note on the spectrometric assay of glutamic-oxaloacetic transaminase in human blood serum. *J. Clin. Invest.* **34**, 131–133.
20. Omura, T. and Sato, R. (1964) The Carbon monoxide-binding pigment of liver microsomes. I. Evidence for Its hemoprotein Nature. *J. Biol. Chem.* **239**, 2370–2378.
21. Swanson, M.A. (1955) Glucose 6-phosphatase from liver. Methods Enzymol. 2, 541–543.
22. Chen, P.S., Toribara, T.Y., and Warner, H. (1956) Microdetermination of phosphorus. *Anal. Chem.* **28**, 1756–1758.
23. Halliwell, B. (1995) How to characterize an antioxidant: an update. *Biochem. Soc. Symp.* **61**, 73–101.
24. Nagababu, E. and Lakshmaiah, N. (1992) Inhibitory effect of eugenol on non-enzymatic lipid peroxidation in rat liver mitochondria. *Biochem. Pharmacol.* **43**, 2393–2400.
25. Nagababu, E. and Lakshmaiah, N. (1994) Inhibition of microsomal lipid peroxidation and monooxygenase activities by eugenol. *Free Radic. Res.* **20**, 253–266.
26. Nagababu, E., Sesikeran, B., and Lakshmaiah, N. (1995) The protective effects of eugenol on carbon tetrachloride induced hepatotoxicity in rats. *Free Radic. Res.* **23**, 617–627.
27. Buettner, G.R. (1988) In the absence of catalytic metals ascorbate does not autoxidize at pH 7: Ascorbate as a test for catalytic metals. *J. Biochem. Biophys. Methods* **16**, 27–40.

Section III

Cellular Oxidative Stress

Chapter 11

The Generation of Stable Oxidative Stress-Resistant Phenotypes in Chinese Hamster Fibroblasts Chronically Exposed to Hydrogen Peroxide or Hyperoxia

Douglas R. Spitz and Shannon J. Sullivan

Abstract

With the recognition that chronic exposure to oxidative stress occurs in many disease states and can be deleterious to the human health, great interest has emerged in understanding the mechanisms by which mammalian cells develop resistance to chronic oxidative stress. In order to study the mechanisms of development of resistance to chronic oxidative stress, a model system where Chinese hamster fibroblasts (HA1) are chronically exposed to progressively increasing concentrations of H_2O_2 (50–800 μM) or hyperoxia (80–95% O_2) has been developed. Following >200 d of exposure to H_2O_2 (or 18 months of exposure to hyperoxia), the cells developed stable H_2O_2-resistant (or O_2-resistant) phenotypes that are characterized by increases in total glutathione, antioxidant enzyme activity, heme oxygenase activity, stress protein gene expression, DNA repair pathways, and resistance to a wide variety of other toxic stress known to cause oxidant injury. In addition, these oxidant-resistant cells exhibited amplification of the gene for catalase and constitutively elevated AP-1 DNA binding activity. Further, beyond 240 d genomic instability as evidenced by chromosomal rearrangements and alterations in ploidy was stably maintained upon removal from the chronic oxidative stress conditions. These results demonstrate the capability of mammalian cells to develop stable oxidative stress-resistant phenotypes in response to both exogenous (H_2O_2) as well as endogenous (95% O_2) oxidative stress. The understanding of mechanisms of resistance to oxidative stress and its possible relevance in various disease states are discussed.

Key words: Oxidative stress, hydrogen peroxide, hyperoxia, adaptive response, oxidant injury cellular resistance, mammalian cell phenotype, antioxidant enzymes.

1. Introduction

Chronic exposure to oxidative stress is emerging as an important component in many degenerative diseases associated with aging and exposure to environmental mutagens and carcinogens (1–6). The deleterious effects of chronic oxidative stress can be

R.M. Uppu et al. (eds.), *Free Radicals and Antioxidant Protocols*, Methods in Molecular Biology 610,
DOI 10.1007/978-1-60327-029-8_11,

mitigated by cellular and dietary antioxidants. Further, a wide variety of repair pathways governed by stress-responsive genes have resulted in a great deal of interest in investigations into the mechanisms by which mammalian cells become resistant to oxidative stress. Traditionally, one of the simplest strategies for determining mechanisms for resistance to cytotoxic agents is to carry out stepwise selection using cell culture systems with chronic exposure to the cytotoxic agent of interest. This is followed by characterizing stable resistance phenotypes at both the cellular and the molecular levels. By manipulating the biochemical pathway of interest, the causal relationships between resistance mechanisms and alterations in the resistant phenotype are determined. Using this strategy, stable oxidative stress-resistant cell lines have been developed from HA1 Chinese hamster fibroblasts. A method of stepwise selection to chronic oxidative stress, both exogenous (H_2O_2) and endogenous (hyperoxia), was used (7–12). The present report deals with the procedures involved in selecting the resistant cell lines, characterizing the phenotypes, and describing tests to establish the main biochemical pathway(s) responsible in the development of resistance to chronic oxidative stress.

2. Materials

1. Water: use nanopure water with resistance $\geq$18.2 MΩ/cm.
2. Eagle's minimal essential media (EMEM).
3. Trypsin–EDTA (0.1%) from GIBCO.
4. Fetal calf serum FCS (Hyclone, Logan, UT).
5. Penicillin–streptomycin.
6. Hydrogen peroxide (H_2O_2, 30% w/v).
7. Coomassie blue stain: dissolve 4 g of coomassie brilliant blue G250 (Sigma) in 2 L of methanol/acetic acid/water (5/1/4, v/v/v), filter, and can be used 4–5 times filtering out particulates after each usage. To prepare the stain, dissolve Coomassie blue in methanol first and then add acetic acid and water.

3. Methods

3.1. Cells and Culture Conditions

Maintain HA1 (13) Chinese hamster fibroblasts in EMEM supplemented with 10% (v/v) FCS and 100 units/mL–0.1 mg/mL penicillin–streptomycin (*see* **Note 1**).

3.2. Selection of H_2O_2-Resistant Cells (see Note 2)

1. Stock cultures of exponentially growing control HA1 cells in T-25 culture flasks are quickly (5 s) rinsed once with 2 mL of 0.1% trypsin–EDTA solution and then trypsinized in 1 mL of trypsin for 5–10 min at room temperature. Inactivate trypsin by the addition of 4 mL of complete medium.
2. Count the cells in a Coulter counter. Passage 100,000 cells per T-25 tissue culture flasks in 5 mL of freshly prepared complete media every 3–4 d when confluent.
3. Dilute 30% ($\approx$10 M) H_2O_2 1:750 with water and determine the concentration based on absorbance at 240 nm ($\varepsilon = 39.4/M/cm$) and filter using a sterile filter (0.2 μm). Place 25 μL of *ca.* 10 mM H_2O_2 solution on the tip of a T-25 culture flask and wash it down into the flask with 5 mL of complete medium containing 250,000 freshly trypsinized HA1 cells (final concentration of H_2O_2 equals 50 μM). This first selection step (passage 1 in 50 μM H_2O_2) is quite toxic and surviving colonies will take 10–14 d to become clearly visible.
4. Once colonies are clearly visible and growing, rinse quickly with 1 mL of 0.1% trypsin–EDTA and then trypsinize with 1 mL of trypsin for 5–10 min. Inactivate trypsin with 4 mL of complete medium (see above) and incubate 250,000 cells in a T-25 culture flask with 50 μM H_2O_2.
5. Repeat this procedure for 7–10 passages (writing the passage number on the flask each time with the H_2O_2 concentration) until the cells grow into a monolayer within 4 d. (Sham-treated control cell cultures are being passaged at the same time in the absence of H_2O_2.)
6. Once the H_2O_2-resistant cells are growing as a monolayer within 4 d in 50 μM H_2O_2, then one flask is continued to be passaged in 50 μM of H_2O_2. Transfer 250,000 cells to a flask with 100 μM H_2O_2 and repeat same procedure until the cells are growing into a monolayer within 4 d. When the cells at 100 μM H_2O_2 have grown into a monolayer within 4 d (~7 passages) the dose is again doubled (to 200 μM H_2O_2) and the same procedure is repeated.
7. Once the cells in 200 μM H_2O_2 are growing as a monolayer within 4 d (about five passages) double the dose again (to 400 μM H_2O_2) and repeat the same procedure. When the cells at 400 μM H_2O_2 are growing into a monolayer within 4 d (about five passages), double the dose again to 800 μM H_2O_2 and repeat the same procedure.
8. Once the cells in 800 μM H_2O_2 are growing as a monolayer within 4 d (about five passages), the selection process is complete because attempts to increasing the dose beyond

800 μM H_2O_2 have proven to be unsuccessful. The five selected cell populations (growing in 50, 100, 200, 400, and 800 μM H_2O_2) and sham-treated control are then maintained with passage in their respective concentrations of H_2O_2 for a total of 240 d to allow them to develop a stable H_2O_2-resistant phenotype. Then, remove the cell populations from H_2O_2 selection, test for mycoplasma, and 10 aliquots of 2,000,000 cells from each group (including control) are frozen as described under cell freezing protocol. The populations are also passaged out of H_2O_2 for 240 d to allow unstable phenotypic characteristics to dissipate at which time they are again tested for mycoplasma and aliquots of each are frozen. The passage numbers in H_2O_2 selection as well as the passage numbers out of H_2O_2 selection are recorded on the flasks and also on the cryo-vials containing frozen cells.

3.3. Characterization of H_2O_2-Resistant Cells

1. Expose the HA1 cells chronically to 50–800 μM H_2O_2 as described above. The cell lines (designated 50, 100, 200, 400, and 800 μM H_2O_2-adapted cells lines) are tested in acute (1 h) H_2O_2 challenge experiments following passage for 60–200 d out of the H_2O_2 selection protocol and assayed for maintenance of reproductive integrity using the clonogenic cell survival assay (7–13) (*See* **Note 3**).
2. Place 150,000 cells from each cell line into 60-mm tissue culture plates in 4 mL of complete media and grow asynchronously at 37°C incubation in 5% CO_2 for 2 d at which time there are approximately 1,000,000 exponentially growing cells/dish. Expose control HA1 cells and H_2O_2-resistant cell lines to increasing concentrations of H_2O_2 (0.1–2.0 mM for control cells and 0.5–10.0 mM for H_2O_2-adapted cells) added directly to fresh complete medium (immediately before placing the medium on the cells) for 1 h at 37°C in a 5% CO_2 incubator. Include sham-treated cultures from each cell line as controls.
3. After treatment for 1 h, rinse the cells with 4 mL of phosphate-buffered saline, trypsinize, and suspend in 4 mL of complete medium. Dilute serially in two successive 1:10 dilutions with complete medium. The cells in the originally trypsinized dish are considered to be the 01 dilution and are at a concentration of ~200,000 cells/mL. The cells in the first 1:10 dilution are considered to be the 02 dilution and are at a concentration of ~20,000 cells/mL, and the cells in the second 1:10 dilution are considered to be the 03 dilution and are at a concentration of ~2,000 cells/mL.
4. At this point, transfer 1 mL of the 02 dilution into 9 mL of isoton (Coulter counting solution) and cells are counted

on a Coulter counter. Then increasing numbers of cells depending on the severity of the acute H_2O_2 challenge dose (~200 for control-untreated HA1 cells or HA1 cells treated with 0.1 mM H_2O_2, 2,000 for 0.5 mM H_2O_2-treated HA1 cells, 20,000 for 1 mM H_2O_2-treated HA1 cells, or 200,000 for >1 mM H_2O_2-treated HA1 cells) are placed into each of three cloning dishes (60 mm) containing 4 mL of complete medium sitting on a stainless steel incubator tray and spread evenly on the bottom of the cloning dish by vertically and horizontally sliding the tray gently on the counter top in the tissue culture hood. Fewer H_2O_2-treated cells need to be plated in cloning dishes from the H_2O_2-selected populations at a given acute challenge dose since they are more resistant. If there is a question as to the survival level of the treated cells (for instance the first time the experiment in accomplished), it is better to plate more dilutions (200–200,000) from each treated dish to ensure that at least one set of cloning dishes will have the required number of colonies to be counted easily (20–200 colonies/dish).

5. Once the cells are evenly spread, the cloning dishes are gently placed into a cloning incubator at 37°C and 5% CO_2 for 9 d to allow surviving clones to develop. Care is taken not to disturb the cloning dishes during the 9-d period to avoid the formation of satellite colonies. At the end of the cloning period, the dishes are gently removed from the incubator, rinsed once with 2 mL of 70% ethanol, and stained 5 min with 2–3 mL of Coomassie blue stain (*see* **Materials**). Count the colonies under a dissecting microscope and only colonies with at least 50 cells are scored as survivors.
6. Once the colony counts are obtained, surviving fractions in the treated and untreated dishes from each cell line are calculated by the following expression: colonies counted from each cloning dish/cells plated in the cloning dish = surviving fraction. The surviving fraction of untreated dishes from each cell line represents the plating efficiency of that cell line (typically 60–90%) and is used to normalize the calculation of the surviving fraction from the H_2O_2 treatments using the following expression: surviving fraction of the cells exposed to the test dose of H_2O_2/surviving fraction of the untreated cells from the same cell line = normalized surviving fraction for the given cell line and dose of H_2O_2 (10).
7. The data are then plotted on a log/linear scale with the normalized surviving fraction on the *y*-axis ($\log_{10}$ scale) and dose of H_2O_2 on the *x*-axis (in mol H_2O_2 present at time 0 of the exposure/cell in the dish at time 0 of exposure).

8. Once full survival curves for the cell lines are obtained, the resistance of the cell lines can be compared using a calculation of dose modifying factors at 20% survival ($DMF_{20\%}$). $DMF_{20\%}$ can be calculated by the following expression: dose to reach 20% survival in the resistant cell line/dose to reach 20% survival in the HA1 parental cell line = $DMF_{20\%}$ for that particular H_2O_2-resistant cell line (7, 10). When this analysis was applied to the cell lines obtained following chronic exposure to 50–800 μM H_2O_2 between 80 and 200 d following the last H_2O_2 treatment in the chronic selection protocol, the cell line exposed to 50 μM H_2O_2 demonstrated DMFs of 5–8, the cell line exposed to 100 μM H_2O_2 demonstrated DMFs of 8–14, the cell line exposed to 200 μM H_2O_2 demonstrated DMFs of 20–29, the cell line exposed to 400 μM H_2O_2 demonstrated DMFs of 20–23, and the cell line exposed to 800 μM H_2O_2 demonstrated DMFs of 26–36 (7). This method of doing acute H_2O_2 challenge of cells for the purpose of characterizing H_2O_2-resistant phenotypes has been utilized in a number of publications (7–13) and cell lines selected for H_2O_2 resistance derived from HA1 have been shown to maintain a stable resistant phenotype for at least 240 d following the selection protocol using this method (7, 8).
9. Studies of cell biological and biochemical characterizations of H_2O_2-resistant cell lines have been accomplished (14–25) and highlights will be summarized. The H_2O_2-resistant cells are larger, have more protein/cell, demonstrated somewhat slower growth rates, increased catalase activity, and increases in aneuploid/tetraploid cells within the population as the H_2O_2 dose to which they were selected increased (7). When clones were obtained from the H_2O_2-resistant cell population selected in 800 μM H_2O_2, the tetraploid clones are found to be, on average, 3-fold more resistant to H_2O_2 than are the diploid clones and the prevalence of tetraploid cells increased with the increasing selection doses above 200 μM H_2O_2 (7, 8). This led to the hypothesis that increases in ploidy during chronic exposure to oxidative stress contributed to the adaptation process leading to the H_2O_2-resistant phenotypes at the higher doses of H_2O_2 (8). In addition to increases in ploidy, many gross chromosomal rearrangements are noted in the 800 μM H_2O_2-selected cells as well as catalase gene amplification (that was not seen in populations exposed to one severe H_2O_2 challenge dose) showing that chronic exposure to oxidative stress mediated by H_2O_2 is capable

of inducing genomic instability and gene amplification in mammalian cells (21).

10. In addition to these cell biology changes, many biochemical changes are noted in H_2O_2-resistant cells. In addition to catalase activity, glutathione peroxidase activity, glutathione transferase activity, CuZn superoxide dismutase activity, heme oxygenase activity, aldose reductase activity, total glutathione content and the expression of several stress-responsive genes including GADD153, c-Jun, and c-Fos, as well as AP-1 DNA binding activity are all found to be increased in H_2O_2-resistant cells indicating a global response to chronic oxidative stress (9–11, 17–20, 22–24). Using various chemical inhibitors, catalase, heme oxygenase, aldose reductase, total glutathione, and AP-1 DNA binding activity are all found to significantly contribute to the H_2O_2-resistant phenotype (11, 15, 17, 20, 22, 24). In addition the H_2O_2-resistant cells are also found to be cross-resistant to cell killing induced by moderate hyperthermia, heat-induced radiosensitization, cisplatin, aldehydic byproducts of lipid peroxidation, O_2 toxicity, xanthine oxidase, and nitric oxide, indicating a great deal of overlap between mechanisms of resistance to oxidative stress induced by numerous agents (9–11, 14–17, 22, 25). Finally, the H_2O_2-resistant cells were found to be resistant to mitochondrial DNA damage induced by exposure to xanthine oxidase-mediated oxidative stress as well as demonstrating enhanced mitochondrial DNA damage repair and increased levels of apurinic/apyrimidinic endonuclease (APE) activity (25). Overall, these H_2O_2-resistant cell lines derived from the parental HA1 cell line following chronic exposure to H_2O_2 provide a valuable resource for studying molecular mechanisms of oxidative stress resistance in mammalian cells. However, it is also worth noting that H_2O_2-resistance mechanisms that arise following chronic exposure to oxidative stress are likely to vary with the cell types being studied at the level of both tissue origin and species, so many patterns of resistant phenotypes and mechanisms of resistance are likely to be obtained (26–38).

3.4. Selection of O_2-Resistant Cells

1. To begin the selection of O_2-resistant cells seed 200,000 HA1 cells into two T-25 tissue culture flasks with 5 mL of complete medium and grow at ambient O_2 tension in a humidified 5% CO_2 incubator at 37°C for 2 d at which time one flask of the cells is placed in an incubator at 37°C with 80% O_2/5% CO_2/15% N_2 and the medium changed every 24 h (*see* **Notes 4** and **5**).

2. Leave the control flask in the ambient O_2 chamber and passage each time when it becomes confluent (every 3–4 d).
3. During the first 100 h of this first exposure to hyperoxia the cells go through a crisis where approximately 50% detach from the flask and are discarded each day when the media is changed (12, 39).
4. After 144 h of exposure using this protocol, trypsinize the surviving attached cells, count, and passage approximately 250,000 into a fresh T-25 flask with 5 mL complete medium and place in 80% O_2/5% CO_2/15% N_2.
5. Change the medium every 24 h. During this second passage into hyperoxia, the cells begin to proliferate and become confluent by 144 h.
6. Repeat the passaging and feeding protocol every 4–5 d when the cells become confluent for 280 d at which time the cells are moved to a humidified incubator at 37°C with 95% O_2/5% CO_2.
7. Repeat the process for another 150 d at which time the cells are capable of stable growth in 95% O_2. At this point in the selection protocol expand the cell populations (designated O_2R95) into T-75 flasks (750,000 cells per passage in 15 mL medium) and test for mycoplasma.
8. Ten (10) aliquots of 2,000,000 cells from each group (including controls passaged the same length of time in 5% CO_2 and air) are frozen down as described below under cell freezing protocol.
9. The O_2R95 cell populations are also passaged out of hyperoxia in 5% CO_2 and air for 75 d to allow unstable phenotypic characteristics to dissipate at which time they are again tested for mycoplasma and aliquots of each frozen down and designated O_2R95-P21%O_2. The passage numbers in O_2 selection as well as the passage numbers out of O_2 selection are recorded on the flasks and on the cryo-vials containing frozen aliquots of cells.

3.5. Characterization of O_2-Resistant Cells

1. In order to quantitate the O_2-resistant phenotype of HA1 cells chronically exposed to hyperoxia, the cell lines (designated O_2R95 and O_2R95-P21%O_2) are tested in acute 95% O_2 challenge experiments (compared to HA1 controls), and assayed for maintenance of reproductive integrity using the clonogenic cell survival assay (12, 40).
2. Because O_2-resistant cells in culture that are fed fresh medium daily do not experience cell death due to O_2 toxicity, the acute O_2 challenge experiments are done in unfed HA1 control and O_2-resistant cultures exposed to 95%

O_2/5% CO_2 in humidified 37°C incubation and surviving fraction using clonogenic cell survival assay plotted as a function of time in hyperoxia (12, 40).

3. Both the controls and the O_2-resistant cells must be at the same cell density (cell number/dish) at the beginning of acute 95% O_2 exposure because O_2 toxicity in this model system has been shown to be highly cell density dependent (10). This method of doing the acute 95% O_2 challenge experiments and expressing the data has been shown to eliminate the artifacts associated with cell density and was used to characterize the cross-resistance of H_2O_2-resistant cells to O_2 toxicity (10).
4. To accomplish the acute O_2 challenge experiments, place 150,000 cells from each cell line into 60 mm tissue culture plates in 4 mL of complete medium and grow them asynchronously at 37°C incubation in 5% CO_2 for 2 d at which time there are approximately 1,000,000 exponentially growing cells per dish in every group.
5. Feed control HA1 cells and O_2-resistant cell lines with fresh medium and move the cultures to 95% O_2/5% CO_2 incubation at 37°C with no further feeding.
6. Sham-treated cultures in an ambient O_2 incubator from each cell line are also included as controls.
7. Remove the cells from hyperoxia after various times of exposure to 95% O_2 (24–96 h), rinsed with 4 mL of phosphate-buffered saline, trypsinize, and suspend in 4 mL complete medium, serially dilute in two successive 1:10 dilutions with complete medium.
8. The cells in the originally trypsinized dish are considered to be the 01 dilution and are at a concentration of approximately 200,000–400,000 cells/mL, the cells in the first 1:10 dilution are considered to be the 02 dilution and are at a concentration of approximately 20,000–40,000 cells/mL, and the cells in the second 1:10 dilution are considered to be the 03 dilution and are at a concentration of approximately 2,000–4,000 cells/mL.
9. At this point, place 1 mL of the 02 dilution into 9 mL of isoton (Coulter counting solution) and count on a Coulter counter. Then increasing numbers of cells depending on the severity of the acute 95% O_2 challenge dose (approximately 200 for control-untreated HA1 cells or HA1 cells treated with 24 h 95% O_2, 200–500 for 48 h O_2-treated HA1 cells, 2,000–10,000 for 60 h O_2-treated HA1 cells, or 10,000–200,000 for $\geq$72 h O_2-treated HA1 cells) are placed into each of three cloning dishes (60 mm) containing 4 mL of complete media sitting on a

stainless steel incubator tray and spread evenly on the bottom of the cloning dish by vertically and horizontally sliding the tray gently on the counter top in the tissue culture hood. O_2-resistant cells challenged with hyperoxia are plated in cloning dishes at lower cell densities (usually only 200–500 total) since they are much more resistant to the acute 95% O_2 challenge (12). If there is a question as to the survival level of the treated cells (for instance the first time the experiment is accomplished) it is better to plate more dilutions (200–200,000) from each treated dish to ensure that at least one set of cloning dishes will have the required number of colonies to be counted easily (20–200 colonies/dish).

10. Once the cells are evenly spread, the cloning dishes are gently placed into a cloning incubator at 37°C and 5% CO_2 for 9–10 d to allow surviving clones to develop. Care is taken not to disturb the cloning dishes during the cloning period to avoid the formation of satellite colonies.
11. At the end of the cloning period the dishes are gently removed from the incubator, rinsed once with 2 mL of 70% EtOH, and stained 5 min with 2–3 mL of Coomassie blue stain. Colonies are counted under a dissecting microscope and only colonies with at least 50 cells are scored as survivors.
12. Once the colony counts are obtained, surviving fractions for treated and untreated dishes from each cell line are calculated by the following expression: colonies counted from each cloning dish/cells plated in the cloning dish = surviving fraction. The surviving fraction of untreated dishes from each cell line represents the plating efficiency of that cell line and is used to normalize the calculation of the surviving fraction from the O_2 treatments using the following expression: surviving fraction of the cells exposed to a given amount of time at 95% O_2/surviving fraction of the untreated cells from the same cell line at the same time point = normalized surviving fraction for the given cell line and time of O_2 exposure (12).
13. The data are then plotted on a log/linear scale with the normalized surviving fraction on the *y*-axis ($\log_{10}$ scale) and exposure time in 95% O_2 on the *x*-axis.
14. Once full survival curves for the entire cell lines are obtained, the resistance of the cell lines can be compared using a calculation of dose modifying factors at 50% isosurvival ($DMF_{50\%}$). $DMF_{50\%}$ can be calculated by the following expression: time of 95% O_2 exposure to reach 50% survival in the resistant cell line/time to reach 50% survival

in the HA1 parental cell line = $DMF_{50\%}$ for that particular O_2-resistant cell line (12). When this analysis was applied to the HA1 and O_2R95-P21%O_2 cell lines DMFs of 2–3 were obtained and essentially all the HA1 cells were clonogenically inactivated at 96 h of 95% O_2, while approximately 50% of the O_2R95-P21%O_2 cell line survived at this time point (12). This method of doing acute O_2 challenge of cells for the purpose of characterizing mechanisms associated with the O_2-resistant phenotype has been utilized in a number of publications (12, 18–20, 40) and cell lines selected for O_2 resistance derived from HA1 have been shown to maintain a stable resistant phenotype for at least 75 d following the selection protocol (12).

15. Many further studies of cell biological and biochemical characteristics of these O_2-resistant cell lines have been accomplished (12, 18–21, 23, 25, 40), and highlights of those studies will be briefly summarized. The O_2-resistant cells were approximately the same size, had approximately the same protein/cell, demonstrated slower growth rates (relative to HA1) while being passaged in the feeding protocol in 95% O_2 (28 h vs. 14 h) as well as while being passaged in ambient O_2 (18 h vs. 14 h), demonstrated increased catalase activity, and demonstrated no apparent increase in ploidy (12, 21). Gross chromosomal rearrangements were noted in the O_2 selected cells as well as catalase gene amplification (that was not seen in HA1 populations exposed to one severe O_2 challenge dose) showing that chronic exposure to oxidative stress mediated by 95% O_2 was capable of inducing genomic instability and gene amplification in mammalian cells (21).
16. In addition to these cell biology changes, many biochemical changes were noted in O_2-resistant cells. In addition to catalase activity, glutathione peroxidase activity, glutathione transferase activity, CuZn superoxide dismutase activity, MnSOD activity, heme oxygenase activity, aldehyde reductase activity, total glutathione content, and the expression of several stress-responsive genes including GADD153, c-Jun, and a UV-resistance-associated gene (AA490771) were all found to be increased in O_2-resistant cells, indicating a global response to chronic oxidative stress (20, 12, 18, 23). Using various chemical inhibitors catalase, heme oxygenase, and total glutathione were all found to significantly contribute to the O_2-resistant phenotype (20, 40).
17. In addition, the O_2-resistant cells were also found to be cross-resistant to cell killing induced by aldehydic byproducts of lipid peroxidation, hydrogen peroxide, and xanthine oxidase indicating significant overlap between mechanisms

of resistance to oxidative stress induced by these agents (12, 18, 20, 40). Finally, the O_2-resistant cells were found to be resistant to mitochondrial DNA damage induced by exposure to xanthine oxidase-mediated oxidative stress as well as demonstrating enhanced mitochondrial DNA damage repair and increased levels of APE activity (25). Overall, these O_2-resistant cell lines provide a valuable resource for studying mechanisms of oxidative stress resistance in mammalian cells. However, it is again worth noting that O_2-resistance mechanisms that arise following chronic exposure to hyperoxia are likely to vary with the cell types being studied at the level of both tissue origin and species, so many patterns of resistant phenotypes and mechanisms of resistance are likely to be obtained (41–44).

3.6. Cell Freezing Protocol

1. 1.0 N potassium hydroxide – 625 μL 8 N KOH; 4.38 mL nanopure water.
2. Sterile DMSO – Autoclave DMSO on liquid cycle for 30 min at 275°C.

To make	100 mL	200 mL	400 mL
KH_2PO_4	0.68 g	0.136 g	0.272 g
KOH, 1.0 N	2.5 mL	5.0 mL	1.0 mL
NaCl	0.175 g	0.350 g	0.700 g
Lactic acid	0.180 g	0.360 g	0.720 g
Dextrose	0.090 g	0.180 g	0.360 g
$MgCl_2$ $6H_2O$	0.010 g	0.020 g	0.040 g
Sorbitol	3.640 g	7.280 g	14.569 g

3.7. Procedure for Making Freezing Solution (200 mL)

1. Add the remaining reagents in order, beginning with KH_2PO_4. Mix gently.
2. Check pH and final pH should be 7.2–7.4. Adjust with KOH or HCl.
3. Bring final volume to 160 mL with water.
4. Sterile filter solution with bottle top 0.22 μm filter.
5. Filter 80 mL into two sterile, 200 mL-capacity glass bottles.
6. To one bottle add 20 mL sterile water. Label DMSO-free cell freezing solution.
7. To the second bottle add 20 mL sterile DMSO. Label DMSO cell freezing solution. Do not filter this solution. The DMSO leaches particles from the filter.
8. Final concentrations of ingredients in the freezing solution are 5 mM [K^+] as KH_2PO_4; 25 mM [K^+] as KOH; 30 mM

$[Na^+]$ as NaCl; 20 mM lactate as L(+) lactic acid; 5 mM glucose as dextrose (D-glucose); 0.5 mM $[Mg^{2+}]$ as $MgCl_2$; and 0.2 M sorbitol.

3.8. Preparation of Cells for Freezing

Every step should be done within sterile tissue culture hood.

1. Trypsinize cells and inactivate with complete media.
2. Count the cells using a Coulter counter.
3. Transfer 20×10^6 cells to a sterile centrifuge tube (capacity: 15 mL).
4. Centrifuge at $400\times g$ in a refrigerated tabletop centrifuge.
5. In the tissue culture laminar flow hood sterilely remove the supernatant.
6. Add 5 mL of 4°C DMSO-free freezing solution, mix, and incubate in the dark for 7 min at 4°C.
7. Add 5 mL of 20 % (v/v) DMSO containing freezing solution. Note final concentration of DMSO in freezing tube equals 10% (v/v).
8. Gently finger vortex the freezing tube to mix solution.
9. Triturate gently and then aliquot 2×10^6 cells (1 mL) to each sterile cryo-vial.
10. Immediately (within 2–3 min) place vials in –80°C freezer or liquid N_2.
11. Tubes should be labeled with cell type, freeze date, and passage number.

3.9. Thawing Cells

1. Bring vial out of freezer.
2. Thaw as rapidly as possible (hold in 37°C water bath).
3. Add to T-25 flask with 5 mL complete media and place in 5% CO_2 37°C humidified incubator allowing for free gas exchange by loosening the cap of the flask slightly.
4. Once cells attach (usually 6–24 h) change media to remove any residual DMSO.

3.10. Removal of DMSO More Rapidly to Facilitate Recovery from Freezing (Optional)

1. After vial has thawed, add contents to a sterile 15 mL centrifuge tube.
2. Add 10 mL complete media and mix gently.
3. Centrifuge at 1,200 rpm ($400\times g$) for 5 min.
4. Discard supernatant.
5. Add 5 mL complete media and resuspend the pellet.
6. Transfer to a T-25 flask and place in 5% CO_2 37°C humidified incubator allowing for free gas exchange by loosening the cap of the flask slightly.

3.11. Overall Summary

The protocols for selecting O_2- and H_2O_2-resistant cell lines outlined in this report using HA1 cells represent robust and reproducible methods that can be applied to generating stable oxidative stress-resistant cell populations for cellular and molecular studies of mechanisms of resistance to chronic oxidative stress. As mentioned earlier these resistant phenotypes can vary from cell line to cell line and from cell lines from different species. Many research groups have used similar protocols for generating chronic oxidative stress-resistant phenotypes from HeLa cells (41, 43), other hamster fibroblast cell lines (31–34), mouse nerve cells (36), L929 cells (35), murine L1210 cells (30), monkey kidney cells (37), and human HL-60 cells (27–30). While mechanisms governing the chronic oxidative stress-resistant phenotypes that arise vary widely between cell lines and species, they do appear to fall into several distinct categories. These categories include increased expression of antioxidant capacity (6–12, 21, 24, 27, 30, 37, 42), increased oxidative damage repair pathways (15, 18, 25, 37), decreased prooxidant production pathways (28, 43, 44), and increased stress-responsive signaling and gene expression pathways (18, 22, 23, 26). Many times coordinated regulation of more than one of these pathways leading to resistance to chronic oxidative appears to be involved with the resistant phenotypes that arise and cross-resistance to several forms of oxidative stress usually ensues. The diversity of oxidative stress-resistant phenotypes that appear to develop in different cell types and different species exposed to chronic oxidative stress seems to support the hypothesis that the specific oxidative metabolism that is predominate in each cell type at the time of exposure to the selection pressure seems to govern the types of specific protective mechanisms that emerge. This hypothesis would predict that coordinated redox-sensitive signaling circuitry that allows for the sensing of changes in flux of electrons within various metabolic compartments of the cell governs the specific biochemical changes that emerge as contributing to each particular chronic oxidative stress-resistant phenotype. Given the plethora of modern molecular biological tools that now exist to manipulate these various pathways and dissect the specific linkages between the oxidative metabolic pathways that are being disrupted and the protective pathways that up regulated, these cellular model systems promise to yield useful information in the future that can be utilized to modify mechanisms of resistance to chronic oxidative stress that are relevant to human health.

4. Notes

1. Periodically test the cells for mycoplasma contamination following passage for 12 d in antibiotic free complete media and always make sure they are negative.
2. Maintain the cells for chronic exposure to H_2O_2 in 5% CO_2 and room air in a humidified tri-gas incubator at 37°C.
3. Because cells in culture rapidly metabolize H_2O_2 from the cell culture media (7), the cytotoxic response of cells in culture to acute H_2O_2 challenge experiments is plotted as surviving fraction following a 1 h exposure to H_2O_2 as a function of the number of moles of H_2O_2 added to the culture per cell in the culture at the time of exposure (13). This method of expressing the data has previously been shown to eliminate the artifacts associated with cell density and is used to plot all the H_2O_2 survival data we have obtained (13).
4. The concentration of CO_2 is determined using a CO_2 medical gas analyzer (Beckman LB-2) calibrated with certified gas (5.0 ± 0.1% CO_2). Oxygen concentrations are determined using an O_2 detector (Beckman OM-11) calibrated with certified gas (95 ± 0.1% O_2).
5. When the hyperoxia chamber is opened during the course of cell culture work, the desired chamber O_2 concentration is rapidly reestablished by purging the chamber with 95% O_2/5% CO_2 for up to 15 min through an access port in the side of the chamber while monitoring the chamber O_2 concentration with an O_2 gas analyzer.

Acknowledgments

DRS is supported by NIH R01-CA100045, DOE DE-FG02-02ER63447, and NIEHS P42 ES013661. SJS is supported by the Department of Pediatrics at the University of Iowa.

References

1. Harman, D. (1957) Aging: a theory based on free radical and radiation chemistry. *J. Gerontol.* 2, 298–300.
2. Oberley, L.W. and Buettner, G.R. (1979) Role of superoxide dismutase in cancer: a review. ***Cancer Res.*** **39**, 1141–1149.
3. Ames, B.N. (1983) Dietary carcinogens and anticarcinogens: oxygen radicals and degenerative diseases. *Science* **221**, 1256–1262.
4. Cerutti, P.A. (1985) Prooxidant states and tumor promotion. *Science* **227**, 375–381.

5. Finkel, T. and Holbrook, N.J. (2000) Oxidants, oxidative stress and the biology of ageing. *Nature* **408**, 239–247.
6. Spitz, D.R., Azzam, E.I., Li, J.J., and Gius, D. (2004) Metabolic oxidation/reduction reactions and cellular responses to ionizing radiation: a unifying concept in stress response biology. *Cancer Metastasis Rev.* **23**, 311–322.
7. Spitz, D.R., Li, G.C., McCormick, M.L., Sun, Y., and Oberley, L.W. (1988) Stable H_2O_2-resistant variants of Chinese hamster fibroblasts demonstrate increases in catalase activity. *Radiat. Res.* **114**, 114–124.
8. Spitz, D.R., Mackey, M.A., Li, G.C., Elwell, J.H., McCormick, M.L.,, and Oberley, L.W. (1989) Relationship between changes in ploidy and stable cellular resistance to hydrogen peroxide. *J. Cell Physiol.* **139**, 592–598.
9. Spitz, D.R., Malcolm, R.R., and Roberts, R.J. (1990) Cytotoxicity and metabolism of 4-hydroxynonenal and 2-nonenal in H_2O_2-resistant cell lines: Do aldehydic by-products of lipid peroxidation contribute to oxidative stress?. *Biochem. J.* **267**, 453–459.
10. Spitz, D.R., Elwell, J.H., Sun, Y., Oberley, L.W., Oberley, T.D., Sullivan, S.J., and Roberts, R.J. (1990) Oxygen toxicity in control and H_2O_2-resistant Chinese hamster fibroblast cell lines. *Arch. Biochem. Biophys.* **279**, 249–260.
11. Spitz, D.R., Adams, D.T., Sherman, C.M., and Roberts, R.J. (1992) Mechanisms of cellular resistance to hydrogen peroxide, hyperoxia and 4-hydroxy-2-nonenal toxicity: The significance of increased catalase activity in H_2O_2-resistant fibroblasts. *Arch. Biochem. Biophys.* **292**, 221–227.
12. Sullivan, S.J., Oberley, T.D., Roberts, R.J., and Spitz, D.R. (1992) A stable O_2-resistant cell line: Role of lipid peroxidation by-products in O_2-mediated injury. *Am. J. Physiol. (Lung Cell Mol. Physiol.)* **262**, 748–756.
13. Spitz, D.R., Dewey, W.C., and Li, G.C. (1987) Hydrogen peroxide or heat shock induces resistance to hydrogen peroxide in Chinese hamster fibroblasts. *J. Cell. Physiol.* **131**, 364–373.
14. Spitz, D.R. and Li, G.C. (1990) Heat-induced cytotoxicity in H_2O_2-resistant Chinese hamster fibroblasts. *J. Cell. Physiol.* **142**, 255–260.
15. Spitz, D.R., Phillips, J.W., Adams, D.T., Sherman, C.M., Deen, D.F., and Li, G.C. (1993) Cellular resistance to oxidative stress is accompanied by resistance to cisplatin: The significance of increased catalase activity and total glutathione in H_2O_2-resistant fibroblasts. *J. Cell. Physiol.* **156**, 72–79.
16. Walker, M.W., Kinter, M.T., Roberts, R.J., and Spitz, D.R. (1995) Nitric oxide induced cytotoxicity: involvement of cellular resistance to oxidative stress and the role of glutathione in protection. *Pediat. Res.* **37**, 41–49.
17. Spitz, D.R., Kinter, M.T., and Roberts, R.J. (1995) The contribution of increased glutathione content to mechanisms of oxidative stress resistance in hydrogen peroxide resistant hamster fibroblasts. *J. Cell. Physiol.* **165**, 600–609.
18. Guyton, K.Z., Spitz, D.R., and Holbrook, N.J. (1996) Expression of stress response genes GADD153, c-*jun*, and heme oxygenase-1 in H_2O_2- and O_2-resistant fibroblasts. *Free Radic. Biol. Med.* **20**, 735–741.
19. Dennery, P.A., Wong, H.E., Sridhar, K.J., Rodgers, P., Sim, J.E., and Spitz, D.R. (1996) Differences in basal and hyperoxia associated heme oxygenase expression in oxidant resistant hamster fibroblasts. *Am. J. Physiol. (Lung Cell. Mol. Physiol.)* **271**, 672–679.
20. Dennery, P.A., Sridhar, K.J., Lee, C.S., Wong, H.E., Shokoohi, V., Rodgers, P.A., and Spitz, D.R. (1997) Heme oxygenase-mediated resistance to oxygen toxicity in hamster fibroblasts. *J. Biol. Chem.* **272**, 14937–14942.
21. Hunt, C.R., Sim, J.E., Featherstone, T., Golden, W., Von Kapp-Herr, C., Hock, R.A., Gomez, R.A., Parsian, A.J., and Spitz, D.R. (1998) Genomic instability and catalase gene amplification induced by chronic exposure to oxidative stress. *Cancer Res.* **58**, 3986–3992.
22. Bradbury, C.M., Locke, J.E., Wei, S.J., Rene, L.M., Karimpour, S., Hunt, C., Spitz, D.R., and Gius, D. (2001) Increased activator protein 1 activity as well as resistance to heat-induced radiosensitization, hydrogen peroxide, and cisplatin are inhibited by indomethacin in oxidative stress-resistant cells. *Cancer Res.* **61**, 3486–3492.
23. Suzuki, T., Spitz, D.R., Gandhi, P., Lin, H.Y., and Crawford, D.R. (2002) Mammalian resistance to oxidative stress: a comparative analysis. *Gene Expr.* **10**, 179–191.
24. Keightley, J.A., Shang, L., and Kinter, M. (2004) Proteomic analysis of oxidative stress resistant cells: a specific role for aldose reductase overexpression in cytoprotection. *Mol. Cell Proteomics* **3**, 165–175.
25. Grishko, V.I., Rachek, L.I., Spitz, D.R., Wilson, G.L., and LeDoux, S. (2005) Contribution of mitochondrial DNA repair to cell resistance from oxidative stress. *J. Biol. Chem.* **280**, 8901–8905.

26. Bojes, H.K., Suresh, P.K., Mills, E.M., Sim, J.E., Spitz, D.R., Sim, J.E., and Kehrer, J.P. (1998) Bcl-2 and Bcl-xL in peroxide resistant A549 and U87MG cells. *Toxicol. Sci.* **42**, 109–116.
27. Kasugai, I. and Yamada, M. (1992) High production of catalase in hydrogen peroxide-resistant human leukemia HL-60 cell lines. *Leuk. Res.* **16**, 173–179.
28. Yamada, M., Hashinaka, K., Inazawa, J., and Abe, T. (1991) Expression of catalase and myeloperoxidase genes in hydrogen peroxide-resistant HL-60 cells. *DNA Cell Biol.* **10**, 735–742.
29. Kasugai, I. and Yamada, M. (1989) Adaptation of human leukemia HL-60 cells to hydrogen peroxide as oxidative stress. *Leuk. Res.* **13**, 757–762.
30. Lin, F., Jackson, V.E., and Girotti, A.W. (1995) Amplification and hyperexpression of the catalase gene in selenoperoxidase-deficient leukemia cells. *Arch. Biochem. Biophys.* **317**, 7–18.
31. Cantoni, O., Sestili, P., Palomba, L., Guidarelli, A., Cattabeni, F., and Murray, D. (1996) Isolation and preliminary characterization of a Chinese hamster ovary cell line with high-degree resistance to hydrogen peroxide. *Biochem. Pharmacol.* **51**, 1021–1029.
32. Vallis, K.A. and Wolf, C.R. (1996) Relationship between the adaptive response to oxidants and stable menadione-resistance in Chinese hamster ovary cell lines. *Carcinogenesis* **17**, 649–654.
33. Cantoni, O., Guidarelli, A., Sestili, P., Mannello, F., Gazzanelli, G., and Cattabeni, F. (1993) Development and characterization of hydrogen peroxide-resistant Chinese hamster ovary cell variants-I. Relationship between catalase activity and the induction/stability of the oxidant-resistant phenotype. *Biochem. Pharmacol.* **45**, 2251–2257.
34. Martins, E.A., Mori, L., Birnboim, H.C., and Meneghini, R. (1992) Menadione-resistant Chinese hamster cell variants are cross resistant to hydrogen peroxide and exhibit stable chromosomal and biochemical alterations. *Mol. Cell Biochem.* **118**, 181–189.
35. Park, Y.M., Anderson, R.L., Spitz, D.R., and Hahn, G.M. (1992) Hypoxia and resistance to hydrogen peroxide confer resistance to tumor necrosis factor in murine L929 cells. *Radiat. Res.* **131**, 162–168.
36. Sagara, Y., Dargusch, R., Chambers, D., Davis, J., Schubert, D., and Maher, P. (1998) Cellular mechanisms of resistance to chronic oxidative stress. *Free Radic. Biol. Med.* **24**, 1375–1389.
37. Goligorsky, M.S., Morgan, M.A., Lyubsky, S., Gross, R.W., Adams, D.T., and Spitz, D.R. (1993) Establishment of a hydrogen peroxide resistant variant of renal tubular epithelial cells: Role of calcium-independent phospholipase A_2 in cell damage. *Arch. Biochem. Biophys.* **301**, 119–128.
38. Laszlo, A., Davidson, T., Harvey, A., Sim, J.E., Malyapa, R.S., Spitz, D.R., and Roti, J.L. (2006) Alterations in heat-induced radiosensitization accompanied by nuclear structure alterations in Chinese Hamster cells. *Int. J. Hyperthermia.* *22*, 43–60.
39. Sullivan, S.J., Roberts, R.J., and Spitz, D.R. (1991) Replacement of media in cell culture alters oxygen toxicity: Possible role of lipid aldehydes and glutathione transferases in O_2 toxicity. *J. Cell. Physiol.* **147**, 427–433.
40. Spitz, D.R., Sullivan, S.J., Kinter, M.T., Adams, D.T., Sherman, C.M., and Roberts, R.J. (1995) Mechanisms of resistance to oxidative stress in O_2-resistant cells. In: *The Oxygen Paradox* (Davies, K.J.A. and Ursini, F., Eds.), CLEUP University Press, Padova, Italy, pp. 405–412.
41. Joenje, H., Gille, J.J., Oostra, A.B., and Van der Valk, P. (1985) Some characteristics of hyperoxia-adapted HeLa cells. A tissue culture model for cellular oxygen tolerance. *Lab. Invest.* **52**, 420–428.
42. van der Valk, P., Gille, J.J., Oostra, A.B., Roubos, E.W., Sminia, T., and Joenje, H. (1985) Characterization of an oxygen-tolerant cell line derived from Chinese hamster ovary. Antioxygenic enzyme levels and ultrastructural morphometry of peroxisomes and mitochondria. *Cell Tissue Res.* **239**, 61–68.
43. Campian, J.L., Qian, M., Gao, X., and Eaton, J.W. (2004) Oxygen tolerance and coupling of mitochondrial electron transport. *J. Biol. Chem.* **279**, 46580–46587.
44. Li, J., Gao, X., Qian, M., and Eaton, J.W. (2004) Mitochondrial metabolism underlies hyperoxic cell damage. *Free Radic. Biol. Med.* **36**, 1460–1470.

Chapter 12

A Simple Method for Effective and Safe Removal of Membrane Cholesterol from Lipid Rafts in Vascular Endothelial Cells: Implications in Oxidant-Mediated Lipid Signaling

Michelle A. Kline, E.S. O'Connor Butler, Adam Hinzey, Sean Sliman, Sainath R. Kotha, Clay B. Marsh, Rao M. Uppu, and Narasimham L. Parinandi

Abstract

Lipid raft-associated cholesterol has been identified as a pivotal player among membrane lipids in regulating cellular functions. Cholesterol of the vascular endothelial cell (EC) membranes is also being recognized as an important element in the vascular EC signaling. However, methods utilized in studying the important role of lipid raft-associated cholesterol in cell signaling involve removal of the raft cholesterol with the aid of chemical agents called cyclodextrins. Caution should be exercised in using cyclodextrins to remove the cellular lipid raft-associated cholesterol as the cyclodextrins cause adverse effects on cells such as loss of cell viability or induction of cytotoxicity. Therefore, the choice of a cyclodextrin to remove the cellular lipid raft-associated cholesterol is extremely important in order to ensure effective and safe removal of cholesterol from the cellular lipid rafts. In order to achieve this, here, we have selected the bovine pulmonary artery endothelial cells (BPAECs) and subjected them to the removal of cholesterol using two different β-cyclodextrin compounds, methyl-β-cyclodextrin (MβCD) and hydroxypropyl-β-cyclodextrin (HPCD). Phospholipase D (PLD), which generates one of the most potent bioactive lipid signal mediators (phosphatidic acid), is activated by oxidants. Therefore, we examined the effects of cholesterol removal by utilizing our current methods on the hydrogen peroxide (H_2O_2)-activated PLD in BPAECs. Differences in the loss of cholesterol and the resulting effects on the cell membrane, cell viability, morphology, and the extent of oxidant-induced PLD activation were determined. The results revealed that both MβCD and HPCD caused loss of cholesterol, loss of cell viability, and altered cell morphology in the chosen EC system. It was also determined that the HPCD compound caused far less extensive damage to the cells than the MβCD, therefore making the HPCD compound a safer tool for EC cholesterol removal.

Key words: Lipid rafts, cholesterol, vascular endothelial cell, oxidant-induced signaling, phospholipase D.

R.M. Uppu et al. (eds.), *Free Radicals and Antioxidant Protocols*, Methods in Molecular Biology 610,
DOI 10.1007/978-1-60327-029-8_12,

1. Introduction

Cell membranes including the plasma membrane host highly organized microdomains called lipid rafts which contain diverse signal transduction mediators (1). Lipid raft-associated cholesterol is a pivotal player among membrane lipids in modulating the cellular signaling cascades, which regulates normal cellular functions and pathophysiological events (2). Although the distribution of cholesterol in various cellular membranes has been found to be heterogeneous, it has been estimated that 80–90% of the entire cellular cholesterol is localized in the plasma membrane (3). Endothelial cells (ECs) of blood vessels are critical in the maintenance of the vascular tone, homeostasis of the blood vessel, and proper function of the cardiovascular system. Cholesterol of the vascular ECs is being recognized as an important element in the vascular EC signaling that dictates cellular events including replication and cell death (4, 5). Among several key cellular signaling enzymes, phospholipase D (PLD) has been recognized as one of the important players in cellular signaling events (6). PLD in vascular endothelial cells (ECs) has been shown to be activated by oxidants and oxidative stress (6). Hence, we selected the widely used ECs in culture, the bovine pulmonary artery endothelial cells (BPAECs), subjected them to removal of cholesterol using established cholesterol-depleting agents, and established a safer and effective method to deplete the lipid raft-associated cholesterol in order to study its role in the oxidant-induced PLD activation in ECs. However, it is increasingly evident that the lipid signal enzymes are regulated by the lipid raft-associated cholesterol. Needless to mention, the methods of depletion of the raft-associated cholesterol have not been standardized in the vascular ECs in order to study the role of raft-associated cholesterol on the oxidant-induced PLD activation in ECs. The removal of the raft-associated cholesterol molecule from cells using the widely used cyclodextrins was shown to be effective (7).

Our study focused on two of the β-cyclodextrin compounds, methyl-β-cyclodextrin (MβCD) and hydroxypropyl-β-cyclodextrin (HPCD) (**Fig. 12.1**). The ECs were treated for 60 min with 2% (wt/vol) concentration of the chosen cyclodextrins. Differences in the loss of cholesterol, cell viability, morphology, and oxidant (H_2O_2)-induced PLD activation were determined. The cholesterol levels were determined by utilizing the spectrofluorometric method. The lactic dehydrogenase (LDH) released into the medium was assayed by the spectrophotometric method to determine cytotoxicity and ultimately cell viability. Oxidant-induced PLD activation was assayed by determining the PLD-derived phosphatidylbutanol in BPAECs pre-labeled with

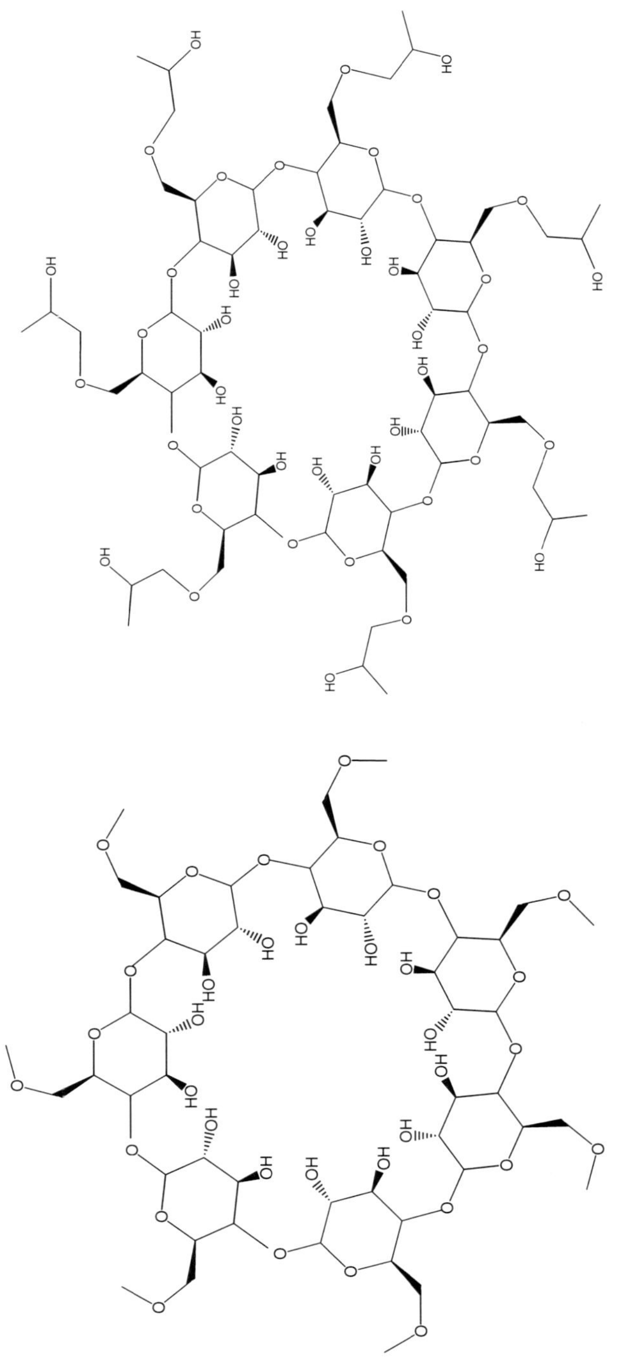

Methyl-β-cyclodextrin

Hydroxypropyl-β-cyclodextrin

Fig. 12.1. MβCD and HPCD example compounds. Above are two proposed structures for both the methyl-β-cyclodextrin (MβCD) and the hydroxypropyl-β-cyclodextrin (HPCD).

[^{32}P] orthophosphoric acid (5 μCi/dish) by utilizing thin layer chromatography and liquid scintillation counting (6). The results revealed that both MβCD and HPCD caused loss of cholesterol, loss of cell viability, and altered cell morphology in the chosen EC system. However, MβCD appeared to cause greater loss of cholesterol, cell viability, and cell morphology as compared to the extent of the same caused by HPCD, under identical conditions, in the chosen EC system. Also, the study demonstrated that removal of cholesterol by MβCD caused enhanced activation of PLD in EC, which was further potentiated by the oxidant (100 μM H_2O_2) treatment. However, depletion of cholesterol by HPCD alone did not cause the activation of PLD in ECs but resulted in the oxidant (100 μM H_2O_2)-induced PLD activation. In comparison with HPCD, cholesterol depletion by MβCD resulted in a greater extent of the oxidant-induced PLD activation in BPAECs. Nevertheless, our study also demonstrated that the MβCD treatment caused greater adverse cellular effects than those caused by HPCD treatment, although both MβCD and HPCD were effective in removing the raft-associated cholesterol including the plasma membrane cholesterol in ECs.

In conclusion, the results of this study showed that the cholesterol was an important player in maintaining the cell viability and morphology in the vascular ECs. Removal of cholesterol by cyclodextrin (especially MβCD) treatment apparently caused loss of fluidity of the cell membrane and leakage of vital cellular components and thus caused loss of cell morphology in ECs. Also, the study offered a safer method of removal of cholesterol by utilizing HPCD, without causing extensive loss of cell viability as seen with the MβCD treatment, for studies to investigate the role of lipid raft-associated cholesterol in cellular functions. Overall, this study underscores the importance of membrane cholesterol in vascular EC integrity.

2. Materials

1. BPAECs (passage 4) were commercially obtained from VEC Technologies (NY, USA).
2. Minimal essential medium (MEM), nonessential amino acids, trypsin, fetal bovine serum (FBS), penicillin/streptomycin, DMEM phosphate-free-modified medium, phosphate-buffered saline (PBS), and Amplex Red Cholesterol Assay Kit were acquired from Invitrogen Corporation (Grand Island, NY).

3. Methyl-β-cyclodextrin (MβCD), 2-hydroxypropyl-β-cyclodextrin (HPCD), and LDH assay kit were purchased from Sigma-Aldrich (St. Louis, MO).
4. Phosphatidylbutanol (PBt) was obtained from Avanti Polar Lipids (Alabaster, AL).
5. [^{32}P]-Orthophosphoric acid was procured from Perkin Elmer (Shelton, CT).
6. Endothelial cell growth factor was purchased from Upstate (Millipore) (Charlottesville, VA).
7. Hydrogen peroxide (3% or 0.88 M) was obtained from Fisher Scientific (Pittsburgh, PA).
8. Uniplate, Silica Gel-H-coated, thin layer chromatography (TLC) plates were purchased from Analtech (Newark, DE).

3. Methods

3.1. Cell Culture

1. BPAECs were grown to confluence (~95%) in MEM supplemented with 10% fetal bovine serum, 100 units/mL penicillin and streptomycin, 5 μg/mL endothelial cell growth factor, and 1% nonessential amino acids at 37°C in a humidified environment of 95% air–5% CO_2 as described earlier in (6, 8).
2. BPAECs, from passages 4–15, were used in the experiments.
3. BPAECs from each primary T-75 cm flask were detached with 0.05% trypsin, re-suspended in fresh medium, and subcultured in 35-mm sterile dishes in complete medium to the required level of confluence under a humidified environment containing 95% air–5% CO_2 at 37°C for treatments with the desired agents.

3.2. Phospholipase D (PLD) Activation in Intact ECs

1. BPAECs in 35-mm dishes (5×10^5 cells/dish) were prelabeled with [^{32}P]-orthophosphoric acid (5 μCi/mL) in DMEM phosphate-free medium containing 2% fetal bovine serum for 12–14 h (6, 8).
2. Cells were washed with basal MEM and treated under the determined conditions with the chosen cyclodextrin compound.
3. Cells were then incubated at 37°C in 1 mL of basal MEM containing 0.05% butanol in absence and presence of H_2O_2 (100 μM) for the desired length of time under a humidified environment of 95% air–5% CO_2.

4. The treatments were terminated by the addition of methanol/concentrated HCl (100/1, v/v).
5. Lipids were extracted essentially according to the method of Bligh and Dyer procedure as described previously (6, 8).
6. [^{32}P]-labeled phosphatidylbutanol (PBt) formed from the PLD activation and transphosphatidylation reaction, as an index of PLD activity in intact cells, was separated by the TLC (6, 8).
7. Radioactivity associated with the [^{32}P]-PBt was determined by the liquid scintillation counting, and data were expressed as DPM normalized to 10^6 counts in the total cellular lipid extract.

3.3. Lactic Dehydrogenase (LDH) Assay of Cytotoxicity

1. BPAECs in 35-mm dishes were grown up to 90% confluence.
2. Cells were treated with basal MEM alone or basal MEM containing the chosen β-cyclodextrin (2% MβCD or HPCD, wt/vol) for 1 h at 37°C in a humidified environment of 5% CO_2–95% air.
3. At the end of the incubation, the medium was collected.
4. The experiment was terminated with 1 N HCl.
5. LDH release was measured by the spectrophotometric method according to the manufacturer's recommendations (Sigma-Aldrich, St. Louis, MO) (9).

3.4. Membrane Cholesterol Depletion and Determination

1. BPAECs in 35-mm dishes were grown up to 90% confluence.
2. Cells were treated with basal MEM alone or basal MEM containing the chosen β-cyclodextrin (2% MβCD or HPCD, wt/vol) for 1 h at 37°C in a humidified environment of 5% CO_2–95% air.
3. At the end of the incubation, the medium was removed and 1 mL of PBS was added to each dish containing the cells.
4. The cells were then detached with a cell scrapper and transferred into an Eppendorf tube and centrifuged for 10 min at 15,000 rpm.
5. After centrifugation, the supernatant was removed without disturbing the cell pellet.
6. The cholesterol amount in the cells was then determined spectrofluorometrically according to the manufacturer's recommendations (Molecular Probes – Invitrogen Detection Technologies, Grand Island, NY).

3.5. Statistical Analysis

1. Standard deviation (SD) for each data point was calculated from triplicate determinations under identical conditions.
2. Data were subjected to one-way analysis of variance and pairwise multiple comparisons were done by Dunnett's method with the significance set at $p < 0.05$.

3.6. Results

Cholesterol is an important constituent of the mammalian cell membranes, mainly the plasma membrane. Being heterogeneously distributed among different membrane locations in the cell, the plasma membrane hosts up to 80–90% of the total cellular cholesterol (3), wherein it plays a major role in the organization and function of cellular lipid rafts which take part in signal transduction events (10). In order to study the role of lipid raft-associated cholesterol in cellular signaling events, the β-cyclodextrins, which offer the utility for solubilizing, have been widely utilized to remove or deplete cholesterol from mammalian cells (11, 12). Among the β-cyclodextrins, MβCD appears to be a popular cyclodextrin that is being widely utilized in removing cholesterol from the plasma membrane and lipid rafts of the mammalian cells. Up to 2% (wt/vol) or 10 mM

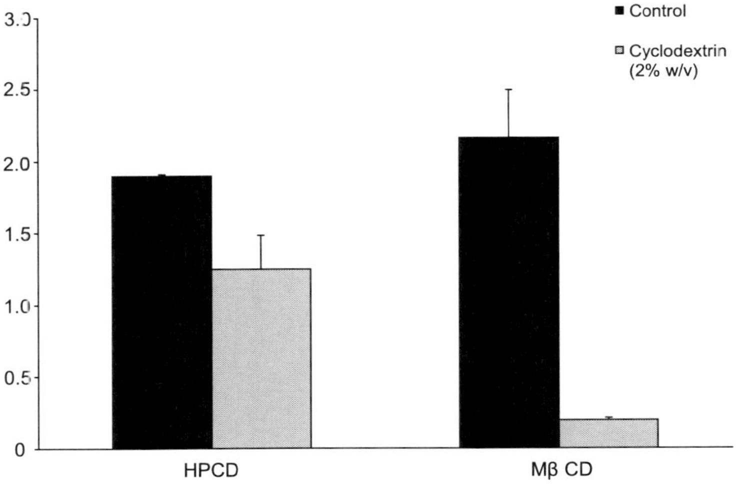

Fig. 12.2. Comparison of cholesterol depletion in BPAECs using HPCD and MβCD. Confluent BPAECs (5×10^5 cells/35-mm dish) were treated for 60 min with basal MEM or basal MEM containing HPCD or MβCD (2% w/v). The amount of cholesterol (μg) in cells following the treatments was determined spectrofluorometrically as described in the Materials section. This figure illustrates how both β-cyclodextrin compounds are effective in removing the cholesterol from the BPAECs; however, the MβCD removes far more cholesterol in comparison with HPCD. Data represents mean ± SD of three independent experiments. *Significantly different at $p < 0.05$ as compared to the control cells treated with basal MEM alone.

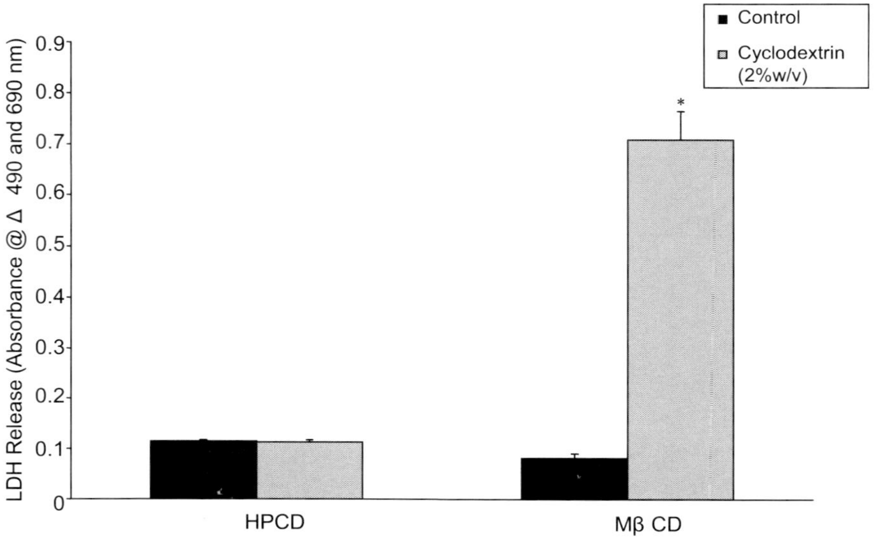

Fig. 12.3. Comparison of LDH released from BPAECs with both HPCD and MβCD. Confluent BPAECs (5×10^5 cells/35-mm dish) were treated for 60 min with basal MEM or basal MEM containing HPCD or MβCD (2% w/v). LDH released is directly proportional to the extent of cytotoxicity and indicative of the cell viability. The extent of LDH release was analyzed spectrophotometrically using the lactic dehydrogenase assay kit as described in the Materials section. This figure demonstrates the cytotoxicity caused by MβCD only after 1 h of treatment of cells. Data represent mean ± SD of three independent experiments. *Significantly different at $p < 0.05$ as compared to the control cells treated with basal MEM alone.

concentration of MβCD has been used in several reported studies to remove cholesterol from different mammalian cell models (13–15). Moreover, in several of these documented studies cells have been treated with MβCD for prolonged time periods of 4 h and beyond. MβCD has been identified to be a toxic compound (11). MβCD has been shown to disrupt lipid rafts and cause apoptotic cell death in keratinocytes (13). In the current study, although MβCD was shown to be effective in depleting cholesterol from BPAECs (**Figs. 12.2** and **12.3**), its cytotoxicity was inevitable (*see* **Note 1**). On the other hand, the less widely used cyclodextrin, HPCD, was demonstrated to cause no cytotoxicity to BPAECs while effectively removing cholesterol from BPAECs (**Figs. 12.2** and **12.3**). Whether the effects of MβCD are solely due to its potential to remove cholesterol from cells or due to other cellular biochemical alterations are not clear. Nevertheless, it is emerging that MβCD causes other cellular biochemical alterations such as the removal of membrane phospholipids, fatty acids, and proteins (Hinzey and Parinandi, unpublished observations), in addition to depleting EC cholesterol. In comparison to MβCD, HPCD appeared to be safer in not causing adverse cellular effects and efficiently removing cholesterol from the ECs (*see* **Notes 2** and **3**).

PLD is an important cell signal membrane phospholipid hydrolase which has been shown to be activated by oxidants (6, 8). The enzyme, which exists in two isoforms, PLD_1 and PLD_2, in mammalian cells including the vascular endothelial cells, forms phosphatidic acid, the bioactive lipid signal indicator, upon its action on the membrane phospholipids, especially, phosphatidylcholine. However, cholesterol (plasma membrane and raft-associated) is known to modulate the structure and function of the biological membranes, and hence the cellular signaling and functions. PLD, a signaling phospholipase, is not an exception to this. As demonstrated in this study, cholesterol removal by both HPCD and MβCD caused significant enhancement of the H_2O_2-induced PLD activation in ECs, wherein MβCD exerted a greater effect (**Fig. 12.4**) (*see* **Notes 4** and **5**). However, MβCD alone, in comparison with HPCD, caused significant activation of PLD as compared to that in the control cells. Nonetheless, our study also showed that MβCD, although removed cholesterol from cells, caused cytotoxicity in ECs as compared to HPCD, which also effectively removed cellular cholesterol without causing

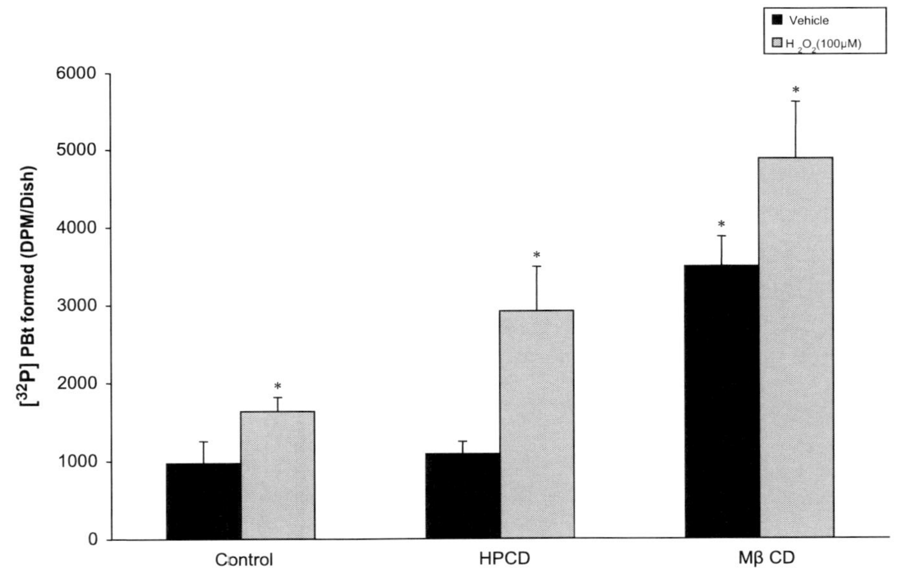

Fig. 12.4. Oxidant-induced PLD activation response to cyclodextrin-assisted cholesterol removal in ECs. Confluent BPAECs (5×10^5 cells/35-mm dish) were treated for 60 min with basal MEM or basal MEM containing HPCD or MβCD (2% w/v). Following cholesterol removal from cells the treatments were removed and the cells were treated with basal MEM only or basal MEM containing 100 μM H_2O_2 for another 60 min and the PLD activity was determined as the formation of [^{32}P]-PBt in intact cells as described in Section **3**. These results indicated that the H_2O_2-induced PLD activation was enhanced by the cellular cholesterol removal assisted by MβCD and HPCD compounds where in the effect of former cyclodextrin was greater than the effect of the latter. In direct comparison of both β-cyclodextrins compounds, the activation increased even greater when more cellular cholesterol was removed with MβCD. Data represent mean ± SD of three independent experiments. *Significantly different at $p < 0.05$ as compared to the control cells treated with basal MEM alone.

any cytotoxicity. Therefore, the enhancement of H_2O_2-induced PLD activation by MβCD could be attributed not only to its cholesterol-depleting (removing) property but also to its cytotoxic effects as shown by others and in this work. Furthermore, as revealed from the current study that the enhancement of the H_2O_2-induced PLD activation in ECs could very well be due to the removal of cholesterol (plasma membrane and lipid raft-associated) by HPCD but not merely due to any adverse effects of HPCD on ECs.

In conclusion, cyclodextrins can be used for the removal of cellular cholesterol (plasma membrane and raft-associated) in studies aimed at the role of lipid raft-associated cholesterol on the oxidant-activated lipid signaling enzymes in ECs, provided the chosen cyclodextrin is not cytotoxic and only is effective in cellular cholesterol removal.

4. Notes

1. Solutions of β-cyclodextrins (MβCD and HPCD) should be freshly prepared in the basal cell culture medium (MEM) for treatment of cells to remove membrane cholesterol. Solutions of β-cyclodextrins should not be prepared ahead and stored either at room temperature or at 4°C.
2. While treating the cells at higher confluence (90–100%), following the addition of β-cyclodextrins to the cells, a gentle swirling for 15 s may be necessary during treatment of cells at 37°C in the tissue culture incubator for effective cholesterol removal from the cells.
3. Any concentration above 1% (w/v) of β-cyclodextrins, especially of MβCD, should be avoided for cholesterol removal from the cells to avoid or minimize cytotoxicity. Also, treatment of the cells beyond 2 h with β-cyclodextrins will also cause adverse cellular effects and should be avoided.
4. Medium containing the radiolabel ($[^{32}P]$-orthophosphate) following pre-labeling of the cells should be carefully removed prior to treating the cells with β-cyclodextrins to remove cellular cholesterol. Avoid treating the cells with β-cyclodextrins while labeling the cells with $[^{32}P]$-orthophosphate as the β-cyclodextrins interfere with the labeling of cellular phospholipids.
5. Solutions of H_2O_2 in the desired medium should be freshly prepared right before treating the cells with oxidants.

Acknowledgments

Support provided toward this work by the Dorothy M. Davis Heart and Lung Research Institute and the Division of Pulmonary, Allergy, Critical Care and Sleep Medicine at The Ohio State University College of Medicine, and the ARCH Program at Southern University-Baton Rouge (NIH grant number ES10018) is acknowledged.

References

1. Zajchowski, L.D. and Robbins, S.M. (2002) Lipid rafts and little caves, compartmentalized signaling in membrane microdomains. *Eur. J. Biochem.* **269**, 737–752.
2. Mishra, S. and Preeti, G.J. (2007) Lipid raft heterogeneity: an enigma. *J. Neurochem.* **103**, 135–142.
3. Mukherjee, S., Zha, X., Tabas, I., and Maxfield, F.R. (1998) Cholesterol distribution in living cells: Fluorescence imaging using dehydroergosterol as a fluorescent cholesterol analog. *Biophys. J.* **75**, 1915–1925.
4. Li, P.L., Zhang, Y., and Yi, F. (2007) Lipid raft redox signaling platforms in endothelial dysfunction. *Antioxid. Redox Signal* **9**, 1457–1470.
5. Xia, M., Wang, Q., Zhu, H., Ma, J., Hou, M., Tang, Z., Li, J., and Ling, W. (2007) Lipid rafts regulate cellular CD40 receptor localization in vascular endothelial cells. *Biochem. Biophys. Res. Commun.* **3**, 768–774.
6. Varadharaj, S., Steinhour, E., Hunter, M.G., Watkins, T., Baran, C.P., Magalang, U., Kuppusamy, P., Zweier, J.L., Marsh, C.B., Natarajan, V., and Parinandi, N.L. (2006) Vitamin C-induced activation of phospholipase D in lung microvascular endothelial cells: regulation by MAP kinases. *Cell Signal* **9**, 1396–1407.
7. Zidovetzki, R. and Levitan, I. (2007) Use of cyclodextrins to manipulate plasma membrane cholesterol content: evidence, misconceptions and control strategies. *Biochim. Biophys. Acta* **6**, 1311–1324.
8. Hagele, T.J., Mazerik, J.N., Gregory, A., Kaufman, B., Magalang, U., Kuppusamy, M.L., Marsh, C.B., Kuppusamy, P., and Parinandi, N.L. (2007) Mercury activates vascular endothelial cell phospholipase D through thiols and oxidative stress. *Int. J. Toxicol.* **1**, 57–69.
9. Mazerik, J.N., Hagele, T., Sherwani, S., Ciapala, V., Butler, S., Kuppusamy, M.L., Hunter, M., Kuppusamy, P., Marsh, C.B., and Parinandi, N.L. (2007) Phospholipase A_2 activation regulates cytotoxicity of methylmercury in vascular endothelial cells. *Int. J. Toxicol.* **26**, 553–569.
10. Crane, J.M. and Tamm, L.K. (2004) Role of cholesterol in the formation and nature of lipid rafts in planar and spherical model membranes. *Biophys. J.* **86**, 2965–2979.
11. Stella, V.J. and He, Q. (2008) Cyclodextrins. *Toxicol. Pathol.* **36**, 30–42.
12. Ostrom, R.S. and Liu, X. (2007) Detergent and detergent-free methods to define lipid rafts and caveolae. *Methods Mol. Biol.* **400**, 459–468.
13. Bang, B., Gniadecki, R., and Gajkowaska, B. (2005) Disruption of lipid rafts causes apoptotic cell death in HaCaT keratinocytes. *Exp. Dermatol.* **14**, 266–272.
14. Pottosin, I.I., Valencia-Cruz, G., Bonales-Alatorre, E., Shabala, S.N., and Dobrovinskaya, O.R. (2007) Methyl-β-cyclodextrin reversibility alters the gating of lipid rafts-associated Kv1.3 channels in Jurkat T lymphocytes. *Eur. J. Physiol.* **454**, 235–244.
15. Larbi, A., Douziech, N., Khalil, A., Dupuis, G., Gherari, S., Guérard, K.P., and Fülöp, T., Jr. (2004) Effects of methyl-β-cyclodextrin on T lymphocytes lipid rafts with aging. *Exp. Gerontol.* **39**, 551–558.

Chapter 13

Superoxide Dismutase – A Target for Gene Therapeutic Approach to Reduce Oxidative Stress in Erectile Dysfunction

W. Deng, T.J. Bivalacqua, H.C. Champion, W.J. Hellstrom, Subramanyam N. Murthy, and Philip J. Kadowitz

Abstract

Erectile dysfunction (ED) is defined as the inability to attain and/or maintain penile erection sufficient for satisfactory sexual performance. Oxidative stress has been demonstrated to be involved in the pathophysiology of age- or diabetes-related ED. Superoxide dismutase (SOD), an antioxidant enzyme catalyzing the conversion of superoxide anion ($O_2^{\bullet -}$) to hydrogen peroxide (H_2O_2) and molecular oxygen (O_2), is a promising therapeutic target for ED. In vivo gene therapy and adult stem cell-based ex vivo gene therapy are two attractive current gene therapies for the treatment of ED. In this chapter we describe the use of two potent gene transfer techniques to deliver the therapeutic gene extracellular superoxide dismutase (ecSOD) into the penis of aged or diabetic rats for therapy of ED: adenoviral-mediated intracavernosal ecSOD gene transfer for gene therapy of ED and ecSOD gene-modified marrow stromal cells, also known as mesenchymal stem cells, based stem cell and gene therapy.

Key words: Oxidative stress, superoxide dismutase, penis, erectile dysfunction, gene therapy, adenovirus, marrow stromal cells, mesenchymal stem cells, stem cell therapy.

1. Introduction

Erectile dysfunction (ED), a medical condition formerly termed impotence, is defined as the inability to attain and/or maintain penile erection sufficient for satisfactory sexual performance (1, 2). ED is a highly prevalent health problem with considerable impact on the quality of life of men and their partners (3–7). According to the Massachusetts Male Aging Study, the combined prevalence of minimal, moderate, and complete ED

R.M. Uppu et al. (eds.), *Free Radicals and Antioxidant Protocols*, Methods in Molecular Biology 610,
DOI 10.1007/978-1-60327-029-8_13,

in men 40–70 years old was 52% and the prevalence of complete impotence tripled from 5 to 15% between subject ages 40 and 70 years (3). Age is the variable most strongly associated with ED and, at age 40–70 years, the percentage of men with ED increased from 40 to 67%. After adjustment of age, a higher probability of ED is directly correlated with heart disease, hypertension, diabetes mellitus, and other diseases (3). ED affects as many as 30 million men in the United States and 152 million men in the world (1, 3, 8, 9). It is estimated that the worldwide prevalence of ED in the year of 2025 will be 322 million, due to the normal aging process of the world population and the accumulation of age-related risk factors contributing to ED (5, 9, 10).

Penile erection, referred to as engorgement of the penis with blood, is a complex neurovascular phenomenon mainly requiring the relaxation of cavernosal smooth muscle (11). Nitric oxide ($^{\bullet}NO$) is the principal mediator for cavernosal smooth muscle relaxation and thus penile erection (12, 13). Recent studies indicate that accumulation of superoxide anion ($O_2^{\bullet-}$) in penis is involved in the pathogenesis of both age- and diabetes-related ED (14–18). Through interaction with $^{\bullet}NO$, $O_2^{\bullet-}$ cannot only scavenge $^{\bullet}NO$ within vessel walls or corporal sinusoids but also form the highly cytotoxic molecule peroxynitrite ($ONOO^-$) (19, 20). Superoxide dismutase (SOD), a metalloenzyme catalyzing the dismutation of $O_2^{\bullet-}$ to hydrogen peroxide (H_2O_2) and molecular oxygen (O_2), is an important part of the cellular antioxidant defense system (21). Extracellular superoxide dismutase (ecSOD), one of the three identified SOD isoforms, is the primary extracellular antioxidant enzyme for $O_2^{\bullet-}$ (22–24). Therefore, ecSOD is an attractive target for gene therapeutic approach for ED because it can relieve oxidative stress in ED. This chapter reviews the use of two potent gene transfer techniques: adenoviral gene transfer of ecSOD to the penis of aged or diabetic rats for gene therapy of ED and ecSOD-modified marrow stromal cells, also known as mesenchymal stem cells, based stem cell and gene therapy.

2. Materials

2.1. Adenoviral Vectors

1. Ad5CMVntlacZ: a replication-deficient recombinant adenoviral vector carrying the nuclear-targeted β-galactosidase gene ntlacZ under the control of the cytomegalovirus (CMV) promoter (17, 18, 25, 26). Prepared at the Gene Transfer Vector Core facility at the University of Iowa (Iowa City, Iowa, IA), the vector has a titer of 1×10^{10} plaque-forming unit (pfu)/mL. The vectors are suspended in 3% (w/v) sucrose/PBS (SPBS). Store at –80°C until use.

2. Ad5CMVecSOD: a replication-deficient recombinant adenoviral vector carrying the human extracellular superoxide dismutase (ecSOD) gene under the control of CMV promoter (17, 18, 27, 28). Prepared at the Gene Transfer Vector Core facility at the University of Iowa, the vector has a titer of 1×10^{10} pfu/mL. The vectors are suspended in SPBS. Store at –80°C until use.

2.2. Experimental Animals

1. Male, 6-wk-old, Sprague Dawley rats (body weight 50–75 g).
2. Male, 6-wk-old, brown Norway rats (body weight 50–75 g).
3. Male, 12-wk-old, Sprague Dawley rats (body weight 325–350 g).
4. Male, 80-wk-old, brown Norway rats (body weight 400–525 g; *see* **Note 1**).

All these rats can be obtained from National Institutes of Health (NIH)/National Institute on Aging colony at Harlan Sprague Dawley (San Diego, CA). The rats must be maintained under controlled temperature and lighting and treated according to NIH regulations.

2.3. Induction of Diabetes-Related ED

1. Water: pyrogen-free ultrapure water (Sigma, St. Louis, MO).
2. Citrate (0.1 M)–phosphate (0.2 M) buffer (CPB), pH 7.0: prepare this buffer by dissolving 21 g of citric acid monohydrate along with 14 g of sodium phosphate (dibasic, anhydrous) in 850 mL of water. Adjust the pH 7.0 and then make up the volume 1 L with water.
3. Streptozotocin (STZ) solution: dissolve STZ (Sigma) in CPB at a concentration of 100 mg/mL. Check the pH and adjust to 7.0 if necessary.
4. Give intraperitoneal (i.p.) injection of STZ to 12-wk-old Sprague Dawley rats a concentration of 60 mg/kg body weight.
5. Measure blood glucose before and after the administration of STZ. Rats develop diabetes-related ED typically in about 8 wk after the administration of STZ. The rats are now ready for use in gene or stem cell therapeutic studies (17, 29, 30).

2.4. Isolation of Rat Marrow Stromal Cells (rMSCs)

1. Male, 6-wk-old, brown Norway or Sprague Dawley rats can be used for isolation of rMSCs (31–33) (*see* **Note 2**).
2. Cell strainer with 70-μm nylon mesh (BD Biosciences, Bedford, MA).
3. T75 culture flask (Fisher Scientific, Pittsburgh, PA).
4. Centrifuge tube, 15 and 50 mL capacity (Fisher Scientific).

5. Culture dish, 10 cm (Fisher Scientific).
6. Needle, 21 gauge (Fisher Scientific).
7. Syringe, 10 mL (Fisher Scientific).
8. Scissors (Fisher Scientific).
9. Forceps (Fisher Scientific).
10. Scalpel (Fisher Scientific).

2.5. Western Blot Analysis for ecSOD Expression in Rat Penis

1. Phosphate-buffered saline (PBS), pH 7.2, 1× (GIBCO Invitrogen, Grand Island, NY).
2. Sodium pentobarbital solution: dissolve pentobarbital sodium salt (Sigma) in PBS at a concentration of 30 mg/mL.
3. Polytron (Brinkmann Instruments, Westbury, NY).
4. Tissue lysing buffer: 1% NP-40, 0.5% sodium deoxycholate, 0.1% SDS, 0.1 mg/mL leupeptin, and 574 μM phenylmethylsulfonyl fluoride (Sigma). Prepare this buffer in PBS.
5. Centrifuge tube, 1.5 mL capacity (Fisher Scientific).
6. Refrigerated benchtop microcentrifuge (Fisher Scientific).
7. BCA protein assay kit (Pierce, Rockford, IL).
8. SDS–PAGE gel, 8% (Fisher Scientific).
9. A protein electrophoresis system, a power supply, and a protein transferring system (Bio-Rad Laboratories, Hercules, CA).
10. Nitrocellulose membrane (Amersham Bioscience, Piscataway, NJ).
11. Rabbit polyclonal anti-human ecSOD antibody (a gift from Dr. James Crapo of National Jewish Medical and Research Center in Denver, CO).
12. Horseradish peroxidase conjugated to goat anti-rabbit IgG (Santa Cruz Biotechnology, Santa Cruz, CA).
13. Enhanced chemiluminescence (ECL) Western blotting detection reagents (Amersham Bioscience).

2.6. In Vivo Measurement of Rat Erectile Responses

1. Sodium pentobarbital solution: dissolve pentobarbital sodium salt (Sigma) in PBS at a concentration of 30 mg/mL.
2. PE-50 tubing (BD Biosciences, Franklin Lakes, NJ).
3. Needle, 25 gauge (Fisher Scientific).
4. Viggo-Spectramed transducer (Viggo-Spectramed, Oxnard, CA).
5. Data acquisition system (Biopac Systems, Santa Barbara, CA).
6. Square wave stimulator (Grass Instruments, Quincy, MA).

2.7. Isolation and Ex Vivo Expansion of rMSCs

1. MEM-α, with ribonucleosides and deoxyribonucleosides (GIBCO Invitrogen).
2. Fetal bovine serum (FBS) (GIBCO Invitrogen).
3. Antibiotic–antimycotic, 100× solution: 10,000 units/mL penicillin, 10,000 μg/mL streptomycin, 2,500 ng/mL amphotericin B (Atlanta Biologicals, Atlanta, GA).
4. L-Glutamine, 100× solution: 200 mM L-Glutamine (GIBCO Invitrogen).
5. Culture medium for rMSCs: MEM-α, 20% FBS, 100 units/mL penicillin, 100 μg/mL. streptomycin, 25 ng/mL amphotericin B, and 2 mM L-glutamine.
6. Phosphate-buffered saline (PBS), pH 7.2, 1× (GIBCO Invitrogen).
7. Trypsin–EDTA solution: 0.25% (w/v) trypsin and 1 mM EDTA (Atlanta Biologicals).
8. T75 culture flask (Fisher Scientific).
9. Centrifuge tube, 15 and 50 mL capacity (Fisher Scientific).
10. Culture plates, 6-well (Fisher Scientific).

2.8. X-Gal Staining for β-Galactosidase Activity in Ad5CMVntlacZ-transduced rMSCs

1. Fixing solution: 2% (w/v) formaldehyde (Sigma) and 0.2% (w/v) glutaraldehyde (Sigma), prepare the solution in PBS.
2. Staining solution: 5 mM potassium ferricyanide (Sigma), 5 mM potassium ferrocyanide (Sigma), 2 mM $MgCl_2$ (Sigma), and 1 mg/mL X-gal (Sigma), prepare the solution in PBS.

2.9. Measurement of SOD Activity In Vitro

A SOD activity assay kit (Cayman Chemical Company, Ann Arbor, MI) can be used to measure SOD activity in the culture supernatant of Ad5CMVecSOD-transduced rMSCs. The kit contains a quality-controlled SOD standard for the generation of a standard curve. SOD activity is assessed by measuring the dismutation of $O_2^{\bullet-}$ generated by xanthine oxidase and hypoxanthine in a convenient 96-well format. The detection of $O_2^{\bullet-}$ is based on the reduction of tetrazolium salt to a formazan dye that absorbs at 450 nm.

2.10. In Vitro Differentiation of rMSCs into Osteoblasts and Adipocytes

1. Osteogenic medium: 1×10^{-5} mM dexamethasone (Sigma), 0.2 mM ascorbic acid (Sigma), and 10 mM β-glycerol phosphate (Sigma). Prepare this in the culture medium for rMSCs.
2. Adipogenic medium: 0.5 μM hydrocortisone (Sigma), 500 μM isobutylmethylxanthine (Sigma), and 60 μM

indomethacin (Sigma). Prepare the medium in culture medium for rMSCs.

3. Methanol (Sigma).
4. Alizarin red S solution: dissolve Alizarin red S (Sigma) in water at a concentration of 2% (w/v). Check the pH and adjust to 4.1 if necessary.
5. Formalin, 10% (Sigma).
6. Oil red O stock solution: dissolve Oil red O (Sigma) in isopropanol (Sigma) at a concentration of 0.5% (w/v).
7. Oil red O working solution: prepare this solution by mixing three parts of the Oil red O stock solution with two parts of water and filtering through a 0.45 μm pore-size filter (Fisher Scientific).

3. Methods

3.1. Adenoviral-Mediated ntlacZ or ecSOD Gene Delivery to Corpora Cavernosa

1. Anesthetize rats with pentobarbital sodium (30 mg/kg, i.p.) and place them in a supine position on a temperature-regulated surgical table.
2. Under sterile conditions, expose the penis and inject 20 μL of vehicle (3% SPBS), Ad5CMVntlacZ, or Ad5CMVecSOD into the corpus cavernosum using a 30-gauge needle attached to a Hamilton microliter syringe (17, 18, 34).
3. Immediately before injection, halt the blood drainage via the dorsal veins by circumferential compression at the base of the penis with an elastic band.
4. Release compression 30 s after the injection of 20 μL vehicle or adenovirus.
5. After time periods ranging from 1 to 2 d following intracavernosal injection of the adenoviral vector, measure erectile responses in the aged or diabetic rats with ED.

3.2. Measurement of Rat Erectile Responses In Vivo

1. Anesthetize rats with sodium pentobarbital (30 mg/kg, i.p.) and place them on a thermoregulated surgical table.
2. Cannulate the trachea using PE 240 tubing.
3. Cannulate a carotid artery using PE 50 tubing and measure mean systemic arterial pressure (MAP) using a Viggo-Spectramed transducer.
4. Expose the bladder and the prostate. Identify the cavernosal nerve. Place an electrical stimulator with a stainless steel bipolar hook around the cavernosal nerve.
5. Give an incision to the skin overlying the penis to expose the right crura. Insert a 25-gauge needle filled with normal

saline containing 250 units/mL of heparin and connected to PE-50 tubing into the right crura and connected to a Viggo-Spectramed transducer to measure intracavernosal pressure (ICP).

6. Connect both pressure transducers to a data acquisition system to continuously measure MAP and ICP.
7. Stimulate the cavernosal nerve with a square wave stimulator. Subject each rat to cavernosal nerve stimulation at a frequency of 15 Hz and pulse width of 30 s. The applications of 2.5, 5, and 7.5 V are used to induce erectile responses with various intensities. The stimulation duration is 1 min with a rest period of 3 min between stimulations.
8. Total erectile response, i.e., total ICP, is determined by the area under the erectile curve (AUC) from the beginning of cavernosal nerve stimulation until the ICP pressure returns to baseline or pre-stimulation pressure. The ratio between maximal ICP and MAP at the peak of erectile response is calculated and used to determine erectile function (17, 18, 31).

3.3. Western Blot Analysis of ecSOD Expression in Corpora Cavernosa

1. Kill rats with an overdose of sodium pentobarbital (100 mg/kg, i.p.) 1 day after intracavernosal injection of Ad5CMVecSOD, Ad5CMVntlacZ, or vehicle, collect rat cavernosal tissue.
2. Homogenize rat cavernosal tissue in ice-cold tissue lysing buffer using a Polytron homogenizer. Incubate for 45 min on ice, centrifuged at $15{,}000\times g$ for 20 min, collect the supernatant, and determine its protein concentration using the BCA protein assay kit.
3. Use 70-μg protein, i.e., cavernosal homogenate, to run protein electrophoresis on an 8% SDS–PAGE gel.
4. Transfer the fractionated cavernosal homogenate onto a nitrocellulose membrane.
5. Use rabbit anti-ecSOD polyclonal antibody (1:5,000 dilution) and horseradish peroxidase conjugated to goat anti-rabbit IgG (1:20,000 dilution) to conduct immunodetection.
6. Process the nitrocellulose membrane with the ECL Western blotting detection reagents to determine the expression of ecSOD protein in rat cavernosal tissue (17).

3.4. Isolation and Ex Vivo Expansion of rMSCs

1. Euthanize 6-wk-old rats with CO_2.
2. Under sterile conditions, remove femurs and tibias and place them in culture medium for rMSCs.

3. Cut both ends of the femur and tibia and flush out the bone marrow with a 21-gauge needle attached to a 10-mL syringe.
4. Filter the bone marrow cells through a cell strainer with 70-μm nylon mesh.
5. Culture the cells from one rat in two T75 flasks at 37°C with 5% humidified CO_2.
6. Add and replace fresh culture medium for rMSCs every 2–3 days to remove non-adherent cells. rMSCs from the bone marrow of rat long bones are isolated by their tight adherence to tissue culture plastic (31, 32; *see* **Note 3**).
7. Grow the adherent rMSCs to 90% confluency over about 7 days.
8. Harvest the cells with trypsin–EDTA solution, diluted 1:3 per passage, re-plate the cells in T75 flasks, again grow the cells to 90% confluency for further ex vivo expansion (*see* **Fig. 13.1**).

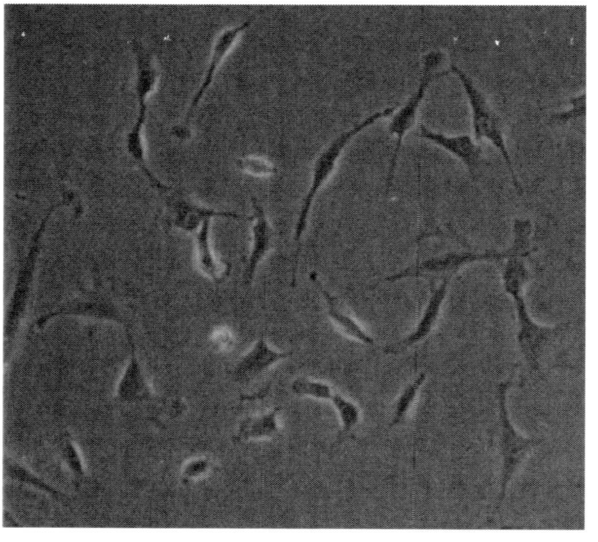

Fig. 13.1. Phase-contrast microscopy of ex vivo expanded rMSCs. Magnification: ×250.

3.5. Adenoviral Gene Transfer of ntlacZ into Ex Vivo Expanded rMSCs

1. Plate rMSCs at a density of 10,000 cells/cm^2 in 6-well plates and culture the cells overnight (*see* **Note 4**).
2. Count the cells, expose the cells to fresh culture medium for rMSCs containing Ad5CMVntlacZ at 300 multiplicity of infection (MOI, defined as pfu per cell), or 3% SPBS for 48 h, and remove the virus-containing supernatant (31, 35; also *see* **Note 5**).

3. Wash the cells with PBS, treat the cells with fixing solution for 5 min, and rinse the cells twice with PBS.
4. Incubate the cells in X-gal staining solution at 37°C for 3 h in the dark and then wash the cells with PBS.
5. Evaluate the expression of ntlacZ transgene in Ad5CMVntlacZ-transduced rMSCs under a phase-contrast microscope through identifying the blue nuclear-targeted β-galactosidase-positive cells (*see* **Fig. 13.2**).

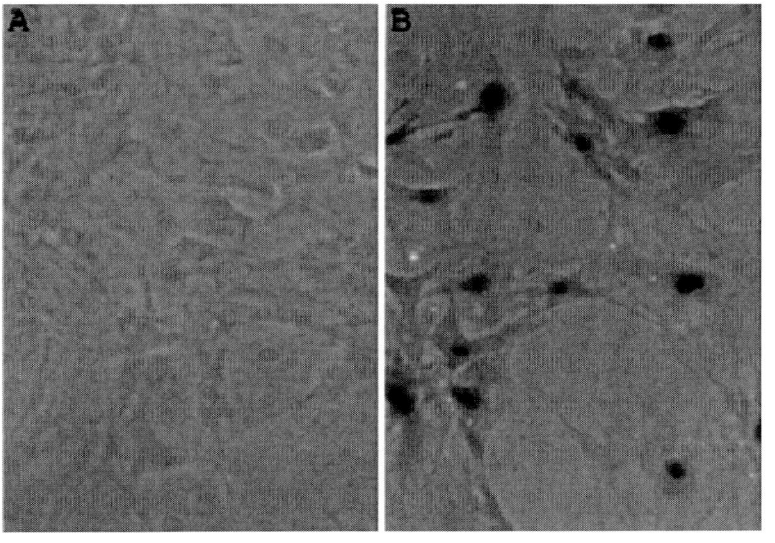

Fig. 13.2. Photomicrograph showing β-galactosidase-positive blue nuclei in Ad5CMVntlacZ-transduced rMSCs. rMSCs were transduced with Ad5CMVntlacZ at MOI 300 for 48 h. The cells were then X-gal stained for nuclear-targeted β-galactosidase activity: (**a**) Control rMSCs and (**b**) rMSCs transduced with Ad5CMVntlacZ at MOI 300. Magnification: ×250.

3.6. Adenoviral Gene Transfer of ecSOD into Ex Vivo Expanded rMSCs

1. Plate rMSCs at a density of 10,000 cells/cm^2 in 6-well plates and culture the cells overnight.
2. Count the cells, expose the cells to fresh culture medium for rMSCs containing Ad5CMVecSOD at 300 MOI, Ad5CMVntlacZ at MOI 300, or 3% SPBS for 48 h, and remove the virus-containing supernatant.
3. Wash the cells three times with PBS, add fresh culture medium for rMSCs to culture the transduced cells for 48 h, and collect the culture supernatant.
4. Measure SOD activity in the culture supernatant of Ad5CMVecSOD-transduced rMSCs, Ad5CMVntlacZ-transduced rMSCs, or naive rMSCs with a SOD activity assay kit. This helps to determine if Ad5CMVecSOD-transduced rMSCs secrete biologically active ecSOD.

3.7. Cell Proliferation Assay for Adenoviral-Transduced rMSCs

1. Plate rMSCs at a density of 10,000 cells/cm^2 in 6-well plates and culture the cells overnight.
2. Count the cells, expose the cells to fresh culture medium for rMSCs containing Ad5CMVntlacZ at MOI 300, Ad5CMVecSOD at 300 MOI, or 3% SPBS for 48 h, and remove the virus-containing supernatant.
3. Wash the cells three times with PBS, add fresh culture medium for rMSCs to further culture the adenoviral-transduced rMSCs for 48 h, and count the cells (32).
4. Compare the final cell number of Ad5CMVntlacZ-transduced rMSCs, Ad5CMVecSOD-transduced rMSCs, and naive rMSCs to determine if Ad5CMVecSOD transduction can enhance the proliferation capability of rMSCs (*see* **Fig. 13.3**).

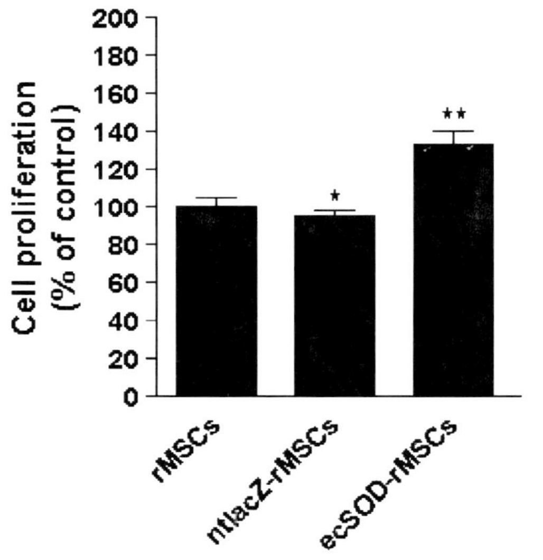

Fig. 13.3. Bar graph showing an increase in cell proliferation capability in Ad5CMVecSOD-transduced rMSCs. rMSCs were transduced with Ad5CMVntlacZ or Ad5CVecSOD at MOI 300 for 48 h. The virus-containing supernatant was removed, the cells were washed three times with PBS, and further incubated in fresh culture medium for 2 days. Cell proliferation was assessed by counting the cells and expressing the data as percentage of control. Each value represents mean ± SEM ($n = 3$) and the data are analyzed by one-way ANOVA followed by post hoc analysis with Tukey test. $^{*}p > 0.05$ versus wild-type rMSCs. $^{**}p < 0.01$ versus wild-type rMSCs or Ad5CMVntlacZ-transduced rMSCs.

3.8. Cell Multipotential Differentiation Assay for Adenoviral-Transduced rMSCs

1. Plate rMSCs at a density of 10,000 cells/cm^2 in 6-well plates and culture the cells overnight.
2. Count the cells, expose the cells to fresh culture medium for rMSCs containing Ad5CMVntlacZ at MOI 300,

Ad5CMVecSOD at 300 MOI, or 3% SPBS for 48 h, and remove the virus-containing supernatant.

3. Wash the cells with PBS, further culture the adenoviral-transduced rMSCs with osteogenic medium or adipogenic medium (31, 32). Change the differentiation medium every 3 days until day 21.
4. To assess mineral deposition, rinse the cells with PBS, fix the cells with cold methanol (–20°C) for 10 min, and wash the cells twice with water. Stain the cells with Alizarin red S solution for 15 min, wash the cells five times with water, and evaluate osteogenic differentiation under a phase-contrast microscope through identifying the red mineral-positive differentiated osteoblasts. Compare the percentage of differentiated osteoblasts of Ad5CMVntlacZ-transduced rMSCs, Ad5CMVecSOD-transduced rMSCs, and naive rMSCs to determine if Ad5CMVecSOD transduction can enhance the osteogenic differentiation capability of rMSCs (*see* **Fig. 13.4**).
5. To assess lipid droplet formation, rinse the cells with PBS, fix the cells with 10% formalin for 1 h, and wash the cells twice with water. Stain the cells with a freshly prepared Oil red O solution for 15 min, wash the cells with water, and evaluate adipogenic differentiation under a phase-contrast microscope through identifying the red lipid droplet-positive differentiated adipocytes. Compare the percentage of differentiated adipocytes of Ad5CMVntlacZ-transduced rMSCs, Ad5CMVecSOD-transduced rMSCs, and naive rMSCs to determine if Ad5CMVecSOD transduction can enhance the adipogenic differentiation capability of rMSCs (*see* **Fig. 13.4**).

3.9. Intracavernosal Injection of Adenovirus-Transduced rMSCs

1. Plate rMSCs at a density of 10,000 cells/cm^2 in T75 flasks and culture the cells overnight.
2. Count the cells and expose the cells to fresh culture medium for rMSCs containing Ad5CMVntlacZ or Ad5CMVecSOD at 300 MOI for 48 h.
3. Remove the virus-containing supernatant, wash the cells three times with PBS, harvest the cells with trypsin–EDTA solution, wash the cells again with PBS, prepare a cell suspension at a concentration of 12,500 cells/μL in PBS, and keep the cells on ice.
4. Inject 40 μL cells, i.e., 500,000 cells total, into the rat corpus cavernosum with a 25-gauge needle, usually within 60 min after cell preparation (31, 35).

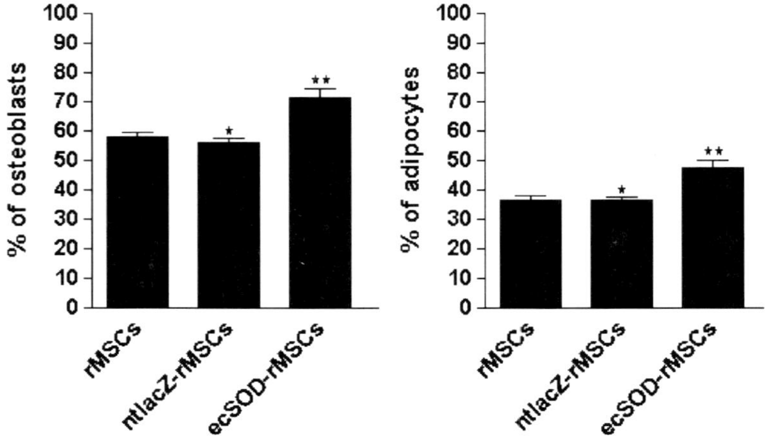

Fig. 13.4. Bar graphs showing an increase in multipotential differentiation capability in Ad5CMVecSOD-transduced rMSCs. rMSCs were transduced with Ad5CMVntlacZ or Ad5CMVecSOD at MOI 300 for 2 days. The cells were then induced to differentiate into osteoblast or adipocyte lineage for 21 days and stained with Alizarin red S for mineral deposition or Oil red O for lipid droplet formation. The percentage of differentiated (**a**) osteoblasts or (**b**) adipocytes in control rMSCs, Ad5CMVntlacZ-transduced rMSCs, and Ad5CMVecSOD-transduced rMSCs were counted. Each value represents mean ± SEM ($n = 3$), and the data are analyzed by one-way ANOVA followed by post hoc analysis with Tukey test. $^{*}p > 0.05$ versus wild-type rMSCs. $^{**}p < 0.01$ versus wild-type rMSCs or Ad5CMVntlacZ-transduced rMSCs.

4. Notes

1. Brown Norway rats are preferred in these studies because the aged animals of this species exhibit a combination of primary and secondary testicular failure that more closely resembles reproductive aging and erectile dysfunction in aged men (31, 35–37).
2. Bone marrow of young animals contains marrow stromal cells (MSCs) with the highest capability for both cell proliferation and multipotential differentiation. Therefore, 6-wk-old rats are ideal for studies involving bone marrow aspiration as performed in this study.
3. Bone marrow contains two types of adult stem cells: hematopoietic stem cells (HSCs) and marrow stromal cells (MSCs). HSCs are suspension cells that can differentiate into hematopoietic cells including erythrocytes, granulocytes, lymphocytes, monocytes, and platelets. MSCs are adherent cells that can differentiate into osteoblasts, chondrocytes, adipocytes, and other cell types. After plating whole bone marrow cells in cell culture plastic, MSCs will rapidly adhere to culture plastic and can be easily separated from non-

adherent HSCs or hematopoietic cells by repeated changing cell culture medium.

4. At a plating density of 10,000 cells/cm^2, rMSCs will be around 50% confluent and the cells will start to rapidly divide within 24 h.
5. At MOI 300, transduction efficiency of adenovirus on rMSCs is almost 100% and the cytotoxicity caused by adenoviral vector transduction on rMSCs is the minimum.

Acknowledgments

The authors would like to thank Drs. Beverley Davidson, Yi Chu, and Donald D. Heistad and the University of Iowa Gene Transfer Vector Core for the preparation of adenoviral vectors. Studies presented in this chapter are supported in part by NIH grant HL62000.

References

1. Consensus Conference., N.I.H. (1993) Impotence. NIH Consensus Development Panel on Impotence. *JAMA* **270**, 83–90.
2. Koskimaki, J., Hakkinen, J., Hakama, M., Huhtala, H., Tammela, T.L., and Shiri, R. (2005) Are questions on both achieving and maintaining an erection needed to define erectile dysfunction?. *Int. J. Impot. Res.* **17**, 335–338.
3. Feldman, H.A., Goldstein, I., Hatzichristou, D.G., Krane, R.J., and McKinlay, J.B. (1994) Impotence and its medical and psychosocial correlates: results of the Massachusetts Male Aging Study. *J. Urol.* **151**, 54–61.
4. Lewis, R.W. (2001) Epidemiology of erectile dysfunction. *Urol. Clin. North. Am.* **28**, 209–216.
5. Seftel, A.D. (2003) Erectile dysfunction in the elderly: Epidemiology, etiology and approaches to treatment. *J. Urol.* **169**, 1999–2007.
6. Shabsigh, R., Perelman, M.A., Lockhart, D.C., Lue, T.F., and Broderick, G.A. (2005) Health issues of men: Prevalence and correlates of erectile dysfunction. *J. Urol.* **174**, 662–667.
7. Pommerville, P. (2003) Erectile dysfunction: An overview. *Can. J. Urol.* **10**(Suppl 1), 2–6.
8. Kinsey, A.C., Pomeroy, W.B., and Martin, C.E. (1948) *Sexual Behavior in the Human Male*, Philadelphia, WB Saunders Co.
9. McKinlay, J.B. (2000) The worldwide prevalence and epidemiology of erectile dysfunction. *Int. J. Impot. Res.* **12**(Suppl 4), S6–S11.
10. Cappelleri, J.C. and Rosen, R.C. (2005) The Sexual Health Inventory for Men (SHIM): A 5-year review of research and clinical experience. *Int. J. Impot. Res.* **17**, 307–319.
11. Shabsigh, R. and Anastasiadis, A.G. (2003) Erectile dysfunction. *Annu. Rev. Med.* **54**, 153–168.
12. Burnett, A.L., Lowenstein, C.J., Bredt, D.S., Chang, T.S.K., and Snyder, S.H. (1992) Nitric oxide: Physiologic mediator of penile erection. *Science* **257**, 401–403.
13. Rajfer, J., Aronson, W.J., Bush, P.A., Dorey, F.J., and Ignarro, L.J. (1992) Nitric oxide as a mediator of relaxation of the corpus cavernosum in response to nonadrenergic, noncholinergic neurotransmission. *N. Engl. J. Med.* **326**, 90–94.
14. Jeremy, J.Y., Angelini, G.D., Khan, M., Mikhailidis, D.P., Morgan, R.J., Thompson, C.S., Bruckdorfer, K.R., and Naseem, K.M. (2000) Platelets, oxidant stress and erectile dysfunction: An hypothesis. *Cardiovasc. Res.* **46**, 50–54.
15. Jones, R.W., Rees, R.W., Minhas, S., Ralph, D., Persad, R.A., and Jeremy, J.Y. (2002) Oxygen free radicals and the penis. *Expert Opin. Pharmacother.* **3**, 889–897.

16. Azadzoi, K.M., Schulman, R.N., Aviram, M., and Siroky, M.B. (2005) Oxidative stress in arteriogenic erectile dysfunction: Prophylactic role of antioxidants. *J. Urol.* **174**, 386–393.
17. Bivalacqua, T.J., Armstrong, J.S., Biggerstaff, J., Abdel-Mageed, A.B., Kadowitz, P.J., Hellstrom, W.J., and Champion, H.C. (2003) Gene transfer of extracellular SOD to the penis reduces $O_2^{\bullet -}$ and improves erectile function in aged rats. *Am. J. Physiol. Heart Circ. Physiol.* **284**, H1408–H1421.
18. Bivalacqua, T.J., Usta, M.F., Kendirci, M., Pradhan, L., Alvarez, X., Champion, H.C., Kadowitz, P.J., and Hellstrom, W.J. (2005) Superoxide anion production in the rat penis impairs erectile function in diabetes: Influence of in vivo extracellular superoxide dismutase gene therapy. *J. Sex. Med.* **2**, 187–197, Discussion 197–198.
19. Gryglewski, R.J., Palmer, R.M., and Moncada, S. (1986) Superoxide anion is involved in the breakdown of endothelium-derived vascular relaxing factor. *Nature* **320**, 454–456.
20. Fukai, T., Folz, R.J., Landmesser, U., and Harrison, D.G. (2002) Extracellular superoxide dismutase and cardiovascular disease. *Cardiovasc. Res.* **55**, 239–249.
21. Squadrito, G.L. and Pryor, W.A. (1998) Oxidative chemistry of nitric oxide: The roles of superoxide, peroxynitrite, and carbon dioxide. *Free Radic. Biol. Med.* **25**, 392–403.
22. Stralin, P., Karlsson, K., Johansson, B.O., and Marklund, S.L. (1995) The interstitium of the human arterial wall contains very large amounts of extracellular superoxide dismutase. *Arterioscler. Thromb. Vasc. Biol.* **15**, 2032–2036.
23. Marklund, S.L. (1982) Human copper-containing superoxide dismutase of high molecular weight. *Proc. Natl. Acad. Sci. USA* **79**, 7634–7638.
24. Levin, E.D. (2005) Extracellular superoxide dismutase (EC-SOD) quenches free radicals and attenuates age-related cognitive decline: Opportunities for novel drug development in aging. *Curr. Alzheimer Res.* **2**, 191–196.
25. Davidson, B.L., Allen, E.D., Kozarsky, K.F., Wilson, J.M., and Roessler, B.J. (1993) A model system for in vivo gene transfer into the central nervous system using an adenoviral vector. *Nat. Genet.* **3**, 219–223.
26. Champion, H.C., Bivalacqua, T.J., D'Souza, F.M., Ortiz, L.A., Jeter, J.R., Toyoda, K., Heistad, D.D., Hyman, A.L., and Kadowitz, P.J. (1999) Gene transfer of endothelial nitric oxide synthase to the lung of the mouse in vivo. Effect on agonist-induced and flow-mediated vascular responses. *Circ. Res.* **84**, 1422–1432.
27. Gunnett, C.A., Lund, D.D., Chu, Y., Brooks, R.M., 2nd., Faraci, F.M., and Heistad, D.D. (2001) NO-dependent vasorelaxation is impaired after gene transfer of inducible NO-synthase. *Arterioscler. Thromb. Vasc. Biol.* **21**, 1281–1287.
28. Chu, Y., Iida, S., Lund, D.D., Weiss, R.M., DiBona, G.F., Watanabe, Y., Faraci, F.M., and Heistad, D.D. (2003) Gene transfer of extracellular superoxide dismutase reduces arterial pressure in spontaneously hypertensive rats: Role of heparin-binding domain. *Circ. Res.* **92**, 461–468.
29. Bivalacqua, T.J., Champion, H.C., Usta, M.F., Cellek, S., Chitaley, K., Webb, R.C., Lewis, R.L., Mills, T.M., Hellstrom, W.J., and Kadowitz, P.J. (2004) RhoA/Rho-kinase suppresses endothelial nitric oxide synthase in the penis: A mechanism for diabetes-associated erectile dysfunction. *Proc. Natl. Acad. Sci. USA* **101**, 9121–9126.
30. Bivalacqua, T.J., Usta, M.F., Champion, H.C., Leungwattanakij, S., Dabisch, P.A., McNamara, D.B., Kadowitz, P.J., and Hellstrom, W.J. (2004) Effect of combination endothelial nitric oxide synthase gene therapy and sildenafil on erectile function in diabetic rats. *Int. J. Impot. Res.* **16**, 21–29.
31. Deng, W., Bivalacqua, T.J., Chattergoon, N.N., Hyman, A.L., Jeter, J.R., Jr., and Kadowitz, P.J. (2003) Adenoviral gene transfer of endothelial nitric oxide synthase: High level expression in ex vivo expanded marrow stromal cells. *Am. J. Physiol. Cell Physiol.* **285**, C1322–C1329.
32. Deng, W., Bivalacqua, T.J., Chattergoon, N.N., Jeter, J.R., Jr., and Kadowitz, P.J. (2004) Engineering ex vivo-expanded marrow stromal cells to secrete calcitonin gene-related peptide using adenoviral vector. *Stem Cells* **22**, 1279–1291.
33. Deng, W., St Hilaire, R.C., Chattergoon, N.N., Jeter, J.R., Jr., and Kadowitz, P.J. (2006) Inhibition of vascular smooth muscle cell proliferation in vitro by genetically engineered marrow stromal cells secreting calcitonin gene-related peptide. *Life Sci.* **78**, 1830–1838.
34. Christ, G.J., Rehman, J., Day, N., Salkoff, L., Valcic, M., Melman, A., and Geliebter, J. (1998) Intracorporal injection of hSlo cDNA in rats produces physiologically relevant alterations in penile function. *Am. J. Physiol.* **275**, H600–H608.
35. Bivalacqua, T.J., Deng, W., Kendirci, M., Usta, M.F., Robinson, C., Taylor, B.K., Murthy, S.N., Champion, H.C., Hellstrom,

W., and Kadowitz, P.J. (2007) Mesenchymal stem cells alone or ex vivo gene modified with endothelial nitric oxide synthase reverse age-associated erectile dysfunction. *Am. J. Physiol. Heart Circ. Physiol.* **292**, H1278–H1290.

36. Greuenewald, D.A., Naai, M.A., Hess, D.L., and Matsumoto, A.M. (1994) The brown Norway rat as a model of reproductive aging: Evidence for both primary and secondary testicular failure. *J. Gerontol. B Psychol. Sci. Soc. Sci.* **49**, B42–B50.

37. Rajasekaran, M., Kasyan, A., Jain, A., Kim, S.W., and Monga, M. (2002) Altered growth factor expression in the aging penis: The Brown-Norway rat model. *J. Androl.* **23**, 393–399.

Chapter 14

Assessing the Reductive Capacity of Cells by Measuring the Recycling of Ascorbic and Lipoic Acids

James M. May

Abstract

Most mammalian cells cannot synthesize vitamin C, or ascorbic acid, and thus must have efficient mechanisms for its intracellular recycling. Ascorbate can be recycled from both its oxidized forms using electrons from several intracellular reducing co-factors, including GSH and the reduced pyridine nucleotides. Methods have been developed to assess the ability of intact cells to recycle ascorbate, which include assay of extracellular ferricyanide reduction and measurement of the ability of the cells to reduce dehydroascorbic acid to ascorbate. Lipoic acid, a disulfide containing medium chain fatty acid, is also taken up by cells and reduced to dihydrolipoic acid, which can be measured upon efflux from the cells using Ellman's reagent. Together, these assays provide an estimate of the ability of different cell types to recycle ascorbate and to generate intracellular reducing equivalents required to maintain the redox status of the cells.

Key words: Ascorbic acid, lipoic acid, ferricyanide, Ellman's reagent, intracellular redox environment, oxidant stress.

1. Introduction

1.1. Ascorbic Acid Recycling

Vitamin C, or ascorbic acid, is required for collagen synthesis, serves as a cofactor for various dioxygenase enzymes, and also contributes to the antioxidant defenses of plasma and cells. Regarding the latter, ascorbate is a sensitive marker of oxidant stress in plasma (1) and cells (2). Ascorbate is synthesized de novo in the liver in mammals and not at all in primates and humans (3). To maintain levels of the vitamin, there are efficient mechanisms for recycling it from its oxidized forms. In contrast to thiols and pyridine nucleotides, which typically donate

R.M. Uppu et al. (eds.), *Free Radicals and Antioxidant Protocols*, Methods in Molecular Biology 610,
DOI 10.1007/978-1-60327-029-8_14, © Humana Press, a part of Springer Science+Business Media, LLC 1998, 2010

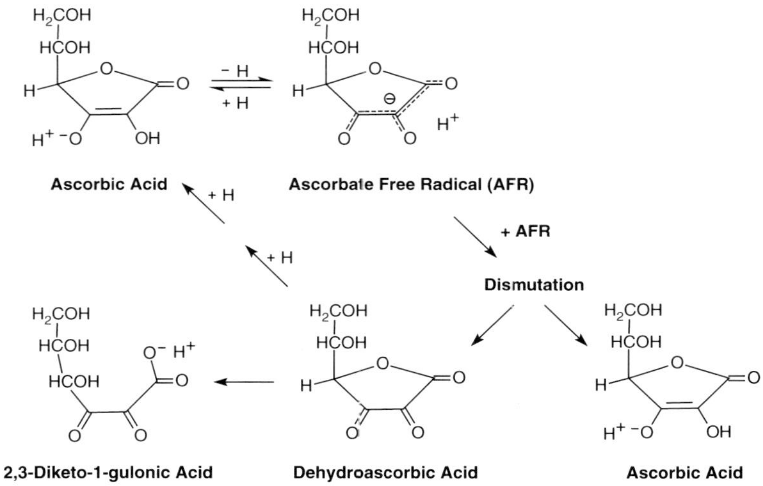

Fig. 14.1. Schematic representation of ascorbic acid recycling.

two electrons, ascorbate functions primarily as a one-electron donor (4), generating the ascorbate free radical (AFR; **Fig. 14.1**). Even though the AFR is more strongly reducing than ascorbate (5), it is relatively stable, sharing the unpaired electron across three vicinal oxygen molecules. Rather than lose another electron to become the two-electron-oxidized form of ascorbate, dehydroascorbic acid (DHA), the AFR dismutates to form one molecule each of ascorbate and dehydroascorbate (6). The AFR can also be reduced to ascorbate by NAD(P)H-dependent reductases within cells (7–10). These enzymes have affinity for the AFR in the low micromolar range. Although their activity can be measured in cell lysates as NAD(P)H-dependent AFR reduction, assays specific for AFR reductases are not feasible in intact cells. In contrast to the AFR, DHA can undergo two-electron reduction back to ascorbate by a variety of mechanisms (**Fig. 14.1**). These include direct reduction by GSH (11) or by GSH-dependent enzymes, such as glutaredoxin or thiol–disulfide isomerase (12, 13). DHA is also reduced by NADPH-dependent enzymes, including a 3α-hydroxysteroid dehydrogenase in liver (14) and thioredoxin reductase (15). Most cells appear to use multiple mechanisms to recycle ascorbate from its oxidized forms, although they may vary in the extent to which they depend on a specific mechanism. For example, despite the presence of GSH in all cells, neither HL-60 cells (16) nor human skin keratinocytes (17) require GSH for DHA reduction.

Since the ability of cells to recycle ascorbate is likely to contribute to their ability to maintain intracellular ascorbate and to withstand oxidant stress, it is useful to assess this capacity, especially in intact cells. Two approaches have been used to do this in either suspended cells or cells in monolayer culture. It was found several decades ago that human erythrocytes reduce ferricyanide to ferrocyanide in an ascorbate-dependent manner (18). Since ferricyanide does not enter cells due to its size and charge (19), and since ascorbate is also trapped within cells, it was postulated that electrons from ascorbate are transferred across the plasma membrane by an oxidoreductase activity (20, 21). Although this putative enzyme or enzyme complex has not been identified, all cells tested thus far can reduce ferricyanide, and this reduction is enhanced by intracellular ascorbate. For example, in freshly prepared human erythrocytes, endogenous ascorbate contributes about two-thirds of basal ferricyanide reductase activity (21). When cells are loaded with increasing ascorbate concentrations, ferricyanide reduction increases in a saturable manner that is limited in part by the ability of the cells to regenerate ascorbate from its oxidized forms (22). Ferricyanide is a one-electron oxidant and has been shown to generate the AFR in human erythrocytes (23, 24). However, because the AFR undergoes rapid dismutation to ascorbate and DHA (6) (**Fig. 14.1**), rates of ferricyanide reduction likely measure the capacity of cells to reduce both the AFR and the DHA. Indeed, ferricyanide reduction is dependent on the GSH status of human erythrocytes (25) and other cells (26, 27), implying that GSH-dependent mechanisms serve to recycle ascorbate from DHA. Rates of ferricyanide reduction also depend on rates of trans-plasma membrane electron transport (28–31) and the extent to which electrons are derived from other intracellular donors, such as pyridine nucleotides. Despite these caveats, assay of ascorbate-dependent ferricyanide reduction provides at least an indirect measure of cellular ascorbate recycling capacity from both the AFR and the DHA. To the extent that ascorbate recycling depends upon both the reserve of both pyridine nucleotides and cellular thiols, it may reflect the redox capacity of the cells in general. The assay, originally developed by Avron and Shavit (32), is easy to perform using aliquots of medium from either suspended cells or monolayer cultures, is non-destructive to the cells, and requires only a spectrophotometer.

A second approach to measurement of ascorbate recycling depends on the ability of cells to take up and reduce DHA to ascorbate. DHA enters cells on the ubiquitous glucose transporter (33) and is then rapidly reduced to ascorbate once inside the cells (25, 34). Ascorbate is not a substrate for transport by these GLUT-type glucose transporters (33). DHA is considered to form an hydrated hemiketal in solution (35) that resembles glucose in its structure and thus is taken up by cells in competition with glucose. In fact, the term "dehydroascorbic acid" is a misnomer, since it is not an acid in solution (35). Once DHA enters cells, it is rapidly reduced to ascorbate by the mechanisms outlined above. The ascorbate content of cells is measured after cell lysis. Although various methods have been developed to measure ascorbate, the most sensitive and specific is by HPLC, either using electrochemical or using UV detection (36). Since reduction of DHA by GSH- and pyridine nucleotide-dependent mechanisms does not involve the AFR (25), this assay specifically measures DHA reduction to ascorbate and not AFR reduction. Depending on the source of donor electrons, it can also provide an estimate of the ability of cells to recycle GSH and of their redox metabolism in general.

1.2. Lipoic Acid Recycling

Lipoic or thioctic acid is a medium chain saturated fatty acid that contains a thiolane ring composed of an internal disulfide between carbons 6 and 8 (**Fig. 14.2**). Lipoic acid is a natural constituent of α-keto reductases, where it contributes to decarboxylation reactions. It has received increasing attention over the last few years for its pharmacologic property as an antioxidant precursor. Although lipoic acid itself can chelate transition metals, it becomes a potent antioxidant when reduced to the dithiol form, dihydrolipoic acid (DHLA). In contrast to ascorbic acid, both lipoic and dihydrolipoic acid can rapidly enter cells by several mechanisms. In some cells it appears to use a multivitamin transporter (37), in most cells it uses a medium chain fatty acid transporter (38), and there is also likely to be simple diffusion of the protonated forms across the plasma membrane. Once inside the cells, lipoic acid is rapidly reduced to dihydrolipoic acid by any of several pyridine nucleotide-dependent oxidoreductases. Because its midpoint reduction potential of –0.34 mV is less than that of GSH (–0.29 mV) (39), it cannot be reduced directly by thiols in the cell unless they form the active site of an enzyme. Lipoic acid reduction has been described only for NADH-dependent lipoamide dehydrogenase in mitochondria (the enzyme responsible for reduction of lipoamide bound to α-keto dehydrogenases), glutathione reductase, and thioredoxin reductase (40, 41). The

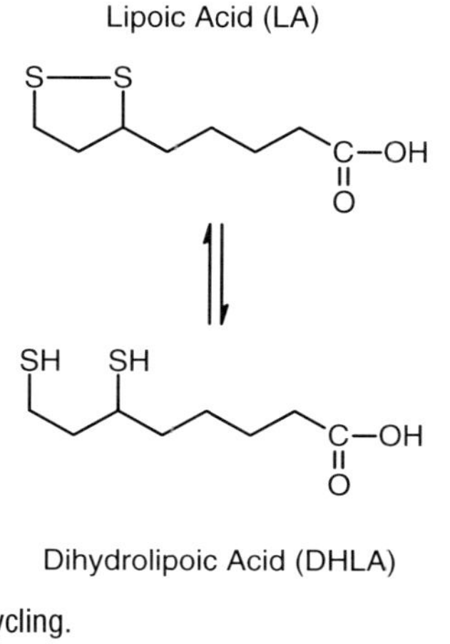

Fig. 14.2. Lipoic acid recycling.

latter two enzymes are NADPH-dependent, so the ability of a cell to reduce lipoic acid or its derivatives reflects both the activities of these enzymes the and NAD(P)H availability. In this regard, it also differs somewhat from recycling of ascorbate from DHA, which may involve reduction by cellular thiols. An important feature of the enzyme-dependent reduction of lipoic acid is that the different enzymes have different rates of reduction for the R- and S-forms of lipoic acid (40, 41), which could provide clues as to the enzymes involved.

In contrast to ascorbic acid, which is trapped in cells, dihydrolipoic acid readily exits cells into the interstitium or culture medium (42, 43). However, because of its high reducing capacity, dihydrolipoic acid is readily oxidized, especially in oxygenated buffer or culture medium. Although lipoic acid recycling by cells can be measured directly as dihydrolipoic acid present in cells or medium, because of the short half-life of dihydrolipoic acid, assays have been developed (44, 45) to measure its appearance in cell buffers by its ability to reduce 5,5'-(dithiobis)-2-nitrobenzoic acid (DTNB), also known as Ellman's reagent. This disulfide has two negative charges due to carboxylic acid groups and does not enter cells. Thus, it detects only lipoic acid that has been released into the incubation medium and does so as soon as this release occurs, thus decreasing the chance for oxidation of dihydrolipoic acid. Reaction of DTNB with a thiol releases the 5-thio-2-nitrobenzoic acid anion, which is a bright yellow, strongly absorbing light at 410 nm, so that it can be detected at visible wavelengths in a spectrophotometer.

This chapter describes approaches for measuring ascorbic and lipoic acid recycling, which can be used to provide time-dependent and non-destructive estimates of the redox capacity of cells in suspension and culture.

2. Materials

2.1. Assay of Ascorbate-Dependent Ferricyanide Reduction

1. Ferricyanide: Analytical grade potassium ferricyanide (potassium hexacyanoferrate (III)) is dissolved in deionized water (dH_2O) or appropriate buffer at a concentration of 100 mM (32.9 mg/mL). This reagent is stable for several months when stored in the dark in a refrigerator.
2. Acetate buffer, 3.0 M: dissolve 40.8 g of sodium acetate trihydrate in ca. 60 mL of dH_2O, adjust the pH to 6.0–6.5, and then make up the volume to 100 mL. Stable in the refrigerator for several months.

3. Citric acid solution, 0.2 M: dissolve 3.84 g of citric acid in 100 mL of dH_2O. Stable several months in the refrigerator.
4. Acetic acid, 0.1 M: dilute 5.75 mL of glacial acetic acid to 1 L with dH_2O.
5. Ferric chloride solution, 3.3 mM: dissolve 53.5 mg of $FeCl_3$ (anhydrous) in 100 mL of 0.1 M acetic acid. This reagent must be made fresh every few days.
6. Phenanthroline derivatives (*see* **Note 1**): The disodium salt of 4,7-diphenyl-1,10-phenanthroline disulfonate (bathophenanthroline disulfonate) is dissolved in dH_2O at a concentration of 6.2 mM (100 mg/30 mL). If 1,10-phenanthroline is to be used, it is prepared by dissolving 100 mg in 10 mL of ethanol. Stable in the refrigerator for several weeks.

2.2. Assay of DHA Reduction Capacity

1. DHA solution, 10 mM: just before use, dissolve DHA in dH_2O or appropriate buffer at a concentration of 1.74 mg/mL (*see* **Note 2**).
2. *Meta*-phosphoric acid: prepare 25% *meta*-phosphoric acid (w/v) by adding 25 g of *meta*-phosphoric acid to 75 mL of dH_2O. After it is dissolved, bring the final volume to 100 mL with dH_2O. This is stable indefinitely at room temperature.
3. Phosphate-EDTA buffer: prepare 100 mM phosphate-EDTA buffer containing 0.05 mM EDTA by dissolving 1.42 g of disodium phosphate (Na_2HPO_4, anhydrous) in 80 mL of dH_2O, adding 1.86 mg disodium EDTA, adjust the pH to 8.0, and then make up the volume to 100 mL with dH_2O.
4. Sodium acetate solution, 1.0 M: dissolve 11.3 g sodium acetate (anhydrous) along with 20.9 mL of glacial acetic acid and bringing the total volume to 500 mL with dH_2O. Store at 3°C.
5. Ion-pair reagent: dissolve 3.79 g of tetrapentylammonium bromide in 100 mL of methanol and store at 3°C (*see* **Note 3**).
6. HPLC mobile phase: mix 80 mL of 1.0 M sodium acetate solution with 10 mL of the tetrapentylammonium bromide solution, 300 mL of HPLC grade methanol, and 610 mL of HPLC grade water.
7. Standard solutions of ascorbic acid: prepare by dissolving 1.76 mg of ascorbic acid in 1 mL of HPLC mobile phase to give an ascorbate concentration of 10 mM. The concentration of this solution can be verified by measuring the absorbance at 265 nm of an aliquot that has been diluted

to 0.1 mM ascorbate. The optical density of this solution should be 0.33, based on a molar extinction coefficient of 3,300/M/cm at this wavelength. Standards of 0.5–4 nmol/mL are prepared by serial dilution of this stock solution with mobile phase. The ascorbate stock and standards should be kept on ice and prepared daily.

2.3. Assay of Lipoic Acid Recycling

1. Krebs–Ringer–Hepes buffer (KRH): prepare a solution in dH_2O of 20 mM N-2-hydroxyethylpiperazine-N′-2-ethanesulfonic acid (Hepes), 128 mM NaCl, 5.2 mM KCl, 1 mM NaH_2PO_4, 1.4 mM $MgSO_4$, and 1.4 mM $CaCl_2$, pH 7.4.
2. Lipoic acid solution: dissolve R,S-α-lipoic acid in KRH at a concentration of 2 mM (0.41 mg/mL).
3. 5,5′-Dithiobis(2-nitrobenzoic acid) (DTNB) solution: prepare a 2 mM solution by dissolving 0.8 mg of DTNB in 0.8 mL of KRH and adjusting the pH to 7.4 by carefully adding 5 μL aliquots of 1 M NaOH. The total volume is brought to 1 mL with KRH after adjusting the pH to 7.4 (*see* **Note 4**).

3. Methods

3.1. Assay of Ascorbate-Dependent Ferricyanide Reduction

1. *Incubation of cells with ferricyanide.* The ferricyanide reduction can be performed using either suspended (e.g., erythrocytes) or in monolayer cell cultures. In either case, it is necessary that all cells in the incubation are exposed to ferricyanide. The cells may be rinsed before the addition of ferricyanide reagent (*see* **Note 5**). Ferricyanide is added by dilution from the stock solution and the incubations continued at 37°C. At the end of the incubation (*see* **Note 6**), cell-free aliquots of the incubation medium are sampled for assay of ferrocyanide. If suspended cells are used, they are pelleted by centrifugation and aliquots of the supernatant used for assay.
2. To start the assay of ferrocyanide, dilute 100:1 of a sample containing 1–100 nmol of ferrocyanide with 0.6 mL of water in a plastic disposable spectrophotometer cuvette.
3. Just before use, prepare a reaction mixture of the following: 10 mL of the sodium acetate buffer, 10 mL of the citric acid solution, 5 mL of the ferric chloride solution, and 5 mL of the solution containing the phenanthroline derivative. This is a volume adequate to assay 100 samples.

4. Add 0.3 mL of the reaction mixture from step 3 (above) to each cuvette, mix, and allow 5 min to allow full development of color (*see* **Note 7**).
5. Read the optical density at 510 nm if 1,10-phenanthroline is used, and at 535 nm if bathophenanthroline disulfonate is used.
6. Calculate the concentration of ferrocyanide in the cuvette based on the molar extinction coefficient of the phenanthroline derivative used for color development (32). For 1,10-phenanthroline this is 10,500/M/cm, and for bathophenanthroline disulfonate it is 20,500/M/cm. Accordingly, an optical density of 1.0 will be generated by a solution of 95.2 μM ferrocyanide in the cuvette using 1,10-phenanthroline; for bathophenanthroline disulfonate, an optical density of 1.0 will be generated by a ferrocyanide concentration of 48.8 μM in the cuvette (*see* **Note 8**).
7. After correcting for the 10-fold dilution of the sample into the cuvette, the amount of ferrocyanide generated over the time of exposure to the cells (*see* **Note 9**) can be normalized to cell number, cell protein, or intracellular water space (*see* **Note 10**).
8. Correction for a reagent blank is unnecessary, but color in the cell medium may require correction using a sample not treated with ferricyanide that is carried through the assay and subtracted before the calculation (*see* **Note 11**). Additional correction may be required for extracellular ascorbate, if present (*see* **Note 12**)

3.2. Assay of DHA Reduction Capacity

1. Cells in culture or in a defined physiologic medium are incubated with freshly prepared DHA. The optimal concentration of DHA must be determined empirically but is usually in the range of 0.1–1 mM. DHA is very unstable at physiologic pH and must be added to the cells quickly after it is dissolved. Uptake and reduction of DHA is usually complete after 15–20 min, and the cells can be taken for assay of intracellular ascorbate. The extent to which intracellular ascorbate increases in different cell types or under different conditions provides an estimate of the ability of the cells to recycle ascorbate from DHA (*see* **Note 13**).
2. Intracellular ascorbate is determined by lysis of the cells and assay of the ascorbate content in the lysate by HPLC. The method of cell lysis varies with the cell type. For cells not containing hemoglobin, after rinsing or centrifugation to remove the incubation buffer, the cell pellet or monolayer of cells (approximately 50–100 μL in volume) is lysed according to the method of Hissin and Hilf (46), which also allows

assay of GSH as another peak in the chromatogram. Cells are treated with 0.1 mL of 25% meta-phosphoric acid, mixed or vortexed, and then partially neutralized with 0.35 mL of phosphate–EDTA buffer. The extract is centrifuged for 5 min in the cold to pellet cell debris, and aliquots of the supernatant are stored at –80°C or taken for assay immediately (*see* **Note 14**). For erythrocytes or samples that contain substantial amounts of iron bound in hemoglobin, an alternative method of lysis is required (*see* **Note 15**).

3. The ascorbate content of cell lysates or ultrafiltrates is determined by HPLC according to the ion-pairing electrochemical method of Pachla and Kissinger (47). Although many HPLC systems are adequate to detect ascorbate (48), including UV detection (*see* **Note 16**), we have found the following method to provide excellent sensitivity and reliability. Samples and standards are prepared in mobile phase and injected in 100 μL volumes. Mobile phase is pumped by an ESA Model 582 Solvent Delivery system (ESA Biosciences, Chelmsford, MA), followed by an in-line pulse dampener to decrease background fluctuations, and by an ESA Guard cell set at 0.5 V to decrease background in the mobile phase. Samples are separated on a 10 × 1 cm Waters RadialPak C18 column (BondaPak, 5 μm particle size), although similar results are obtained with other C18 columns, such as an Alltech Absorbosphere C18 column (4.6 × 100 mm, 5 μm particle size). The main column is preceded by a 4 mm guard column of the same packing material. Detection is accomplished with either an ESA Model 5100A or a Coulochem II detector using an ESA Model 5010 analytical cell with the first electrode set at 0.4 V. Peaks are analyzed using either an ESA 501 data analysis package or a Shimadzu C-R5A Chromatopac integrator. The sensitivity of detection for ascorbate is 10 pmol/sample with this system (2).

3.3. Assay of Lipoic Acid Recycling

1. Cultured cells (*see* **Note 17**) in 6-well plates (*see* **Note 18**) are subjected to pretreatments in culture and rinsed three times in 2 mL of KRH buffer and incubated at 37 EC with gentle mixing in 2 mL of KRH containing additives as desired (e.g., 5 mM D-glucose), 0.1 mM lipoic acid, and 0.2 mM DTNB (*see* **Note 19**).
2. After 30 min, a 1 mL aliquot of the supernatant is transferred to a 1 mL disposable plastic cuvette and the absorbance at 412 nm is determined in spectrophotometer.
3. The concentration of 2-nitro-5-thiobenozoate anion of DTNB is calculated based on a molar extinction coefficient of 13,600/M/cm (49). Results are corrected for absorbance of a blank containing the amounts of lipoic acid and DTNB

used in the experiment without cells. This blank is typically near zero, indicating that the LA preparations used do not contain contaminants that can reduce DTNB.

4. Results are expressed relative to the cell protein present in each well.

4. Notes

1. Ferricyanide has an absorption maximum at 420 nm, so its disappearance from the incubation medium could be simply followed with time. However, this approach is only about 5–10% as sensitive as assay of the complex formed between ferrocyanide and phenanthroline derivatives developed by Avron and Shavit (32). Although use of bathophenanthroline disulfonate increases sensitivity of the assay about 2-fold, this is usually not a problem and we typically use 1,10-phenanthroline.
2. Whereas commercial DHA (Sigma-Aldrich, cat. No. 261556) is adequate for most determinations, it is contaminated with about 0.5% ascorbate, which will directly react with ferricyanide outside the cells. Although we have not found it necessary for this type of assay, DHA may be generated directly from ascorbate using the bromine oxidation method. Dissolve ascorbic acid to a concentration of 50 mM in deionized water (8.8 mg/mL) in a 12 × 75 mm glass culture tube. To this add about 2 μl of liquid bromine in the hood (the amount is not critical, it is in large excess). Vortex vigorously for 1–2 min to allow the bromine to dissolve. The solution should turn yellow. Then pass a stream of nitrogen gas over the surface of the solution for several minutes until the solution loses its color, but not so long as to cause evaporation of the water. The final solution is diluted just before use into the appropriate buffer and added to the cell assay. This method results in an acidic solution of DHA that is relatively stable. On the other hand, care must be taken that the medium used for cell incubations has adequate buffering capacity to neutralize the residual hydrobromic acid after dilution of into cells. This can be checked directly with a pH meter.
3. Any of several ion-pair reagents work well. We have also used tridecylamine and tetraoctylammonium bromide. These reagents bind the hydrophobic column material because of their short-chain fatty acids and bind the ascorbate anion due to their positive quaternary ammonium charge.

4. DTNB is only sparingly soluble in water. As it is neutralized, it will gradually dissolve. However, overshooting the pH will result in hydrolysis of the DTNB and a bright yellow color, so the additions must be done slowly. This solution must be made just before use, since the intensity of the yellow color will increase with time.
5. Although ferricyanide is a weak oxidant, it will react with ascorbate and thiols if present outside the cells. It is important to remove such interfering substances if they are present by rinsing the cells before adding ferricyanide, or to test for their presence by comparing results in rinsed and non-rinsed cells, which should have similar rates of reduction.
6. If a single time point (typically 30 min) is to be used, a time course of ferricyanide reduction by cells must be determined to document that the reaction is linear. The main reasons for lack of a linear response are depletion of extracellular ferricyanide below 50% of the starting concentration, or exhaustion of intracellular reducing equivalents. The former can be avoided by using extracellular ferricyanide concentrations as high as 5 mM, or by decreasing cell number. The latter may be the outcome of interest but can be minimized by shorter times of incubation with ferricyanide.
7. The color of the reaction mixture is stable at room temperature for up to 2 h if the cuvettes containing the reaction mixtures are stored in the dark. As long as the pH range of the reaction mixture is between 2.5 and 6.5, the color of the reaction remains unchanged (32).
8. The assays are linear with ferrocyanide up to 100 μM ferrocyanide for bathophenanthroline disulfonate as the color indicator, and up to 200 μM ferricyanide for 1,10-phenanthroline as the color indicator (32).
9. The assay is typically linear for at least 30 min of treating cells with ferricyanide, but this should be determined for each cell type. Using the linear phase, results can be expressed as a rate, or mol ferrocyanide generated per min.
10. For erythrocytes, which vary in volume and surface area from different species, rates of ferricyanide reduction have been typically expressed as a function of the intracellular water space, which is about 70% of the packed cell volume of human erythrocytes (18). For cultured cells, the rate of ferricyanide reduction is usually expressed per mg cell protein (50).
11. Examples of colored media requiring correction might be a small amount of hemoglobin resulting from erythrocyte

lysis or colored cell culture medium. In general, conditions should be adjusted so that such a blank is less than 5% of the basal reading.

12. In some experiments evaluating ascorbate-dependent ferricyanide reduction, cells may be loaded with ascorbate using DHA. If commercial DHA used, its content of ascorbate (~0.5%) will raise the apparent "background" by reducing ferricyanide directly. In most instances there is a small effect that can be corrected for in by using a blank containing the loading DHA concentration.
13. Conditions that have been used stress or modify the ability of the cells to reduce DHA include removing glucose from the medium, or depleting the cells of endogenous GSH by various agents.
14. Cells can also be lysed using 5–10% (w/v) *meta*-phosphoric acid alone, followed by the centrifugation step. The acid precipitation method has the advantage that ascorbate is stable at a low pH and can be stored at –80°C for analysis at a later date. It has the disadvantage that the low pH may eventually damage the HPLC column. Alternatively, cells in suspension or monolayer culture may also be lysed in 60–90% methanol (v/v) containing saturating amounts of EDTA (51, 52) and incubated for 10 min on ice. After centrifugation to pellet cell debris, aliquots of the supernatant are taken for assay of ascorbate.
15. In erythrocytes or cells that contain heme proteins, acidic or methanolic lysis denatures hemoglobin and releases ferric iron that immediately oxidizes ascorbate in the sample. To avoid this problem, the ultrafiltration method of Iheanacho et al. (53) for hemoglobin-containing plasma was modified for use with lysed cells (2). In this method, erythrocytes at a 25% packed cell volume are frozen in a mixture of dry ice and acetone and allowed to thaw on ice. The hemolysate is transferred to a Centricon-YM10 filter apparatus (Millipore Corporation, cat. No. 4241) and centrifuged at 3–4°C for 30 min at 5,000×*g*. This results in a clear ultrafiltrate that can be diluted as needed with mobile phase for assay of ascorbate.
16. UV detection of ascorbate at 265 nm, although slightly less sensitive than coulometric detection, may be useful if there are overlapping peaks on the chromatogram.
17. Suspended cells such as erythrocytes can also be used in this assay. Erythrocytes are suspended in 1 mL of phosphate-buffered saline (140 mM NaCl, 12.5 mM NaH_2PO_4, pH 7.4) to a packed cell volume of 1%. All other incubation conditions are the same as described for cultured cells,

except that erythrocytes are pelleted by centrifugation at the end of the assay, and the results are expressed per mL of packed cells.

18. Twelve-well plates can also be used by decreasing all volumes by 50%.
19. The concentrations of lipoic acid and DTNB are determined empirically, but for cultured cells and erythrocytes, 0.1 mM lipoic acid and 0.2 mM DTNB are useful starting points.

Acknowledgments

This work was supported by RO1 DK050435.

References

1. Frei, B., England, L., and Ames, B.N. (1989) Ascorbate is an outstanding antioxidant in human blood plasma. *Proc. Natl. Acad. Sci. USA* **86**, 6377–6381.
2. Mendiratta, S., Qu, Z.-C., and May, J.M. (1998) Erythrocyte ascorbate recycling: Antioxidant effects in blood. *Free Radic. Biol. Med.* **24**, 789–797.
3. Chatterjee, I.B., Majumder, A.K., Nandi, B.K., and Subramanian, N. (1975) Synthesis and some major functions of vitamin C in animals. *Ann. N. Acad. Sci.* **258**, 24–47.
4. Njus, D. and Kelley, P.M. (1991) Vitamins C and E donate single hydrogen atoms in vivo. *FEBS Lett.* **284**, 147–151.
5. Bielski, B.H., Richter, H.W., and Chan, P.C. (1975) Some properties of the ascorbate free radical. *Ann. NY Acad. Sci.* **258**, 231–237.
6. Bielski, B.H., Allen, A.O., and Schwarz, H.A. (1981) Mechanism of disproportionation of ascorbate radicals. *J. Am. Chem. Soc.* **103**, 3516–3518.
7. Iyanagi, T. and Yamazaki, I. (1969) One-electron-transfer reactions in biochemical systems. 3. One- electron reduction of quinones by microsomal flavin enzymes. *Biochim. Biophys. Acta* **172**, 370–381.
8. Lumper, L., Schneider, W., and Staudinger, H. (1967) Untersuchungen zur Kinetik der mikrosomalen NADH: semidehydroascorbat-oxydoreduktase. *Hoppe Seylers. Z. Physiol. Chem.* **348**, 323–328.
9. Schulze, H.-U., Gallenkamp, H., and Staudinger, H. (1970) Untersuchungen zum mikrosomalen NADH-abhängigen elektronentransport. *Hoppe Seylers. Z. Physiol. Chem.* **351**, 809–817.
10. May, J.M., Cobb, C.E., Mendiratta, S., Hill, K.E., and Burk, R.F. (1998) Reduction of the ascorbyl free radical to ascorbate by thioredoxin reductase. *J. Biol. Chem.* **273**, 23039–23045.
11. Winkler, B.S., Orselli, S.M., and Rex, T.S. (1994) The redox couple between glutathione and ascorbic acid: A chemical and physiological perspective. *Free Radic. Biol. Med.* **17**, 333–349.
12. Wells, W.W., Xu, D.P., Yang, Y.F., and Rocque, P.A. (1990) Mammalian thioltransferase (glutaredoxin) and protein disulfide isomerase have dehydroascorbate reductase activity. *J. Biol. Chem.* **265**, 15361–15364.
13. Park, J.B. and Levine, M. (1996) Purification, cloning and expression of dehydroascorbic acid- reducing activity from human neutrophils: Identification as glutaredoxin. *Biochem. J.* **315**, 931–938.
14. Del Bello, B., Maellaro, E., Sugherini, L., Santucci, A., Comporti, M., and Casini, A.F. (1994) Purification of NADPH-dependent dehydroascorbate reductase from rat liver and its identification with 3a-hydroxysteroid dehydrogenase. *Biochem. J.* **304**, 385–390.
15. May, J.M., Mendiratta, S., Hill, K.E., and Burk, R.F. (1997) Reduction of dehydroascorbate to ascorbate by the selenoenzyme thioredoxin reductase. *J. Biol. Chem.* **272**, 22607–22610.

16. Guaiquil, V.H., Farber, C.M., Golde, D.W., and Vera, J.C. (1997) Efficient transport and accumulation of vitamin C in HL-60 cells depleted of glutathione. *J. Biol. Chem.* **272**, 9915–9921.
17. Savini, I., Duflot, S., and Avigliano, L. (2000) Dehydroascorbic acid uptake in a human keratinocyte cell line (HaCaT) is glutathione-independent. *Biochem. J.* **345**, 665–672.
18. Orringer, E.P. and Roer, M.E. (1979) An ascorbate-mediated transmembrane-reducing system of the human erythrocyte. *J. Clin. Invest.* **63**, 53–58.
19. Székely, M., Mányai, S., and Straub, F.B. (1952) Über den Mechanismus der osmotischen Hämolyse. *Acta Physiol. Acad. Sci. Hung.* **3**, 571–583.
20. Goldenberg, H., Grebing, C., and Löw, H. (1983) NADH-monodehydroascorbate reductase in human erythrocyte membranes. *Biochem. Int.* **6**, 1–9.
21. May, J.M., Qu, Z.-C., and Whitesell, R.R. (1995) Ascorbic acid recycling enhances the antioxidant reserve of human erythrocytes. *Biochemistry* **34**, 12721–12728.
22. May, J.M. and Asard, H. (2004) Ascorbate Recycling. In: *Vitamin C. Functions and biochemistry in animals and plants* (H. Asard, May, J.M., and Smirnoff, N., Eds.), Bios Scientific Publishers, London, UK, pp.139–158.
23. May, J.M., Qu, Z.C., and Cobb, C.E. (2001) Recycling of the ascorbate free radical by human erythrocyte membranes. *Free Radic. Biol. Med.* **31**, 117–124.
24. van Duijn, M.M., Tijssen, K., van Steveninck, J., van den Broek, P.J.A., and van der Zee, J. (2000) Erythrocytes reduce extracellular ascorbate free radicals using intracellular ascorbate as an electron donor. *J. Biol. Chem.* **275**, 27720–27725.
25. May, J.M., Qu, Z.C., Whitesell, R.R., and Cobb, C.E. (1996) Ascorbate recycling in human erythrocytes: Role of GSH in reducing dehydroascorbate. *Free Radic. Biol. Med.* **20**, 543–551.
26. Giblin, F.J., Winkler, B.S., Sasaki, H., Chakrapani, B., and Leverenz, V. (1993) Reduction of dehydroascorbic acid in lens epithelium by the glutathione redox cycle. *Invest. Ophthalmol. Vis. Sci.* **34**, 1298.
27. May, J.M., Qu, Z.C., and Li, X. (2001) Requirement for GSH in recycling of ascorbic acid in endothelial cells. *Biochem. Pharmacol.* **62**, 873–881.
28. Grebing, C., Crane, F.L., Löw, H., and Hall, K. (1984) A transmembranous NADH-dehydrogenase in human erythrocyte membranes. *J. Bioenerg. Biomembr.* **16**, 517–533.
29. Himmelreich, U. and Kuchel, P.W. (1997) ^{13}C-NMR studies of transmembrane electron transfer to extracellular ferricyanide in human erythrocytes. *Eur. J. Biochem.* **246**, 638–645.
30. Iyanagi, T., Yamazaki, I., and Anan, K.F. (1985) One-electron oxidation-reduction properties of ascorbic acid. *Biochim. Biophys. Acta* **806**, 255–261.
31. May, J.M. (1999) Is ascorbic acid an antioxidant for the plasma membrane? *FASEB J.* **13**, 995–1006.
32. Avron, M. and Shavit, N. (1963) A sensitive and simple method for determination of ferrocyanide. *Anal. Biochem.* **6**, 549–554.
33. Vera, J.C., Rivas, C.I., Fischbarg, J., and Golde, D.W. (1993) Mammalian facilitative hexose transporters mediate the transport of dehydroascorbic acid. *Nature* **364**, 79–82.
34. Schweinzer, E., Mao, Y., Krajnik, P., Getoff, N., and Goldenberg, H. (1996) Reduction of extracellular dehydroascorbic acid by K562 cells. *Cell Biochem. Funct.* **14**, 27–31.
35. Tolbert, B.M. and Ward, J.B. (1982) Dehydroascorbic acid. In: *Ascorbic Acid: Chemistry, Metabolism, and Uses* (P.A.Seib and Tolbert, B.M., Eds.), American Chemical Society, Washington, DC, pp. 101–123.
36. Levine, M., Wang, Y.H., and Rumsey, S.C. (1999) Analysis of ascorbic acid and dehydroascorbic acid in biological samples. *Methods Enzymol.* **299**, 65–76.
37. Prasad, P.D., Wang, H., Kekuda, R., Fujita, T., Fei, Y.J., Devoe, L.D., Leibach, F.H., and Ganapathy, V. (1998) Cloning and functional expression of a cDNA encoding a mammalian sodium- dependent vitamin transporter mediating the uptake of pantothenate, biotin, and lipoate. *J. Biol. Chem.* **273**, 7501–7506.
38. Adkison, K.D. and Shen, D.D. (1996) Uptake of valproic acid into rat brain is mediated by a medium-chain fatty acid transporter. *J. Pharmacol. Exp. Ther.* **276**, 1189–1200.
39. Jocelyn, P.C. (1967) The standard redox potential of cysteine-cystine from the thiol-disulfide exchange reaction with glutathione and lipoic acid. *Eur. J. Biochem.* **2**, 327–331.
40. Pick, U., Haramaki, N., Constantinescu, A., Handelman, G.J., Tritschler, H.J., and Packer, L. (1995) Glutathione reductase and lipoamide dehydrogenase have opposite stereospecificities for a-lipoic acid enantiomers. *Biochem. Biophys. Res. Commun.* **206**, 724–730.
41. Arnér, E.S.J., Nordberg, J., and Holmgren, A. (1996) Efficient reduction of lipoamide and lipoic acid by mammalian thioredoxin

reductase. *Biochem. Biophys. Res. Commun.* **225**, 268–274.

42. Constantinescu, A., Pick, U., Handelman, G.J., Haramaki, N., Han, D., Podda, M., Tritschler, H.J., and Packer, L. (1995) Reduction and transport of lipoic acid by human erythrocytes. *Biochem. Pharmacol.* **50**, 253–261.
43. Jones, W., Li, X., Perriott, L.M., Whitesell, R.R., and May, J.M. (2002) Uptake, recycling, and antioxidant functions of a-lipoic acid in endothelial cells. *Free Radic. Biol. Med.* **33**, 83–93.
44. Biaglow, J.E., Donahue, J., Tuttle, S., Held, K., Chrestensen, C., and Mieyal, J. (2000) A method for measuring disulfide reduction by cultured mammalian cells: relative contributions of glutathione-dependent and glutathione-independent mechanisms. *Anal. Biochem.* **281**, 77–86.
45. May, J.M., Qu, Z.C., and Nelson, D.J. (2006) Cellular disulfide-reducing capacity: an integrated measure of cell redox capacity. *Biochem. Biophys. Res. Commun.* **344**, 1352–1359.
46. Hissin, P.J. and Hilf, R. (1976) A fluorometric method for determination of oxidized and reduced glutathione in tissues. *Anal. Biochem.* **74**, 214–226.
47. Pachla, L.A. and Kissinger, P.T. (1976) Determination of ascorbic acid in foodstuffs, pharmaceuticals, and body fluids by liquid chromatography with electrochemical detection. *Anal. Chem.* **48**, 364–367.
48. Washko, P.W., Welch, R.W., Dhariwal, K.R., Wang, Y., and Levine, M. (1992) Ascorbic acid and dehydroascorbic acid analyses in biological samples. *Anal. Biochem.* **204**, 1–14.
49. Ellman, G.L. (1959) Tissue sulfhydryl groups. *Arch. Biochem. Biophys.* **82**, 70–77.
50. May, J.M., Qiu, Z., and Whitesell, R.R. (2003) Generation of oxidant stress in cultured endothelial cells by methylene blue: protective effects of glucose and ascorbic acid. *Biochem. Pharmacol.* **66**, 777–784.
51. Farber, C.M., Kanengiser, S., Stahl, R., Liebes, L., and Silber, R. (1983) A specific high-performance liquid chromatography assay for dehydroascorbic acid show an increased content in CLL lymphocytes. *Anal. Biochem.* **134**, 355–360.
52. Bergsten, P., Amitai, G., Kehrl, J., Dhariwal, K.R., Klein, H.G., and Levine, M. (1990) Millimolar concentrations of ascorbic acid in purified human mononuclear leukocytes. Depletion and reaccumulation. *J. Biol. Chem.* **265**, 2584–2587.
53. Iheanacho, E.N., Stocker, R., and Hunt, N.H. (1993) Redox metabolism of vitamin C in blood of normal and malaria-infected mice. *Biochim. Biophys. Acta* **1182**, 15–21.

Chapter 15

Biomarkers of Oxidative Stress: Methods and Measures of Oxidative DNA Damage (COMET Assay) and Telomere Shortening

Muthuswamy Balasubramanyam, Antonysunil Adaikalakoteswari, Zaheer Sameermahmood, and Viswanathan Mohan

Abstract

Oxidative stress is fast becoming the nutritional and medical buzzword for the twenty-first century. The theoretical importance of oxidative stress in diabetes is highlighted by its potential double impact on metabolic dysfunction on one hand and the vascular system on the other hand. The new concept of oxidative stress, being an important trigger in the onset and progression of diabetes and its complications, emphasizes the need for measurement of markers of oxidation to assess the degree of oxidative stress. While we have been routinely measuring biomarkers in our molecular epidemiology projects, here we discuss the utility of two assays, (a) DNA damage assessment by COMET measurement and (b) telomere length measurement. As DNA damage is efficiently repaired by cellular enzymes, its measurement gives a snapshot view of the level of oxidative stress. The protocol allows for measurement of oxidative DNA damage (FPG-sensitive DNA strand breaks). Telomere length measured by Southern blotting technique allows one to estimate the chronic burden of oxidative stress at the molecular level and is now considered as biomarker of biological aging.

Key words: Oxidative stress, telomere shortening, comet assay, DNA damage, biomarker, diabetes.

1. Introduction

The new concept of oxidative stress, being an important trigger in the onset and progression of diabetes and its complications (1), is often challenged because intervention studies with classic antioxidants, such as vitamin E, failed to demonstrate any convincing beneficial effects on cardiovascular outcomes (2). However, these

R.M. Uppu et al. (eds.), *Free Radicals and Antioxidant Protocols*, Methods in Molecular Biology 610,
DOI 10.1007/978-1-60327-029-8_15, © Humana Press, a part of Springer Science+Business Media, LLC 1998, 2010

studies point out several flaws in the trials among which much emphasis is on the need for measurement of markers of oxidation to assess the degree of oxidative stress (3). The absence of epidemiologic data on oxidative damage in human populations represents a serious gap in our knowledge about the distribution, correlates, and causative factors of oxidative damage. As DNA damage is efficiently repaired by cellular enzymes, its measurement gives a snapshot view of the level of oxidative stress, in contrast to measurement of oxidation of other biomolecules which are not repaired and/or have a slow turnover, such as lipids or proteins.

Since direct measurement of free radical generation is not always practical, it can be assessed indirectly by measurement of oxidative products. Oxidative changes to DNA occur by oxidative modifications of the nucleotide bases or sugars or by forming cross-links, thus resulting in (i) base oxidation and fragmentation products, (ii) single- and double-strand breaks, (iii) interstrand/intrastrand cross-links, and (iv) DNA protein cross-links and sugar fragmentation products (4).

Comet assay otherwise called as single cell gel electrophoresis (SCGE) is used for detecting DNA damage and for determination of oxidized purines on DNA by using damage-specific repair endonucleases such as endonuclease III and formamidopyrimidine glycosylase (FPG) (5–7). This rapid, sensitive fluorescence microscopic method for detection of damage in individual cell has been increasingly employed for monitoring DNA damage in molecular epidemiological studies. Comet assay in combination with the FPG recognizes ring-opened purines and 8-oxoguanine and represents a reliable test for detecting oxidative base damage. This method is also less prone to the artifact of the additional oxidation and is particularly suitable for small amounts of biological material (8, 9).

Telomeres are specialized DNA-protein structures found at the ends of eukaryotic chromosomes. Telomeric DNA consists of small, tandemly repeated DNA sequences (e.g., human repeat sequence, TTAGGG). These G-rich sequences are highly conserved during evolution. They appear to be involved in determining the proliferative capacity and lifespan of both normal and malignant cells. Determinations of telomere length may provide important information about normal cell aging, as well as assisting investigation of disease processes. Accumulation of oxidative damage is also thought to play an important role in aging and associated diseases. Studies report that telomeric DNA sequences are particularly prone to chromosomal breakage and their GGG triplets are a favorable target for reactive oxygen species (ROS) (10). Telomeres are also considered to fulfill a function as stress sensors or sentinels for the risk of genomic damage due to low physiological levels of cumulative oxidative damage (11). Several studies have proposed that telomere shortening is a marker of biological aging and age-associated diseases (12–16). These studies indicate the importance of a long-term biomarker (such as shortening of telomeres) to identify those individuals who are at particularly high risk of developing disease, or to signal the disease-onset at an early stage of development, or to identify those individuals who will benefit from the intervention. So far, at least four techniques have been described for telomere length measurement: Southern-blot, Q-FISH, Flow-FISH, and Quantitative Real-Time PCR (Tel-PCR). Here, we describe the Southern blot-based telomere length measurement, which is a conventional and gold standard technique.

2. Materials

2.1. Comet Assay

1. Freshly isolated lymphocytes or other nucleated cells from culture (*see* **Note 1**).
2. Phosphate-buffered saline (PBS): dissolve 8 g of NaCl, 0.2 g of KCl, 1.44 g of Na_2HPO_4, and 0.24 g of KH_2PO_4 in 800 mL of double-distilled water (ddH_2O). Adjust the pH to 7.4 with dilute HCl and then make up the volume to 1 L with ddH_2O. Dispense the solution into aliquots and sterilize by autoclaving for 20 min at 15 psi on liquid cycle. Store at room temperature.
3. Normal melting agarose (NMA), 1% (w/v): dissolve 1 g of NMA in 100 mL of sterile PBS.
4. Low melting agarose (LMA), 0.5% (w/v): dissolve 0.5 g of LMA in 100 mL of sterile PBS.
5. Lysis buffer: 10 mM Tris–HCl, 2.5 M NaCl, 100 mM EDTA–Na_2H_2, pH adjusted to 10.0. Immediately before use, add 1 mL of Triton X-100 and 10 mL of DMSO per 100 mL of the Tris buffer at pH 10.0.
6. Electrophoresis solution: 0.3 M NaOH containing 1 mM EDTA–Na_2H_2.
7. Neutralizing buffer: 0.4 M Tris–HCl, pH 7.5.
8. Ethidium bromide (EB) staining solution: prepare EB in ddH_2O at 20 μg/mL concentration.
9. Enzyme reaction buffer: 40 mM HEPES, 0.1 M KCl, 0.5 mM EDTA–Na_2H_2, pH adjusted to 8.0 with KOH. Dissolve BSA at a concentration of 0.2 mg/mL buffer.
10. FPG enzyme: dilute 1,000-fold with enzyme reaction buffer (*see* **Note 2**).

2.2. Telomere Length Assay

1. Cell culture, biopsy material, and other biological samples.
2. Restriction enzymes: *Hinf*I and *Rsa*I (NEBL, UK).
3. TAE buffer: 0.04 M Tris base, 1 mM EDTA–Na_2H_2, pH adjusted 8.0 with acetic acid.
4. Maleic acid buffer (10×): 0.1 M maleic acid, 0.15 M NaCl, pH 7.5.
5. Washing buffer (10×): dissolve Tween 20 in 10× maleic acid buffer at a concentration of 0.3% (v/v).
6. Washing buffer, 1×: dilute an appropriate volume of 10× washing buffer 10-fold with sterile (autoclaved) ddH_2O.
7. Detection buffer (10×): 0.1 M Tris–HCl, 0.1 M NaCl, pH 9.5.
8. Detection buffer (1×): dilute an appropriate volume of 10× detection buffer, 10-fold with sterile ddH_2O.
9. Blocking solution (1×): dilute an appropriate volume of 10× blocking buffer, 1:10 with 1× maleic acid buffer.
10. HCl, 0.25 M: for a 200 cm^2 blot about 250 mL of solution is needed.
11. Denaturation solution: 0.5 M NaOH, 1.5 M NaCl. For a 200 cm^2 blot about 500 mL of solution are needed.
12. Neutralization buffer: 0.5 M Tris–HCl, 3.0 M NaCl, pH 7.5. For a 200 cm^2 blot about 500 mL of solution are needed.
13. SSC, 20× solution: 3.0 M NaCl, 0.3 M sodium citrate, pH 7.0.
14. Sodine-sodium citrate SSC, 2× solution: dilute 20× SSC 1:10 with sterile ddH_2O.
15. DIG Easy Hyb: reconstitute the granules with 65 mL sterile ddH_2O and incubate at 37°C until complete reconstitution. Prepare the solution several hours before use.
16. Stringent wash buffer I: 2× SSC with 0.1% (w/v) SDS.
17. Stringent wash buffer II: 0.2× SSC with 0.1% (w/v) SDS.
18. Anti-DIG-AP working solution: dilute an appropriate volume of anti-DIG-AP with 1× blocking solution to a final concentration of 75 mU/mL (1:10,000) (*see* **Note 3**).

3. Methods

3.1. Comet Assay

The Comet assay also called single cell gel electrophoresis is a rapid and very sensitive fluorescent microscopic method to assess DNA damage and repair in individual, nucleated cells (*see* **Table 15.1**). The assay is based on the measurement of the

Table 15.1
Troubleshooting guide for comet assay

Problem	Cause	Action
Majority of cells in untreated control sample have large comet tails	➢Unwanted damage to cells occurred in culture or in sample preparations ➢Electrophoresis solution too hot ➢Intracellular activity ➢LMAgarose too hot	➢Check morphology of cells to ensure health appearance ➢Handle cells or tissues gently to avoid physical damage ➢Control temperature by recirculating the electrophoresis solution or performing the assay at less than 20°C ➢Keep cells on ice and prepare cell samples immediately before combining with molten LMA ➢Cool LMAgarose to 42°C before adding cells
Majority of cells in untreated control sample have small to medium comet tails	➢Endogenous oxidative damage or endonuclease activity after sample preparation is damaging DNA	➢Ensure that lysis solution was chilled before use ➢Add DMSO to any cell sample that may contain heme groups ➢Ensure that PBS used is calcium and magnesium free ➢Work under dimmed light conditions or under yellow light
In positive control (e.g., 100 μM hydrogen peroxide for 30 min on ice) no evidence of comet tail	➢No damage to DNA ➢Sample was not processed correctly	➢Use fresh hydrogen peroxide to induce damage ➢Ensure that each step in protocol was performed correctly. Failure to lyse, denature in alkali (optional), or to properly perform electrophoresis may generate poor results
Comet tails present but not significant in positive control	➢Insufficient denaturation in alkaline solution ➢Insufficient electrophoresis time	➢Increase time in alkaline solution up to 1 h ➢Increase time of electrophoresis up to 20 min for TBE and up to 1 h for alkaline electrophoresis. Increase time of electrophoresis when running at cold temperature

Table 15.1 (continued)

Problem	Cause	Action
Cells in LM agarose did not remain attached to the microscopic slide	➢Electrophoresis solution too hot ➢Cells were not washed to remove medium before combining with LMAgarose ➢Agarose percentage was too low ➢LMAgarose was not fully set before samples were processed ➢LMAgarose unevenly set on the slide	➢Control temperature by recirculating the electrophoresis solution or performing the assay at less than 20°C ➢The pH of medium and carry over serum proteins, etc., can reduce the adherence of the agarose. Resuspend cells in 1× PBS ➢Do not increase ratio of cells to molten agarose by more than 1–10 ➢Ensure that 0.5 mm dried rings due to agarose disc retraction is seen at the edge of the slide area ➢Spread the agarose with the slide of pipette tip to ensure uniformity of agarose disc and better adherence

ability of denatured, cleaved DNA fragments to migrate out of the cell under the influence of an electric field, whereas undamaged DNA migrates slower and remains within the confines of the nucleoid when a current is applied. Evaluation of the DNA "comet" tail shape and migration pattern allows for assessment of DNA damage. In this assay, cells are immobilized in a bed of low melting point agarose, followed by gentle cell lysis, treated with alkali to unwind and denature the DNA, and hydrolyze sites of damage. The samples are then submitted to electrophoresis and stained with a fluorescent DNA intercalating dye. The sample is then visualized by epifluorescence microscopy.

The causality and specificity of oxidative DNA damage, i.e., the DNA strand breaks specific to 8-oxo-2′-deoxyguanosine can be assessed by incorporating a digestion step of nucleoid DNA with a lesion-specific endonuclease, formamidopyrimidine glycosylase (FPG). Therefore, comet assay in combination with the enzyme (FPG) recognizes DNA damage due to ring-opened purines and 8-oxoguanine. After lysis of cells, the nucleoids are incubated with FPG, where additional breaks are formed at sites of such lesions and the amount of DNA in the tail of the comet is increased.

Cell samples should be prepared immediately before starting the assay, although success has been achieved using cryopreserved cells (*see* **Note 4**). Handling of cell samples should be performed under dimmed or yellow light to prevent DNA damage from ultraviolet light. Buffers should be chilled to 4°C or on ice to inhibit endogenous damage occurring during sample preparation and to inhibit repair in the unfixed cells. Phosphate-buffered saline must be free of calcium and magnesium to inhibit endonuclease activities. The appropriate controls should also be included (*see* **Notes 5** and **6**).

3.1.1. Assay Protocol

1. Pre-cote microscopic slides by layering 1% (w/v) NMA and drying them at 37°C.
2. Suspend control and treated cells in PBS at 2×10^4/mL.
3. Mix 100 μL of cell suspension with 200 μL of 0.5% (w/v) LMA. Layer the solution onto the pre-coated slide, allow the gel to solidify with a cover slip over the gel, and chill the slide to 4°C for 5 min.
4. Remove the cover slip, add 200 μL of 1% (w/v) NMA onto the slide, and incubate for 5 min at 4°C with a cover slip over the gel.
5. Remove the cover slip and place the slide in the lysis buffer (10 mM Tris–HCl, pH 10.0, containing 2.5 M NaCl, 100 mM EDTA–Na_2H_2, 1% (v/v) Triton X-100, and 10% (v/v) DMSO) for 1 h at 4°C.
6. Remove the slide from the lysis buffer. Place the slide in the electrophoresis tank containing pre-chilled electrophoresis solution (0.3 M NaOH containing 1 mM EDTA–Na_2H_2) and incubate for 40 min at 4°C before beginning electrophoresis.
7. Set the run for 30 min at 25 V with the current adjusted to 300 mA.
8. Remove the slide after the run and wash three times with neutralization buffer (0.5 M Tris–HCl, 3.0 M NaCl, pH 7.5) for 5 min each at 4°C.
9. Place 75 μL of 20 μg/mL EB solution onto the slide and examine under fluorescent microscope set at excitation and emission wavelengths of 515 and 535 nm, respectively.

3.1.2. Data Analysis and Results

Fluorescence microscopy. When excited, the DNA-bound ethidium bromide emits red light. In healthy cells the fluorescence is confined to the nucleoid: undamaged DNA is supercoiled and thus does not migrate very far of the nucleoid under the influence of an electric current (**Fig. 15.1**). In cells that have accrued damage to the DNA, the alkali treatment unwinds the DNA, releasing fragments that migrate from the cell when subjected to an electric field. The negatively charged DNA migrates toward the anode and the extrusion length reflects increasing relaxation of

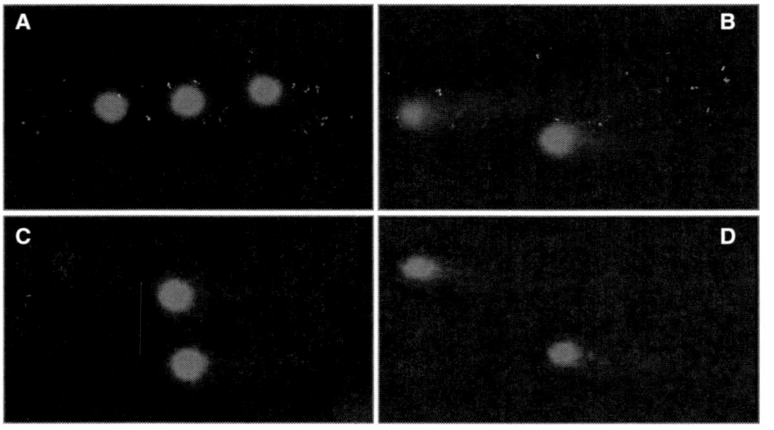

Fig. 15.1. Typical results from a comet assay: (**a**) basal DNA damage from control subject, (**b**) basal DNA damage from diabetic subject, (**c**) FPG-sensitive DNA damage in control, and (**d**) FPG-sensitive DNA damage in diabetics.

supercoiling, which is indicative of damage. When using alkaline electrophoresis conditions, the distribution of DNA between the tail and the head of the comet should be used to evaluate the degree of DNA damage. The characteristics of the comet tail including length, width, and DNA content may also be useful in assessing qualitative differences in the type of DNA damage.

Reproducibility. To test the reproducibility of comet assay, we measured percent DNA in the tail of eight subjects on two different occasions. For this, blood samples were taken twice from the same subject on two different occasions and the respective samples used for the comet assay assessment were referred to as basal DNA damage-1 and basal DNA damage-2. As shown in **Fig. 15.2**, the two values correlated well ($r = 0.87$; $p < 0.001$), indicating that comet assay is a reproducible measure.

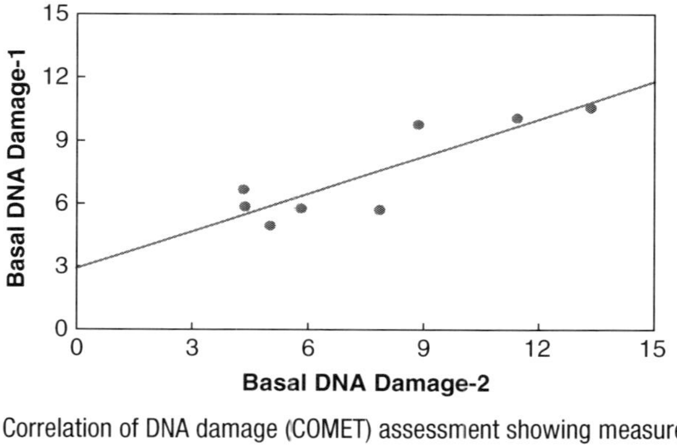

Fig. 15.2. Correlation of DNA damage (COMET) assessment showing measure of reproducibility. Blood sample were collected twice from the same subjects (n=8) on two different occasions and the respective samples used for the comet assay were referred to as basal DNA damage-1 and basal DNA damage-2.

Qualitative analysis. The comet tail can be scored according to DNA content (intensity). The untreated, control cells should be used to determine the characteristics of data for a healthy cell. Scoring can then be made according to nominal, medium, or high intensity tail DNA content. At least 75 cells should be scored per sample. Score each comet on a scale of 0 (no tail) to 4 (almost all DNA in the tail, insignificant head). Each slide can then be given an arbitrary score from 0 to 400.

Quantitative analysis. There are several image analysis systems that are suitable for quantitation of data on comet assay. The more sophisticated systems include the microscope, camera, and computer analysis package. These systems can be set up to establish the length of DNA migration, image length, nuclear size, and calculate the tail moment. At least 75 randomly selected cells should be analyzed from each slide avoiding those comets on the edge of the gel as these are prone to artifacts. Subtracting the percent of DNA in the tail without enzyme incubation from the percent of DNA in the tail with enzyme incubation gives the net amount of damage represented by FPG-sensitive sites (*see* **Fig. 15.1**).

Troubleshooting. **Table 15.1** gives an outline of the problems generally encountered while performing the comet assay, their cause, and the action plan to proceed with the experiment.

3.2. Telomere Length Assay

Various methods have been described to detect telomeres and to measure telomere length (*see* **Chart 15.1** and **Table 15.2**). The standard method to assess telomere length utilizes Southern blot analysis of terminal restriction fragments (TRFs) obtained by digestion of genomic DNA using frequently cutting restriction enzymes. The TRFs obtained contain DNA with uniform telomeric (TTAGGG) repeats as well as degenerate repeats other than at the distal end of the chromosome (sub-telomeric region). After digestion, the DNA fragments are separated by gel electrophoresis, blotted, and TRFs are visualized directly or indirectly by hybridization with labeled oligonucleotides complementary to the telomeric repeat sequence. Finally, the size distribution of the TRFs can be compared to a DNA length standard.

3.2.1. Assay Protocol

1. Isolation of genomic DNA: follow standard protocols or use a commercial kit such as the one marketed by Roche Applied Science (*see* **Note 7**).
2. Digestion of DNA (*see* **Note 8**):
   (a) Prepare *Hin*fI/*Rsa*I by mixing equal volumes of *Hin*fI (20 units/μL per sample) and *Rsa*I (20 units/μL per sample).

Chart 1. Flow Chart of the Telomere Length Assay

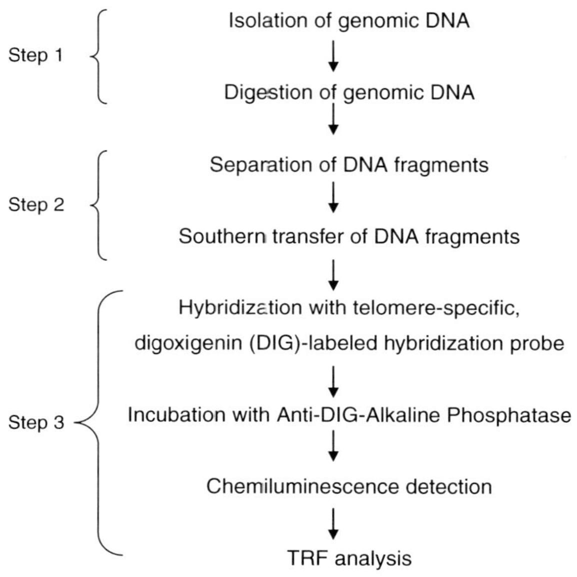

(b) Use 1 μL of the enzyme mixture for each sample to be tested and for the positive (control-DNA high) and negative controls (control-DNA low).

(c) Dilute DNA (1 μg; in 10 μL) from the positive as well as the negative control samples with 7 μL of nuclease-free water.

(d) Dilute 1–2 μg of purified genomic DNA similarly with nuclease-free water to a final volume of 17 μL per sample. Add 2 μL of 10× digestion buffer, and 1 μL of *Hin*fI/*Rsa*I enzyme mixture. Mix and incubate for 2 h at 37°C.

(e) At the end of incubation period, stop the reaction by adding 4 μL of 5× gel electrophoresis loading buffer and tap spin.

3. Gel electrophoresis (*see* **Note 9**):
   (a) Cast 0.8% agarose gel (about 15 cm in length) in 1× TAE buffer. Use highly pure, nucleic acid grade agarose.
   (b) Mix 4 μL DIG molecular weight marker with 12 μL nuclease-free water and 4 μL 5× loading buffer.
   (c) Load digested DNA samples (1–2 μg/lane) in respective lanes with 10 μL of DIG molecular weight marker and run the gel at 5 V/cm in 1× TAE buffer (*see* **Note 10**).
4. Southern blotting (*see* **Note 11**):

Table 15.2
Troubleshooting guide for Telomere Length Assay

No.	Problem	Possible cause	Recommendation
1	High membrane background		• Check that there is sufficient solution of the detection reagents • Reduce concentration of the telomere probe • Quick-spin Anti-DIG-AP before use • Check diluted working solutions for bacterial contamination
2	Weak signals or none at all	• Check that DIG molecular weight marker is clearly visible, meaning there is no transfer problem • DIG molecular weight marker is hardly or not at all visible. Inefficient transfer of the restricted DNA to the membrane occurred	• Increase the DNA amount loaded onto the gel (up to 7.5 μg/lane) • Repeat the complete experiment with the control-DNA supplied with the kit
3	Multiple bands above telomere signal		• Check if there are multiple bands observed also with the control-DNA supplied with the kit • Inadequate purity of the genomic DNA • Prolong digestion of the genomic DNA
4	Multiple bands below the telomere signal		• Check temperature for stringent wash step II carefully • Increase time for stringent washing of the membrane during detection
5	DIG marker shows more bands than expected	Marker has been incubated at 65°C for 10 min	Do not heat DIG molecular weight marker prior to loading to the agarose gel

(a) Submerge the gel in 0.25 N HCl for 5–10 min with agitation at ambient temperature. Make sure that the bromophenol blue stain changes its color to yellow.

(b) Rinse the gel in ddH_2O twice and submerge in denaturation solution for 2 × 15 min at ambient temperature followed by rinsing in ddH_2O again two times.

(c) Submerge the gel in neutralization solution for 2 × 15 min at ambient temperature.

   (d) Blot the digested DNA from the gel to nylon membrane by capillary transfer at ambient temperature using 20× SSC (*see* **Notes 12** and **13**).

   (e) Fix the transferred DNA onto the wet blotting membrane by UV-cross-linking (120 mJ) followed by washing the blotting membrane with 2× SSC (*see* **Note 14**).

5. Hybridization (*see* **Note 15**):

   (a) Prewarm approximately 25 mL of DIG Easy Hyb to 42°C.

   (b) Submerge the blot in 18 mL of pre-warmed DIG Easy Hyb and incubate for 30–60 min at 42°C with gentle agitation (pre-hybridization).

   (c) Prepare the hybridization solution by mixing 1 μL of telomere probe per 5 mL of fresh, pre-warmed DIG Easy Hyb.

   (d) Discard the pre-hybridization solution and incubate the blot in hybridization solution for 3 h at 42°C with gentle agitation.

   (e) At the end of the incubation period, discard the hybridization solution and proceed with the washing procure (*see* below).

6. Washing:

   (a) Wash the membrane with sufficient stringent wash buffer I for 5 min at 15–25°C followed by stringent wash buffer II for 15–20 min at 50°C in a heated water bath with gentle agitation.

   (b) Incubate the membrane with 100 mL of freshly prepared 1× blocking solution for 30 min at 15–25°C with gentle agitation.

   (c) After blocking, treat the membrane with 50–100 mL anti-DIG-AP solution for 30 min at 15–25°C with gentle agitation.

   (d) Incubate the membrane with 100 mL Anti-DIG-AP working solution for 30 min at 15–25°C with gentle agitation.

   (e) Wash the membrane with 200 mL of 1× washing buffer 2 × 15 min at 15–25°C with gentle agitation.

   (f) Finally, incubate the membrane with 100 mL of 1× detection buffer for 2–5 min at 15–25°C with gentle agitation.

7. Chemiluminescence detection:

   (a) Discard the detection buffer. Place the wet membrane on a hybridization tray with DNA side facing up. Add approximately 3 mL (40 drops) of substrate solution quickly onto the membrane.

(b) Cover the membrane with a sheet of hybridization bag to spread of the substrate solution uniformly. Care must be taken to avoid air bubbles over the membrane.

(c) Incubate the membrane for 5 min at 15–25°C, then squeeze out excess substrate solution, seal off edges of the hybridization bag, and then expose the whole setup to X-ray film for 5–20 min at 15–25°C (*see* **Note 16**).

3.2.2. Data Analysis – Quantification of Terminal Restriction Fragments (TRFs) Length

After exposing the blot to an X-ray film, an estimate of the mean TRF length can be obtained by visually comparing the mean size of the smear to the molecular weight marker. However, for quantitative measurements of mean TRF length, the chemiluminescence signals should be scanned using a densitometer or other imaging system. To obtain reliable results, the signal strength must be within the linear range of the X-ray film.

1. Scan the exposed X-ray film with a densitometer.
2. Overlay each sample lane of the scanned image with a grid (**Fig. 15.2**). The vertical size of the individual squares of the grid defines the resolution in determining the TRF length. Typically, >30 squares per lane are recommended.
3. For background subtraction, select several boxes in each lane where no telomere-specific signal is found and which are representative for the background of the corresponding lane. Signals of these boxes should be averaged and subtracted from each grid box.
4. For each square, determine the signal (OD_i) and the corresponding length L_i where OD_i is the total signal intensity within the grid box and L_i is the molecular weight at the midpoint of the corresponding box.
5. Calculate the mean TRF length using the formula

$$\mathrm{TRF} = \frac{\sum(\mathrm{OD}_i)}{\sum(\mathrm{OD}_i/L_i)}$$

where OD_i is the chemiluminescent signal and L_i is the length of the TRF fragment at position *i*. The calculation takes into account the higher signal intensity from larger TRF fragments because of multiple hybridization of the telomere-specific hybridization probe.

3.2.3. Typical Results

To show the reproducibility of our method, we measured telomere lengths of eight subjects on two different occasions. For this blood samples were taken twice from the same subjects on

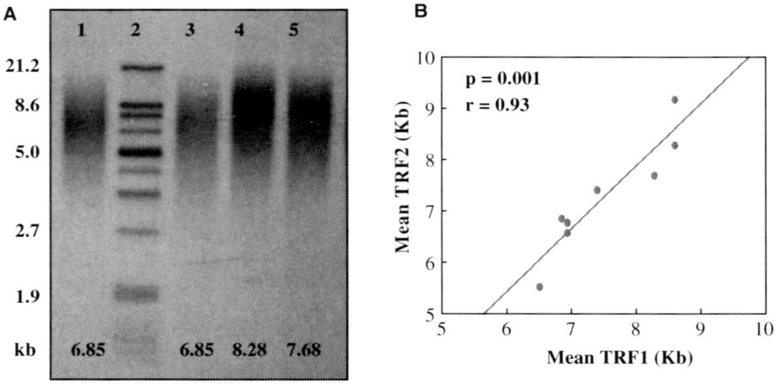

Fig. 15.3. (**a**) Reproducibility illustration of Southern blot of TRF from two subjects. Lanes 1 and 3 refer to the telomere length of subject 1 determined on two occasions. Lanes 4 and 5 refer to the telomere length of subject 2, and lane 2 represents the molecular weight (0.8–21.2 Kb) marker used. (**b**) Correlation between the TRF values eight subjects determined on two different occasions The first and second sets of TRF values are referred to as TRF 1 and TRF 2, respectively.

two different occasions and the respective DNA used for the TRF length measurements were referred to as TRF1 and TRF2. **Figure 15.3a** is the reproducibility illustration of Southern blot of TRF from two subjects showing reproducibility. As shown in **Fig. 15.3b**, the two values correlated well ($r = 0.93$; $p < 0.001$), indicating that white blood cell mean TRF length is a reproducible measure.

4. Notes

1. A slight modification in the comet assay is needed to detect oxidized bases using the repair endonucleases FPG. Follow the comet assay procedure till the lysis step, do the enzyme treatment, and continue with electrophoresis.
2. Alternatively, one can use hOGG1 (human 8-oxoguanine DNA glycosylase) or endonuclease III (for oxidized pyrimidines) or AlkA (for alkylation damage).
3. Spin the vial for 5 min at 13,000 rpm before use. This helps to reduce background due the aggregated antibody, if any, present in the sample.
4. The comet assay can be performed using cryopreserved cells. For this, centrifuge cell suspensions at 2,000 rpm for 5 min at 4°C. Re-suspend the cell pellet in fetal calf serum/DMSO (9/1, v/v) at 1×10^7 cells/mL. Using a controlled rate freezer. cool the cell suspensions at a rate of 1°C per min between +4 and –30°C, 2°C per min between

–30 and –70°C, and then transfer them to liquid nitrogen. Just before the comet assay, cells can be revived and resuspended in ice-cold PBS.

5. A sample of untreated cells should always be processed as a control for endogenous levels of damage within cells and for damage that may occur during sample preparation. Both control cells and treated cells should be handled in an identical manner.
6. For positive controls, cells are treated with either H_2O_2 (100 μM) or $KMnO_4$ (25 μM) for 20 min at 4°C. This treatment will generate significant oxidative damage in the majority of cells, thereby providing a positive control for each step in the comet assay. Note that the dimensions and characteristics of the comet tail, as a consequence of H_2O_2 or $KMnO_4$ treatment, may be different to those agents inducing the damage under investigation.
7. For maximum efficiency in the digestion of genomic DNA, spin the DNA solution obtained after preparation in a centrifuge for 5 min at maximum speed. Transfer DNA-containing supernatant to a fresh tube and continue with digestion of genomic DNA.
8. Handling of the solutions and pipetting should be done on ice.
9. Separation of digested DNA is done by agarose gel electrophoresis following standard protocols.
10. To obtain valid quantitative analysis of telomere length, each sample should contain the same amount of DNA.
11. Southern transfer of the digested DNA can be done by both capillary transfer and vacuum transfer using 20× SSC or alkaline transfer buffer. However, maximum transfer efficiency and sensitivity are obtained with positively charged nylon membranes using capillary transfer with 20× SSC. All incubation steps should be performed with gentle agitation.
12. It is recommended that unpowdered rubber gloves be worn and that the membrane be handled with forceps only at the edges.
13. Transfer for 6 h is generally sufficient in most applications. However, for maximum sensitivity and reproducibility of the results, overnight blotting is recommended.
14. If not used immediately for the hybridization and chemiluminescence detection step, air-dry the blotting membrane and store at 2–8°C.

15. The volumes recommended in the hybridization step are based on a membrane size of 200 cm^2. The volumes should be adjusted accordingly if membranes of other sizes are used. It is highly recommended to control the hybridization and stringent wash temperatures exactly.
16. Luminescence continues for at least 24 h and signal intensity will increase during the first few hours.

Acknowledgments

This work was supported by research grants from the Department of Science and Technology (DST & DST-FIST), the Department of Biotechnology (DBT), and the Indian Council of Medical Research (ICMR), Government of India.

References

1. Ceriello, A. and Motz, E. (2004) Is oxidative stress the pathogenic mechanism underlying insulin resistance, diabetes, and cardiovascular disease? The common soil hypothesis revisited. *Arterioscler. Thromb. Vasc. Biol.* **24**, 816–823.
2. Balasubramanyam, M. (2006) Antioxidants and cardiovascular disease – Where do we stand? *Asian J. Diabetol.* **8**, 1–5.
3. Evans, J.L., Maddux, B.A., and Goldfine, I.D. (2005) The molecular basis for oxidative stress-induced insulin resistance. *Antioxid. Redox. Signal* **7**, 1040–1052.
4. Evans, M.D. and Cooke, M.S. (2004) Factors contributing to the outcome of oxidative damage to nucleic acids. *Bioessays* **26**, 533–542.
5. Singh, N.P., McCoy, M.T., Tice, R.R., and Schneider, E.L. (1988) A simple technique for quantitation of low levels of DNA damage in individual cells. *Exp. Cell. Res.* **175**, 184–191.
6. Collins, A.R., Raslova, K., Somorovska, M., Petrovska, H., Ondrusova, A., Vohnout, B., Fabry, R., and Dusinska, M. (1998) DNA damage in diabetes: correlation with a clinical marker. *Free Radic. Biol. Med.* **25**, 373–377.
7. Adaikalakoteswari, A., Rema, M., Mohan, V., and Balasubramanyam, M. (2007). Oxidative DNA damage and augmentation of Poly (ADP-ribose) polymerase/Nuclear factor-kappa B signaling in patients with Type 2 diabetes and microangiopathy. *Int. J. Biochem. Cell Biol.* **39**,1673–1684.
8. Collins, A.R. (2004) The comet assay for DNA damage and repair: principles, applications, and limitations. *Mol. Biotechnol.* **26**, 249–261.
9. Cadet, J., Bellon, S., Berger, M., Bourdat, A.G., Douki, T., Duarte, V., Frelon, S., Gasparutto, D., Muller, E., Ravanat, J.L., and Sauvaigo, S. (2002) Recent aspects of oxidative DNA damage: guanine lesions, measurement and substrate specificity of DNA repair glycosylases. *Biol. Chem.* **383**, 933–943.
10. Kawanishi, S. and Oikawa, S. (2004) Mechanism of telomere shortening by oxidative stress. *Ann. NY Acad. Sci.* **1019**, 278–284.
11. Von Zglinicki, T. and Martin-Ruiz, C.M. (2005) Telomeres as biomarkers for ageing and age-related diseases. *Curr. Mol. Med.* **5**, 197–203.
12. Samani, N.J., Boultby, R., Butler, R., Thompson, J.R., and Goodall, A.H. (2001) Telomere shortening in atherosclerosis. *Lancet* **358**, 472– 473.
13. Jeanclos, E., Schork, N.J., Kyvik, K.O., Kimura, M., Skurnick, J.H., and Aviv, A. (2000) Telomere length inversely correlates with pulse pressure and is highly familial. *Hypertension* **36**, 195–200.
14. Adaikalakoteswari, A., Balasubramanyam, M., and Mohan, V. (2005) Telomere shortening occurs in Asian Indian type 2 diabetic patients. *Diabet. Med.* **22**, 1151–1156.
15. Adaikalakoteswari, A., Balasubramanyam, M., Ravikumar, R., Deepa, R., and Mohan, V. (2007) Association of telomere shorten-

ing with impaired glucose tolerance and diabetic macroangiopathy. *Atherosclerosis* **195**, 83–89.

16. Brouilette, S.W., Moore, J.S., McMahon, A.D., Thompson, J.R., Ford, I., Shepherd, J., et al. (2007) West of Scotland Coronary Prevention Study Group. Telomere length, risk of coronary heart disease, and statin treatment in the West of Scotland Primary Prevention Study: a nested case-control study. *Lancet* **369**, 107–114.

Chapter 16

Simultaneous Analysis of Expression of Multiple Redox-Sensitive and Apoptotic Genes in Hypothalamic Neurons Exposed to Cholesterol Secoaldehyde

K. Sathishkumar, Achuthan C. Raghavamenon, Karunakaran Ganeshkumar, Rameshwari Telaprolu, Narasimham L. Parinandi, and Rao M. Uppu

Abstract

Oxidative stress and apoptotic cell death are two important processes that occur under several disease states and in conditions of toxicant insult. Traditionally, investigators have chosen a variety of analytical methods to detect and/or quantify oxidative stress and apoptosis. The approach has proven less satisfying, however, when applied to complex systems with many unknown influences. Such areas of study could benefit from the development and application of new and more powerful analytical tools. Microarray-based approach has been developed for analyzing various cellular phenomena at the level of gene expression. These gene arrays are hybridization chips that are capable of simultaneous analysis of the expression of thousands of genes. Often, this approach warrants examining a multitude of unrelated genes which can greatly impede the interpretation of results. The real-time RT-PCR-based methodology presented here allows simultaneous detection and analysis of as many as 84 well-characterized genes associated with either oxidative stress or apoptosis in hypothalamic neuronal cells exposed to cholesterol secoaldehyde, an "ozone-/singlet oxygen-specific" oxidation product of cholesterol that has been shown to be present at the inflammatory sites including the arterial plaque and the brain specimens of patients with Alzheimer's disease. This pathway-specific analysis of the expression of the well-defined chosen set of genes offers ways of convenient and reliable interpretation of results that often corroborate well with the results obtained from other standard biochemical analytical approaches.

Key words: Oxidative stress, apoptosis, cholesterol secoaldehyde, microarray gene expression, RT-PCR.

1. Introduction

Living cells under normal physiological states constantly encounter various intrinsic and extrinsic stimuli that, often, results

R.M. Uppu et al. (eds.), *Free Radicals and Antioxidant Protocols*, Methods in Molecular Biology 610,
DOI 10.1007/978-1-60327-029-8_16, © Humana Press, a part of Springer Science+Business Media, LLC 1998, 2010

in oxidative stress and apoptotic cell death. The affected cells respond by altering the expression of several genes/proteins. Among the most common defense mechanisms activated in these cells, the production of endogenous radical scavengers (1), anti-apoptotic proteins (2), and an increased synthesis of antioxidants or repair enzymes play an important role. There is a fine regulation of the antioxidant effects and if this is disturbed, the rescue process is hampered resulting in the oxidative stress and can lead to apoptotic cell death (3–6). Accordingly, oxidative stress and apoptosis play a major role in a number of pathologies including cancer (7–8), neurodegenerative diseases (9–10), and cardiovascular diseases (11–12). Therefore, a comprehensive analytical system that reliably facilitates the early detection of these pathological processes is an imminent choice. Both qualitative and quantitative measurements of gene induction/expression (mRNA transcripts) provide an insight of various pathways associated with the altered cellular physiology for arriving at reasonable conclusions. With varying degrees of reliability and sensitivity, the modern microarray platforms are fine tools to assess the expression of several thousands of specific mRNA (13–14). However, identification of the expression of all the genes that are present in the target organ/cell will be a tedious process as this warrants examining a multitude of unrelated genes which often might impede in the interpretation of results (15). Application of the real time RT-PCR methodology as presented in this chapter allows simultaneous detection and analysis of pathway-specific expression of gene(s) that would be highly useful to describe the mechanism of toxicity or stress. A battery of well-characterized genes associated with oxidative stress and apoptosis were studied in murine GT1-7 hypothalamic neurons exposed to cholesterol secoaldehyde (ChSeco or 3β-hydroxy-5-oxo-5,6-secocholestan-6-al), an "ozone-/singlet oxygen-specific" oxidation product of cholesterol identified in the arterial plaque (16) and in the brain specimens of patients with Alzheimer's disease (17). ChSeco has been shown to be cytotoxic to several mammalian cells including cultured 16-HBE cells (18), GT1-7 hypothalamic neurons (19), J744 and RAW264.7 macrophages (20, 21), and H9c2 cardiomyoblasts (22), mostly through eliciting cellular oxidative stress (23). The approach of pathway-specific gene expression presented in this chapter provides corroborative evidence that goes well with routine biochemical techniques.

2. Materials

2.1. Cell Culture

1. Water (*see* **Note 1**).
2. Murine GT1-7 hypothalamic neurons are a gift from Dr. Pam Mellon, University of California, San Diego, CA.

3. Dulbecco's Modified Eagle's Medium (DMEM; Sigma, St. Louis, MO) containing glutamine (2 mM), glucose (4.5 g/L), sodium pyruvate (1.0 mM), penicillin (100 units/mL), streptomycin (100 μg/mL), and $NaHCO_3$ (1.5 g/L).
4. Supplement DMEM with 10% fetal bovine serum (FBS; Atlanta Biologicals, Lawrenceville, GA).
5. Solution of trypsin (0.25%) and ethylenediaminetetraacetic acid (EDTA; 1 mM) from Sigma.
6. ChSeco, 20 mM: prepare in ethanol and store at −70°C in aliquots of 50–100 μL each.

2.2. RNA Isolation, cDNA Synthesis, and Real-Time (RT) PCR

1. Tri Reagent® and bromochlorophenol from Molecular Research Center (Cincinnati, OH).
2. RNeasy Mini Kit consisting of wash buffers I and II and an extraction buffer containing guanidine isothiocyanate (GITC) from Qiagen (Valencia, CA). Just before use, add 10 μL of β-mercaptoethanol (Sigma) per milliliter of GITC-containing buffer.
3. SYBR green master mix (2x), cDNA synthesis kit and RT^2 profiler PCR array for rat oxidative stress and apoptosis from SuperArray (Frederick, MD).
4. RNase-free DNase set containing a diluent buffer for reconstitution from Qiagen (Valencia, CA).
5. Ethanol (70%) from Aaper (Shelbyville, KY).
6. Cell scrapers from BD Falcon (Bedford, MA).
7. ABI Prism 7900HT sequence detection system from Applied Biosystems (Foster City, CA).

3. Methods

3.1. Synthesis of Cholesterol Secoaldehyde

ChSeco can be synthesized by ozonation of cholesterol in mixed solvents of dichloromethane (DCM)/methanol (1/1, v/v) and subsequent reduction of the concentrated product mixture with Zn/acetic acid containing traces (up to 5%, v/v) of water, as described by Wang et al. (24) and Sathishkumar et al. (22). A typical synthesis of ChSeco (~400 mg) can be accomplished as follows:

1. Dissolve 0.39 g (~1 mmol) of cholesterol (purity >99%; Sigma) in 100–150 mL of DCM/methanol (1/1, v/v) in a bubbler (*see* **Note 2**).
2. Using a Sander 200 ozonizer (Erwin Sander, Uetze-Eltze, Germany), generate ozone (4 mmol/h) in a stream of oxygen (300 mL/min) (*see* **Note 3**).

3. Pass the ozone in O_2 through the solution of cholesterol for 20 min (*see* **Notes 4** and **5**).
4. Concentrate the product mixture under vacuum using a rotary evaporator (Buchi Rotavapor R205; Brinkmann Instruments, Westbury, NY).
5. Reconstitute the residue in 20 mL of glacial acetic acid/water (19/1, v/v).
6. Add 600 mg of zinc dust and stir the mixture for 2 h at room temperature. Filter.
7. Dilute the filtrate with 80 mL of water and extract the mixture with 100 mL of DCM using a separatory funnel with Teflon stopper.
8. Wash the organic (DCM) layer with 3 × 1 volume of 1 M NaCl in water followed by 1 volume of water.
9. Dry the DCM layer over anhydrous Na_2SO_4 and concentrate the product under vacuum using a flash evaporator.
10. Test the purity of the final product by RP-HPLC (*see* **Note 5**).
11. Prepare stock solutions of ChSeco in DMSO or ethanol at a concentration of 20 mM. Store in aliquots of 50–200 μL each at −80°C until use.

3.2. Cell Culture and Treatment with ChSeco

1. Grow Murine GT1-7 hypothalamic neurons in T-75 flasks with DMEM containing 10% FBS at 37°C in 5% CO_2/95% air.
2. Disperse the cells with trypsin–EDTA, briefly wash the cells, resuspend in the culture medium, and transfer 2 mL to each well in six-well plates and allow to grow overnight.
3. At sub-confluence, remove the culture medium and add fresh DMEM containing 2% FBS.
4. Add 2 μL of 20 mM ChSeco to obtain a final concentration of 20 μM. Expose the control cells to an equal volume of vehicle (2 μL ethanol).
5. After 8 h of incubation at 37°C, remove the medium by centrifugation and isolate total RNA as described in **Section 3.3**.

3.3. Isolation of RNA

3.3.1. Extraction Using Trizol

1. Remove the culture medium from treated and control wells, wash the cells once with 1 mL of PBS, and then add 1 mL of Tri reagent (*see* **Note 6**).
2. Dissociate cells from plate using a scraper and transfer to a sterile tube (~5 mL).
3. Homogenize with a handheld homogenizer for 1–2 min (*see* **Note 7**).

4. Incubate the homogenate at room temperature for 5 min and add 0.1 mL of bromochlorophenol (BCP) reagent and mix vigorously.
5. Keep the tubes for 15 min at room temperature and centrifuge at 12,000×*g* for 15 min at 4°C to facilitate separation of phases.
6. Transfer the aqueous phase into a new tube and add 0.5 mL of chilled isopropanol.
7. Mix and allow standing at room temperature for 5 min, thereafter centrifuge the samples at 12,000×*g* for 8 min at 4°C to pellet the total RNA.
8. Discard the supernatant carefully without losing the pellet.
9. Wash the pellet with 1 mL of 70% ethanol, centrifuge at 7,500×*g* for 5 min at 4°C to collect the pellet. Repeat the step twice.
10. Allow the pellets to air dry (*see* **Note 8**).
11. Dissolve the pellet in 100 μL of RNase-free water and purify using a Qiagen column.

3.3.2. RNA Cleanup Using RNeasy Kit

1. Mix the RNA sample (*see* above) thoroughly with 350 μL of GITC-containing buffer.
2. Add 250 μL of ethanol (200% proof) and mix the contents gently.
3. Transfer the sample onto a silica gel-based spin column and centrifuge at 8,000×*g* for 15 s. Discard the eluent.
4. Add 350 μL of wash buffer I to the column and centrifuge at 8,000×*g* for 15 s.
5. Perform digestion of DNA by adding 80 μL of DNase (*see* **Note 9**) directly onto the silica gel column and incubating at room temperature (20–30°C) for 15 min.
6. At the end of incubation, add 350 μL of wash buffer I onto the column and centrifuge at 8,000×*g* for 15 s.
7. Wash the column with 2 × 500 μL of wash buffer II, centrifuging each time at 8,000×*g* for 2 min. Discard the flow-through.
8. Transfer the column to a microfuge tube (capacity: 1.5 mL). Add 50 μL of RNase-free water to the column and elute the RNA by centrifuging at 8,000×*g* for 1 min.

3.3.3. RNA Quality and Quantity Assessment

The most important prerequisite for real-time RT-PCR analysis is to have a high-quality RNA. Traces of proteins, salts, or other contaminants will decrease the performance of PCR. The quantity and quality of RNA obtained can be examined by UV spectrophotometry and agarose gel electrophoresis.

3.3.3.1. Quantitation of RNA

1. Dilute the RNA sample 1:200 with RNase-free 10 mM Tris–HCl buffer, pH 8.0 (typically, 5 μL of RNA sample is added to 995 μL of the buffer).
2. Measure the absorbance at 230, 260, and 280 nm using UV spectrophotometer.
3. Determine the A_{260}:A_{280} and A_{260}:A_{230} ratios (*see* **Note 10**).
4. The following formula can be used to determine the concentration of RNA
   Total RNA (μg/mL) = 40 × 260 × 200 (200 is the dilution factor).

3.3.3.2. Determination of Quality of RNA

1. Prepare a 2% formalin agarose gel in 0.04 M Mopes buffer, pH 8.0.
2. Load approximately 500 ng of RNA samples per well and run at 5 V/cm for 45 min using Mopes buffer.
3. Stain the gel with ethidium bromide dye.
4. Two sharp bands corresponding to 18S and 28S ribosomal RNA should be visible, which indicates a good-quality RNA. A streaking or a smearing band indicates degradation of RNA and should be discarded.

3.4. cDNA Synthesis

1. Prepare an annealing mixture in a 0.2-μL PCR tube by adding up to 3.0 μg of RNA and 1 μL of random primer. Bring the final volume to 10 μL with RNase-free water.
2. Mix the contents, heat denature at 70°C for 3 min in a thermal cycle, and then cool to room temperature.
3. While the mixture is in thermal cycle, prepare reverse transcriptase (RT) reaction mix by adding 4 μL of 5× RT-buffer to 4 μL of RNase-free water, 1 μL of RNase inhibitor, and 1 μL of RT enzyme.
4. Mix together 10 μL of RT mix and the annealing mix thoroughly. Do not vortex.
5. Incubate the reaction mixture at 37°C for 60 min in a thermal cycle followed by an enzyme inactivation step at 95°C for 5 min.
6. Make up the final volume to 100 μL using RNase-free water. cDNA can be stored at –20°C.

3.5. Analysis of Redox-Sensitive and Apoptotic Gene Expression by RT-PCR

The RT^2-profiler array contains a panel of 96 primer sets for 84 relevant, pathway- or disease-specific genes, plus five housekeeping genes, and three RNA and PCR quality controls. The quantitative RT-PCR is performed in an ABI Prism 7900HT sequence detection system. Super Array's primer sets are optimized for SYBR green real-time RT-PCR detection.

1. Prepare the master mix #1 by adding 1,225 μL of 2× SYBR green PCR master mix, 98 μL of cDNA synthesis reaction mix, and 1,127 μL of water. The final volume should be 2,550 μL.
2. Add 25 μL of master mix #1 (the above mixture) to each well from A1 to H5 of 96-well array plate preferably with an eight-channel pipette.
3. Prepare a second control master mix (master mix #2) by adding 100 μL of 2× SYBR green PCR master mix and 100 μL of water.
4. Prepare a five 10-fold serial dilution of master mix #1 in a 0.2-mL strip tubes. To make this, add 3 μL of master mix #1 to 27 μL of master mix #2 in the first well of the eight-well strip. Aspirate 3 μL from the mix in the first well and add to the second well which already contains 27 μL of master mix #2; similarly make three more 10-fold serial dilutions.
5. Add 25 μL of these diluted samples sequentially from wells H6 to H10.
6. Add 24 μL of master mix #2 and 1 μL of 1:100 dilution of the original RNA to the well H11 (no reverse transcription control). This will allow testing for any contamination of the original RNA with genomic DNA.
7. Add 25 μL of master mix #2 to well H12 (no template control needed). This is to make sure that there is no DNA contamination in the PCR system.
8. Centrifuge the PCR plate at 1,000×*g* for 1 min to get the contents to the bottom of the well.
9. Seal the plate tightly with optical thin-wall eight-cap strips.
10. Perform real-time PCR with ABI Prism 7900HT sequence detection system (SDS) using standard thermal setting (95°C for 10 min and 40 cycles of 95°C for 15 s and 60°C for 1 min) (*see* **Note 11**).
11. After completion of run, analyze the data using the Applied Biosystems SDS software to get the C_t values (threshold values) and export to a blank Excel sheet for further analysis.

3.6. Data Analysis

1. The threshold cycle values of the control wells (H11 and H12) are examined to check for any DNA contamination. A value of 35 or greater indicates the absence of any DNA contamination. When the samples are analyzed in replicates, the average C_t value for each gene across the array is calculated.

2. Calculate the ΔC_t for each gene by normalizing with the housekeeping genes
   ΔC_t = Average C_t values – average of housekeeping genes in the same PCR plate.
3. Now calculate the $\Delta \Delta C_t$ for each gene across the control and treatment groups
   $\Delta\Delta C_t = \Delta C_t$ (20 μM ChSeco treatment) – ΔC_t (control).
4. The fold change in gene expression in treatment group is calculated as $2^{-\Delta\Delta Ct}$. If the fold change is greater than 1, the gene is considered to be upregulated. If the fold change is less than 1, the gene is considered to be downregulated.

3.7. Results

Table 16.1 shows the results of the expression of various genes related to oxidative stress in GT1-7 hypothalamic neuronal cells exposed to ChSeco for 8 h. It was found that several genes related to oxidative metabolism are downregulated. These include catalase (*Cat*), soluble and intracellular superoxide dismutase (*SOD1* and *2*), NADPH oxidase 4 (*Nox4*) and NADPH oxidase organizer (*Noxo1*), and GSH peroxidases 1 (*Gpx1*), 3 (*Gpx3*), 4 (*Gpx4*), 6 (*Gpx6*), and 7 (*Gpx7*). These results are in agreement with our earlier biochemical studies which showed depletion of GSH associated with marked increase in ROS production in neuronal cells exposed to ChSeco for 6 h (19). The upregulation of NADPH oxidase 1(*Nox1*) and NADPH oxidase activator (*Noxa1*) (**Table 16.1**) suggests the involvement of the plasma membrane-bound NADPH oxidase system in the observed ROS production in neuronal cells. Prostaglandin synthases 1 and 2 (*Ptgs1* and *Ptgs2*) are also upregulated presumably to counter the effects of lipid peroxidation by ROS.

Table 16.2 shows the results of the expression of genes associated with the apoptotic pathway(s). Effector caspases 8 and 9 (*Casp8* and *Casp9*) showed a marginal increase of 3.55–5.33-fold, whereas the executor caspase (*Casp3*) showed a 11-fold increase. In addition, genes for DNA fragmentation factors alpha (*Dffa*) and beta (*Dffb*) are upregulated by 1.22–1.44-fold. These findings corroborate well with our previous biochemical studies which showed that the mode of neuronal cell death in ChSeco exposures confirms to an apoptotic process (19). The findings that *Fas*, *Fasl*, and *Fadd* are upregulated confirm that death receptor pathway plays a significant role in the observed cytotoxicity of ChSeco.

In general, the gene expression analysis performed using real time RT-PCR gives reliable information on the pathway-specific genes. In addition to providing additional or supporting information to the existing biochemical evidence, the approach can be highly exploratory and can be used for prescreening before undertaking laborious and often time-consuming functional analysis.

Table 16.1
Effect of cholesterol secoaldehyde (ChSeco) on the expression of genes related to oxidative stress[a]

		AVG ΔC_t=(Avg C_t – Avg C_t of HKG)		$2^{-\Delta Ct}$		Fold up or downregulation
Symbol	Description	Test sample	Control sample	Test sample	Control sample	Test sample/control sample
2310016C16Rik	RIKEN cDNA 2310016C16 gene	24.20	18.34	5.2E–08	3.0E–06	–57.83
Aass	Aminoadipate-semialdehyde synthase	18.34	34.32	3.0E–06	4.7E–11	64703.81
Als2	Amyotrophic lateral sclerosis 2 (juvenile) homolog (human)	33.89	20.72	6.3E–11	5.8E–07	–9213.44
Apc	Adenomatosis polyposis coli	20.66	19.79	6.0E–07	1.1E–06	–1.82
Apoe	Apolipoprotein E	19.25	24.46	1.6E–06	4.3E–08	36.95
Aqr	Aquarius	25.76	19.79	1.8E–08	1.1E–06	–62.65
Atr	Ataxia telangiectasia and rad3 related	19.36	20.07	1.5E–06	9.1E–07	1.64
Cat	Catalase	20.14	19.14	8.7E–07	1.7E–06	–2.00
Ccs	Copper chaperone for superoxide dismutase	18.83	19.73	2.1E–06	1.1E–06	1.87
Cmya1	Cardiomyopathy associated 1	19.73	32.74	1.1E–06	1.4E–10	8226.54
Ctsb	Cathepsin B	32.30	16.37	1.9E–10	1.2E–05	–62485.88
Cyba	Cytochrome b-245, alpha polypeptide	15.89	18.42	1.6E–05	2.8E–06	5.77
Cygb	Cytoglobin	18.07	25.34	3.6E–06	2.3E–08	154.39
Dnm2	Dynamin 2	25.10	19.00	2.8E–08	1.9E–06	–68.89
Duox1	Dual oxidase 1	18.67	26.02	2.4E–06	1.5E–08	163.60
Ehd2	EH-domain containing 2	25.94	17.81	1.6E–08	4.4E–06	–279.59
Epx	Eosinophil peroxidase	17.75	30.23	4.5E–06	8.0E–10	5690.55
Ercc2	Excision repair cross-complementing rodent repair deficiency, complementation group 2	30.30	21.10	7.6E–10	4.4E–07	–586.72
Ercc6	Excision repair cross-complementing rodent repair deficiency, complementation group 6	21.26	22.90	4.0E–07	1.3E–07	3.11

Table 16.1 (continued)

Symbol	Description	AVG ΔC_t=(Avg C_t− Avg C_t of HKG)		$2^{-\Delta Ct}$		Fold up or downregulation
		Test sample	Control sample	Test sample	Control sample	Test sample/control sample
Fancc	Fanconi anemia, complementation group C	23.11	20.83	1.1E–07	5.4E–07	−4.87
Fmo2	Flavin containing monooxygenase 2	20.52	26.83	6.7E–07	8.4E–09	79.69
Gab1	Growth factor receptor-bound protein 2-associated protein 1	25.53	19.25	2.1E–08	1.6E–06	−77.75
Gpx1	Glutathione peroxidase 1	19.16	15.04	1.7E–06	3.0E–05	−17.39
Gpx2	Glutathione peroxidase 2	15.05	26.80	3.0E–05	8.5E–09	3459.99
Gpx3	Glutathione peroxidase 3	25.79	25.77	1.7E–08	1.8E–08	−1.02
Gpx4	Glutathione peroxidase 4	26.46	16.54	1.1E–08	1.1E–05	−968.54
Gpx5	Glutathione peroxidase 5	15.96	35.00	1.6E–05	2.9E–11	540132.00
Gpx6	Glutathione peroxidase 6	34.59	33.19	3.9E–11	1.0E–10	−2.64
Gpx7	Glutathione peroxidase 7	32.40	18.94	1.8E–10	2.0E–06	11257.35
Gsr	Glutathione reductase 1	18.84	17.30	2.1E–06	6.2E–06	−2.90
Gstk1	Glutathione *S*-transferase kappa 1	26.69	29.59	9.3E–09	1.2E–09	7.50
Hbq1	Hemoglobin, theta 1	29.22	31.56	1.6E–09	3.1E–10	5.08
Idh1	Isocitrate dehydrogenase 1 (NADP+), soluble	32.06	17.31	2.2E–10	6.2E–06	−27514.95
Ift172	Intraflagellar transport 172 homolog (Chlamydomonas)	17.26	32.91	6.4E–06	1.2E–10	51437.36
Il19	Interleukin 19	32.17	29.52	2.1E–10	1.3E–09	−6.27
Il22	Interleukin 22	29.69	31.51	1.2E–09	3.3E–10	3.54
Kif9	Kinesin family member 9	30.71	35.00	5.7E–10	2.9E–11	19.63
Lpo	Lactoperoxidase	35.00	35.00	2.9E–11	2.9E–11	−1.00

Table 16.1 (continued)

Symbol	Description	AVG ΔC_t=(Avg C_t− Avg C_t of HKG)		$2^{-\Delta Ct}$		Fold up or downregulation
		Test sample	Control sample	Test sample	Control sample	Test sample/control sample
Mb	Myoglobin	33.25	30.88	9.8E–11	5.1E–10	−5.17
Mpo	Myeloperoxidase	29.59	30.74	1.2E–09	5.6E–10	2.22
Mpp4	Membrane protein, palmitoylated 4 (MAGUK p55 subfamily member 4)	31.22	35.00	4.0E–10	2.9E–11	13.76
Ncf2	Neutrophil cytosolic factor 2	34.06	25.50	5.6E–11	2.1E–08	−376.39
Ngb	Neuroglobin	25.89	27.88	1.6E–08	4.0E–09	3.99
Nos2	Nitric oxide synthase 2, inducible, macrophage	27.02	26.35	7.4E–09	1.2E–08	−1.58
Nox1	NADPH oxidase 1	25.59	30.35	2.0E–08	7.3E–10	27.13
Nox4	NADPH oxidase 4	29.79	24.88	1.1E–09	3.2E–08	−30.02
Noxa1	NADPH oxidase activator 1	24.84	30.64	3.3E–08	6.0E–10	55.79
Noxo1	NADPH oxidase organizer 1	30.62	22.36	6.1E–10	1.9E–07	−305.78
Nqo1	NAD(P)H dehydrogenase, quinone 1	22.12	28.01	2.2E–07	3.7E–09	59.06
Nudt15	Nudix (nucleoside diphosphate linked moiety X)-type motif 15	28.10	20.82	3.5E–09	5.4E–07	−155.77
Nxn	Nucleoredoxin	21.07	19.62	4.5E–07	1.2E–06	−2.75
Park7	Parkinson disease (autosomal recessive, early onset) 7	19.99	17.36	9.6E–07	5.9E–06	−6.17
Ppp1r15b	Protein phosphatase 1, regulatory (inhibitor) subunit 15b	17.30	18.13	6.2E–06	3.5E–06	1.77
Prdx1	Peroxiredoxin 1	17.86	19.21	4.2E–06	1.7E–06	2.54

Table 16.1 (continued)

Symbol	Description	AVG ΔC_t=(Avg C_t− Avg C_t of HKG)		$2^{-\Delta Ct}$		Fold up or downregulation
		Test sample	Control sample	Test sample	Control sample	Test sample/control sample
Prdx2	Peroxiredoxin 2	19.04	23.05	1.9E–06	1.2E–07	16.14
Prdx3	Peroxiredoxin 3	23.14	17.38	1.1E–07	5.8E–06	−54.06
Prdx4	Peroxiredoxin 4	17.11	19.79	7.1E–06	1.1E–06	6.44
Prdx5	Peroxiredoxin 5	19.52	20.86	1.3E–06	5.3E–07	2.52
Prdx6	Peroxiredoxin 6	20.99	16.86	4.8E–07	8.4E–06	−17.59
Prdx6-rs1	Peroxiredoxin 6, related sequence 1	16.84	15.93	8.5E–06	1.6E–05	−1.88
Prnp	Prion protein	15.94	17.89	1.6E–05	4.1E–06	3.86
Psmb5	Proteasome (prosome, macropain) subunit, beta type 5	17.63	16.67	4.9E–06	9.6E–06	−1.95
Ptgs1	Prostaglandin–endoperoxide synthase 1	16.25	18.47	1.3E–05	2.8E–06	4.65
Ptgs2	Prostaglandin–endoperoxide synthase 2	17.55	19.35	5.2E–06	1.5E–06	3.47
Rag2	Recombination activating gene 2	17.53	27.51	5.3E–06	5.2E–09	1010.47
Recql4	RecQ protein-like 4	28.07	19.86	3.5E–09	1.1E–06	−296.40
Scd1	Stearoyl-Coenzyme A desaturase 1	19.18	16.74	1.7E–06	9.1E–06	−5.40
Serpinb1b	Serine (or cysteine) peptidase inhibitor, clade B, member 1b	16.24	26.40	1.3E–05	1.1E–08	1140.05
Slc38a1	Solute carrier family 38, member 1	26.18	17.96	1.3E–08	3.9E–06	−299.82
Slc41a3	Solute carrier family 41, member 3	17.53	22.88	5.3E–06	1.3E–07	40.86
Sod1	Superoxide dismutase 1, soluble	22.26	18.42	2.0E–07	2.8E–06	−14.33
Sod2	Superoxide dismutase 2, mitochondrial	18.41	16.82	2.9E–06	8.6E–06	−3.01
Sod3	Superoxide dismutase 3, extracellular	17.27	24.57	6.3E–06	4.0E–08	157.70
Srxn1	Sulfiredoxin 1 homolog (*Saccharomyces cerevisiae*)	25.23	18.13	2.5E–08	3.5E–06	−137.66

Table 16.1 (continued)

Symbol	Description	AVG ΔC_t=(Avg C_t− Avg C_t of HKG)		$2^{-\Delta Ct}$		Fold up or downregulation
		Test sample	Control sample	Test sample	Control sample	Test sample/control sample
Tmod1	Tropomodulin 1	18.26	22.96	3.2E–06	1.2E–07	26.04
Tpo	Thyroid peroxidase	22.98	34.78	1.2E–07	3.4E–11	3554.97
Txnip	Thioredoxin interacting protein	35.00	18.30	2.9E–11	3.1E–06	−106761.06
Txnrd1	Thioredoxin reductase 1	18.39	16.24	2.9E–06	1.3E–05	−4.47
Txnrd2	Thioredoxin reductase 2	16.16	20.98	1.4E–05	4.8E–07	28.13
Txnrd3	Thioredoxin reductase 3	20.71	21.80	5.8E–07	2.7E–07	2.12
Ucp3	Uncoupling protein 3 (mitochondrial, proton carrier)	22.13	28.80	2.2E–07	2.1E–09	101.53
Vim	Vimentin	29.68	12.67	1.2E–09	1.5E–04	−131726.40
Xpa	Xeroderma pigmentosum, complementation group A	12.91	22.24	1.3E–04	2.0E–07	643.02
Zmynd17	Zinc finger, MYND domain containing 17	21.20	23.04	4.1E–07	1.2E–07	3.56
Gusb	Glucuronidase, beta	28.03	27.15	3.6E–09	6.7E–09	−1.85
Hprt1	Hypoxanthine guanine phosphoribosyl transferase 1	18.52	17.15	2.7E–06	6.9E–06	−2.58
Hspcb	Heat-shock protein 1, beta	17.11	12.92	7.1E–06	1.3E–04	−18.28
Gapdh	Glyceraldehyde-3-phosphate dehydrogenase	12.77	12.63	1.4E–04	1.6E–04	−1.10
Actb	Actin, beta, cytoplasmic	23.61	23.35	7.8E–08	9.3E–08	−1.20

[a]Murine GT1-7 hypothalamic neurons were exposed to ChSeco (10 μM) for 8 h and the gene expression profile was studied against untreated controls. RNA isolation, cDNA synthesis, real-time RT-PCR, and data analysis were done as mentioned in the text. Fold difference expressed in negative values indicate the downregulation, whereas those represented in positive values indicate the upregulation of gene expression.

Table 16.2
Effect of cholesterol secoaldehyde (ChSeco) on the expression of genes related to apoptosis[a]

Symbol	Description	AVG ΔC_t=(Avg C_t of control – Avg C_t of HKG)		$2^{-\Delta C_t}$		Fold up- or downregulation
		Test sample	Control sample	Test sample	Control sample	Test sample/Control sample
Akt1	Thymoma viral proto-oncogene 1	2.62	2.54	1.60E–01	1.70E–01	–1.06
Apaf1	Apoptotic peptidase activating factor 1	6.18	8.94	1.40E–02	2.00E–03	6.77
Api5	Apoptosis inhibitor 5	4.3	3.88	5.10E–02	6.80E–02	–1.33
Atf5	Activating transcription factor 5	3.96	5.73	6.40E–02	1.90E–02	3.4
Bad	Bcl-associated death promoter	7.43	7.45	5.80E–03	5.70E–03	1.01
Bag1	Bcl2-associated athanogene 1	2.56	2.68	1.70E–01	1.60E–01	1.08
Bag3	Bcl2-associated athanogene 3	5.55	5.36	2.10E–02	2.40E–02	–1.14
Bak1	BCL2-antagonist/killer 1	16.46	15.5	1.10E–05	2.20E–05	–1.94
Bax	Bcl2-associated X protein	3.81	7.05	7.10E–02	7.50E–03	9.45
Bcl10	B-cell leukemia/lymphoma 10	5.06	6	3.00E–02	1.60E–02	1.92
Bcl2	B-cell leukemia/lymphoma 2	9.44	7.83	1.40E–03	4.40E–03	–3.06
Bcl2l1	Bcl2-like 1	5.93	5.58	1.60E–02	2.10E–02	–1.28
Bcl2l10	Bcl2-like 10	16.7	16.78	9.40E–06	8.90E–06	1.06
Bcl2l2	Bcl2-like 2	8.03	8.79	3.80E–03	2.30E–03	1.69
Bid	BH3 interacting domain death agonist	5.62	5.58	2.00E–02	2.10E–02	–1.03
Birc1a	Baculoviral IAP repeat-containing 1a	19.67	19.58	1.20E–06	1.30E–06	–1.07
Birc1b	Baculoviral IAP repeat-containing 1b	8.86	9.42	2.20E–03	1.50E–03	1.47
Birc2	Baculoviral IAP repeat-containing 2	7.39	7.59	6.00E–03	5.20E–03	1.15
Birc3	Baculoviral IAP repeat-containing 3	8.87	9.64	2.10E–03	1.30E–03	1.7
Birc4	Baculoviral IAP repeat-containing 4	6.31	6.23	1.30E–02	1.30E–02	–1.06
Birc5	Baculoviral IAP repeat-containing 5	5.56	6.05	2.10E–02	1.50E–02	1.4

Table 16.2 (continued)

Symbol	Description	AVG ΔC_t=(Avg C_t of control – Avg C_t of HKG)		2–ΔC_t		Fold up- or downregulation
		Test sample	Control sample	Test sample	Control sample	Test sample/Control sample
Bnip2	BCL2/adenovirus E1B interacting protein 1, NIP2	3.93	4.27	6.60E–02	5.20E–02	1.26
Bnip3	BCL2/adenovirus E1B interacting protein 1, NIP3	5.56	6.23	2.10E–02	1.30E–02	1.59
Bnip3l	BCL2/adenovirus E1B interacting protein 3-like	3.48	3.92	9.00E–02	6.60E–02	1.37
Bok	Bcl-2-related ovarian killer protein	6.44	6.23	1.20E–02	1.30E–02	–1.16
Card10	Caspase recruitment domain family, member 10	9.83	9.49	1.10E–03	1.40E–03	–1.27
Card4	Caspase recruitment domain 4	8.65	8.49	2.50E–03	2.80E–03	–1.12
Card6	Caspase recruitment domain family, member 6	11.18	11.37	4.30E–04	3.80E–04	1.14
Casp1	Caspase 1	9.97	10.18	1.00E–03	8.60E–04	1.16
Casp12	Caspase 12	7.07	7.35	7.40E–03	6.10E–03	1.21
Casp14	Caspase 14	14.04	13.98	5.90E–05	6.20E–05	–1.05
Casp2	Caspase 2	6.84	6.39	8.70E–03	1.20E–02	–1.36
Casp3	Caspase 3	3.25	6.71	1.00E–01	9.50E–03	11
Casp11	Caspase 11, apoptosis-related cysteine peptidase	8.3	9.32	3.20E–03	1.60E–03	2.04
Casp6	Caspase 6	5.52	5.78	2.20E–02	1.80E–02	1.2
Casp7	Caspase 7	5.09	5.27	2.90E–02	2.60E–02	1.14
Casp8	Caspase 8	5.4	7.79	2.40E–02	4.50E–03	5.23
Casp9	Caspase 9	7.08	8.91	7.40E–03	2.10E–03	3.55

Table 16.2 (continued)

Symbol	Description	AVG ΔC_t=(Avg C_t of control – Avg C_t of HKG)		$2\text{–}\Delta C_t$		Fold up- or downregulation
		Test sample	Control sample	Test sample	Control sample	Test sample/Control sample
Cflar	CASP8 and FADD-like apoptosis regulator	6.43	6.74	1.20E–02	9.30E–03	1.24
Cidea	Cell death-inducing DNA fragmentation factor, alpha subunit-like effector A	20.01	21.02	9.50E–07	4.70E–07	2.01
Cideb	Cell death-inducing DNA fragmentation factor, alpha subunit-like effector B	11.28	12	4.00E–04	2.40E–04	1.65
Cradd	CASP2 and RIPK1 domain containing adaptor with death domain	8.31	9	3.20E–03	1.90E–03	1.62
Dad1	Defender against cell death 1	3.08	3.55	1.20E–01	8.50E–02	1.39
Dapk1	Death-associated protein kinase 1	18.89	17.35	2.10E–06	6.00E–06	–2.9
Dffa	DNA fragmentation factor, alpha subunit	6	6.3	1.60E–02	1.30E–02	1.22
Dffb	DNA fragmentation factor, beta subunit	8.32	8.83	3.10E–03	2.20E–03	1.42
Tsc22d3	TSC22 domain family 3	6.37	7.86	1.20E–02	4.30E–03	2.83
Fadd	Fas (TNFRSF6)-associated via death domain	8.34	8.92	3.10E–03	2.10E–03	1.5
Fas	Fas (TNF receptor superfamily member)	10.96	11.87	5.00E–04	2.70E–04	1.89
Fasl	Fas ligand (TNF superfamily, member 6)	17.06	18.51	7.30E–06	2.70E–06	2.73

Table 16.2 (continued)

Symbol	Description	AVG ΔC_t=(Avg C_t of control – Avg C_t of HKG)		$2{-}\Delta C_t$		Fold up- or downregulation
		Test sample	Control sample	Test sample	Control sample	Test sample/Control sample
Hells	Helicase, lymphoid specific	4.37	4.78	4.80E–02	3.60E–02	1.33
Il10	Interleukin 10	19.16	21.02	1.70E–06	4.70E–07	3.63
Lhx4	LIM homeobox protein 4	18.07	17.95	3.60E–06	4.00E–06	–1.09
Ltbr	Lymphotoxin B receptor	6.71	7.11	9.50E–03	7.20E–03	1.31
Mcl1	Myeloid cell leukemia sequence 1	10.63	10.01	6.30E–04	9.70E–04	–1.53
Nfkb1	Nuclear factor of kappa light chain gene enhancer in B-cells 1, p105	4.77	4.63	3.70E–02	4.00E–02	–1.1
Nme5	Expressed in non-metastatic cells 5	16.98	18.53	7.70E–06	2.60E–06	2.93
Nol3	Nucleolar protein 3 (apoptosis repressor with CARD domain)	6.93	7.34	8.20E–03	6.20E–03	1.32
Pak7	P21 (CDKN1A)-activated kinase 7	19.29	19.66	1.60E–06	1.20E–06	1.29
Pim2	Proviral integration site 2	14.43	9.01	4.50E–05	1.90E–03	–42.83
Polb	Polymerase (DNA directed), beta	6.56	6.87	1.10E–02	8.60E–03	1.24
Prdx2	Peroxiredoxin 2	9.07	9.05	1.90E–03	1.90E–03	–1.01
Pycard	PYD and CARD domain containing	10.37	10.9	7.60E–04	5.20E–04	1.44
Ripk1	Receptor (TNFRSF)-interacting serine–threonine kinase 1	5.88	6.59	1.70E–02	1.00E–02	1.64
Rnf7	Ring finger protein 7	5.59	6.12	2.10E–02	1.40E–02	1.45

Table 16.2 (continued)

Symbol	Description	AVG ΔC_t=(Avg C_t of control – Avg C_t of HKG)		$2-\Delta C_t$		Fold up- or downregulation
		Test sample	Control sample	Test sample	Control sample	Test sample/Control sample
Sphk2	Sphingosine kinase 2	5.97	6.92	1.60E–02	8.20E–03	1.93
Tnf	Tumor necrosis factor	17.15	18.82	6.90E–06	2.20E–06	3.17
Tnfrsf10b	Tumor necrosis factor receptor superfamily, member 10b	5.89	5.96	1.70E–02	1.60E–02	1.06
Tnfrsf11b	Tumor necrosis factor receptor superfamily, member 11b (osteoprotegerin)	19.88	19.9	1.00E–06	1.00E–06	1.01
Tnfrsf1a	Tumor necrosis factor receptor superfamily, member 1a	4.75	5.04	3.70E–02	3.00E–02	1.23
Cd40	CD40 antigen	18.54	19.46	2.60E–06	1.40E–06	1.9
Tnfsf10	Tumor necrosis factor (ligand) superfamily, member 10	18	17.14	3.80E–06	6.90E–06	–1.81
Tnfsf12	Tumor necrosis factor (ligand) superfamily, member 12	6.21	6.29	1.30E–02	1.30E–02	1.06
Cd40lg	CD40 ligand	20.01	21.02	9.50E–07	4.70E–07	2.01
Tnfsf7	Tumor necrosis factor (ligand) superfamily, member 7	20.01	21.02	9.50E–07	4.70E–07	2.01
Traf1	Tnf receptor-associated factor 1	8.45	8.76	2.90E–03	2.30E–03	1.23
Traf2	Tnf receptor-associated factor 2	4.82	5.07	3.60E–02	3.00E–02	1.19

Table 16.2 (continued)

Symbol	Description	AVG ΔC_t=(Avg C_t of control – Avg C_t of HKG)		$2-\Delta C_t$		Fold up- or downregulation
		Test sample	Control sample	Test sample	Control sample	Test sample/Control sample
Traf3	Tnf receptor-associated factor 3	6.77	6.95	9.20E–03	8.10E–03	1.14
Trp53	Transformation-related protein 53	2.19	2.62	2.20E–01	1.60E–01	1.35
Trp53bp2	Transformation-related protein 53 binding protein 2	6.5	6.43	1.10E–02	1.20E–02	–1.05
Trp53inp1	Transformation-related protein 53 inducible nuclear protein 1	7.21	7.85	6.70E–03	4.30E–03	1.55
Trp63	Transformation-related protein 63	17.94	18.22	4.00E–06	3.30E–06	1.22
Trp73	Transformation-related protein 73	17.42	18.76	5.70E–06	2.30E–06	2.52
Zc3hc1	Zinc finger, C3HC type 1	12.78	6.25	1.40E–04	1.30E–02	–92.57
Gusb	Glucuronidase, beta	5.69	5.21	1.90E–02	2.70E–02	–1.39
Hprt1	Hypoxanthine guanine phosphoribosyl transferase 1	3.55	3.68	8.50E–02	7.80E–02	1.09
Hspcb	Heat-shock protein 1, beta	–0.62	–0.34	1.50E+00	1.30E+00	1.21
Gapdh	Glyceraldehyde-3-phosphate dehydrogenase	–0.59	–0.51	1.50E+00	1.40E+00	1.06

[a] Murine GT1-7 hypothalamic neurons were exposed to ChSeco (10 μM) for 8 h and the gene expression profile was studied against untreated controls. RNA isolation, cDNA synthesis, real-time RT-PCR, and data analysis were done as mentioned in the text. Fold difference expressed in negative values indicates the downregulation and that in positive values indicates the upregulation of gene expression.

4. Notes

1. Unless specified otherwise, throughout this procedure "water" refers to autoclaved, ultra pure water with resistance ≥18.2 MΩ or water from Sigma.
2. A general setup for carrying out ozone-mediated oxidations and the design of bubbler are given in Ref. (25). The bubbler described in this article (25) can be modified for better performance and safety. For instance, a Teflon stopper can be placed at the bottom of the bubbler for easy sampling. Also, it is a good idea to have a jacket around the bubbler and circulate coolant at 5–10°C to minimize evaporation of solvents.
3. All experiments involving ozone should be performed behind a safety shield in a fume hood. No direct contact or exposure to ozone is allowed during experimentation. Excess unreacted ozone must be absorbed and destroyed in a solution of 5% (w/v) KI in 0.067 M phosphate buffer, pH 7.0 (25).
4. It is important to monitor the course of ozone reaction with cholesterol (22). One easy way to avoid oxidation of cholesterol beyond the level of ChSeco or the peroxide intermediates (18, 26) is to stop ozonation at various intervals of time and check for the presence of unreacted cholesterol.
5. The presence of cholesterol can be tested by reversed-phase (RP) HPLC on a Discovery HS C18 column (4.6 × 250 mm) using 40% isopropanol in acetonitrile as the mobile phase at a flow of 1 mL/min and monitoring the eluent at 210 nm (22).
6. Approximately 1×10^5 cells are present in each well in a six-well plate having a growth area of 9.5 cm^2 and 1 mL of Tri reagent is sufficient for cells up to $5–10 \times 10^6$.
7. To shear the high molecular weight genomic DNA and other cellular components, the lysate can be passed 5–10 times through a 20-gauge needle using a sterile plastic syringe.
8. The pellet should not dry for longer time, as solubility will decrease.
9. Dilute stock DNase solution eightfold with a diluent buffer before adding to the column. This digestion step will ensure complete removal of contaminating DNA from the RNA sample.

10. Measuring absorbance at 230, 260, and 280 nm will help to determine any protein or salt contamination. A ratio ≤ of 1.8–2.0 for A_{260}/A_{280} and ≥1.7 for A_{260}/A_{230} are considered suitable for PCR analysis.
11. As the primers are already optimized, an additional analysis of dissociation (melting) curve is not necessary.

Acknowledgments

We thank Dr. Pam Mellon (University of California, San Diego) for providing murine GT1-7 hypothalamic neurons. This publication was made possible by National Institutes of Health (NIH) Grants P20 RR16456 (from the BRIN Program of the National Center for Research Resources) and ES10018 (from the ARCH Program of the National Institute of Environmental Health Sciences), and the US Department of Education (Title III, Part B – Strengthening Historically Black Graduate Institutions, HBGI; grant number: PO31B040030). Its contents are solely the responsibility of authors and do not necessarily represent the official views of the NIH, the NSF, or the US Department of Education.

References

1. Cadenas, E.P.L. (1996) Handbook of Antioxidants, Marcel Dekker, INC, New York.
2. Videla, L.A., Fernandez, V., Tapia, G., and Varela, P. (2007) Thyroid hormone calorigenesis and mitochondrial redox signaling: upregulation of gene expression. *Front Biosci.* **12**, 1220–1228.
3. Esposito, F., Cuccovillo, F., Russo, L., Casella, F., Russo, T., and Cimino, F. (1998) A new p21waf1/cip1 isoform is an early event of cell response to oxidative stress. *Cell Death Differ.* **5**, 940–945.
4. Perrice, F., Floccari, F., Nostro, L., Caccamo, C., Belghity, N., Mantuano, S., Romeo, A., Barilla, A., Aloisi, C., Ruello, A., Frisina, N., and Buemi, M. (2006) Oxidative stress, sister chromatid exchanges and apoptosis in the pathogenesis of lymphocytopenia in ESRD patients. *J. Nephrol.* **19**, 613–620.
5. Polyak, K., Xia, Y., Zweier, J.L., Kinzler, K.W., and Vogelstein, B. (1997) A model for p53-induced apoptosis. *Nature* **389**, 300–305.
6. Haddad, J.J. (2004) Redox and oxidant-mediated regulation of apoptosis signaling pathways: immuno-pharmaco-redox conception of oxidative siege versus cell death commitment. *Int. Immunopharmacol.* **4**, 475–493.
7. Birch-Machin, M.A. (2006) The role of mitochondria in ageing and carcinogenesis. *Clin. Exp. Dermatol.* **31**, 548–552.
8. Migliaccio, E., Giorgio, M., and Pelicci, P.G. (2006) Apoptosis and ageing: role of p66Shc redox protein. *Antioxid. Redox Signal* **8**, 600–608.
9. Follezou, J.Y., Emerit, J., and Bricaire, F. (1999) Neuro-degenerative diseases: role of reactive oxygen species and of apoptosis. *Presse. Med.* **28**, 1661–1666.
10. Blomgren, K. and Hagberg, H. (2006) Free radicals, mitochondria, and hypoxia–ischemia in the developing brain. *Free Radic. Biol. Med.* **40**, 388–397.
11. Tsutsui, H. (2006) Mitochondrial oxidative stress and heart failure. *Intern. Med.* **45**, 809–813.

12. Ferrari, R., Guardigli, G., Mele, D., Percoco, G.F., Ceconi, C., and Curello, S. (2004) Oxidative stress during myocardial ischaemia and heart failure. *Curr. Pharm. Des.* **10**, 1699–1711.
13. Waring, J.F. and Ulrich, R.G. (2000) The impact of genomics-based technologies on drug safety evaluation. *Annu. Rev. Pharmacol. Toxicol.* **40**, 335–352.
14. Aardema, M.J. and MacGregor, J.T. (2002) Toxicology and genetic toxicology in the new era of "toxicogenomics": impact of "-omics" technologies. *Mutat. Res.* **499**, 13–25.
15. Baker, S.G. and Kramer, B.S. (2006) Identifying genes that contribute most to good classification in microarrays. *BMC Bioinformatics* 7, 407.
16. Wentworth P, Jr., Nieva, J., Takeuchi, C., Galve, R., Wentworth, A.D., Dilley, R.B., DeLaria, G.A., Saven, A., Babior, B.M., Janda, K.D., Eschenmoser, A., and Lerner, R.A. (2003). Evidence for ozone formation in human atherosclerotic arteries. *Science* **302**, 1053–1056.
17. Zhang, Q., Powers, E.T., Nieva, J., Huff, M.E, Dendle M.A, Bieschke, J., Glabe, C.G., Eschenmoser, A., Wentworth, P. Jr., Lerner, R.A., and Kelly, J.W. (2004) Metabolite-initiated protein misfolding may trigger Alzheimer's disease. *Proc. Natl. Acad. Sci. USA* **101**, 4752–4757.
18. Pulfer, M.K. and Murphy, R.C. (2004) Formation of biologically active oxysterols during ozonolysis of cholesterol present in lung surfactant. *J. Biol. Chem.* **279**, 26331–26338.
19. Sathishkumar, K., Xi, X., Martin, R., and Uppu, R.M. (2007) Cholesterol secoaldehyde, an ozonation product of cholesterol, induces amyloid aggregation and apoptosis in murine GT1-7 hypothalamic neurons. *J. Alzheimer's Dis.* **11**, 261–274.
20. Takeuchi, C., Galve, R., Nieva, J., Witter, D.P., Wentworth, A.D., Troseth, R.P., Lerner, R.A., and Wentworth, P, Jr. (2006.) Proatherogenic effects of the cholesterol ozonolysis products, atheronal-A and atheronal-B. *Biochemistry* **45**, 7162–7170.
21. Gao, X., Raghavamenon, A.C., D'Auvergne, O., and Uppu, R.M. (2009) Cholesterol secoaldehyde induces apoptosis in J774 macrophages via mitochondrial pathway but not involving reactive oxygen species as mediators. *Biochem. Biophys. Res. Commun.* **389**, 382–387.
22. Sathishkumar, K., Haque, M., Perumal, T.E., Francis, J., and Uppu, R.M. (2005) A major ozonation product of cholesterol, 3beta-hydroxy-5-oxo-5, 6-secocholestan-6-al, induces apoptosis in H9c2 cardiomyoblasts. *FEBS Lett.* **579**, 6444–6450.
23. Sathishkumar, K., Gao, X, Ragavamenon, A.C., Parinandi, N., Pryor, W.A., and Uppu, R.M. (2009) Cholesterol secoaldehyde induces apoptosis in H9c2 cardiomyoblasts through reactive oxygen species involving mitochondrial and death receptor pathways. Free Radic. Biol. Med. **47**, 548–558.
24. Wang, K., Bermudez, E., and Pryor, W. (1993) The ozonation of cholesterol: separation and identification of 2,4-dinitrophenylhydrazine derivatization products of 3 beta- hydroxyl-5-oxo-5,6-secocholestan-6-al. *Steroids* **58**, 225–229.
25. Uppu, R.M., Squadrito, G.L., Ceuto, R., and Pryor, W.A. (1996) Synthesis of peroxynitrite by azide-ozone reaction. *Meth. Enzymol.* **269**, 311–321.
26. Smitth, L.L., Ezell, E.L., and Jaworski, K. (1996) On the ozonation of cholesterol 3-acetyl esters in protic media. *Steroids* **61**, 401–406.

Chapter 17

Redox Homeostasis and Cellular Stress Response in Aging and Neurodegeneration

Vittorio Calabrese, Carolin Cornelius, Cesare Mancuso, Riccardo Lentile, A.M. Giuffrida Stella, and D. Allan Butterfield

Abstract

Decreased expression and/or activity of antioxidant proteins leads to oxidative stress, accelerated aging, and neurodegeneration. While overwhelming levels and uncontrolled/dysregulated actions of reactive oxygen species (ROS) lead to deleterious effects, tighter regulation of those plays an important role in cell signaling. Mutations causing protein misfolding and the overload of toxic products derived from the free radical oxidation of polyunsaturated fatty acids, cholesterol, and glucose contribute to the disruption of the cellular redox homeostasis. Collectively or individually, these effects create pro-oxidant conditions in cells. Oxidative stress can induce neuronal damage, modulate intracellular signaling, and can ultimately lead to neuronal death by apoptosis or necrosis. Emerging evidence indicates that homocysteine (Hcy), a non-protein amino acid naturally present in the plasma, is implicated as a risk factor for numerous diseases. In particular, increased levels of circulating Hcy have been recognized as an independent risk factor for the development of vascular disease(s). Recent findings emphasize a relationship between elevated Hcy levels and neurodegeneration, which can be observed in Alzheimer's and Parkinson's diseases. An integrated response exists in the brain to detect and control diverse forms of stress. This is accomplished by a complex network of the so-called longevity assurance processes, which are controlled by several genes termed "vitagenes." Among these, the heat-shock proteins (HSPs) form a highly conserved system that is responsible for the preservation and repair of the correct protein conformation. Recent studies have shown that the heat-shock response (HSR) contributes to cytoprotection in a number of human diseases including inflammation, cancer, aging, and neurodegenerative disorders. Given the broad cytoprotective properties of the HSR, interest mounts currently among investigators toward discovering and developing pharmacological agents capable of inducing HSR. L-Acetylcarnitine (LAC) is proposed as a therapeutic agent for several neurodegenerative disorders and also current evidence suggests that the compound may play a critical role in the modulation of cellular stress response in health and disease conditions. Here, we review the emerging salient concepts highlighting the pathways of neurodegeneration and the role of LAC in modulating the redox-dependent mechanisms responsible for the upregulation of vitagenes in brain that leads to the enhancement of stress tolerance in brain.

Key words: Acetylcarnitine, heme oxygenase, homocysteine, mitochondria, neurodegenerative disorders, oxidative stress, redox regulation, vitagenes.

R.M. Uppu et al. (eds.), *Free Radicals and Antioxidant Protocols*, Methods in Molecular Biology 610,
DOI 10.1007/978-1-60327-029-8_17,

1. Introduction

In the resting state, 2% of oxygen upon entry into living cells is converted into reactive oxygen species (ROS) (1). McCord and Fridovich first described superoxide dismutase implying a potential physiological role for superoxide (2), which was subsequently confirmed in numerous studies (3). In contrast to its physiological functions, it has been shown that ROS exert adverse effects (pathological actions) at an order of magnitude higher concentrations. To protect from the ROS-induced oxidative attack, the living cell has evolved a wide battery of efficient antioxidant defense systems which include the enzymatic antioxidants (e.g., superoxide dismutase, catalase, peroxidases, and heme oxygenase) and the non-enzymatic redox-regulating antioxidants (e.g., glutathione and vitamin C). Therefore, under normal physiological conditions, the cell is well equipped with the antioxidant defenses to cope up with the production of ROS. Indeed, ROS generated continuously at low concentrations induce the expression of antioxidant enzymes and related defense mechanisms. Alzheimer's disease (AD), Parkinson's disease (PD), and Huntington's diseases (HD), amyotrophic lateral sclerosis (ALS), and Friedreich's ataxia (FRDA) belong to a class of the "protein conformational diseases" (4). Under normal conditions, cells employ mechanisms such as the unfolded protein response, which generally balance the protein delivery to the endoplasmic reticulum (ER) and the capacity for protein folding (5). Furthermore, if proteins are misfolded, the unfolded protein response either (i) breaks up abnormal protein aggregates and facilitates protein refolding or (ii) delivers them to the proteasome to be recycled if proteins cannot be rescued by refolding (5). Disorders or diseases arising from/associated with the abnormalities in protein conformations are conditions that arise from the dysfunctional aggregation of proteins in non-native conformations. One consequence of the formation of these intracellular abnormal protein aggregates is ER dysfunction (stress) that leads to mitochondrial dysfunction, excessive production of ROS, and oxidative stress. The ability of a cell to combat the elevated levels of ROS and reactive nitrogen species (RNS) involves the activation of pro-survival pathways as well as the production of molecules endowed with antioxidant and anti-apoptotic actions. Heme oxygenase-1 (HO-1), also referred to as heat-shock protein 32 (Hsp32), belongs to the Hsp family of proteins and protects brain cells from oxidative stress by degrading toxic heme to free iron, carbon monoxide (CO), and biliverdin. Biliverdin is then reduced by biliverdin reductase (BVR) to bilirubin (BR), a linear tetrapyrrole with antioxidant and anti-nitrosative properties (6–10)

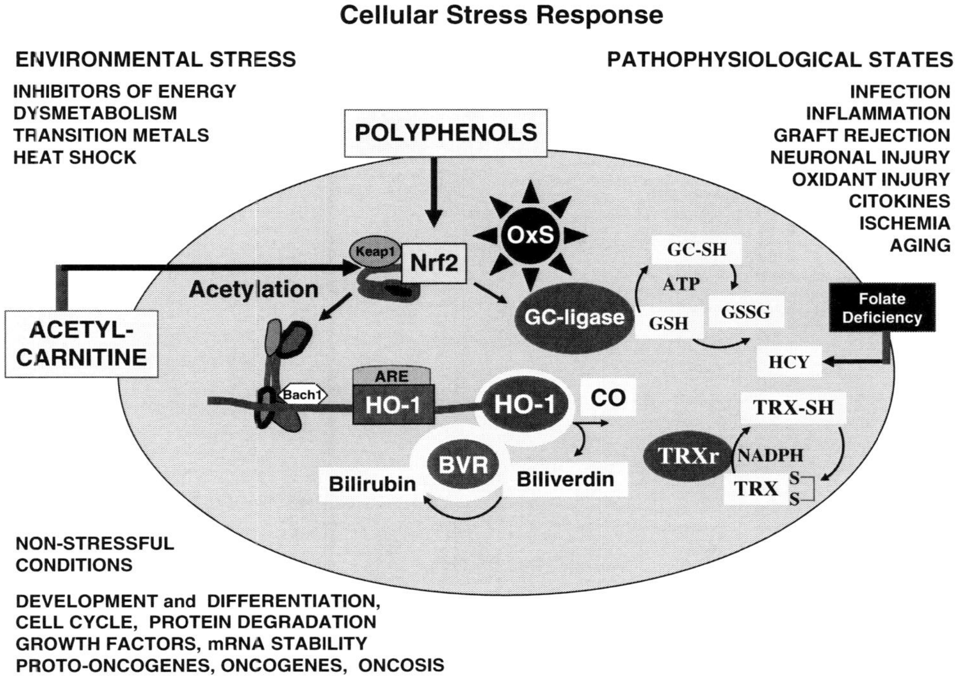

Fig. 17.1. Heme oxygenase and the pathway of cellular stress response. Environmental stress factors, such as heavy metals, cytokines, heat shock, conditions of metabolic dysregulation, inhibitors of energy generation, and pathophysiological conditions perturbing the oxidant–antioxidant balance such as inflammation, graft rejections, neuronal damage, ischemia, and brain aging are situations associated with the induction of cellular stress response. Hsp response is also involved in cellular homeostasis during various physiological conditions, such as brain development and differentiation, cell cycle, apoptosis and oncosis, oncogene and action of growth factors, as well as mRNA and protein turnover. The HO represents an integrated system for cellular stress tolerance and can be activated by non-noxious stimuli also, such as nutritional antioxidants or L-acetylcarnitine (LAC). LAC, through activation (via acetylation) of the redox-sensitive transcription factor Nrf2 and its consequent binding to the antioxidant-responsive element (ARE) in the *ho* gene, up-regulates HO-1, glutathione (GSH) synthesis, and TRXr, thus counteracting nitrosative stress and NO-mediated neurotoxicity. In the same figure are indicated the respective roles of the transcription factors such as Bach1 and Keap1 in the Nrf2 activation, the redox cycling between bilirubin and biliverdin through the enzyme biliverdin reductase (BVR), as well as the role of folate deficiency in inducing neurotoxicity via increasing levels of the neurotoxin homocysteine (Hcy) and the consequent GSH depletion associated with mitochondrial and DNA damage.

(**Fig. 17.1**). Recently, BR has also been shown to effectively counteract nitrosative stress through its properties to bind and inactivate nitric oxide (NO) and RNS (11–13).

There are several mechanisms by which ROS may be generated including the aerobic respiration, nitric oxide synthesis, and NADPH oxidase pathways during inflammation. In aerobic respiration, the mitochondrial respiratory chain produces ROS as it transfers electrons during the reduction of molecular oxygen to water. During this process, some electrons escape the electron transport chain and activate molecular oxygen into superoxide, hydrogen peroxide, and hydroxyl radical (14). Activated neutrophils can also release the enzyme myeloperoxidase which produces the highly reactive oxidant hypochlorous acid (HOCl) from hydrogen peroxide and chloride ions.

The nitric oxide synthase (NOS) enzymes produce NO via deamination of L-arginine to L-citrulline. NO is not highly reactive per se but can react with other species like oxygen, superoxide, and transition metals leading to the formation of the highly reactive products which affect the functionality of macromolecules (15–18). Recently the term "nitrosative stress" has been introduced to describe the cellular damage elicited by the RNS including NO and its congeners peroxynitrite, N_2O_3, nitroxyl anion, and nitrosonium (19–22). From a molecular point of view, the cell is able to counteract the oxidant stress by utilizing its antioxidant ammunition such as the antioxidant vitamins (A, C, and E), bioactive molecules (GSH, thioredoxin, flavonoids), antioxidant enzymes (heme oxygenase, superoxide dismutase, catalase, glutathione peroxidases, thioredoxin reductase), and the redox-sensitive transcriptional factors (AP-1, NFκB, Nrf2, HSF). The Hsps are one of the widely studied defense systems which are known to offer protection against the cellular damage.

In this chapter we describe the recent reports on the biochemical changes occurring in the central nervous system (CNS) when brain cells are exposed to chronic oxidative insult as well as the key role played by the HSR, particularly the heme oxygenase (Hsp32) and Hsp70 pathways. Mounting evidence underscores the potential of the Hsp system as a target for new neuroprotective strategies, especially those aimed at minimizing deleterious consequences associated with the oxidative stress encountered during neurodegenerative disorders and brain aging. Here, we also present the evidence in favor of the emerging role of homocysteine in the pathogenesis of neurodegenerative damage as well as the role of L-acetylcarnitine (LAC) in modulating the redox-dependent mechanisms leading to the upregulation of vitagenes in brain toward the enhancement of stress tolerance in brain.

2. Heme Oxygenase-1

Heme oxygenase-1 (HO-1), also referred to as Hsp32, is a redox-sensitive and inducible isoform of the HO family of antioxidant enzymes. HO-1 is a microsomal enzyme and catalyzes the degradation of heme in a multistep energy-requiring system. The reaction catalyzed by HO-1 is the α-specific oxidative cleavage of heme moieties to form equimolar amounts of ferrous iron, CO, and biliverdin. Biliverdin is then reduced by the cytosolic enzyme biliverdin reductase to bilirubin (BR) which is then conjugated with glucuronic acid and excreted (23).

Accumulating evidence suggests that the *HO-1* gene is redox regulated (**Fig. 17.2**) and contains the antioxidant-responsive element (ARE) in its promoter region, similar to other antioxidant enzymes (24). In fact, HO-1 can be induced by several stimuli including oxidative and nitrosative stresses, ischemia, heat shock, lipopolysaccharide (LPS), hemin, and the neuroprotective agent neotrofin (25–27). HO-1 induction, one of the earlier cellular responses to tissue damage, is responsible for the rapid transformation of the pro-oxidant heme into CO and BR, the two molecules with important physiological and pathophysiological functions (28–30).

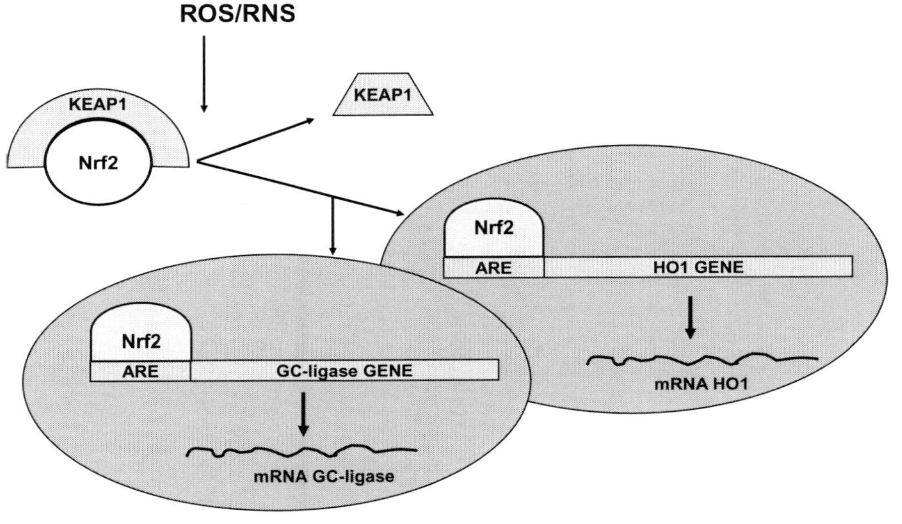

Fig. 17.2. Regulation of Nrf2-related genes. Nuclear factor-erythroid 2-related factor 2 (Nrf2) is a transcription factor responsible for the induction of several genes related to the cellular stress response, including HO-1 and GC-ligase. Under normal conditions, Nrf2 is sequestered in the cytoplasm by an actin-binding protein, Kelch-like ECH-associating protein 1 (Keap1), but upon exposure of cells to oxidative stress or nitrosative stress, Nrf2 dissociates from Keap1, translocates to the nucleus, binds to antioxidant-responsive elements (AREs), and activates HO-1 and GC-ligase genes.

HO-1 gene is induced by a variety of stimuli (**Fig. 17.1**), including metalloporphyrins, hemin, ultraviolet A (UVA) irradiation, hydrogen peroxide, pro-oxidants, and inflammation (26, 31, 32). The induction of HO-1 gene (*ho-1*) strictly relies on its configuration: the 6.8-kilobase gene is organized into four introns and five exons. A promoter sequence is located approximately 28 base pairs upstream from the transcriptional site of initiation. In addition, different transcriptional enhancer elements, such as heat-shock element and metal regulatory element, reside in the flanking 5′ region. Also, inducer-responsive sequences have been identified in the proximal enhancer located upstream of the promoter and, more distally, in two enhancers located 4 and 10 kbp upstream of the initiation site (33). The molecular mechanism that confers inducible expression of *ho-1* in response to numerous and diverse conditions has remained elusive. One important clue has recently emerged from a detailed analysis of the transcriptional regulatory mechanisms controlling the mouse and human *ho-1* genes. The induction of *ho-1* is regulated principally by two upstream enhancers E1 and E2 (34). Both enhancer regions contain multiple stress (or antioxidant)-responsive elements (StRE, also called ARE) that also conform to the sequence of the Maf recognition element (MARE) (35) with a consensus sequence (GCnnnGTA) similar to that of other antioxidant enzymes (36). There is now evidence to suggest that heterodimers of NF-E2-related factor 2 (Nrf2) and one or another of the small Maf proteins (i.e., MafK, mafF, and MafG) are directly involved in the induction of *ho-1* through these MAREs (35). A possible model, which centers around Nrf2 activity, suggests that the *ho-1* locus is situated in a chromatin environment that is permissive for activation. Since the MARE can be bound by various heterodimeric basic leucine zipper (bZip) factors including NF-E2, as well as several other NF-E2-related factors (Nrfl, Nrf2, and Nrf3), Bach, Maf, and AP-1 families (34), random interaction of activators with the *ho-1* enhancers would be expected to cause spurious expression. This raises a paradox as to how cells reduce transcriptional noise from the *ho-1* locus in the absence of metabolic or environmental stimulation. This problem could be reconciled by the activity of repressors that prevent non-specific activation. One possible candidate is the heme protein Bach1, a transcriptional repressor endowed with DNA-binding activity, which is negatively regulated upon binding with heme. Bach1–heme interaction is mediated by evolutionarily conserved heme regulatory motifs (HRM), including the cysteine–proline dipeptide sequence in Bach1. Hence, a plausible model accounting for the regulation of *ho-1* expression by Bach1 and heme is that expression of *ho-1* gene is regulated through antagonism between transcription activators and the repressor Bach1 (**Figs. 17.1 and 17.2**). While under normal physiological conditions expression of *ho-1* is repressed by Bach1/Maf complex, increased levels of heme displace Bach1 from the enhancers and allow activators, such as heterodimers of Maf or Keap with NF-E2-related activators (Nrf2), to the transcriptional promotion of *ho-1* gene (34) (**Fig. 17.2**). To our knowledge, the Bach1–*ho-1* system is the first example in higher eukaryotes that involves a direct regulation of a transcription factor for an enzyme gene by its substrate. Thus, regulation of *ho-1* involves a direct sensing of heme levels by Bach1 (by analogy to *lac* repressor sensitivity to lactose), generating a simple feedback loop whereby the substrate effects repressor–activator antagonism.

The promoter region also contains two metal-responsive elements, similar to those found in metallothionein-1 gene, which respond to heavy metals (cadmium and zinc) only after recruitment of another fragment located upstream, between –3.5 and 12 kbp (CdRE). In addition, a 163-bp fragment containing two binding sites for HSF-1, which mediates the HO-1 transcription, is located 9.5 kbp upstream of the initiation site (37). The distal enhancer regions are important in regulating HO-1 in inflammation, since, as has been demonstrated, they are responsive to endotoxin. In the promoter region also resides a 56-bp fragment which responds to the STAT-3 acute-phase response factor, involved in the downregulation of *HO-1* gene induced by glucocorticoid (38, 39).

2.1. HO-1, Oxidative Stress, and Neurodegenerative Disorders

The mechanisms responsible for neuronal death are not completely elucidated, even if many studies suggest that ROS are primarily involved in the genesis of neurodegenerative disorders (14, 15, 40–42). Due to its strong antioxidant properties and wide distribution within the CNS, HO-1 has been proposed as a key enzyme in the prevention of brain damage (25, 26, 43, 44). Panahian et al., using transgenic mice over-expressing HO-1 in neurons, demonstrated the neuroprotective effect of this enzyme in an experimental model of ischemic brain damage (45). The neuroprotective effects of over-expressed HO-1 can be attributed to (i) increase in cGMP and bcl-2 levels in neurons; (ii) inactivation of p53, a protein involved in promoting cell death; (iii) increase in antioxidant sources; and (iv) increase in the iron-sequestering protein ferritin (45). Particularly interesting is the role played by HO-1 in AD, a neurodegenerative disorder which involves a chronic inflammatory response associated with both oxidative brain injury and β-amyloid-associated pathology. Significant increases in the levels of HO-1 have been observed in AD brains in association with neurofibrillary tangles, and HO-1 mRNA was also found increased in AD neocortex and cerebral vessels (46, 47). HO-1 increase was not only in association with neurofibrillary tangles but also co-localized with senile plaques and glial fibrillary acidic protein-positive astrocytes in AD brains (48). It is plausible that the dramatic increase in HO-1 in AD may be a direct response to an increase in free heme concentrations, associated with neurodegeneration, and can be considered as an attempt of brain cells to convert the highly toxic heme into the antioxidants CO and BR. The protective role played by HO-1 and its products in AD raised new possibilities regarding the possible use of natural substances, which are able to increase HO-1 levels, as potential drugs for the treatment of AD. In this light, very promising are the polyphenolic compounds contained in some herbs and spices, e.g., curcumin (49–51). Curcumin is the active antioxidant in *Curcuma longa*, a coloring agent and food additive commonly used in Indian culinary preparations. This polyphenolic substance has the potential to inhibit lipid peroxidation and to effectively intercept and neutralize ROS and RNS (52). In addition, curcumin has been shown to significantly increase HO-1 in astrocytes and vascular endothelial cells (50, 53). This latter effect on HO-1 can explain, at least in part, the antioxidant properties of curcumin, in particular keeping in mind that HO-1-derived BR has the ability to scavenge both ROS and RNS (11–13, 28, 54). Epidemiological studies suggested that curcumin, as one of the most prevalent nutritional and medicinal compounds used by the Indian population, is responsible for the significantly reduced (4.4-fold) prevalence of AD in India as compared to the United States (55). Based on these findings, Lim and colleagues have provided convincing evidence that dietary curcumin given to an Alzheimer transgenic APPSw mouse model (Tg2576) for 6 months resulted in a suppression of indices of inflammation and oxidative damage in the brain of these mice (56). Furthermore, in a human neuroblastoma cell line, it has recently been shown that curcumin inhibits nuclear factor-κB (NFκB) activation, efficiently preventing neuronal cell death (52).

Although it is generally agreed that *ho-1* over-expression is a common feature during oxidative stress, recent reports have demonstrated that *ho-1* can be repressed following pro-oxidant conditions. In particular human and rodent cells exposed to cigarette smoke condensate, β-carotene, hypoxia, thermal stress, and interferon-γ treatment showed a marked *ho-1* repression (57–61). The importance of *ho-1* repression has been corroborated by the discovery of Bach1/Bach2 as heme-regulated transcription factors for *ho-1* gene (62, 63). In fact, Bach1 is broadly expressed in mice and human tissues and, in human cells, it is induced by the same stimuli which are able to repress *ho-1* gene (57, 59–61). The reason why the cell should react to an oxidant stress by repressing *ho-1* gene is strictly related to the maintenance of a good metabolic balance during stressful conditions. The current hypothesis suggests that *ho-1* repression is useful for the cell because it (i) decreases the energy costs necessary for heme degradation, (ii) reduces the accumulation of CO and BR which can become toxic if produced in excess, and (iii) increases the intracellular content of heme necessary for the preservation of vital functions such as respiration and defense (61, 64, 65).

2.2. Carbon Monoxide and Stress Response

Carbon monoxide (CO) is the gaseous product of HO and it has been found to play a role in several biological phenomena, including hippocampal long-term potentiation, non-adrenergic and non-cholinergic gastrointestinal relaxation, and vasodilatation and is currently regarded as a neuromodulator in the peripheral and central nervous system (43, 44). Evidence from the in vitro and in vivo studies suggests that the HO–CO pathway is involved in the modulation of the neuroendocrine mechanism of stress. Thus, increased CO generation is clearly associated with the inhibition of K^+-stimulated arginine vasopressin (AVP) and oxytocin release from rat hypothalamic explants, whereas the inhibition of HO activity significantly potentiates the LPS-induced increase in AVP circulating levels while reducing the hypothalamic content of this neuropeptide (66–68). With regard to corticotropin-releasing hormone (CRH), the effects of CO on the release of this hormone are contradictory, since increases in CO generation induced by two HO substrates, hematin and hemin, were associated with reduced or enhanced CRH release, respectively, in two different in vitro models (69, 70). As far as the intracellular mechanism(s) by which CO exerts its biological functions is concerned, it is generally agreed that this gas activates the cytosolic form of guanylyl cyclase (sGC) which in turn increases intracellular cGMP levels (26). However, during the last 10 years, many studies arose in the literature demonstrating that CO signals through the activation of alternative intracellular signal transduction pathways. As demonstrated by Mancuso et al., the activation of another hemoprotein, cyclooxygenase (COX), plays a significant role in CO signaling in the rat hypothalamus (71). In these studies, hemin, the precursor of CO via HO, dose-dependently increased PGE_2 production from rat hypothalamus in vitro and this effect was specifically due to CO because it is counteracted by the HO inhibitor Sn-mesoporphyrin-IX and oxyhemoglobin, the latter being a well-known scavenger for CO (71). The direct evidence about the stimulatory role of CO in PG production was obtained by incubating hypothalami directly in CO saturated solutions and measuring significantly increased PGE_2 levels with respect to control tissue (29). Recently Jaggar and colleagues, in a very elegant paper, demonstrated that exogenously or endogenously produced CO dilates cerebral arterioles by directly activating large-conductance Ca^{2+}-activated K^+ (K_{Ca}) channels primarily by increasing the coupling ratio and amplitude relationship between Ca^{2+} sparks and K_{Ca} channels (72). Although CO is a potent and effective activator of K_{Ca} channels, the gas does not dilate arterioles in the absence of Ca^{2+} sparks. Therefore, CO appears to act by priming K_{Ca} channels for activation by Ca^{2+} sparks, and this ultimately leads to arteriole dilation via membrane hyperpolarization (72). Finally, Otterbein and colleagues have shown that in organs and tissues different from brain, CO exerts anti-inflammatory and anti-apoptotic effects depending on the modulation of the p38 mitogen-activated protein kinase (p38-MAPK) signaling pathway (30). By virtue of these effects, CO confers protection in oxidative lung injury models and likely plays a role in HO-1-mediated tissue protection (73).

2.3. Heat-Shock Protein-70

The 70-kDa family of stress proteins is one of the most extensively studied. Included in this family are Hsc70 (heat-shock cognate, the constitutive form), Hsp70 (the inducible form, also referred to as Hsp72), and GRP-75 (a constitutively expressed glucose-regulated protein found in the endoplasmic reticulum) (74–76).

The availability of transgenic animals and gene transfer allowed us to over-express the gene encoding for Hsp70, thus demonstrating that overproduction of this protein leads to protection in several different models of nervous system injury (77, 78). Following focal cerebral ischemia, Hsp70 mRNA is synthesized in most ischemic cells except in areas of very low blood flow, due to the scarcity of ATP levels. Hsp70 proteins are produced mainly in endothelial cells, in the core of infarcts in the cells that are most resistant to ischemia, in glial cells at the edges of infarcts, and in neurons outside the areas of infarction (79). It has been suggested that this neuronal expression of Hsp70 outside an infarct can be used to define the ischemic penumbras, which means the zone of protein denaturation in the ischemic areas (79).

As mentioned above, Hsps are induced in many neurodegenerative disorders mainly in the view of its cytoprotective function. Hsp72 was over-expressed in the postmortem cortical tissue of AD patients and an increase in Hsp70 mRNA was found in the cerebellum hippocampus and the cortex of AD patients during the axonal phase of the disease (80–82). Recently, Kakimura et al. demonstrated that Hsp70 induces IL-6 and tumor necrosis factor-α (TNF-α) in microglial cells and this event is associated with an increased phagocytosis and clearance of Aβ peptides (83). The same authors hypothesize that Hsps could activate microglial cells through NFκB- and p-38 MAPK-dependent pathways (83).

A large body of evidence now suggests a correlation between mechanisms of nitrosative stress and Hsp induction. We have demonstrated in astroglial cell cultures that cytokine-induced nitrosative stress is associated with an increased synthesis of Hsp70 stress proteins. The molecular mechanisms regulating the NO-induced activation of heat-shock signal seem to involve cellular oxidant/antioxidant balance, mainly represented by the glutathione status and the antioxidant enzymes (84, 85).

2.4. Acetylcarnitine and Homocysteine

Mitochondria are cellular organelles involved in many crucial metabolic processes such as the pyruvate oxidation, the tricarboxylic acid cycle, the fatty acid β-oxidation, and the final pathway of oxidative phosphorylation which generates most of the cellular energetic source, ATP. It has been proposed that accumulation of mitochondrial DNA (mtDNA) during life is a major cause of age-related disease and this is because of its high mutagenic propensity. The lack of introns and protective histones, limited nucleotide excision and recombination DNA repair mechanisms, and the location in proximity of the inner mitochondrial membrane which exposes it to an enriched free radical milieu are all factors contributing to a 10-fold higher mutation rate occurring in the mtDNA than in the nuclear DNA. Relevant to mitochondrial bioenergetics, in fact, is the finding of a significant decrease in state 3/state 4 ratio, which has been observed to occur in brain as a function of age (86). Since this ratio relates to the coupling efficiency between electron flux through the electron transport chain and ATP production, an increase in state 4 would result in a more reductive state of mitochondrial complexes and, consequently, to an increase in free radical species production. A decrease in state 3/state 4 respiration during aging has been found associated with a significant decrease in cardiolipin content in brain mitochondria (87). This loss could play a critically important role in the age-related decrements in mitochondrial function and appears to be associated with both quantitative and qualitative region-specific protein changes which are parallel to structural changes, such as decreased inner membrane surface, smaller as well as sparser *cristae*, decreased fluidity, and increased fragility. Modifications in cardiolipin composition are recognized to accompany functional changes in brain mitochondria which include all proteins of the inner mitochondrial membrane that generally require interaction with cardiolipin for optimal catalytic activity (88).

Acetylcarnitine (LAC) is an ester of the trimethylated amino acid, L-carnitine, and is synthesized in the brain, liver, and kidney by the enzyme LAC-transferase. LAC facilitates the uptake of acetyl-CoA into the mitochondria during fatty acid oxidation, enhances acetylcholine production, and stimulates protein and membrane phospholipid synthesis (89). At present, studies have shown that LAC is a compound of great interest for its wide clinical application in various neurological disorders: it may be of benefit in treating Alzheimer's dementia, chronic fatigue syndrome, depression in the elderly, HIV infection, diabetic neuropathies, ischemia and reperfusion of the brain, cognitive impairment of alcoholism, and aging (90–92). The neuroprotective benefits of this compound have been observed in the hippocampus, the prefrontal cortex, the substantia nigra, and muscarinic receptor portions of the brain (93). These include antioxidant activity, improved mitochondrial energetics, stabilization of intracellular membranes, and cholinergic neurotransmission (94). Promising therapeutic applications of LAC are derived from observations that this compound crosses the blood–brain barrier through a saturable process in a sodium-dependent manner and improves neuronal energetic and repair mechanisms, while modifying acetylcholine production in the CNS (95). LAC treatment restores the altered neurochemical abnormalities, cerebral energy metabolites in ischemia and aging, and, in particular, ammonia-induced cerebral energy depletion (94). In addition, it increases the responsiveness of aged neurons to neurotrophic factors in the CNS and has preventive and corrective effects on diabetic neuropathology. Its beneficial effects have also been observed on EEG, evoked potentials, and long-term synaptic potentiation in aged animals (96). Moreover, LAC is also commonly used for the treatment of painful neuropathies: it exerts a potent analgesic effect by upregulating metabotropic glutamate receptors (97). There are experimental data that LAC improves memory function in Alzheimer's patients and influences attention, learning, and memory in the rat (98). Chronic treatment enhances spatial acquisition in a novel environment of rats with behavioral impairments and has a slight effect on retention of the spatial discrimination in a familiar environment (99). Furthermore, recent evidences have reported that it influences the expression of glyoxylase 1, a gene involved in the detoxification of metabolic byproducts, and increases p75-mRNA in mutant transgenic mouse model Tg2576 of Alzheimer's disease (100). Recently, by using suppressive subtractive hybridization (SSH) strategy, a PCR-based cDNA subtraction procedure particularly efficient for obtaining expressed transcripts often obscured by more abundant ones, it has been reported that LAC modulates specific genes in the rat CNS, such as the *hsp72* gene, the gene for the isoform of 14-3-3 protein, and that encoding for the precursor mitochondrial P3 of ATP synthase lipid-binding protein (101).

LAC fed to old rats increases cardiolipin levels as compared to that fed to young rats and also restores protein synthesis in the inner mitochondrial membrane, as well as cellular oxidant/antioxidant balance, suggesting that administration of this compound may improve cellular bioenergetics in aged rats (102). Fascinatingly, caloric restriction, a dietary regimen that extends life span in rodents, maintains the levels of 18:2 acyl side chains and inhibits the cardiolipin compositional changes (103). In addition, caloric restriction has shown to retard the aging-associated changes in oxidative damage, the mitochondrial oxidant generation, and the antioxidant defenses observed during aging (104).

Interestingly, we have recently demonstrated that LAC treatment of astrocytes induces HO-1 in a dose- and time-dependent manner and that this effect is associated with the upregulation of other Hsps as well as the high expression of the redox-sensitive transcription factor Nrf2 in the nuclear fraction of treated cells. In addition, we have shown that addition of LAC to astrocytes, prior to LPS and INF-γ-induced nitrosative stress, prevents changes in mitochondrial respiratory chain complex activity, protein nitrosation, and antioxidant status induced by inflammatory cytokine insult (89). Very importantly, this new envisioned role of LAC as a molecule endowed with the capability of enhancing the cellular stress response pathways appears to provide an alternative therapeutic approach for those pathophysiological conditions where stimulation of the HO pathway is warranted (7). Although clinical application of compounds potentiating the action of stress-responsive genes should be fully considered, a better understanding of how HO mediates its action will guide therapeutic strategies to enhance or suppress HO effects. Remarkably, the recent envisioned role of Hsp70 as a vehicle for intra-cytoplasmic and intra-nuclear delivery of fusion proteins or DNA to modulate gene expression (105, 106) along with the evidence that binding of HO protein to HO-1 DNA modifies HO expression via non-enzymatic signaling events associated with CO and p38-dependent induction of Hsp70 (107) opens intriguing perspectives, as it is possible to speculate that synergy between these two systems might represent a possible important target(s) for the LAC action, with possible impact on cell survival during times of oxidative stress, activation of cell life programs, and to the extent of cellular stress tolerance.

Homocysteine (Hcy) is a non-protein amino acid naturally occurring in the plasma, which has been implicated as a risk factor for numerous diseases. In particular, increased levels of circulating Hcy have been recognized as an independent risk factor for the development of pre-eclampsia, diabetes, and vascular disease, and recent findings emphasize a relationship between elevated Hcy and neurodegeneration, which can be observed in some neurodegenerative diseases such as Alzheimer's disease (AD) and Parkinson's disease (PD) (108–112).

Multiple enzymes and cofactors are involved in Hcy metabolism. Indeed, metabolic pathways consist of (i) re-methylation to methionine, which requires methylenetetrahydrofolate reductase (MTHFR)/methionine synthase (MS) or betaine Hcy methyltransferase (BHMT), and folic acid and vitamin B_{12} as co-factors; (ii) trans-sulfuration to cysteine, which requires cystathionine-β-synthase (CBS) and pyridoxal-5′-phosphate, the vitamin B6 coenzyme (112). Since MTHFR and BHMT are rate-limiting enzymes, vitamin deficiency may lead to increased plasma Hcy levels (113). The other major determinants of hyperhomocysteinemia are the C677T and A1298C missense polymorphisms of the *MTHFR* gene. Carriers of either TT677/AA1298 or CT677/AC1298 diplotype can have dramatically elevated Hcy levels, especially if stores of folate or vitamin B_{12} are depleted (114).

Plasma membrane carrier-mediated uptake is present in neurons and glial cells, suggesting that Hcy can be imported from the plasma into the brain (115). Particularly, high levels of Hcy in the CSF have been observed in several neuropathological conditions and have paralleled those in serum; however, serum concentrations are 20–100-fold higher than concentrations in the CSF (116–118).

Therefore, it appears that mild to moderate increases in Hcy levels can be a risk factor for neuronal degeneration and thus hyperhomocysteinemia can be involved in neuronal damage. Additionally, it is noteworthy that plasma levels of Hcy increase with age and have been implicated in vascular disease as well as in brain atrophy (119–121). Although in vivo and in vitro studies have suggested a causal role for Hcy in the neurodegenerative diseases, clinical studies should be carried out to further evaluate the relationship between high levels of Hcy and cognitive decline, dementia, and AD. The overstimulation of glutamate receptors can be considered as a common feature in several neurodegenerative conditions and alterations in intracellular calcium homeostasis mediate the toxicity of glutamate and NMDA for neurons. Consistently, among excitatory amino acids, Hcy is toxic to human and murine neuronal cell in vitro (122), and its neurotoxicity is partially dependent on overstimulation of NMDA subtype glutamate receptors which promote increases in calcium influx. Moreover, recently, attention has been focused on the cell damage produced in different cell types by Hcy exposure. It has been reported that Hcy impairs DNA repair in hippocampal neurons and sensitizes them to oxidative stress. Hcy also potentiates the β-amyloid-induced increase in cytosolic calcium and apoptosis in differentiated human neuroblastoma cells (123). Additionally, in cerebellar granule cells, Hcy-mediated cell death can be prevented by co-administrating superoxide dismutase (SOD) and catalase or by administrating catalase alone, suggesting that the formation of hydrogen peroxide contributes to the Hcy-mediated cell death. Hyperhomocysteinemia can be a consequence of folic acid deficiency that contributes to metabolic changes during nervous system development as well as in aging process and is involved in the pathogenesis of various diseases. In light of the data implicating increased DNA damage in neurons that degenerate in AD, it is noteworthy that folic acid deficiency and Hcy can impair DNA repair in neural and non-neuronal cells. However, folic acid deficiency and elevated Hcy levels increase the vulnerability of cultured hippocampal neurons to β-amyloid-induced neuronal degeneration and death (124). Both oxidative stress and DNA damage have been documented in neurons associated with β-amyloid-containing plaques in the brains of AD patients (for review, *see* 125). All these suggest that folic acid deficiency and Hcy promote the accumulation of DNA damage in neurons by impairing DNA repair mechanisms. In this context, postmitotic cells have been shown to be more vulnerable to DNA damage than are mitotic cells, probably because of the lack of efficacy of DNA repair. In fact, it was shown that neurons and neuroblastoma cells become extremely UV sensitive after terminal differentiation. Therefore, in mature neurons the mechanisms involved in the cell cycle activation can be an important component of the mechanisms associated with DNA damage leading to cell death. The exposure of cultured neurons to β-amyloid induces caspase activation (126) and increases production of Par-4 and Bax, each of which appears to play an important role in the cell death process (127). Apoptosis triggered by DNA damage typically involves activation of PARP and induction and activation of the tumor suppressor protein p53 (128). Increased PARP activity and p53 levels have been documented in association with degenerating neurons in AD patients and in cultured neurons exposed to β-amyloid. Moreover, pifithrin-α, a p53 transactivation inhibitor, can protect neurons against β-amyloid toxicity, suggesting that, in DNA damage-responsive cell death, this pathway might play a key role in the pathogenesis of AD (129). This evidence indicates that, by impairing the DNA repair capacity in neurons, folic acid deficiency and elevated Hcy levels lower the threshold level of DNA damage that is required to trigger neuronal death. Consistently, folic acid deficiency and elevated Hcy levels accelerate the accumulation of DNA damage that is promoted by age-related increases in oxidative stress and by the accumulation of β-amyloid (123). Neurons are more vulnerable to DNA damage than are the non-neuronal cells, suggesting that the brain may be particularly sensitive to diets deficient in folic acid and other nutritional and genetic factors associated with one-carbon metabolism (123).

In humans, the normal range of Hcy concentrations in plasma is 5–13 μM and that in CSF and brain tissue is 0.5–10 μM (130). Plasma folic acid levels decrease and Hcy levels increase with age, and to an even greater extent in patients with AD and PD (131). However, in long-term patients with AD, or following acute lesions such as stroke, neurons will degenerate and axons will not be replaced. Therefore, vitamin supplementation does not improve brain function (132). However, the involvement of Hcy as a risk factor for stroke has been confirmed by many investigators and recent data demonstrated an improvement in stroke mortality after folate fortifications (133). To summarize, different treatments are used to lower Hcy levels. The efficacy can be improved when the intervention starts at an early stage of the disease. Several findings suggest a causal link between hyperhomocysteinemia and neurological disorders. However, available results from treatment studies are very limited and further studies are required to demonstrate the efficacy of treatments against hyperhomocysteinemia conditions. Nevertheless, on the basis of the available results, useful efforts can be made to prevent more than reduce the effects of hyperhomocysteinemia and early intervention can play a crucial role to be beneficial. Therefore, increasing vitamin intake can be considered a neuroprotective strategy in age-related disease.

3. Conclusions and Perspectives

Modulation of endogenous cellular defense mechanisms via the stress response signaling represents an innovative approach to therapeutic intervention of diseases causing tissue damage, such as neurodegeneration. Efficient maintenance and repair processes seem to be crucial for survival, physical well-being, and quality of life. This is accomplished by a complex network of the so-called longevity assurance processes, which are composed of several genes called "vitagenes" (**Table 17.1**). Consistently, maintaining or recovering the activity of vitagenes can possibly delay the aging process and decrease the occurrence of age-related diseases, resulting in the prolongation of a healthy life span (6, 7, 65). One of the most important neurodegenerative disorders, AD is a progressive disorder with cognitive and memory decline, loss of speech, personality changes, and loss of synapse. Among the increasingly aging population of the United States, the number of AD patients who will face the lack of effective interventions is predicted to reach 14 million in the mid twenty-first century (2, 49). This will pose an immense economic and personal burden on the people of this country. Similar considerations apply worldwide, except in sub-Saharan Africa, where HIV infection rates seem to be leading as opposed to the decreased incidence of AD. There is now strong evidence to suggest that factors such as oxidative stress and disturbed protein metabolism and their interactions in a vicious cycle are central to AD pathogenesis. Brain-targeted antioxidants, potentially, may provide the means of implementing this therapeutic strategy of delaying the onset of AD, and more in general all the degenerative diseases associated with oxidative stress (51, 55). As one potentially successful approach, the potentiation of endogenous secondary antioxidant systems can be achieved by interventions which target the HO-1/BR and/or Hsp70 systems. In this review, the importance of the stress response signaling and, in particular, the central role of HO-1 together with the redox-dependent mechanisms involved in cytoprotection are highlighted. The beneficial effects of HO-1 induction result from the heme degradation and cytoprotective regulatory functions of biliverdin/bilirubin redox cycling. Thus, HO-1 can amplify intracellular cytoprotective mechanisms against a variety of insults. Consequently, the induction of HO-1/BVR, by increasing CO and BR availability, can be of clinical relevance.

Very importantly, HO-1 and CO can suppress the development of atherosclerotic lesions associated with chronic rejection of transplanted organs (134). Consistently, LAC, as a molecule endowed with the capability of enhancing the cellular stress

Table 17.1
Homeodynamic vitagene network involved in cellular stress tolerance

Biological functions involved in molecular fidelity maintenance and repair

Molecular control	*Cellular control*
Antioxidant defense	Cell proliferation
DNA repair systems	Cell differentiation
Transfer of genetic information	Stability of cell membrane
Stress protein synthesis	Stability of intracellular milieu
Proteasomal function	Macromolecular turnover
Tissue and organ control	*Physiological and redox control*
Removal/detoxification of toxic chemicals	Neuronal response and synaptic plasticity
Tissue regeneration	Hormonal response
Tumor suppression	Thermoregulation
Cell death and cell replacement	HO-1/CO, BVR/BR, UCP, Hsp70, Hsp27, TRXr/TRX

response pathways, appears to offer similar protective action, thereby providing an alternative therapeutic approach valuable for all those pathophysiological conditions where stimulation of the HO pathway becomes a primary target.

Presented here is a strong evidence that a cross talk between stress response genes is critical for cell stress tolerance, highlighting the compelling reason for a renewed effort to understand the central role of this most extraordinary defense system in biology and medicine. All of the above evidences also support the notion that the stimulation of maintenance and repair pathways through exogenous intervention, such as mild stress, or compounds targeting the heat-shock signal pathway, such as LAC, may have biological significance as a novel approach to delay the onset of various age-associated alterations in cells, tissues, and organisms. Hence, by maintaining or recovering the activity of vitagenes it can be possible to delay the aging process and decrease the occurrence of age-related diseases which result in the prolongation of a healthy life span.

Acknowledgments

This work was supported by grants of FIRB RBRN07BMCT, and INBB, and by NIH grants [AG-05119; AG-10836].

References

1. Boveris, A., Oshino, N., and Chance, B. (1972) The cellular production of hydrogen peroxide. *Biochem. J.* **128**, 617–630.
2. McCord, J.M. and Fridovich, I. (1988) Superoxide dismutase: the first twenty years (1968–1988). *Free Radic. Biol. Med.* **5**, 363–369.
3. Vina, J., Borras, C., Gomez-Cabrera, M.C., and Orr, W.C. (2006) Part of the series: from dietary antioxidants to regulators in cellular signalling and gene expression. Role of reactive oxygen species and (phyto)oestrogens in the modulation of adaptive response to stress. *Free Radic. Res.* **40**, 111–119.
4. Tabner, B.J., Turnbull, S., El-Agnaf, O., and Allsop, D. (2001) Production of reactive oxygen species from aggregating proteins implicated in Alzheimer's disease, Parkinson's disease and other neurodegenerative diseases. *Curr. Top. Med. Chem.* **1**, 507–517.
5. Zhang, K. and Kaufman, R.J. (2006) The unfolded protein response: a stress signaling pathway critical for health and disease. *Neurology* **66**, S102–S109.
6. Maines, M.D. (2005) The heme oxygenase system: update 2005. *Antioxid. Redox Signal* **7**, 1761–1766.
7. Maines, M.D. (2005) New insights into biliverdin reductase functions: linking heme metabolism to cell signaling. *Physiology* **20**, 382–389.
8. Baranano, D.E., Rao, M., Ferris, C.D., and Snyder, S.H. (2002) Biliverdin reductase: a major physiologic cytoprotectant. *Proc. Natl. Acad. Sci. USA* **99**, 16093–16098.
9. Ewing, J.F. and Maines, M.D. (2006) Regulation and expression of heme oxygenase enzymes in aged-rat brain: age related depression in HO-1 and HO-2 expression and altered stress-response. *J. Neural. Trans.* **13**, 439–454.
10. Maines, M.D., Ewing, J.F., Huang, T.J., and Panahian, N. (2001) Nuclear localization of biliverdin reductase in the rat kidney: response to nephrotoxins that induce heme oxygenase-1. *J. Pharmacol. Exp. Ther.* **296**, 1091–1097.
11. Kaur, H., Martin, N.H., Collin, J.G., Patrick, N., Roberta, F., and Roberto, M. (2003) Interaction of bilirubin and biliverdin with reactive nitrogen species. *FEBS Lett.* **543**, 113–119.
12. Mancuso, C., Bonsignore, A., Di Stasio, E., Mordente, A., and Motterlini, R. (2003) Bilirubin and S-nitrosothiols interaction: evidence for a possible role of bilirubin as a scavenger of nitric oxide. *Biochem. Pharmacol.* **66**, 2355–2363.
13. Mancuso, C., Bonsignore, A., Capone, C., Di Stasio, E., and Pani, G. (2006) Albumin-bound bilirubin interacts with nitric oxide by a redox mechanism. *Antioxid. Redox Signal* **8**, 487–494.
14. Pappolla, M.A., Chyan, Y.J., Omar, R.A., Hsiao, K., Perry, G., Smoth, M.A., and Bozner, P. (1998) Evidence of oxidative stress and in vivo neurotoxicity of beta-amyloid in a transgenic mouse model of Alzheimer's disease: a chronic oxidative paradigm for testing antioxidant therapies in vivo. *Am. J. Pathol.* **152**, 871–877.
15. Smith, M.A., Hirai, K., Hsiao, K., Pappolla, M.A., Harris, P.L., Seidlak, S.I., Tabaton, M., and Perry, G. (1998) Amyloid-beta deposition in Alzheimer transgenic mice is associated with oxidative stress. *J. Neurochem.* **70**, 2212–2215.
16. Butterfield, D.A., Drake, J., Pocernich, C., and Castegna, A. (2001) Evidence of oxidative damage in Alzheimer's disease brain: central role for amyloid beta-peptide. *Trends Mol. Med.* 7, 548–554.
17. Butterfield, D.A. and Lauderback, C.M. (2002) Lipid peroxidation and protein oxidation in Alzheimer's disease brain: potential causes and consequences involving amyloid β-peptide-associated free radical oxidative stress. *Free Radic. Biol. Med.* **32**, 1050–1060.
18. Mattson, M.P. (2004) Pathways towards and away from Alzheimer's disease. *Nature* **430**, 631–639.
19. Drew, B. and Leeuwenburgh, C. (2002) Aging and the role of reactive nitrogen species. *Ann. NY Acad. Sci.* **959**, 66–81.
20. Kroncke, K.D. (2003) Nitrosative stress and transcription. *Biol. Chem.* **384**, 1365–1377.
21. Ridnour, L.A., Thomas, D.D., Maneardi, D., Espey, M.G., Miranda, K.M., et al. (2004) The chemistry of nitrosative stress induced by nitric oxide and reactive nitrogen oxide species. Putting perspective on stressful biological situations. *Biol. Chem.* **385**, 1–10.
22. Forman, H.J., Fukuto, J.M., and Torres, M. (2004) Redox signaling: thiol chemistry defines which reactive oxygen and nitrogen species can act as second messengers. *Am. J. Physiol. Cell Physiol.* **287**, C246–256.
23. Maines, M.D. (1988) Heme oxygenase: function, multiplicity, regulatory mechanisms, and clinical applications. *FASEB J.* **2**, 2557–2568.

24. Prestera, T., Talalay, P., Alam, J., Ahn, Y.I., Lee, P.J., and Choi, A.M. (1995) Parallel induction of heme oxygenase-1 and chemoprotective phase 2 enzymes by electrophiles and antioxidants: regulation by upstream antioxidant-responsive elements (ARE). *Mol. Med.* **1**, 827–837.
25. Maines, M.D. (Ed.) Heme Oxygenase: Clinical Applications and Functions, CRC Press, Boca Raton, FL, 1992.
26. Maines, M.D. (1997) The heme oxygenase system: a regulator of second messenger gases. *Annu. Rev. Pharmacol. Toxicol.* **37**, 517–554.
27. Wang, X., Hauptmann, N., Taylor, E., Foreman, M., Khawli, L.A., and Maines, M.D. (2003) Neotrofin increases heme oxygenase-1 selectively in neurons. *Brain Res.* **962**, 1–14.
28. Stocker, R., Yamamoto, Y., McDonagh, A.F., Glazer, A.N., and Ames, B.N. (1987) Bilirubin is an antioxidant of possible physiological importance. *Science* **235**, 1043–1046.
29. Mancuso, C., Tringali, G., Grossman, A., Preziosi, P., and Navarra, P. (1998) The generation of nitric oxide and carbon monoxide produces opposite effects on the release of immunoreactive interleukin-1beta from the rat hypothalamus in vitro: evidence for the involvement of different signaling pathways. *Endocrinology* **139**, 1031–1037.
30. Otterbein, L.E., Baach, F.H., Allam, J., Soares, M., Lu, H., Wysk, M., Davis, R.J., Flavell, R.A., and Choi, A.M. (2000) Carbon monoxide has anti-inflammatory effects involving the mitogen-activated protein kinase pathway. *Nat. Med.* **6**, 422–428.
31. Turner, C.P., Panter, S.S., and Sharp, F.R. (1999) Anti-oxidants prevent focal rat brain injury as assessed by induction of heat shock proteins (HSP70, HO-1/HSP32, HSP47) following subarachnoid injections of lysed blood. *Brain Res. Mol. Brain Res.* **65**, 87–102.
32. Motterlini, R., Forestry, R., Bassi, R., Calabrese, V., Clark, J.F., and Green, C.J. (2000) Endothelial heme oxygenase-1 induction by hypoxia. Modulation by inducible nitric-oxide synthase and S-nitrosothiols. *J. Biol. Chem.* **275**, 13613–13620.
33. Hill-Kapturczak, N., Sikorski, E., Voakes, C., Garcia, J, Nick, H.S., and Agarwal, A. (2003) An internal enhancer regulates heme and cadmium-mediated induction of human heme oxygenase-1. *Am. J. Physiol. Renal Physiol.* **285**, F515–F523.
34. Sun, J., Hoshino, H., Takaku, K., Nakajima, O., Muto, A., Suzuki, H., et al. (2002) Hemoprotein Bach1 regulates enhancer availability of heme oxygenase-1 gene. *EMBO J.* **21**, 5216–5224.
35. Stewart, D., Killeen, E., Naquin, R., Alam, S., and Alam, J. (2003). Degradation of transcription factor Nrf2 via the ubiquitin-proteasome pathway and stabilization by cadmium. *J. Biol. Chem.* **278**, 2396–2402.
36. Scapagnini, G., D'Agata, V., Calabres, V., Pascale, A., Colombrita, C., Alkon, D., and Cavallaro, S. (2002) Gene expression profiles of heme oxygenase isoforms in the rat brain. *Brain Res.* **954**, 51–59.
37. Balogun, E., Hoque, M., Gong, P., Killeen, E., Green, C.J., et al. (2003) Curcumin activates the haem oxygenase-1 gene via regulation of Nrf2 and the antioxidant-responsive element. *Biochem. J.* **371**, 887–895.
38. Liu, N., Wang, X., Mc Coubrey, W.K., and Maines, M.D. (2000) Developmentally regulated expression of two transcripts for heme oxygenase-2 with a first exon unique to rat testis: control by corticosterone of the oxygenase protein expression. *Gene* **241**, 175–183.
39. Raju, V.S., McCoubrey, W.K., and Maines, M.D. (1997) Regulation of heme oxygenase-2 by glucocorticoids in neonatal rat brain: characterization of a functional glucocorticoid response element. *Biochim. Biophys. Acta* **1351**, 89–104.
40. Simonian, N.A. and Coyle, J.T. (1996) Oxidative stress in neurodegenerative diseases. *Annu. Rev. Pharmacol. Toxicol.* **36**, 83–106.
41. Sayre, L.M., Smith, M.A., and Perry, G. (2001) Chemistry and biochemistry of oxidative stress in neurodegenerative disease. *Curr. Med. Chem.* **8**, 721–738.
42. Andersen, J.K. (2004) Oxidative stress in neurodegeneration: cause or consequence? *Nat. Med.* **10**, S18–S25.
43. Mancuso, C. (2004) Heme oxygenase and its products in the nervous system. *Antioxid. Redox Signal* **6**, 878–887.
44. Calabrese, V., Butterfield, D.A., Scapagini, G., Stella, A.M., and Maines, M.D. (2006) Redox regulation of heat shock protein expression by signaling involving nitric oxide and carbon monoxide: relevance to brain aging, neurodegenerative disorders, and longevity. *Antioxid. Redox Signal* **8**, 444–477.
45. Panahian, N., Yoshiura, M., and Maines, M.D. (1999) Overexpression of heme oxygenase-1 is neuroprotective in a model of permanent middle cerebral artery

occlusion in transgenic mice. *J. Neurochem.* **72**, 1187–1203.
46. Takeda, A., Perry, G., Abraham, N.G., Dwyer, B.E., Kutty, R.K., Laitinen, J.T.,Petersen, R.B., and Smith, M.A. (2000) Overexpression of heme oxygenase in neuronal cells, the possible interaction with Tau. *J. Biol. Chem.* **275**, 5395–5399.
47. Premkumar, D.R., Smith, M.A., Richey, P.L., Petersen, R.B., Castellani, R., et al. (1995) Induction of heme oxygenase-1 mRNA and protein in neocortex and cerebral vessels in Alzheimer's disease. *J. Neurochem.* **65**, 1399–1402.
48. Schipper, H.M. (2000) Heme oxygenase-1: role in brain aging and neurodegeneration. *Exp. Gerontol.* **35**, 821–830.
49. Calabrese, V., Butterfield, D.A., and Giuffrida Stella, A.M. (2003) Nutritional antioxidants and the heme oxygenase pathway of stress tolerance: novel targets for neuroprotection in Alzheimer's disease. *Ital. J. Biochem.* **52**, 72–76.
50. Scapagnini, G., Foresti, R., Calabrese, V., Giuffrida Stella, A.M., Green, C.J., and Otterlini, R. (2002) Caffeic acid phenethyl ester and curcumin: a novel class of heme oxygenase-1 inducers. *Mol. Pharmacol.* **61**, 554–561.
51. Scapagnini, G., Butterfield, D.A., Colombrita, C., Sultana, R., Pascale, A., and Calabrese, V. (2004) Ethyl ferulate, a lipophilic polyphenol, induces HO-1 and protects rat neurons against oxidative stress. *Antioxid. Redox Signal* **6**, 811–818.
52. Butterfield, D., Castegna, A., Pocernich, C., Drake, J., Scapagini, G., and Calabrese, V. (2002) Nutritional approaches to combat oxidative stress in Alzheimer's disease. *J. Nutr. Biochem.* **13**, 444–461.
53. Motterlini, R., Foresti, R., Bassi, R., and Green, C.J. (2000) Curcumin, an antioxidant and anti-inflammatory agent, induces heme oxygenase-1 and protects endothelial cells against oxidative stress. *Free Radic. Biol. Med.* **28**, 1303–1312.
54. Baranano, D.E., Rao, M., Ferris, C.D., and Synder, S.H. (2002) Biliverdin reductase: a major physiologic cytoprotectant. *Proc. Natl. Acad. Sci. USA* **99**, 16093–16098.
55. Ganguli, M., Chandra, V., Kamboh, M.I., Johnston, J.M., Dodge, H.H., et al. (2000) Apolipoprotein E polymorphism and Alzheimer disease: The Indo-US Cross-National Dementia Study. *Arch. Neurol.* **57**, 824–830.
56. Lim, G.P., Yang, F., Chu, T., Chen, P., Beech, W., Teter, B., Tran, T., Ubeda, O., et al. (2000) Ibuprofen suppresses plaque pathology and inflammation in a mouse model for Alzheimer's disease. *J. Neurosci.* **20**, 5709–5714.
57. Kitamuro, T., Takahashi, K., Ogawa, K., Udono-Fujimori, R., Takeda, K., et al. (2003) Bach1 functions as a hypoxia-inducible repressor for the heme oxygenase-1 gene in human cells. *J. Biol. Chem.* **278**, 9125–9133.
58. Nakayama, M., Takahashi, K., Kitamuro, T., Yasumoto, K., Katayose, D., et al. (2000) Repression of heme oxygenase-1 by hypoxia in vascular endothelial cells. *Biochem. Biophys. Res. Commun.* **271**, 665–671.
59. Okinaga, S., Takahashi, K., Tkeda, K., Yoshizawa, M., Fujita, H., Sasaki, H., and Shibahara, S. (1996) Regulation of human heme oxygenase-1 gene expression under thermal stress. *Blood* **87**, 5074–5084.
60. Takahashi, K., Nakayama, M., Tkeda, K., Fujia, H., and Shibahara, S. (1999) Suppression of heme oxygenase-1 mRNA expression by interferon-gamma in human glioblastoma cells. *J. Neurochem.* **72**, 2356–2361.
61. Palozza, P., Serini, S., Curro, D., Calviello, G., Igarashi, K., and Mancuso, C. (2006) Beta-Carotene and cigarette smoke condensate regulate heme oxygenase-1 and its repressor factor Bach1: relationship with cell growth. *Antioxid. Redox Signal* **8**, 1069–1080.
62. Oyake, T., Itoh, K., Motohashi, H., Hyashi, N., Hoshino, H., Nishizawa, M., Yamamoto, M., and Igarashi, K. (1996) Bach proteins belong to a novel family of BTB-basic leucine zipper transcription factors that interact with MafK and regulate transcription through the NF-E2 site. *Mol. Cell. Biol.* **16**, 6083–6095.
63. Sun, J., Hoshino, H., Takaku, K., Nakajima, O., Muto, A., Suzuki, H., et al. (2002) Hemoprotein Bach1 regulates enhancer availability of heme oxygenase-1 gene. *EMBO J.* **21**, 5216–5224.
64. Shibahara, S. (2003) The heme oxygenase dilemma in cellular homeostasis: new insights for the feedback regulation of heme catabolism. *Tohoku J. Exp. Med.* **200**, 167–186.
65. Calabrese, V., Boyd-Kimball, D., Scapagnini, G., and Butterfield, D.A. (2004) Nitric oxide and cellular stress response in brain aging and neurodegenerative disorders: the role of vitagenes. *In Vivo* **18**, 245–268.
66. Kostoglou-Athanassiou, I., Forsling, M.L., Navarra, P., and Grossman, A.B. (1996) Oxytocin release is inhibited by the

generation of carbon monoxide from the rat hypothalamus – further evidence for carbon monoxide as a neuromodulator. *Brain Res. Mol. Brain Res.* **42**, 301–306.

67. Mancuso, C., Kostoglou-Athanassiou, I., Forsling, M.L., Grossman, A.B., Preziosi, P., Navarra, P., and Minotti, G. (1997) Activation of heme oxygenase and consequent carbon monoxide formation inhibits the release of arginine vasopressin from rat hypothalamic explants. Molecular linkage between heme catabolism and neuroendocrine function. *Brain Res. Mol. Brain Res.* **50**, 267–276.
68. Mancuso, C., Ragazzoni, E., Trigalli, G., Librale, I., Preziosi, P., Grossman, A., and Navarra, P.(1999) Inhibition of heme oxygenase in the central nervous system potentiates endotoxin-induced vasopressin release in the rat. *J. Neuroimmunol.* **99**, 189–194.
69. Parkes, D., Kasckow, J., and Vale, W. (1994) Carbon monoxide modulates secretion of corticotropin-releasing factor from rat hypothalamic cell cultures. *Brain Res.* **646**, 315–318.
70. Pozzoli, G., Mancuso, C., Mirtella, A., Preziozi, P., Grossman, A.B., and Navarra, P. (1994) Carbon monoxide as a novel neuroendocrine modulator: inhibition of stimulated corticotropin-releasing hormone release from acute rat hypothalamic explants. *Endocrinology* **135**, 2314–2317.
71. Mancuso, C., Pistritto, G., Tringali, G., Grossman, A.B., Prezıozi, P., and Navarra, P. (1997) Evidence that carbon monoxide stimulates prostaglandin endoperoxide synthase activity in rat hypothalamic explants and in primary cultures of rat hypothalamic astrocytes. *Brain Res. Mol. Brain Res.* **45**, 294–300.
72. Jaggar, J.H., Leffler, C.W., Cheranov, S.Y., Techeranova, D, and Cheng, X. (2002) Carbon monoxide dilates cerebral arterioles by enhancing the coupling of Ca^{2+} sparks to Ca^{2+}-activated K^+ channels. *Circ. Res.* **91**, 610–617.
73. Ryter, S.W., Otterbein, L.E., Morse, D., and Choi, A.M.(2002) Heme oxygenase/carbon monoxide signaling pathways: regulation and functional significance. *Mol. Cell. Biochem.* **234**, 249–263.
74. Calabrese, V., Stella, A.M., Butterfield, D.A., and Scapagnini, G. (2004) Redox regulation in neurodegeneration and longevity: role of the heme oxygenase and Hsp70 systems in brain stress tolerance. *Antioxid. Redox Signal* **6**, 895–913.
75. Calabrese, V., Scapagnini, G., Ravagna, A., Colombrita, C., Spadaro, F., Butterfield D.A., and Giuffrida Stella, A.M. (2004) Increased expression of heat shock proteins in rat brain during aging: relationship with mitochondrial function and glutathione redox state. *Mech. Age Dev.* **125**, 325–335.
76. Calabrese, V., Scapagnini, G., Colombrita, C., Ravagna, A., Pennisi, G., Giuffrida Stella, A.M., Galli, F., and Butterfield, D.A. (2003) Redox regulation of heat shock protein expression in aging and neurodegenerative disorders associated with oxidative stress: A nutritional approach. *Amino Acids* **25**, 437–444.
77. Yenari, M.A.,Giffard, R.G., Sapolski, R.M., and Steinberg, G.K. (1999) The neuroprotective potential of heat shock protein 70 (HSP70). *Mol. Med. Today* **5**, 525–531.
78. Yenari, M.A. (2002) Heat shock proteins and neuroprotection. *Adv. Exp. Med. Biol.* **513**, 281–299.
79. Hata, R., Maeda, K., Hermann, D., Mies, G., and Hossmann, K.A. (2000) Dynamics of regional brain metabolism and gene expression after middle cerebral artery occlusion in mice. *J. Cereb. Blood Flow Metab.* **20**, 306–315.
80. Perez, N., Sugar, J., Charya, S., Johnson, G., Merril, C., Bierer, L., Perl, D., Haroutunian, V., and Wallace, W. (1991) Increased synthesis and accumulation of heat shock 70 proteins in Alzheimer's disease. *Brain Res. Mol. Brain Res.* **1**, 249–254.
81. Yoo, B.C., Seidl, R.,Cairns, N., and Lubec, G. (1999) Heat-shock protein 70 levels in brain of patients with Down syndrome and Alzheimer's disease. *J. Neural. Transm. Suppl.* **57**, 315–322.
82. Morrison-Bogorad, M., Zimmerman, A.L., and Pardue, S. (1995) Heat-shock 70 messenger RNA levels in human brain: correlation with agonal fever. *J. Neurochem.* **64**, 235–246.
83. Kakimura, J., Kitamura, Y., Takata, K., Umeki, M., Suzuki, S., Shibagaki, K., et al. (2002) Microglial activation and amyloid-beta clearance induced by exogenous heat-shock proteins. *FASEB J.* **16**, 601–603.
84. Calabrese, V., Copani, A., Testa, D., Ravagna, A., Spadaro, F., et al. (2000) Nitric oxide synthase induction in astroglial cell cultures: effect on heat shock protein 70 synthesis and oxidant/antioxidant balance. *J. Neurosci. Res.* **60**, 613–622.
85. Calabrese, V., Testa, G., Ravagna, A., Bates, T.E., and Stella, A.M. (2000) HSP70 induction in the brain following ethanol administration in the rat: regulation by glutathione

redox state. *Biochem. Biophys. Res. Commun.* **269**, 397–400.

86. Calabrese, V., Bates, T.E., and Giuffrida Stella, A.M. (2000) NO synthase and NO-dependent signal pathways in brain aging and neurodegenerative disorders: the role of oxidant/antioxidant balance. *Neurochem. Res.* **65**, 1315–1341.
87. Calabrese, V., Scapagnini, G., Stella, A.M., Bates, T.E., and Clark, J.B. (2000) Mitochondrial involvement in brain function and dysfunction: relevance to aging, neurodegenerative disorders and longevity. *Neurochem. Res.* **26**, 739–764.
88. Iverson, S.L. and Orrenius, S. (2004) The cardiolipin-cytochrome c interaction and the mitochondrial regulation of apoptosis. *Arch. Biochem. Biophys.* **423**, 37–46.
89. Calabrese, V., Ravagna, A., Colombrita, C., Scapagnini, G., Guagliano, E., et al. (2005) Acetylcarnitine induces heme oxygenase in rat astrocytes and protects against oxidative stress: involvement of the transcription factor Nrf2. *J. Neurosci. Res.* **79**, 509–521.
90. Rai, G., Wright, G., Scott, L, Beston, B., Rest, J., and Exon-Smth, A.N. (1990) Double-blind, placebo controlled study of acetyl-L-carnitine in patients with Alzheimer's dementia. *Curr. Med. Res. Opin.* **11**, 638–647.
91. Patrick, L. (2000) Nutrients and HIV: part three - N-acetylcysteine, alpha-lipoic acid, L-glutamine, and L-carnitine. *Altern. Med. Rev.* **5**, 290–305.
92. Calabrese, V., Scapagnini, G., Ravagna, A., Bella, R., Butterfield, D.A., Calvani, M., Pennisi, G., and Giuffrida Stella, A.M. (2003) Disruption of thiol homeostasis and nitrosative stress in the cerebrospinal fluid of patients with active multiple sclerosis: evidence for a protective role of acetylcarnitine. *Neurochem. Res.* **28**, 1321–1328.
93. Albertini, P., Amenta, F., Bossoni, G., Cavellotti, C, Felici, L., Ferrante, F., and Gaviraghi, G. (1989) Effect of acetyl-L-carnitine treatment on the density of muscarinic receptors in the brain of methylazoxymethanol-microencephalic rats. *Drugs Exp. Clin. Res.* **15**, 421–427.
94. Rao, K.V., Mawal, Y.R., and Qureshi, I.A. (1997) Progressive decrease of cerebral cytochrome C oxidase activity in sparse-fur mice: role of acetyl-L-carnitine in restoring the ammonia-induced cerebral energy depletion. *Neurosci. Lett.* **224**, 83–86.
95. Carta, A., Calvani, M., Bravi, D., and Bhuachalla, S.N. (1993) Acetyl-L-carnitine and Alzheimer's disease: pharmacological considerations beyond the cholinergic sphere. *Ann. N.Y. Acad. Sci.* **695**, 324–326.
96. Ando, S., Kobayashi, S., Waki, H., Kon, K., Fukui, F., Tadenuma, T., et al. (2002) Animal model of dementia induced by entorhinal synaptic damage and partial restoration of cognitive deficits by BDNF and carnitine. *J. Neurosci. Res.* **70**, 519–527.
97. Chiechio, S., Caricasole, A., Barletta, E., Storto, M., Catania, M.V., et al. (2002) L-Acetylcarnitine induces analgesia by selectively up-regulating mGlu2 metabotropic glutamate receptors. *Mol. Pharmacol.* **61**, 989–996.
98. Spagnoli, A., Lucca, U., Menasce, G., Bandera, L., Cizza, G., Forloni, G., et al. (1991) Long-term acetyl-L-carnitine treatment in Alzheimer's disease. *Neurology* **41**, 1726–1732.
99. Caprioli, A., Markowska, A.L., and Olton, D.S. (1995) Acetyl-L-Carnitine: chronic treatment improves spatial acquisition in a new environment in aged rats. *J. Gerontol.* **50**, 232–236.
100. Chauhan, N.B. and Siegel, G.J. (2003) Effect of PPF and ALCAR on the induction of NGF- and p75-mRNA and on APP processing in Tg2576 brain. *Neurochem. Int.* **43**, 225–233.
101. Calabrese, V., Scapagnini, G., Latteri, S., Colombrita, C., Ravagna, A., Catalano, C., et al. (2002) Long-term ethanol administration enhances age-dependent modulation of redox state in different brain regions in the rat: protection by acetyl carnitine. *Int. J. Tissue React.* **24**, 97–104.
102. Hagen, T.M., Ingersoll, R.T., Wehr, C.M., Lykkesfeldt, J., Vinarsky, V., Bartholomew, J.C., Song, M.H., and Ames, B.N. (1998) Acetyl-L-carnitine fed to old rats partially restores mitochondrial function and ambulatory activity. *Proc. Natl. Acad. Sci. USA* **95**, 9562–9566.
103. Wolff, R.L., Combe, N.A., and Entressangles, B. (1985) Positional distribution of fatty acids in cardiolipin of mitochondria from 21-day-old rats. *Lipids* **20**, 908–914.
104. Armeni, T., Principato, G., Quiles, J.L., Pieri, C., Bompadre, S., and Battino, M. (1998) Mitochondrial dysfunctions during aging: vitamin E deficiency or caloric restriction – two different ways of modulating stress. *J. Bioenerg. Biomembr.* **35**, 181–191.
105. Wheeler, D.S., Dunsmore, K.E., and Wong, H.R. (2003) Intracellular delivery of HSP70 using HIV-1 Tat protein transduction domain. *Biochem. Biophys. Res. Commun.* **301**, 54–59.

106. Pocernich, C.B., Sultana, R., Hone, E., Turchan, J., Martins, R.N., Calabrese, V., Nath, A., and Butterfiled, D.A. (2004) Effects of apolipoprotein E on the human immunodeficiency virus protein tat in neuronal cultures and synaptosomes. *J. Neurosci. Res.* 77, 532–539.
107. Weng, Y.H., Yang, G., Weiss, S., Dennery, P.A., et al. (2003) Interaction between heme oxygenase-1 and 2 proteins. *J. Biol. Chem.* **278**, 50999–61005.
108. Seshadri, S., Beiser, A., Selhub, J., Jacques, P.E., Rosenberg, I.H., et al. (2002) Plasma homocysteine as a risk factor for dementia and Alzheimer's disease. *N. Engl. J. Med.* **346**, 476–483.
109. Postuma, R.B. and Lang, A.E. (2004) Homocysteine and levodopa: should Parkinson disease patients receive preventative therapy? *Neurology* **63**, 886–891.
110. D'Anna, R., Baviera, G., Corrado, F., Ientile, R., Granese, D., and Stella, N.C. (2004) Plasma homocysteine in early and late pregnancies complicated with preeclampsia and isolated intrauterine growth restriction. *Acta Obstet. Gynecol. Scand.* **83**, 155–158.
111. Russo, G.T., DiBendedetto, A., Giorda, C., Alessi, E., Crisafulli, G., Ientile, R., et al. (2004) Correlates of total homocysteine plasma concentration in type 2 diabetes. *Eur. J. Clin. Invest.* **34**, 197–204.
112. Morris, M.S. (2003) Homocysteine and Alzheimer's disease. *Lancet Neurol.* **2**, 425–428.
113. Schnyder, G., Roffi, M., Flammer, Y., Pin, R., and Hess, O.M. (2002) Effect of homocysteine-lowering therapy with folic acid, vitamin B(12), and vitamin B(6) on clinical outcome after percutaneous coronary intervention: the Swiss Heart study: a randomized controlled trial. *JAMA* **288**, 973–979.
114. Van Der Put, N.M., Gabreels, F., Stevens, E.M., Smeitink, J.A., Trijebels, F.J., et al. (1998) A second common mutation in the methylenetetrahydrofolate reductase gene: an additional risk factor for neural tube defects? *Am. J. Hum. Genet.* **62**, 1044–1051.
115. Grieve, A., Butcher, S.P., and Griffiths, R. (1992) Synaptosomal plasma membrane transport of excitatory sulphur amino acid transmitter candidates: kinetic characterisation and analysis of carrier specificity. *J. Neurosci. Res.* **32**, 60–68.
116. Eto, K., Asada, T., Arima, K., Makifuchi, T., and Kimura, H. (2002) Brain hydrogen sulfide is severely decreased in Alzheimer's disease. *Biochem. Biophys. Res. Commun.* **293**, 1485–1488.
117. Regland, B., Abrahamsson, L., Blenow, K., Grenfeldt, B., and Gottfries, C.G. (2004) CSF-methionine is elevated in psychotic patients. *J. Neural. Transm.* **111**, 631–640.
118. Isobe, C., Murata, T., Sato, C., and Terayama, V. (2005) Increase of total homocysteine concentration in cerebrospinal fluid in patients with Alzheimer's disease and Parkinson's disease. *Life Sci.* 77, 1836–1843.
119. Obeid, R., Schorr, H., Eckert, R., and Herrmann, W. (2004) Vitamin B12 status in the elderly as judged by available biochemical markers. *Clin. Chem.* **50**, 238–241.
120. Herrmann, W., Quast, S., Ullrich, M., Schultze, H., Bodis, M., and Geisel, J. (1999) Hyperhomocysteinemia in high-aged subjects: relation of B-vitamins, folic acid, renal function and the methylenetetrahydrofolate reductase mutation. *Atherosclerosis* **144**, 91–101.
121. den Heijer, T., Vermeer, S.E., Clarke, R., Koudstaal, P.J., et al. (2003) Homocysteine and brain atrophy on MRI of non-demented elderly. *Brain* **126**, 170–175.
122. Kruman, I.I., Culmsee, C., Chan, S.L., Kruman, Y., Guo, Z., et al. (2000) Homocysteine elicits a DNA damage response in neurons that promotes apoptosis and hypersensitivity to excitotoxicity. *J. Neurosci.* **20**, 6920–6926.
123. Ho, P.I., Collins, S.C., Dhitavat, S., Ortis, D., Ashline, D., Rogers, E., et al. (2001) Homocysteine potentiates beta-amyloid neurotoxicity: role of oxidative stress. *J. Neurochem.* **78**, 249–253.
124. Kruman, I.I., Kumaravel, T.S., Lohani, A., Pederson, W.A., Cuttler, R.G., Kruman, Y., et al. (2002) Folic acid deficiency and homocysteine impair DNA repair in hippocampal neurons and sensitize them to amyloid toxicity in experimental models of Alzheimer's disease. *J. Neurosci.* **22**, 1752–1762.
125. Mattson, M.P. (2000) Emerging neuroprotective strategies for Alzheimer's disease: dietary restriction, telomerase activation, and stem cell therapy. *Exp. Gerontol.* **35**, 489–502.
126. Harada, J. and Sugimoto, M. (1999) Activation of caspase-3 in beta-amyloid-induced apoptosis of cultured rat cortical neurons. *Brain Res.* **842**, 311–323.
127. Love, S., Barber, R., and Wilcock, G.K. (1999) Increased poly(ADP-ribosyl)ation of nuclear proteins in Alzheimer's disease. *Brain* **122**, 247–253.

128. Smulson, M.E., Simbulan-Rosenthal, C.M., and Boulares, A.H. (2000) Roles of poly(ADP-ribosyl)ation and PARP in apoptosis, DNA repair, genomic stability and functions of p53 and E2F-1. *Adv. Enzyme Regul.* **40**, 183–215.
129. Culmsee, C., Monning, J., Kemp, B.E., and Matterson, M. (2001) AMP-activated protein kinase is highly expressed in neurons in the developing rat brain and promotes neuronal survival following glucose deprivation. *J. Mol. Neurosci.* **17**, 45–58.
130. Welch, G.N. and Loscalzo, J. (1998) Homocysteine and atherothrombosis. *N. Engl. J. Med.* **338**, 1042–1050.
131. de Lau, L.M.L., Koudstaal, P.J., Witteman, J.C.M., Hofman, A., and Breteler, M.M.B. (2006) Dietary folate, vitamin B12, and vitamin B6 and the risk of Parkinson disease. *Neurology* **67**, 315–318.
132. Stott, D.J., MacIntoswh, G., Lowe, G.D., Rumley, A., McMahaon, A.D., et al. (2005) Randomized controlled trial of homocysteine-lowering vitamin treatment in elderly patients with vascular disease. *Am. J. Clin. Nutr.* **82**, 1320–1326.
133. Yang, Q., Botto, L.D., Erickson, J.D., Berry, R.J., Sambell, C., Johansen, H., and Friedman, J.M. (2006) Improvement in stroke mortality in Canada and the United States, 1990 to 2002. *Circulation* **113**, 1335–1343.
134. Otterbein, L.E., Zuckerbraun, B.S., Haga, M., Liu, F., Song, R., Usheva, A., et al. (2003) Carbon monoxide suppresses arteriosclerotic lesions associated with chronic graft rejection and with balloon injury. *Nat. Med.* **9**, 183–190.

Chapter 18

Gene Therapy Techniques for the Delivery of Endothelial Nitric Oxide Synthase to the Lung for Pulmonary Hypertension

W. Deng, T.J. Bivalacqua, H.C. Champion, W.J. Hellstrom, Subramanyam N. Murthy, and Philip J. Kadowitz

Abstract

Pulmonary hypertension (PH) is a serious, often fatal disease characterized by remodeling of the pulmonary vascular bed, increased pulmonary arterial pressure, and right heart failure. The increased vascular resistance in the pulmonary circulation is due to structural changes and increased vasoconstrictor tone. Although current therapies have prolonged survival, the long-term outcome is not favorable. Nitric oxide (NO) is synthesized by endothelial nitric oxide synthase (eNOS) and is important in regulating vascular resistance and in vascular remodeling in the lung. NO deficiency due to endothelial dysfunction plays an important role in the pathogenesis of PH. Therefore, local *eNOS* gene delivery to the lung is a promising approach for the treatment of PH. Adenoviral-mediated in vivo gene therapy and adult stem cell-based ex vivo gene therapy are two attractive current gene therapies for the treatment of cardiovascular and pulmonary diseases. In this chapter we describe the use of two gene transfer techniques, i.e., adenoviral gene transfer of *eNOS* and *eNOS* gene-modified rat marrow stromal cells, for *eNOS* gene delivery to the lung of laboratory animals for the treatment of PH.

Key words: Gene therapy, pulmonary hypertension, endothelial nitric oxide synthase, adenovirus, marrow stromal cells, mesenchymal stem cells, lung.

1. Introduction

Pulmonary hypertension (PH) is a progressive disorder characterized by increased pulmonary arterial pressure, remodeling of the pulmonary vascular bed, right heart failure, and death. This disorder progresses and usually leads to death in 2–3 years

R.M. Uppu et al. (eds.), *Free Radicals and Antioxidant Protocols*, Methods in Molecular Biology 610,
DOI 10.1007/978-1-60327-029-8_18, © Humana Press, a part of Springer Science+Business Media, LLC 1998, 2010

after diagnosis (1–6). The cause of primary pulmonary hypertension is unknown but the pathogenesis of PH involves pulmonary arterial endothelial dysfunction and pulmonary vasoconstriction (5, 7). Although current therapies such as intravenous prostacyclin, endothelin-receptor antagonists, type 5 phosphodiesterase inhibitor (sildenafil), and lung transplantation can lower pulmonary arterial pressure, improve the quality of life, and prolong survival, however the long-term outcome in PH is not favorable (5, 6, 8, 9). Nitric oxide (NO) is synthesized in a reaction catalyzed by endothelial nitric oxide synthase (eNOS) in endothelial cells and is an important vasodilator for the regulation of vascular smooth muscle tone (10, 11). NO deficiency has been suggested to be involved in pulmonary hypertension (12–15). Therefore, the enhancement of NO delivery to the lung is a promising approach for the treatment of PH. An attractive method of increasing local NO delivery is the use of gene transfer technology to increase *eNOS* gene expression at or near the site of endothelial dysfunction in the lung.

For gene transfer, adenoviral vectors have several advantages over other vectors such as the ability to infect both dividing and non-dividing cells, the capability to accommodate large DNA inserts, the ease of preparation of high-titer viral stock, high-level transgene expression, and a broad host range. Furthermore, as adenoviral DNA does not integrate into the host cell chromosome, there is no risk of insertional mutagenesis (16–19). The major disadvantage is that adenoviral vectors can elicit strong host inflammatory and immune responses after direct in vivo administration, which precludes the possibility of repeated adenoviral administration in the same patient, and can even cause patient death (20–23).

Marrow stromal cells (MSCs), also known as mesenchymal stem cells, are multipotent adult stem cells from bone marrow that have the capability of differentiation into osteoblasts, chondrocytes, adipocytes, myocytes, and other cell types (24–29). Since MSCs are relatively easy to isolate, expand ex vivo, and gene engineer, genetically modified MSCs have been used not only for local delivery of therapeutic gene but also for tissue repair or regeneration in adult stem cell-based ex vivo gene therapy of a variety of diseases (30–35).

To explore alternative approaches for better therapies for PH, here we describe two methods of *eNOS* gene transfer to the lung: intratracheal injection of adenovirus containing *eNOS* to the lung of the mouse and intracheal injection of adenoviral-transduced rat marrow stromal cells (rMSCs) to the lung of the rat (33, 36–38, Kadowtiz et al., unpublished data). In this report, a mouse model of bleomycin-induced PH and a rat model of monocrotaline-induced PH are used and the focus is on gene transfer techniques for the delivery of *eNOS* gene to the lung.

2. Materials

2.1. Adenoviral Vectors

The following two adenoviral vectors were used in the study:

1. Ad5RSVeNOS: a replication-deficient recombinant adenoviral vector carrying the bovine *eNOS* gene under the control of the Rous sarcoma virus (RSV) promoter, and it has a concentration of 1.2×10^{12} viral particles (vp)/mL and a titer of 3.0×10^{10} plaque-forming units (pfu)/mL (33, 36, 39, 40).
2. Ad5RSVntlacZ: a replication-deficient recombinant adenoviral vector carrying the nuclear-targeted β-galactosidase gene *ntlacZ* under the control of the RSV promoter, and it has a concentration of 1.4×10^{12} vp/mL and a titer of 1.0×10^{10} pfu/mL (33, 34, 36, 41).

The above replication-deficient recombinant adenoviral vectors were prepared in the University of Iowa Gene Transfer Vector Core (Iowa city, IA) using the standard procedure and were suspended in 3% sucrose/PBS. After arrival, the adenoviruses were stored at –70°C until used.

2.2. CD-1 Mice with Bleomycin-Induced PH

Male, body weight 22–30 g CD-1 mice were purchased from Harlan (Indianapolis, IN). Bleomycin (Sigma, St. Louis, MO) was dissolved in 0.9% NaCl and used for the induction of PH in CD-1 mice.

2.3. Sprague–Dawley Rats

Male, 6-week- or 3-month-old Sprague–Dawley rats were obtained from Harlan (San Diego, CA). The 6-week-old rats were used for the isolation of rMSCs and the 3-month-old rats were used for the induction of PH via tail vein injection of monocrotaline. Monocrotaline (Sigma) was dissolved in dH_2O and the pH value was adjusted to 7.0.

2.4. Isolation of Rat Marrow Stromal Cells (rMSCs)

Scalpel, scissors, forceps, syringe, needle, cell strainer with 70-μm nylon mesh (BD Bioscience, Bedford, MA), cell culture dish, centrifuge tube, and T75 tissue culture flask.

2.5. Ex Vivo Expansion of rMSCs

Culture medium for rMSCs [α-MEM (GIBCO Invitrogen, Grand Island, NY), 20% fetal bovine serum (GIBCO Invitrogen), 100 units/mL penicillin, 100 μg/mL streptomycin, 25 ng/mL amphotericin B (Atlanta Biologicals, Norcross, GA), and 2 mM L-glutamine (GIBCO Invitrogen)], PBS, 0.25% trypsin/1 mM EDTA, T75 flask, and six-well cell culture plate.

2.6. X-Gal Staining for β-Galactosidase Activity in Ad5RSVntlacZ-Transduced rMSCs

Fixing solution [2% formaldehyde and 0.2% glutaraldehyde (Sigma), prepared in PBS], staining solution [5 mM potassium ferricyanide, 5 mM potassium ferrocyanide, 2 mM $MgCl_2$, and 1 mg/mL X-gal (Sigma), prepared in PBS].

2.7. Immunostaining for eNOS Expression in Ad5RSVeNOS-Transduced rMSCs

Mouse anti-eNOS monoclonal antibody (BD Transduction Laboratories; San Diego, CA), Mouse Immunocruz Staining System (Santa Cruz Biotechnology; Santa Cruz, CA), methanol, hematoxylin (Rowley Biochemical, Danvers, MA).

2.8. X-Gal Staining for β-Galactosidase-Positive Ad5RSVntlacZ-Transduced rMSCs in Rat Lung

Sodium pentobarbital (Sigma), 4% paraformaldehyde in PBS (USB Corp., Cleveland, OH), sucrose (Sigma), OCT compound (Triangle Biomedical Sciences, Durham, NC), Superfrost Plus slides (Fisher Scientific, Pittsburgh, PA).

3. Methods

The methods in the following paragraphs describe (1) intratracheal injection of adenovirus containing *eNOS*; (2) isolation and ex vivo expansion of rMSCs; (3) adenoviral-mediated *ntlacZ* or *eNOS* gene transfer into rMSCs; (4) intratracheal injection of wild-type or adenoviral-transduced rMSCs; (5) X-gal staining for the analysis of the survival and distribution of rMSCs in rat lung. Ultimately it is important to determine whether pulmonary hypertension can be attenuated and endothelial responses can be restored after direct in vivo administration of adenovirus containing *eNOS* to the lung or adenoviral-transduced rMSCs to the lung in laboratory animals with PH. The methods for the measurement of pulmonary hemodynamics have been described previously (36, 37, 42, 43).

3.1. Safe Handling of Adenoviruses in the Laboratory

The biosafety level of replication-deficient recombinant adenoviruses is biosafety level 2. NIH biosafety level 2 practices, equipment, and facilities shall be required for all research conducted using adenoviruses. Adenoviruses obtained from the University of Iowa Gene Transfer Vector Core were sold in 1-mL aliquots and prepared in 3% sucrose/PBS. Therefore, they should be stored at –70°C. When used, adenoviruses should be thawed for the least amount of time possible and then kept on ice. After the first usage, adenoviruses should be aliquoted at the smallest amount per use and stored at –70°C (*see* **Note 1**). Multiple rounds of freeze-thawing should be avoided to insure viral infectivity.

3.2. Intratracheal Injection of Adenovirus Containing eNOS to the Mice with Bleomycin-Induced PH

CD-1 mice were anesthetized with thiopentobarbital (85–95 mg/kg, i.p.)/ketamine (3 mg/kg, i.p.) and put in a supine position on a thermoregulated surgical table. Under sterile conditions, trachea was approached via a midline neck incision and isolated by blunt dissection. Using a 27-gauge needle attached to a 100-μL microsyringe, 50 μL vehicle (3% sucrose in PBS), Ad5RSVntlacZ (1×10^{12} vp/mL), or Ad5RSVeNOS (1×10^{12} vp/mL) was instilled into the trachea. Before instillation of 50 μL of vehicle/adenovirus, a forced expiration was achieved by manual compression of the thorax of the mouse. Immediately after instillation, manual compression was rapidly released. This resulted in a forceful inspiration that facilitated adenoviral dispersion to distal air spaces (*see* **Note 2**). The incision was sutured and the mice were allowed to recover. Five days after Ad5RSVeNOS administration, the mice were anesthetized and 50 μL of bleomycin (4 U/kg) was instilled into the trachea in the same way as above (*see* **Note 3**). The incision was sutured and the mice were allowed to recover. Fourteen days after bleomycin administration, experiments were performed to assess pulmonary arterial pressure, pulmonary vascular responses, and cardiac output and the effectiveness of direct in vivo administration of Ad5RSVeNOS on bleomycin-induced PH in mice was determined (*see* **Note 4**).

3.3. Isolation and Ex Vivo Expansion of rMSCs

rMSCs were isolated by their tight adherence to cell culture plastic as previously described (33, 34; *see* **Note 5**). Briefly, 6-week-old male Sprague–Dawley rats were euthanized with CO_2 and under sterile conditions, femurs and tibias were collected. Both ends of femurs and tibias were removed and the bone marrow was flushed out using a 21-gauge needle attached to a 5-mL syringe filled with the culture medium for rMSCs. The whole bone marrow cells were filtered through a cell strainer with 70-μm nylon mesh and the cells were plated in a T75 flask. The cells were incubated in the culture medium for rMSCs at 37°C with 5% humidified CO_2 and rMSCs were isolated by their tight adherence to cell culture plastic. Fresh culture medium was added and replaced every 2–3 days to remove non-adherent cells. The adherent rMSCs were grown to 90% confluency (passage 0) over about 7 days, harvested with 0.25% trypsin/1 mM EDTA, diluted 1:3, re-plated in T75 flasks, and again grown to 90% confluency (passage 1). The cells were harvested with 0.25% trypsin/1 mM EDTA and diluted 1:3 per passage for further ex vivo expansion (**Fig. 18.1**). rMSCs at passages 1–3 were used for all the experiments.

3.4. Adenoviral-Mediated ntlacZ or eNOS Gene Transfer into rMSCs

rMSCs were plated at a density of 1×10^4 cells/cm^2 in T75 flask or six-well plate and cultured overnight (*see* **Note 6**). The cells were counted and exposed to fresh culture medium

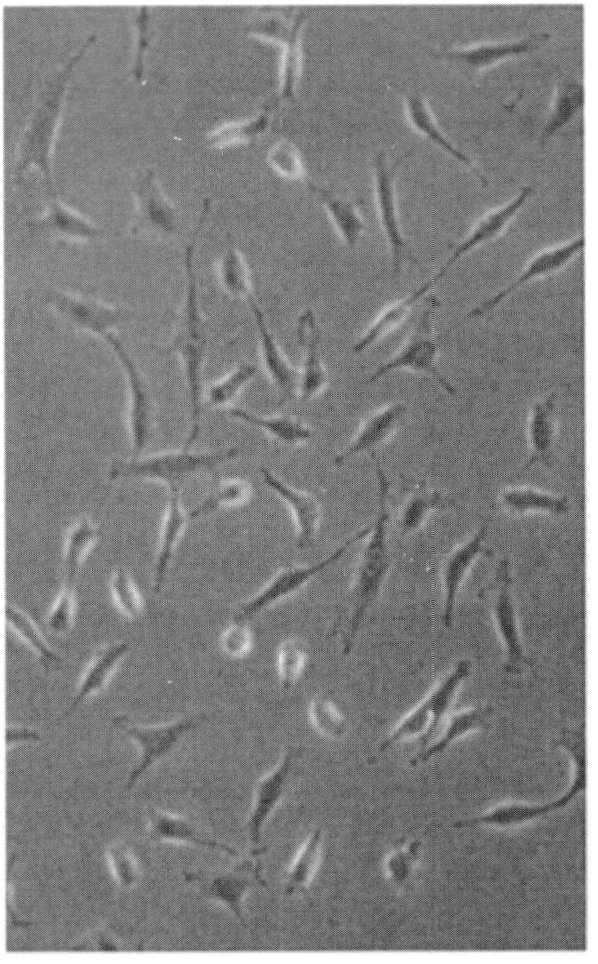

Fig. 18.1. Photomicrograph showing rMSCs at log phase. Magnification, 25×.

containing Ad5RSVntlacZ or Ad5RSVeNOS at 300 multiplicities of infection (MOI, defined as pfu per cell) for 48 h and the virus-containing supernatant was removed. These adenoviral-transduced rMSCs were then analyzed for the expression of β-galactosidase or *eNOS* and the transduction efficiency was determined (*see* **Note 7**). For X-gal cytochemistry of β-galactosidase activity in rMSCs, cells in six-well plate were washed with PBS, fixed for 5 min, and rinsed twice with PBS. Cells were then incubated in X-gal staining solution at 37°C overnight in the dark. Cells were washed with PBS and the expression of *ntlacZ* transgene in rMSCs was evaluated under a phase contrast microscope. The blue nuclear-targeted β-galactosidase-positive Ad5RSVntlacZ-transduced rMSCs found in 2–3 representative microscopic fields were then counted and expressed as a percentage of the total number of cells in these fields (**Fig. 18.2**).

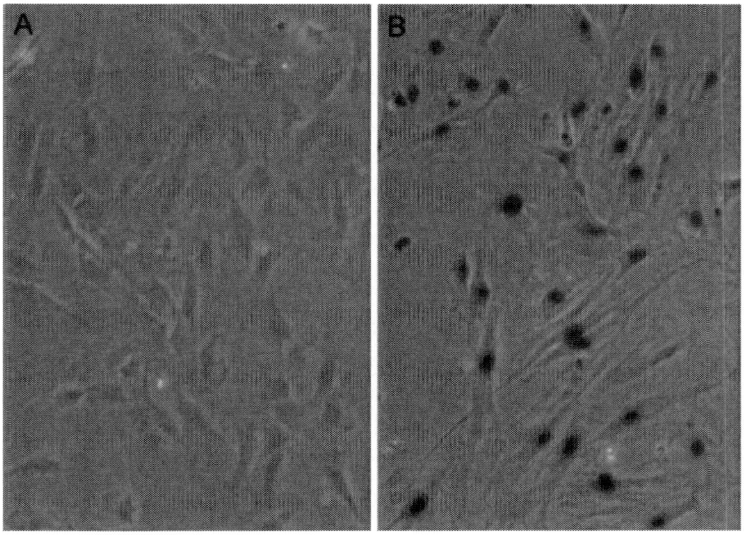

Fig. 18.2. Photomicrograph showing blue β-galactosidase-positive Ad5RSVntlacZ-transduced rMSCs. rMSCs were transduced with Ad5RSVntlacZ at MOI 300 for 48 h and the cells were stained for the nuclear-targeted β-galactosidase activity. (**a**) rMSCs. (**b**) rMSCs transduced with Ad5RSVntlacZ at MOI 300. Magnification, 25×.

For immunocytochemistry of *eNOS* transgene expression in rMSCs, a mouse anti-eNOS monoclonal antibody and a mouse Immunocruz Staining System were used. Briefly, cells were fixed with ice-cold methanol for 5 min, washed with PBS twice, and immunostained with mouse anti-eNOS monoclonal antibody (1:100 dilution) followed by biotinylated secondary antibody, HRP–streptavidin complex, and HRP substrate. The cells were then counterstained with hematoxylin. Expression of *eNOS* transgene in rMSCs was evaluated by phase contrast microscopy. The brown *eNOS*-positive Ad5RSVeNOS-transduced rMSCs found in 2–3 representative microscopic fields were counted and expressed as a percentage of the total number of cells in these fields (**Fig. 18.3**).These adenoviral-transduced rMSCs can also be washed three times with PBS, trypsinized, and used for intratracheal injection for the treatment of PH as described below.

3.5. Intratracheal Injection of Wild-Type or Adenoviral-Transduced rMSCs to the Rats with Monocrotaline-Induced PH

To induce PH, the rats were given monocrotaline in a dose of 60 mg/kg i.v. into the tail vein. Two weeks after monocrotaline administration, the rats were given 3 × 10^6 wild-type or adenoviral-transduced rMSCs for the treatment of PH. In brief, cells were washed three times with PBS, harvested with 0.25% trypsin/1 mM EDTA, washed with PBS again, and a cell suspension at a concentration of 6,000 cells/μL was prepared in PBS. The cell suspension was kept on ice until the rats were anesthetized with sodium pentobarbital (30 mg/kg, i.p.),

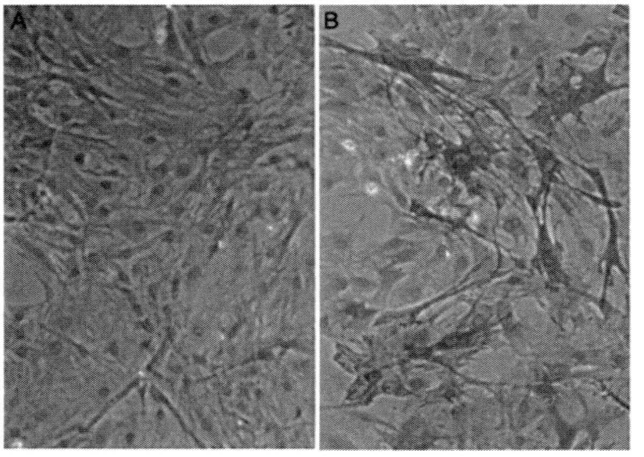

Fig. 18.3. Photomicrograph showing brown eNOS-positive Ad5RSVeNOS-transduced rMSCs. rMSCs were transduced with Ad5RSVeNOS at MOI 300 for 48 h and the cells were immunostained for *eNOS* transgene expression. (**a**) rMSCs and (**b**) rMSCs transduced with Ad5RSVeNOS at MOI 300. Magnification, 25×.

usually within 1 h. Under sterile conditions, the trachea was exposed and 500 μL of cell suspension, i.e., 3×10^6 cells, or 500 μL of PBS was injected into the lung with a 1-mL syringe connected to a 25-gauge needle. Before cell injection, the breath of the rat was halted by circumferential compression of the chest with hand. Immediately after cell injection, the compression was rapidly released to allow the rat to inhale the stem cells (*see* **Note 8**). Three weeks after stem cell administration, the rats were anesthetized with sodium pentobarbital (30 mg/kg, i.p.), experiments were performed for the measurement of pulmonary hemodynamics such as pulmonary arterial pressure, pulmonary vascular responses, and cardiac output, and the effectiveness of intratracheal injection of wild-type or adenoviral-transduced rMSCs on bleomycin-induced PH in rats was determined.

3.6. X-Gal Staining of Rat Lung After Intratracheal Injection of Ad5RSVntlacZ-Transduced rMSCs

rMSCs were transduced with Ad5RSVntlacZ at MOI 300 for 48 h. The virus-containing culture supernatant was removed and the cells were washed three times with PBS. These Ad5RSVntlacZ-transduced rMSCs were harvested with 0.25% trypsin/1 mM EDTA, washed with PBS, and a cell suspension at a concentration of 6,000 cells/μL was prepared in PBS. The cell suspension was kept on ice until the rats were anesthetized with sodium pentobarbital (30 mg/kg, i.p.). The rats were given monocrotaline in a dose of 60 mg/kg i.v. into the tail vein 2 weeks before. Under sterile conditions, the trachea was exposed and 500 μL of cell suspension was injected into the lung in the same way as above. A total of 3×10^6 Ad5RSVntlacZ-transduced rMSCs were injected into each rat. Seven days later, the rats were deeply anesthetized with sodium pentobarbital (80 mg/kg, i.p.)

and intracardiacally perfused with 200 mL PBS. The lung was removed, fixed with 4% paraformaldehyde in PBS for 10 min, washed three times with PBS, and incubated in X-gal staining solution at 37°C overnight in the dark. The lung was washed with PBS, fixed in 4% paraformaldehyde in PBS at 4°C overnight, and transferred to 30% sucrose in PBS at 4°C overnight. The lung tissue was embedded in OCT compound, snap frozen in liquid nitrogen, and stored at –70°C. Ten-micrometer lung sections were prepared with a cryostat and mounted on Superfrost Plus slides. The slides were checked under a phase contrast microscope and the blue β-galactosidase-positive cells in the lung were identified as the transplanted Ad5RSVntlacZ-transduced rMSCs that have survived for 7 days in rat lung with PH (**Fig. 18.4**). Normal rats were used as a control.

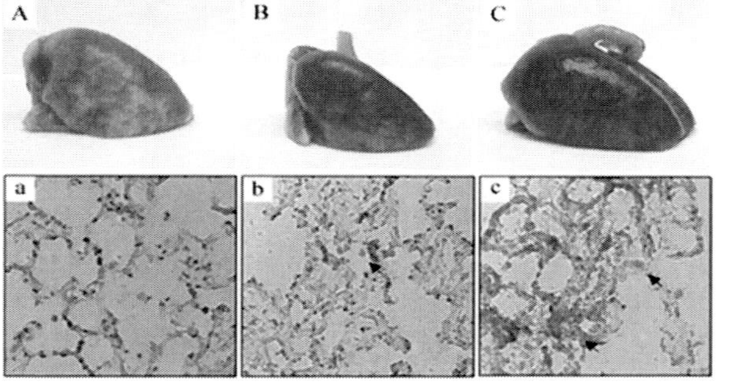

Fig. 18.4. X-gal histochemical analysis for the survival and distribution of rMSCs in rat lung. Seven days after intratracheal injection of 3 × 10^6 Ad5RSVntlacZ-transduced rMSCs, rats were sacrificed and lungs were X-gal stained for β-galactosidase activity. *Top*: Photomacrographs of the lung of a normal rat (**A**), a normal rat that was intratracheally injected with Ad5RSVntlacZ-transduced rMSCs (**B**), and a rat with monocrotaline-induced PH that was intratracheally injected with Ad5RSVntlacZ-transduced rMSCs (**C**). *Bottom*: Ten-micrometer lung cross sections were then prepared and checked under a phase contrast microscope for the blue β-galactosidase-positive Ad5RSVntlacZ-transduced rMSCs. Photomicrographs of the typical lung section of a normal rat (**a**), a normal rat that was intratracheally injected with Ad5RSVntlacZ-transduced rMSCs (**b**), and a rat with monocrotaline-induced PH that was intratracheally injected with Ad5RSVntlacZ-transduced rMSCs (**c**). *Arrows* denote the presence of blue β-galactosidase-positive Ad5RSVntlacZ-transduced rMSCs in rat lung.

4. Notes

1. To facilitate the freeze-thawing process, a 37°C water bath can be used to thaw the adenovirus and after use, the adenovirus can be rapidly frozen using dry ice and ethanol.
2. Before intratracheal injection of adenovirus, a forced expiration must be achieved by manual compression of mouse

thorax. Immediately after adenoviral administration, the compression should be rapidly released so that a forceful inspiration will facilitate adenoviral dispersion in the whole lung.

3. Pilot studies demonstrated that eNOS reached peak expression at 5 days after intratracheal injection of Ad5RSVeNOS, and the expression was maintained for 21–25 days. Therefore, induction of PH via intratracheal injection of bleomycin into the mice was carried out 5 days after the administration of Ad5RSVeNOS.
4. Pilot studies demonstrated that intratracheal injection of bleomycin at 4 U/kg resulted in lung fibrosis and increased pulmonary arterial pressure at 14 days with low mortality. Therefore, pulmonary hemodynamics was assessed in the mice 14 days after bleomycin treatment.
5. There are two types of adult stem cells in the bone marrow: hematopoietic stem cells and non-hematopoietic stem cells (31, 44). Hematopoietic stem cells are suspension cells that give rise to red cells, monocytes, platelets, granulocytes, and lymphocytes. Non-hematopoietic stem cells, usually referred to as marrow stromal cells (MSCs) or mesenchymal stem cells, are adherent cells that are capable of differentiating into osteoblasts, chondrocytes, adipocytes, myocytes, and other cell types. After cells from the whole bone marrow are plated in tissue culture flask, MSCs firmly adhere to the plastic and can be easily separated from the suspension hematopoietic stem cells by repeatedly changing the cell culture medium.
6. Plating density determines the proliferative rate of MSCs. At a plating density of 1×10^4 cells/cm^2, rMSCs will be at log phase and cell condition will be excellent the next day.
7. Transduction efficiency of adenovirus in rMSCs is dose dependent. At MOI 300, the transduction efficiency of Ad5RSVntlacZ and Ad5RSVeNOS is greater than 90% and greater than 60%, respectively. The difference in transduction efficiency of Ad5RSVntlacZ and Ad5RSVeNOS at the same MOI is due to the difference in the sensitivity of the detection method for β-galactosidase and eNOS, respectively. However, adenoviral-induced cytotoxicity in rMSCs, i.e., inhibition of cell proliferation, cell death, and change of multipotentiality, is also dose dependent. At MOI 300, adenoviral-induced cytotoxicity in rMSCs cannot be observed (33, 34). Therefore, MOI 300 is used for adenoviral transduction in rMSCs.
8. Before intratracheal injection of wild-type or adenoviral-transduced rMSCs, the breath of the rat must be halted

by manual compression of rat chest. Immediately after cell injection, the compression should be rapidly released so that a forceful inspiration will facilitate stem cell dispersion in rat lung.

Acknowledgments

The authors would like to thank Drs. Beverly Davidson and Donald D. Heistad and the University of Iowa Gene Transfer Vector Core for the preparation of replication-deficient recombinant adenoviral vectors. This work was supported by NIH R01 HL62000, NIH R01 HL77421, and National Cancer Institute Grant CA-65600.

References

1. D'Alonzo, G.E., Barst, R.J., Ayres, S.M., Bergofsky, E.H., Brundage, B.H., Detre, K.M., Fishman, A.P., Goldring, R.M., Groves, B.M., and Kernis, J.T. (1991) Survival in patients with primary pulmonary hypertension. results from a national prospective registry. *Ann. Intern. Med.* **115**, 343–349.
2. Rubin, L.J. (1997) Primary pulmonary hypertension. *N. Engl. J. Med.* **336**, 111–117.
3. Galie, N., Manes, A., Uguccioni, L., Serafini, F., De Rosa, M., Branzi, A., and Magnani, B. (1998) Primary pulmonary hypertension: insights into pathogenesis from epidemiology. *Chest* **114** (3 Suppl), 184S–194S.
4. Peacock, A.J. (1999) Primary pulmonary hypertension. *Thorax* **54**, 1107–1118.
5. Runo, J.R. and Loyd, J.E. (2003) Primary pulmonary hypertension. *Lancet* **361**, 1533–1544.
6. Humbert, M., Sitbon, O., and Simonneau, G. (2004) Treatment of pulmonary arterial hypertension. *N. Engl. J. Med.* **351**, 1425–1436.
7. Mandegar, M., Fung, Y.C., Huang, W., Remillard, C.V., Rubin, L.J., and Yuan, J.X. (2004) Cellular and molecular mechanisms of pulmonary vascular remodeling: role in the development of pulmonary hypertension. *Microvasc. Res.* **68**, 75–103.
8. Lee, A.J., Chiao, T.B., and Tsang, M.P. (2005) Sildenafil for pulmonary hypertension. *Ann. Pharmacother.* **39**, 869–884.
9. Nathan, S.D. (2005) Lung transplantation: disease-specific considerations for referral. *Chest* **127**, 1006–1016.
10. Palmer, R.M., Ferrige, A.G., and Moncada, S. (1987) Nitric oxide release accounts for the biological activity of endothelium-derived relaxing factor. *Nature* **327**, 524–526.
11. Fleming, I. and Busse, R. (2003) Molecular mechanisms involved in the regulation of the endothelial nitric oxide synthase. *Am. J. Physiol. Regul. Integr. Comp. Physiol.* **284**, R1–R12.
12. Huang, P.L., Huang, Z., Mashimo, H., Bloch, K.D., Moskowitz, M.A., Bevan, J.A., and Fishman, M.C. (1995) Hypertension in mice lacking the gene for endothelial nitric oxide synthase. *Nature* **377**, 239–242.
13. Michelakis, E.D. (2003) The role of the NO axis and its therapeutic implications in pulmonary arterial hypertension. *Heart Fail. Rev.* **8**, 5–21.
14. Nakamoto, T., Harasawa, H., Fuse, D., Amano, H., and Matsuda T. (2004) Nitric oxide and pulmonary hypertension. *Nippon Rinsho* **62** (Suppl 9), 522–526.
15. Farber, H.W. and Loscalzo, J. (2004) Pulmonary arterial hypertension. *N. Engl. J. Med.* **351**, 1655–1665.
16. Wilson, J.M. (1996) Adenoviruses as gene-delivery vehicles. *N. Engl. J. Med.* **334**, 1185–1187.
17. Brenner, M. (1999) Gene transfer by adenovectors. *Blood* **94**, 3965–3967.
18. Nadeau, I. and Kamen, A. (2003) Production of adenovirus vector for

gene therapy. *Biotechnol. Adv.* **20**, 475–489.
19. Lusky, M. (2005) Good manufacturing practice production of adenoviral vectors for clinical trials. *Hum. Gene Ther.* **16**, 281–291.
20. Lozier, J.N., Metzger, M.E., Donahue, R.E., and Morgan, R.A. (1999) Adenovirus-mediated expression of human coagulation factor IX in the rhesus macaque is associated with dose-limiting toxicity. *Blood* **94**, 3968–3975.
21. Knorr, D. (1999) *Serious Event on NIH Human Gene Transfer Protocol 9512-139. A Phase I Study of Adenovector-Mediated Gene Transfer to Liver in Adults With Partial Ornithine Transcarbamylase Deficiency.* Bethesda, MD: Memorandum, National Institutes of Health, Office of Recombinant DNA Activities.
22. St George, J.A. (2003) Gene therapy progress and prospects: adenoviral vectors. *Gene Ther.* **10**, 1135–1141.
23. Tomanin, R. and Scarpa, M. (2004) Why do we need new gene therapy viral vectors? Characteristics, limitations and future perspectives of viral vector transduction. *Curr. Gene Ther.* **4**, 357–372.
24. Friedenstein, A.J., Petrakova, K.V., Kurolesova, A.I., and Frolova, G.P. (1968) Heterotopic of bone marrow. Analysis of precursor cells for osteogenic and hematopoietic tissues. *Transplantation* **6**, 230–247.
25. Prockop, D.J. (1997) Marrow stromal cells as stem cells for nonhematopoietic tissues. *Science* **276**, 71–74.
26. Ferrari, G., Cusella-De Angelis, G., Coletta, M., Paolucci, E., Stornaiuolo, A., Cossu, G., and Mavilio, F. (1998) Muscle regeneration by bone marrow-derived myogenic progenitors. *Science* **279**, 1528–1530.
27. Pittenger, M.F., Mackay, A.M., Beck, S.C., Jaiswal, R.K., Douglas, R., Mosca, J.D., Moorman, M.A., Simonetti, D.W., Craig, S., and Marshak, D.R. (1999) Multilineage potential of adult human mesenchymal stem cells. *Science* **284**, 143–147.
28. Deng, W., Obrocka, M., Fischer, I., and Prockop, D.J. (2001) In vitro differentiation of human marrow stromal cells into early progenitors of neural cells by conditions that increase intracellular cyclic AMP. *Biochem. Biophys. Res. Commun.* **282**, 148–152.
29. Grove, J.E., Bruscia, E., and Krause, D.S. (2004) Plasticity of bone marrow-derived stem cells. *Stem Cells* **22**, 487–500.
30. Cherington, V., Chiang, G.G., McGrath, C.A., Gaffney, A., Galanopoulos, T., Merrill, W., Bizinkauskas, C.B., Hansen, M., Sobolewski, J., Levine, P.H., Greenberger, J.S., and Hurwitz, D.R. (1998) Retroviral vector-modified bone marrow stromal cells secrete biologically active factor IX in vitro and transiently deliver therapeutic levels of human factor IX to the plasma of dogs after reinfusion. *Hum. Gene Ther.* **9**, 1397–1407.
31. Bianco, P., Riminucci, M., Gronthos, S., and Robey, P.G. (2001) Bone marrow stromal stem cells: nature, biology, and potential applications. *Stem Cells* **19**, 180–192.
32. Toma, C., Pittenger, M.F., Cahill, K.S., Byrne, B.J., and Kessler, P.D. (2002) Human mesenchymal stem cells differentiate to a cardiomyocyte phenotype in the adult murine heart. *Circulation* **105**, 93–98.
33. Deng, W., Bivalacqua, T.J., Chattergoon, N.N., Hyman, A.L., Jeter, J.R., Jr., and Kadowitz, P.J. (2003) Adenoviral gene transfer of endothelial nitric oxide synthase: High level expression in ex vivo expanded marrow stromal cells. *Am. J. Physiol. Cell Physiol.* **285**, C1322–C1329.
34. Deng, W., Bivalacqua, T.J., Chattergoon, N.N., Jeter, J.R., Jr., and Kadowitz, P.J. (2004) Engineering ex vivo-expanded marrow stromal cells to secrete calcitonin gene-related peptide using adenoviral vector. *Stem Cells* **22**, 1279–1291.
35. Kassem, M. (2004) Mesenchymal stem cells: biological characteristics and potential clinical applications. *Cloning Stem Cells* **6**, 369–374.
36. Champion, H.C., Bivalacqua, T.J., D'Souza, F.M., Ortiz, L.A., Jeter, J.R., Toyoda, K., Heistad, D.D., Hyman, A.L., and Kadowitz, P.J. (1999) Gene transfer of endothelial nitric oxide synthase to the lung of the mouse in vivo. Effect on agonist-induced and flow-mediated vascular responses. *Circ. Res.* **84**, 1422–1432.
37. Champion, H.C., Bivalacqua, T.J., Greenberg, S.S., Giles, T.D., Hyman, A.L., and Kadowitz, P.J. (2002) Adenoviral gene transfer of endothelial nitric-oxide synthase (eNOS) partially restores normal pulmonary arterial pressure in eNOS-deficient mice. *Proc. Natl. Acad. Sci. USA* **99**, 13248–13253.
38. Ortiz, L.A., Champion, H.C., Lasky, J.A., Gambelli, F., Gozal, E., Hoyle, G.W., Beasley, M.B., Hyman, A.L., Friedman, M., and Kadowitz, P.J. (2002) Enalapril protects mice from pulmonary hypertension by inhibiting TNF-mediated activation of NF-kappaB and AP-1. *Am. J. Physiol. Lung Cell Mol. Physiol.* **282**, L1209–L1221.
39. Nishida, K., Harrison, D.G., Navas, J.P., Fisher, A.A., Dockery, S.P., Uematsu, M.,

Nerem, R.M., Alexander, R.W., and Murphy, T.J. (1992) Molecular cloning and characterization of the constitutive bovine aortic endothelial cell nitric oxide synthase. *J. Clin. Invest.* **90**, 2092–2096.

40. Ooboshi, H., Chu, Y., Rios, C.D., Faraci, F.M., Davidson, B.L., and Heistad, D.D. (1997) Altered vascular function after adenovirus-mediated overexpression of endothelial nitric oxide synthase. *Am. J. Physiol.* **273**, H265–270.

41. Davidson, B.L., Allen, E.D., Kozarsky, K.F., Wilson, J.M., and Roessler, B.J. (1993) A model system for in vivo gene transfer into the central nervous system using an adenoviral vector. *Nat. Genet.* **3**, 219–223.

42. Hyman, A.L., Hao, Q., Tower, A., Kadowitz, P.J., Champion, H.C., Gumusel, B., and Lippton, H. (1998) Novel catheterization technique for the in vivo measurement of pulmonary vascular responses in rats. *Am. J. Physiol.* **274**, H1218–H1229.

43. Baber, S.R., Champion, H.C., Bivalacqua, T.J., Hyman, A.L., and Kadowitz, P.J. (2003) Role of cyclooxygenase-2 in the generation of vasoactive prostanoids in the rat pulmonary and systemic vascular beds. *Circulation* **108**, 896–901.

44. Shizuru, J.A., Negrin, R.S., and Weissman, I.L. (2005) Hematopoietic stem and progenitor cells: clinical and preclinical regeneration of the hematolymphoid system. *Annu. Rev. Med.* **56**, 509–538.

Section IV

DNA Oxidation, Oxidative Lipidomics, and Biomarkers

Chapter 19

A General Method for Quantifying Sequence Effects on Nucleobase Oxidation in DNA

Yelena Margolin and Peter C. Dedon

Abstract

Oxidative damage to DNA has long been associated with aging and disease, with guanine serving as the primary target for oxidation owing to its low ionization potential. Emerging evidence points to a critical role for sequence context as a determinant of the guanine ionization potential and the associated chemical reactivity of the guanine, as well as the spectrum of damage products that arise from oxidation. Recent studies also suggest that the generally accepted model of oxidation hotspots in runs of guanine bases may not hold for biologically relevant oxidants. One of the primary methods used to address these important problems of sequence context utilizes gel electrophoresis to identify the location and quantity of base damage arising in model oligonucleotides. However, this approach has limited study to those agents that produce few strand breaks arising from deoxyribose oxidation, while ionizing radiation, Fenton chemistry and other biologically relevant oxidants produce sizeable proportions of both base and sugar damage. To this end, we have developed a universal method to quantify sequence context effects on nucleobase damage without interference by strand breaks from deoxyribose oxidation.

Key words: Oxidative DNA damage, guanine oxidation, sequence selectivity, sequence context, nitrosoperoxycarbonate, γ-radiation, hydroxyl radical, strand breaks, exonuclease III.

1. Introduction

DNA damage resulting from oxidative stress has been strongly associated with cancer, chronic degenerative diseases, and aging (reviewed in (1 and 2)). While both the nucleobase and the deoxyribose moieties of DNA are targets for oxidation, recent interest in charge transfer and sequence context effects on the location and quantity of damage have focused attention on the bases, with particular attention paid to guanine due to its low ionization potential (3) and the myriad products arising from

R.M. Uppu et al. (eds.), *Free Radicals and Antioxidant Protocols*, Methods in Molecular Biology 610,
DOI 10.1007/978-1-60327-029-8_19, © Humana Press, a part of Springer Science+Business Media, LLC 1998, 2010

its primary and secondary oxidation (4). Sequence context has been shown to play a significant role in modulating the ionization potential of guanine in duplex DNA and, hence, the reactivity of guanine with oxidizing agents. For example, it has been demonstrated that many one-electron oxidants, such as anthraquinones (5), rhodium complexes (6), and riboflavin-mediated photooxidation, selectively damage guanine when the base is located adjacent to other guanines (e.g., GG, GGG). This reactivity has been rationalized on the basis of the low ionization potential conferred to guanine in these sequence contexts and the migration of cationic holes to these sites from guanine radical cations located in sequence contexts conferring higher ionization potentials (7). On the other hand, we have recently demonstrated that nitrosoperoxycarbonate, an oxidant formed by reactive oxygen species during chronic inflammation, is selective for guanines with the highest ionization potentials (8), while hydroxyl radical generated by Fe^{2+}–EDTA and γ-radiation is equally reactive with guanines irrespective of sequence context (Margolin et al., 2008, (9)). We have, therefore, shown that sequence selectivity of guanine oxidation in double-stranded DNA is not only a function of sequence context, as has been previously thought, but also depends on the oxidant identity and its interactions with the DNA. Determination of sequence effects in nucleobase oxidation by various agents can thus provide valuable information on their mechanism of damage induction in DNA and on the relationship between reactivity and the potential to cause mutations (10).

The most widely employed approach to studying sequence context effects on DNA damage involves gel electrophoretic analysis of damage in model oligodeoxynucleotides exposed to oxidizing agents. Base damage in the oligos is converted to strand breaks by treating the DNA with either hot piperidine or DNA repair enzymes such as *Escherichia coli* formamidopyrimidine DNA glycosylase (Fpg) for oxidized purines and *E. coli* endonuclease IV (Nth) for oxidized pyrimidines (11). The strand breaks are then localized on sequencing gels and quantified by autoradiography or phosphorimager analysis. The problem inherent with this approach is that it is limited to oxidizing agents that produce only base damage, but not 2-deoxyribose damage. Oxidation of deoxyribose results in the formation of direct strand breaks and easily hydrolysable abasic sites, creating a background of strand breaks that can interfere with quantification of base-derived strand breaks. Such is the case with the biologically relevant oxidizing agent such as ionizing radiation, peroxynitrite, and Fenton chemistry arising with iron and copper (12).

We have developed a method that obviates the background of deoxyribose oxidation-induced strand breaks. Using relatively inexpensive 3′-phosphorothioate-protected oligodeoxynucleotides, the background of strand breaks is removed from the analysis by digestion of the oxidized oligos with *E. coli* exonuclease III (*Exo*III). Subsequent treatment with hot piperidine or DNA repair enzymes exposes the base damage as strand breaks that can be localized and quantified in sequencing gels. This approach provides a nearly universal method for defining the sequence context effects on oxidative damage to DNA.

2. Materials

2.1. Gel Electrophoresis and Purification of Oligodeoxynucleotides

1. Oligodeoxynucleotides for analysis, as well as their complements, can be ordered from Integrated DNA Technologies (www.idt.com) or any company specializing in custom oligodeoxynucleotide synthesis. As an example, we and others have used the following model oligodeoxynucleotide for studies of sequence context effects on guanine oxidation: 5′-CGTACTCTTTGGTXGYTXGYTTCTTCTAT-3′ (7, 8). This sequence contains consensus portions on both 5′ and 3′ ends, as well as a TGG sequence that is placed at the same position in all oligodeoxynucleotides and acts as a normalization standard. Damage at guanines in the variable sequences XGY (where X and Y are thymine, cytosine, guanine, or adenine) is always normalized to the damage at the central guanine of the TGG sequence (*see* **Notes 1** and **2**).
2. For experiments with agents that induce high levels of direct strand breaks (i.e., deoxyribose oxidation), the oligodeoxynucleotides should contain three consecutive phosphorothioate linkages at their 3′ ends (also available from Integrated DNA Technologies): 5′-CGTACTCTTTGGTXGYTXGYTTCTTC-S-T-S-A-S-T-3′, where S is a phosphorothioate linkage. The complements to these oligodeoxynucleotides should also contain three phosphorothioate linkages at their 3′ ends (*see* **Note 3**).
3. TBE buffer (10×): 0.89 M Tris base, 0.89 M boric acid, 0.02 M EDTA (disodium salt), pH 8.3. Store at ambient temperature.
4. Acrylamide to bis-acrylamide (19:1; 40% w/v) solution (American Bioanalytical). Store at 4°C. Acrylamide is a neurotoxin in unpolymerized form and should be handled with care.
5. *N,N,N,N′*-Tetramethylethylenediamine (TEMED; Sigma-Aldrich).

6. Ammonium persulfate, 10% (w/v) aqueous solution, prepared directly before use.
7. Gel electrophoresis system: Model S2 Sequencing Gel Electrophoresis Apparatus (Lab Repco).
8. Power Pac 3000 power supply with a temperature probe (Bio-Rad).
9. Elution buffer: 0.5 M ammonium acetate, 10 mM magnesium acetate.
10. Ultrafree-MC Centrifugal Filter Devices (Millipore).
11. Sodium acetate, 3.0 M, pH 5.2.
12. Ethanol, 100% (200% proof).
13. Ethanol, 70% (v/v).

2.2. Labeling of Oligodeoxynucleotides

1. γ-[^{32}P]-ATP, 10 mCi/mL, 6,000 Ci/mmol (Perkin Elmer).
2. T4 polynucleotide kinase (PNK) and 10× PNK reaction buffer (New England Biolabs).
3. Sephadex G-25 spin columns (Roche).

2.3. Damage Reactions of Oligodeoxynucleotides and Preparation of Samples for Gel Analysis

1. Piperidine solution, 2 M, in distilled and deionized water.
2. *E. coli* Fpg glycosylase and 10× reaction buffer (New England Biolabs or Trevigen).
3. Glycogen (Roche).
4. Formamide gel loading buffer: 95% (v/v) formamide, 20 mM EDTA, pH 8.0, 0.05% (w/v) bromophenol blue, and 0.05% (w/v) xylene cyanol.

2.4. Sequencing Gel Analysis and Autoradiography

1. *See* **Section 2.1.**
2. Imaging Screen K (Bio-Rad).
3. Phosphorimager Storm 820 model (GE).

2.5. Image Analysis

ImageQuant software (GE)

2.6. Removal of Direct Strand Breaks from Damaged Oligodeoxynucleotides

1. *E. coli Exo*III and 10× exonuclease III reaction buffer (NEbuffer-1) (New England Biolabs).
2. Sephadex G-25 spin columns (Roche).

3. Methods

This method for studying the sequence effects on nucleobase oxidation uses small, synthetic 5′-^{32}P-labeled, double-stranded oligodeoxynucleotides containing guanines in defined sequence contexts. After treatment with a damaging agent, a strand break is introduced at the sites of guanine oxidation by treatment with either hot piperidine or Fpg glycosylase (reviewed in (10)). The relative instability of most of the primary and secondary guanine oxidation products to treatment by either one or both of these agents ensures the complete conversion of most guanine oxidation products to strand breaks. The treated oligodeoxynucleotides are then resolved on a DNA sequencing gel and the strand breaks formed at each oxidized guanine are quantified using standard image analysis software.

For analysis of guanine oxidation by an agent that produces significant amounts of deoxyribose oxidation, a protocol modification is introduced that allows the removal of the background of direct strand breaks. This is accomplished by using preparing double-stranded oligodeoxynucleotides that contain exonuclease-resistant phosphorothioate linkages at their 3′ ends and treating these oligodeoxynucleotides with *Exo*III after the damage reaction is complete (*see* **Note 3**). Phosphorothioate linkages protect the 3′ ends of the parent oligodeoxynucleotides and oligodeoxynucleotides containing only base lesions from digestion by *Exo*III (13). Oligodeoxynucleotides containing strand breaks now have exposed 3′ ends that are substrates for the enzyme. *Exo*III recognizes substrates with 3′-hydroxyl, 3′-phosphate, and 3′-phosphoglycolate termini (14), as well as substrates containing abasic sites that are cleaved endonucleolytically (15) and thus removed from base damage analysis. After the *Exo*III reaction is complete, the only [^{32}P]-labeled oligodeoxynucleotides remaining in solution are the parent molecules or those containing damaged bases that can be revealed as strand breaks following reaction with hot piperidine or Fpg treatment and gel electrophoresis (*see* **Fig. 19.1**). Our control experiments have shown that *Exo*III treatment does not alter the sequence selectivity of guanine oxidation observed in sequence damage experiments (i.e., the presence or absence of *Exo*III does not affect the quantity and location of base damage, as shown in **Fig. 19.2** for the guanine-specific oxidant, nitrosoperoxycarbonate).

The first part of this section describes the steps necessary to determine sequence selectivity of guanine oxidation by agents that selectively oxidize guanines in duplex DNA. The second part describes the modifications of the method that provide for removal of the background of direct strand breaks induced by agents capable of significant deoxyribose oxidation. When working with a new damaging agent it is necessary to measure the amount of direct strand breaks that it causes, in order to determine if the *Exo*III treatment described in the second part of the method should be used. This can often be accomplished with a single dose–response experiment with and without hot piperidine treatment. The same dose–response experiment should be

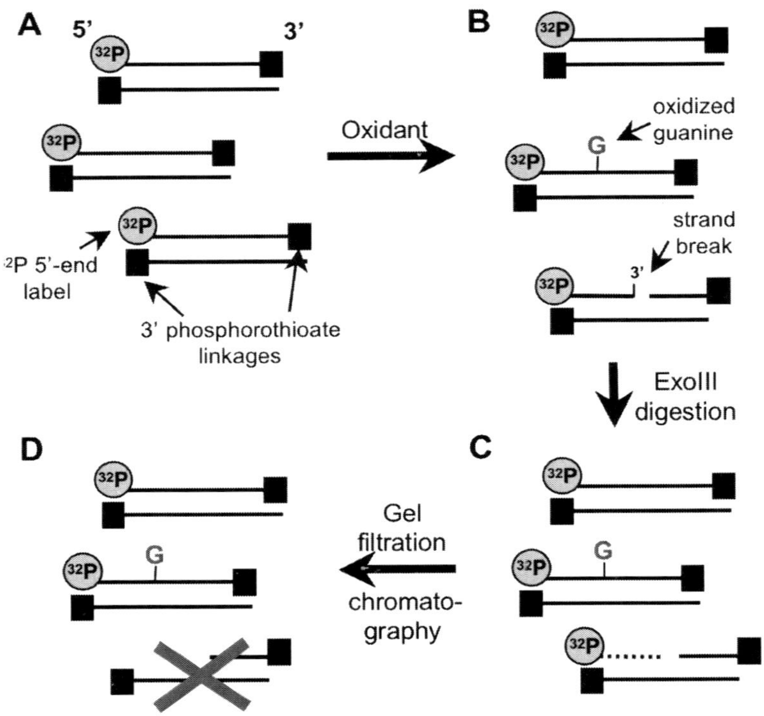

Fig. 19.1. Schematic representation of the *Exo* III method for removing direct strand breaks from analyses of sequence context effects on base damage. (**a**) Oligodeoxynucleotides contain a [^{32}P] label at their 5′ ends and three consecutive phosphorothioate linkages at their 3′ ends. (**b**) Treatment with an agent that oxidizes deoxyribose produces direct strand breaks that possess unprotected 3′ ends. (**c**) *Exo* III recognizes unprotected 3′ ends and hydrolyzes the oligodeoxynucleotide in a 3′-to-5′ direction, releasing the 5′ ^{32}P label. (**d**) After gel filtration chromatography, the oligodeoxynucleotide that contained the direct strand break has lost its 5′ label and is not be detected during subsequent electrophoresis and autoradiography.

conducted to determine a dose of the damaging agent that will be used in all subsequent experiments. A dose that is typically chosen should be high enough to induce statistically significant damage at every guanine of interest, as measured by a paired Student's *t* test. However, it should be low enough to damage less than 30% of the parent oligodeoxynucleotide. According to a Poisson distribution, this low level of damage ensures that each oligodeoxynucleotide sustains an average of one or fewer damage reactions. If these single-hit conditions are violated, sequencing gel quantification of DNA damage becomes impossible due to the inability to quantify a second damage event.

3.1. Purification of Oligodeoxynucleotides by Gel Electrophoresis

All synthetic oligodeoxynucleotides should be purified before use in sequence damage experiments to remove failure sequences and damaged molecules. Gel electrophoresis is the most efficient and reliable method for oligodeoxynucleotide purification

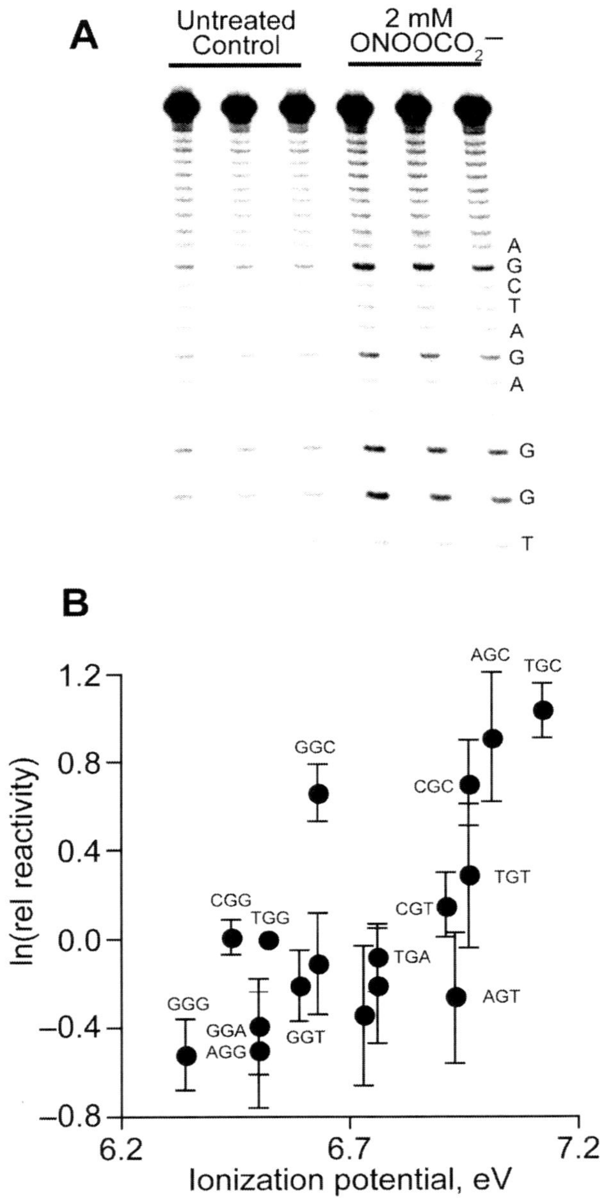

Fig. 19.2. (**a**) A typical autoradiogram of a sequencing gel obtained in an experiment with nitrosoperoxycarbonate, a selective oxidant of guanines (8). An oligonucleotide with the sequence 5′-CGTACTCTTTGGTAGATAGCTTCTTCTAT-3′ was damaged with 0 and 2 mM nitrosoperoxycarbonate and treated with hot piperidine to convert base lesions to strand breaks. The resulting damage products were separated on a sequencing gel. (**b**) Plot of the relative amounts of piperidine-sensitive guanine lesions in different sequence contexts produced by nitrosoperoxycarbonate as a function of sequence-specific guanine ionization potential (7, 8).

(*see* **Note 4**). Due to the high frequency of nucleobase damage occurring in synthesis, all oligodeoxynucleotides should be treated with hot piperidine prior to purification (*see* **Note 5**).

3.1.1. Treatment of Oligodeoxynucleotides by Hot Piperidine

1. Oligodeoxynucleotides are dissolved in TE buffer (10 mM Tris, 0.5 mM EDTA, pH 8.0) to a final concentration of 100–500 pmol/μL.
2. An equal volume of 2.0 M piperidine solution in water is added and the oligodeoxynucleotides are incubated at 90°C for 20 min.
3. Following drying under vacuum (e.g., Speedvac), the samples are again dissolved in TE buffer containing bromophenol blue dye and 20–25% glycerol to a DNA concentration of 100–500 pmol/μL.

3.1.2. Preparative Gel Electrophoresis

1. These instructions assume the use of the S2 gel sequencing apparatus from Gibco/BRL, sold by Lab Repco. It is essential that the glass plates are cleaned in 1 M NaOH followed by thorough rinsing with deionized water and wiping with a small amount of acetone or ethyl acetate to dissolve residual contaminants. Sigmacote is applied to the internal surface of one of the plates before assembly to promote sticking of the gel to only one plate. Large binding clips can be used to hold the gasket in place.
2. Prepare 20% acrylamide, 8.3 M urea gel in 10× TBE by mixing 130 g of urea with 26 mL of 10× TBE buffer, 130 mL of 40% 19:1 acrylamide:bis-acrylamide solution, and water to a total volume of 260 mL. Gently heat the mixture with stirring to dissolve all of urea, followed by filtration through a membrane with 0.45 μ pores to remove any particulate. To this solution, add 600 μL of 10% ammonium persulfate solution and 60 μL of TEMED. Immediately load ~180 mL of this mixture into the assembled plates using a 60 mL syringe to prepare a 1.2 mL thick, 20% acrylamide gel in 1× TBE; tap to remove any bubbles that form between the plates. Insert a comb with 10 mm × 5 mm teeth and secure it with binding clips. The gel should polymerize in ~30 min.
3. Pre-run the gel with 1× TBE running buffer at 70 W until the gel reaches about 39°C (use any type of surface temperature probe).
4. Before the samples are loaded on the gel, the wells should be extensively washed with 1× TBE running buffer using a syringe with a small gauge needle to remove particulate and unpolymerized acrylamide.
5. Load no more than 40 nmol (400 μg) of oligodeoxynucleotide into each lane and run the gel for ~6 h with

1× TBE running buffer at ~70 W to produce a temperature of ~45°C.

3.1.3. Elution of Oligodeoxynucleotides from the Gel

1. Separate the glass plates and wrap the plate attached to the gel in Saran wrap or other thin, clear plastic wrap. Visualize oligodeoxynucleotides in the gel by placing the plate on a white sheet of paper in a darkroom and illuminating it from above with a short-ray UV lamp. Mark the DNA bands on the Saran wrap with a black, fine-tipped marker, then use a razor blade to cut out only the upper half of each band; the lower half may contain poorly resolved failure sequences.
2. Place the gel slices into 1.5 mL capacity tubes and macerate the pieces by repeated rolling of a pipette tip along the tube wall. Add elution buffer to the top of the tube and leave the tubes vortexing overnight at the low setting in a cold room at a temperature of 4°C.
3. Remove the gel particulate using Ultrafree-MC centrifugal filter devices (Millipore). Distribute the filtrate to 1.5 mL tubes in 300 μL aliquots. To each tube add 100 μL of 3.0 M sodium acetate solution and 1 mL of 100% ethanol. Incubate the tubes at –80°C for 1 h and pellet the precipitated DNA at 16,000×*g* for 30 min at 4°C. Carefully wash the DNA pellets several times with 70% cold ethanol and air-dry them. Dissolve the dried oligodeoxynucleotides in the buffer that will be used to carry out damage reactions.

3.2. Preparation of [^{32}P]-Labeled, Duplex Oligodeoxynucleotides

1. Oligodeoxynucleotides with sequences of interest are 5′ end labeled with ^{32}P by transferring a phosphate group from γ-^{32}P-ATP to the oligodeoxynucleotide using T4 polynucleotide kinase (PNK). In a single tube, mix approximately 100 pmol of oligodeoxynucleotide, 20 units of T4 PNK, 5 μL of 10× T4 PNK buffer, 5 μL of γ-^{32}P-ATP, and deionized water to a final volume of 50 μL. Incubate at 37°C for 1 h. Remove unreacted ATP by passing the reaction mixture over a G-25 Sephadex column that has been washed with the buffer in which damage reactions will be carried out (*see* **Note 6**).
2. For annealing of complementary oligodeoxynucleotides, add to the mixture approximately 200 pmol of complementary strand. Incubate the mixture in a heating block at 90–95°C for about 1 min, then turn the heating block off and allow it to cool to 37°C (this takes ~90 min). The resulting solution will contain ~100 pmol of 5′-[^{32}P]-labeled, double-stranded oligodeoxynucleotide, as well as ~100 pmol of unlabeled, single-stranded complementary oligodeoxynucleotide.

3.3. Damage Reactions and Sample Preparation for Sequencing Gel Analysis

1. When working with a new damaging agent, a dose–response study should be performed with several different concentrations of the damaging agent to determine: (a) the optimal dose of the damaging agent to be used in sequence-specific damaging reactions and (b) the level of direct strand breaks produced by the damaging agent. In cases where an agent produces high levels of direct strand breaks (>10% of the total strand breaks revealed after hot piperidine or Fpg treatment) that can interfere with the quantification of damage induced at guanines, *Exo*III should be used (*see* **Section 4**) (*see* **Note 6**).
2. Each damage reaction is performed in triplicate, with triplicate control samples to which only the vehicle is added. Note that, except for the absence of the damaging agent, the control tubes are treated exactly the same way as the damage reactions. Proceed with the damage reactions according to the protocol. Each reaction should contain 5–15 μL of labeled oligodeoxynucleotide for sufficiently strong signal.
3. After the damage reaction is complete, the damaging agent is removed by passing each sample over a G-25 spin column (*see* **Note 7**).
4. Sites of base damage are now converted to strand breaks by treatment either with hot piperidine or with Fpg (10):
   (a) For hot piperidine treatment, aliquot a defined volume of the damaged oligodeoxynucleotide into a separate tube and add equal volume of 2.0 M piperidine solution. Incubate the tubes at 90°C for 20 min, then dry the reaction under vacuum (e.g., Speedvac; *see* **Note 9**). Add 5 μL of formamide gel loading buffer to each tube.
   (b) For Fpg treatment, add the appropriate volume of 10× Fpg buffer and 1–3 units of Fpg to a defined volume of damaged oligodeoxynucleotide (*see* **Note 10**). Incubate at 37°C for 1 h. To precipitate the DNA after the reaction, add 3 M NaoAC (pH 5.2) equal to one-half of the reaction volume, 1 μL of glycogen, and 3.5 volume of 100% ethanol. Incubate at –80°C for 1 h, then pellet the DNA by centrifugation at 16,000×*g* for 30 min at 4°C. Air-dry the pellets and dissolve them in 5 μL of formamide gel loading buffer.

3.4. Separation of Damage Reaction Products on DNA Sequencing Gel and Autoradiography

1. Prepare 0.4 mm thick sequencing gels containing 8.3 M urea and 20% acrylamide sequencing gel in 1× TBE as described in steps 1 and 2 of **Section 3.1.2**, using the 0.4 mm spacers. The sequencing gel is only 0.4 mm thick (compared to 1.2 mm thick preparative gel), so prepare only one-half of

the total gel solution volume described in step 2 of **Section 3.1.2** (i.e., 130 mL).

2. Pre-run the gel at 70 W in 1× TBE until temperature reaches ~39°C.
3. Wash the wells of the gel with 1× TBE as described in step 4 of **Section 3.1.2**.
4. Load 3 μl of each sample in each well and run the gel at ~70 W to achieve a temperature of ~45°C for 3 h (*see* **Note 11**).
5. Separate the plates, taking care that the gel adheres to only one of the plates. Wrap the plate with the gel with Saran wrap. Position the glass plate with the gel facing upward and place the imaging screen on top of the gel. Expose for 2–12 h as needed to provide a strong exposure when damage bands are weak (*see* **Note 12**).
6. Scan the imaging plate according to manufacturer's instructions to obtain a digital image of the gel. An example of a typical result is shown in **Fig. 19.2a**.

3.5. Quantification of Damage at Each Guanine Using Phosphorimager Software

1. Locate the position of each guanine of interest within each lane, using sequencing standards as needed (*see* **Note 11**).
2. Following user directions for the software, determine the amount of radioactivity corresponding to each band as a percentage of total radioactivity in the entire lane (including the band corresponding to intact parent oligodeoxynucleotide).
3. Calculate the average percentage of total radioactivity present in each of the three bands (derived from the TGG and two XGY sites) for the three control and three oxidant-treated samples.
4. Subtract the average values for the controls from the average values for the oxidized samples to obtain the net signal due to guanine oxidation (*see* **Note 13**).
5. Normalize the two signals for guanines in the XGY sequence contexts by dividing their signal values by that for the TGG normalization sequence. An example of a typical result showing the dependence of guanine oxidation by nitrosoperoxycarbonate as a function of guanine ionization potential is shown in **Fig. 19.2b**.

3.6. Removing a Background of Direct Strand Breaks

1. This section describes modifications to the procedure described in **Section 3.2**, in which the background of direct strand breaks produced by some oxidizing agents is removed prior to analysis of guanine oxidation. This is accomplished using oligodeoxynucleotides containing phosphorothioate linkages at their 3′ ends and *Exo*III to remove

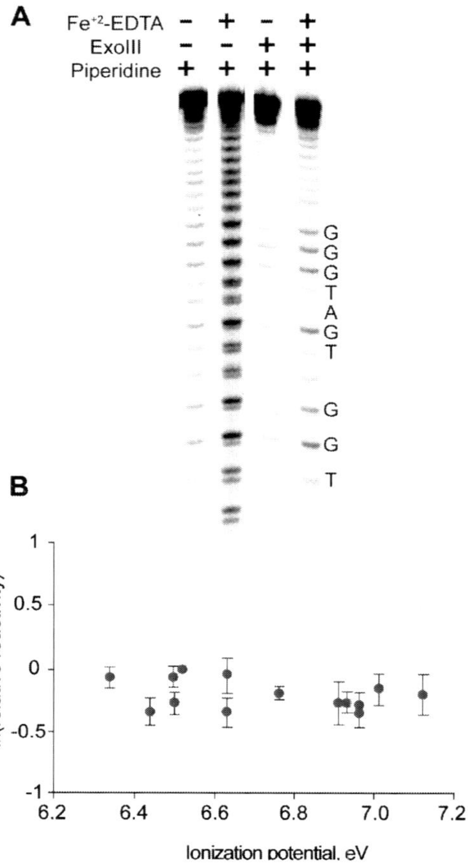

Fig. 19.3. (**a**) Illustration of the *Exo* III method for removing direct strand breaks. The oligodeoxynucleotides employed in **Fig. 19.2** were damaged with Fe^{2+}–EDTA, an oxidant that produces a high proportion of deoxyribose oxidation and thus direct strand breaks. These direct breaks were removed by treatment with *Exo* III, which leaves the intact oligodeoxynucleotides containing guanine base lesions ready for analysis by hot piperidine cleavage. The faster and slower migrating bands in the doublets apparent in lane 2 represent 3′-phosphoglycolate-ended and 3′-phosphate-ended DNA fragments, respectively. Both of these "direct strand break" products are completely removed by *Exo* III, as is apparent in lane 4. (**b**) Plot of the relative amounts of piperidine-sensitive guanine lesions in different sequence contexts produced by Fe^{2+}–EDTA as a function of sequence-specific guanine ionization potential (7).

oligodeoxynucleotide fragments that contain 3′ ends as a result of deoxyribose oxidation. Hot piperidine or Fpg treatment is subsequently used to introduce strand breaks at the sites of guanine oxidation.

2. Follow steps 3.1.1 through 3.2.2, part 3 (*see* **Note 14**).
3. *Exo*III treatment is achieved by adding an appropriate volume of 10× NEbuffer-1 and 1–5 units of *Exo*III to a defined volume of damaged oligodeoxynucleotide. Incubate

the tubes at 37°C for 1 h (shorter incubation times may be sufficient). De-salt each reaction by passing it over a G-25 spin column (*see* **Note 7**).

4. To each of the G-25 eluents, add an equal volume of 2.0 M piperidine solution and incubate at 90°C for 20 min. Dry completely under vacuum (e.g., Speedvac; *see* **Note 15**) and add 5 μL of formamide gel loading buffer to each tube.
5. Proceed with steps described in **Sections 3.2.1** (*see* **Note 16**) and **3.2.2**. An example of the typical result of using *Exo*III to remove the background of direct strand breaks is shown in **Fig. 19.3a**. Relative reactivities of guanines in different sequence contexts with a hydroxyl radical formed by $(Fe\text{-}EDTA)^{2-}$ treatment is shown in **Fig. 19.3b**.

4. Notes

1. It is imperative to have a common normalization sequence in each oligodeoxynucleotide to act as an internal control, as lane-to-lane variations in signal intensity make rigorous quantification impossible without the internal control.
2. It is possible to design longer oligodeoxynucleotides that contain more than two sequence contexts in addition to the internal control. Care should be taken to avoid using shorter oligodeoxynucleotides with low melting temperatures, as they may undergo partial denaturation during incubations at 37°C.
3. The presence of three phosphorothioate linkages completely inhibits the 3′-to-5′ exonuclease activity of the *Exo*III used to remove direct strand breaks induced by the oxidation agents.
4. The oligodeoxynucleotides are received in solid form and contain truncated (i.e., failure) sequences that may interfere with quantification of damage on sequencing gels. It is, therefore, imperative to purify full-length oligodeoxynucleotides away from the truncated sequences. Gel electrophoresis is the most efficient and effective method for oligodeoxynucleotide purification.
5. Synthetic oligodeoxynucleotides also contain detectable amounts of oxidized nucleobases, most notably guanine. This background of base damage may pose serious problems for the analyses by reducing the dynamic range for quantification of guanine oxidation. To remove this background of base oxidation, the oligodeoxynucleotides can be treated with hot piperidine prior to their purification

by gel electrophoresis. If using oligodeoxynucleotides containing various modifications, such as biotinylation, unnatural bases, or phosphorothioate linkages, it is often advisable to keep in mind that these modifications may be affected by the hot piperidine treatment.

6. Usually, washing the column four times with 300 μL of buffer is sufficient for complete buffer exchange.
7. The spin column can be washed with deionized water (2 × 300 μL).
8. It is preferable to perform the [^{32}P] labeling of the oligodeoxynucleotides and the damage reactions on the same day, as it is our experience that storing [^{32}P]-labeled oligonucleotides overnight can result in higher damage background.
9. Since piperidine is a volatile and toxic chemical, incubations should be carried out in an appropriate fume hood in tubes with screw-top caps.
10. Careful attention should be paid to a manufacturer's definition of unit values and concentrations for all commercial enzyme preparations. This is particularly important for Fpg, since unit definitions and concentrations often differ for the various manufacturers.
11. Though the position of each guanine within the oligodeoxynucleotides is known and can usually be determined from the relative migration of the DNA fragments, the identity of the cleavage sites can be verified using Maxam-Gilbert sequencing standards (16) or [^{32}P]-labeled synthetic oligodeoxynucleotides with lengths corresponding to the oligodeoxynucleotides arising from guanine oxidations at the three sites.
12. The time needed for an appropriate exposure of the phosphorimager plate depends on the strength of the signal to be quantified and on the condition of the phosphorimaging plate.
13. The statistical significance of differences between control and oxidized samples can be determined using a paired Student's t test.
14. Depending on the proportion of deoxyribose oxidation-induced strand breaks caused by an oxidant, the radioactive signal for base oxidation will be diminished following *Exo*III digestion. This may necessitate starting with larger amounts of [^{32}P]-labeled oligodeoxynucleotides to produce base damage signals sufficient for quantification. Typically, we use two-fold more of the radioactive oligodeoxynucleotides when analyzing hydroxyl

radical-induced guanine oxidation, as compared to the starting amounts in the experiments with agents producing a higher proportion of base damage (e.g., nitrosoperoxycarbonate).

15. The *Exo*III must be de-activated and/or completely removed before converting guanine base damage to strand breaks with Fpg, since the newly exposed 3′ ends will be susceptible to digestion by residual *Exo*III. Hot piperidine treatment denatures and inactivates *Exo*III, so no additional steps are necessary in the procedure.

16. *Exo*III-digested samples will contain very short, [^{32}P]-labeled oligodeoxynucleotide fragments that will migrate faster than the fragments arising from guanine oxidation and thus will not interfere with the intended analysis. However, the short fragments may run off the gel into the running buffer, so care should be taken in handling and discarding the contaminated running buffer after the experiment. In addition, the gel should be wrapped with at least two layers of Saran or similar plastic wrap to avoid contaminating the phosphorimaging screen with radioactivity from the short fragments that readily elute from the gel during exposure or from residual contaminated running buffer.

References

1. De Flora, S., Izzotti, A., Randerath, K., Randerath, E., Bartsch, H., Nair, J., Balansky, R., van Schooten, F., Degan, P., Fronza, G., Walsh, D., and Lewtas, J. (1996) DNA adducts and chronic degenerative disease. Pathogenetic relevance and implications in preventive medicine *Mutat. Res.* **366**, 197–238.
2. Migliore, L. and Coppede, F. (2002) Genetic and environmental factors in cancer and neurodegenerative diseases. *Mutat. Res.* **512**, 135–153.
3. Steenken, S. and Jovanovic, S.V. (1997) How easily oxidizable is DNA? One-electron reduction potentials of adenosine and guanosine radicals in aqueous solution. *J. Am. Chem. Soc.* **119**, 617–618.
4. Neeley, W.L. and Essigmann, J.M. (2006) Mechanisms of formation, genotoxicity, and mutation of guanine oxidation products. *Chem. Res. Toxicol.* **19**, 491–505.
5. Henderson, P.T., Jones, D., Hampikian, G., Kan, Y., and Schuster, G.B. (1999) Long-distance charge transport in duplex DNA: the phonon-assisted polaron-like hopping mechanism. *Proc. Natl. Acad. Sci. USA* **96**, 8353–8358.
6. Hall, D.B., Holmlin, R.E., and Barton, J.K. (1996) Oxidative DNA damage through long-range electron transfer. *Nature* **383**, 731–735.
7. Saito, I., Nakamura, T., Nakatani, K., Yoshiyoka, Y., Yamaguchi, K., and Sugiama, H. (1998) Mapping of the hot spots for DNA damage by one-electron oxidation: efficacy of GG doublets and GGG triplets as a trap in long-range hole migration. *J. Am. Chem. Soc.* **120**, 12686–12687.
8. Margolin, Y., Cloutier, J.F., Shafirovich, V., Geacintov, N.E., and Dedon, P.C. (2006) Paradoxical hotspots for guanine oxidation by a chemical mediator of inflammation. *Nature Chem. Biol.* **2**, 365–366.
9. Margolin, Y., Shafirovich, V., Geacintov, N.E., DeMott, M.S., Dedon, P.C. (2008) DNA sequence context as a determinant of the quantity and chemistry of guanine oxidation produced by hydroxyl radicals and

one-electron oxidants. *J. Biol. Chem.* **283**, 35569–35578.

10. Tretyakova, N.Y., Burney, S., Pamir, B., Wishnok J.S., Dedon, P.C., Wogan, G.N., and Tannenbaum, S.R. (2000) Peroxynitrite-induced DNA damage in the supF gene: correlation with the mutational spectrum. *Mutat. Res.* **447**, 287–303.
11. Burrows, C.J. and Muller, J.G. (1998) Oxidative nucleobase modifications leading to strand scission. *Chem. Rev.* **98**, 1109–1151.
12. Sutherland, B.M., Bennett, P.V., Sidorkina, O., and Laval, J. (2000) Clustered damages and total lesions induced in DNA by ionizing radiation: oxidized bases and strand breaks. *Biochemistry* **39**, 8026–8031.
13. Putney S.D., Benkovic, S.J., and Schimmel P.R. (1981) A DNA fragment with an alpha-phosphorothioate nucleotide at one end is asymmetrically blocked from digestion by exconuclease III and can be replicated in vivo. *Proc. Natl. Acad. Sci. USA* **78**, 7350–7354.
14. Henner, W.D., Grunberg, S.M., and Haseltine, W.A. (1983) Enzyme action at 3' termini of ionizing radiation-induced DNA strand breaks. *J. Biol. Chem.* **258**, 15198–15205.
15. Takeuchi, M., Lillis, R., Demple, B., and Takeshita, M. (1994) Interactions of Escherichia coli Endonuclease IV and Exonuclease III with abasic sites in DNA. *J. Biol. Chem.* **269**, 21907–21914.
16. Maxam, A. and Gilbert, W. (1980) Sequencing end-labeled DNA with base-specific chemical cleavages. *Meth. Enzymol.* **65**, 499–560.

Chapter 20

Analysis of Urinary 8-oxo-7,8-dihydro-2′-deoxyguanosine by Liquid Chromatography–Tandem Mass Spectrometry

Mark D. Evans, Rajinder Singh, Vilas Mistry, Peter B. Farmer, and Marcus S. Cooke

Abstract

The ability to non-invasively monitor DNA oxidation and its repair has significant utility in large-scale, population-based studies. Such studies could include assessments of the efficacy of antioxidant intervention strategies, pathological roles of DNA oxidation in various disease states and population or inter-individual differences in antioxidant defence and DNA repair. The analysis of urine, or indeed any extracellular matrix, for 8-oxo-7,8-dihydro-2′-deoxyguanosine (8-oxodG), using chromatographic or immunoassay procedures, is by far the most popular method to non-invasively assess oxidative insult to the genome. The actual biological significance of the presence of extracellular 8-oxodG is still a subject for debate however. Studies are slowly ruling out confounding factors such as diet and cell turnover, which would leave endogenous processes, notably repair, as the sole source of extracellular 8-oxodG. The method described herein exploits the non-invasive properties of urine sampling, coupled with efficient extraction of 8-oxodG by a validated solid-phase extraction procedure. Subsequent analysis by liquid chromatography–tandem mass spectrometry has the advantages of sensitivity, internal standardisation and robust peak identification.

Key words: Oxidative stress, DNA damage, DNA repair, chromatography, solid-phase extraction, urine, liquid chromatography, mass spectrometry.

1. Introduction

Methods for high-throughput, non-invasive assessment of DNA oxidation have a potentially wide application in a basic or applied biological science context but would also facilitate transfer of such analyses to a clinical setting. The most widely measured

R.M. Uppu et al. (eds.), *Free Radicals and Antioxidant Protocols*, Methods in Molecular Biology 610,
DOI 10.1007/978-1-60327-029-8_20, © Humana Press, a part of Springer Science+Business Media, LLC 1998, 2010

product of DNA oxidation in urine is 8-oxo-7,8-dihydro-2′-deoxyguanosine (8-oxodG) (1). Although 8-oxodG has been examined in various extracellular matrices, for example, serum, saliva and faecal matter, urine remains by far the most widely examined matrix (2). There are many reports of the analysis of urinary 8-oxodG in various populations and pathologies: these analyses have been performed despite a lack of understanding of the exact sources of this lesion in extracellular matrices (3). This area is still under study, in our laboratory and others, with DNA repair, cell turnover and diet considered the major potential sources of urinary 8-oxodG; however, recent data are increasingly ruling out the significance of the latter two sources (4, 5). In terms of the DNA repair origins for 8-oxodG, the most likely source would be Nudix hydrolase activity [e.g. NUDT1 (MTH1)], whose product, 8-oxodGMP, following dephosphorylation yields 8-oxodG (6). Thus, there is potentially more biological meaning to urinary 8-oxodG to be uncovered, other than simply as a marker of nucleic acid oxidation (7).

Chromatography and immunoassay are the two major analytical approaches that have been used to examine urinary 8-oxodG; each method has its own advantages and disadvantages. The more complex technology, costs, user training and lower throughput of the existing chromatographic procedures are outweighed by their more rigorous separation and compound identification capabilities. Conversely, immunoassay has higher sample throughput potential, requires less user training and capital outlay for equipment but potentially suffers from a lack of specificity and sufficient amounts of appropriately characterised antibodies must be available. The chromatographic analysis of urinary 8-oxodG has, to date, largely been done using HPLC coupled with electrochemical detection, in most cases using column switching devices to provide effective sample clean-up (8). Gas chromatography–mass spectrometry (GC–MS) following HPLC pre-purification of urine has also received increasing attention (9, 10). More recently, HPLC coupled to tandem mass spectrometry (LC-MS/MS) has come to the fore and looks set to be the method of choice for the analysis of 8-oxodG, and other lesions, in urine (11, 12). The procedure reported here is an adaptation of the methodology originally reported by Lin et al. (13), in particular the solid-phase clean-up of urine prior to GC–MS analysis, following derivatisation of 8-oxodG. To simplify the procedure, LC-MS/MS is used, which avoids the derivatisation step, provides a good separation of the urine extract, shows sufficient sensitivity and retains the advantages of mass information for more rigorous sample identification. A more detailed presentation of the work leading to the development and validation of this method has recently been published (14).

2. Materials

2.1. Synthesis of Isotopically Labelled 8-oxodG {[$^{15}N_5$]8-oxodG}

1. Water (*see* **Note 1**).
2. [$^{15}N_5$]-2′-Deoxyguanosine (>98% ^{15}N; Spectra Stable Isotopes, Columbia, MD, USA).
3. Reaction buffer: 20 mM sodium phosphate buffer, pH 7.0, Chelex-treated. For Chelex treatment, add 0.05 g Chelex®-100 resin (Biotechnology grade, 00-20 mesh, sodium form, Bio-Rad, cat. no. 143-2832) per millilitre of buffer, stir for 1 h, allow the resin to settle and then carefully decant the buffer into a fresh container. Alternatively, pellet the resin by centrifugation and decant the buffer. Store the buffer at 4–6°C until needed.
4. Copper sulphate, 20 mM, dissolved in water. This reagent can be stored for several weeks at 4–6°C.
5. Sodium ascorbate, 170 mM, dissolved in water. This reagent should be prepared immediately before use and not stored.
6. Hydrogen peroxide: 30% v/v solution, use straight from the bottle as supplied.
7. Catalase: dissolve 1 mg of catalase (bovine liver; 10,700 units/mg solid; Sigma cat. No. C-40) per 1 mL of Chelex-treated, 20 mM phosphate buffer, pH 7.0. Prepare in sufficient quantities and use it on the same day.

2.2. Purification of [$^{15}N_5$]8-oxodG

1. HPLC system capable of mobile-phase flow rates of at least 5 mL/min and with a UV-visible diode array detection facility.
2. Centrifugal vacuum evaporator (speedvac) or freeze dryer.
3. UV-visible spectrophotometer.
4. HPLC column: Columbus, 5μ C_8 semi-preparative column, 250 × 10 mm (Phenomenex, Macclesfield, UK).
5. Mobile phase: 10% (v/v) methanol in ultrapure water or commercially obtained HPLC-grade water.
6. Standard solutions of 2′-deoxyguanosine (dG) and 8-oxodG (Sigma Chemical Co., Poole, UK, cat. no. D7145 and H5653, respectively), 50 μM prepared in 10% (v/v) methanol in water (mobile phase). Alternatively, more concentrated stocks can be prepared in water and diluted to 50 μM using the mobile phase.

2.3. Urine Collection, Creatinine Analysis and Solid-Phase Extraction

1. Vacuum manifold (*see* **Note 2**), pump and trap.
2. Plate reader for 96-well plates, with filter for measurements at 490 nm.

3. $[^{15}N_5]$8-oxodG – synthesised in-house or obtained commercially (e.g. Cambridge Isotope Laboratories, $[^{15}N_5]$8-oxo-2′-deoxyguanosine, cat no. NLM-6715).
4. Formic acid, 10% (v/v), diluted using water.
5. HPLC-grade methanol.
6. Formic acid, 20 mM, in water (pH ~2.75).
7. Methanol (17.5% v/v) in water containing 20 mM formic acid.
8. Solid-phase extraction columns: Waters Oasis HLB, 3 cm^3, 60 mg (Waters Ltd., Elstree, UK).

2.4. LC-MS/MS Analysis of Urine Extracts

1. LC-Tandem mass spectrometry system (*see* **Note 3**).
2. HPLC column: HyPurity C_{18} (3 μm, 2.1 × 150 mm; Thermo Electron Corporation, Runcorn, UK); Uniguard HyPurity C_{18} (3 μm, 2.1 × 10 mm,) guard cartridge attached to a KrudKatcher™ (Phenomenex, Macclesfield, UK) disposable pre-column (0.5 μm) filter.
3. Mobile phase: 0.1% aqueous acetic acid:methanol (92:8, v/v).
4. Standard 8-oxodG tuning solution: (10 pmol/μL) 8-oxodG dissolved in mobile phase.

3. Methods

The use of isotopically labelled 8-oxodG is critical to this procedure, to account for sample recovery and any instrumental differences in the ionisation of samples from one run to the next during the process of electrospray ionisation. We have synthesised our own isotopically labelled 8-oxodG, based on the method reported by Singh et al. (15); however, this compound can now be obtained commercially (*see* **Section 2.3**). We report the synthesis procedure for the benefit of those laboratories that, for one reason or another, may wish to make their own standard.

3.1. Synthesis of $[^{15}N_5]$8-oxodG

1. Prepare a 1 mg/mL solution of $[^{15}N_5]$dG in Chelex-treated 20 mM sodium phosphate buffer, pH 7.0.
2. To the $[^{15}N_5]$dG solution add, sequentially and rapidly, vortexing between additions, copper solution, sodium ascorbate and hydrogen peroxide (final concentrations, 1.2 mM, 10 mM and 370 mM, respectively).
3. Incubate the reaction mixture at ambient temperature for 20 min (*see* **Note 4**), followed by addition of 300 μL of catalase solution.

4. The reaction mixture can be frozen at –20°C, or preferably –80°C, at this point if required.

3.2. Purification of [$^{15}N_5$]8-oxodG

1. Purification of [$^{15}N_5$]8-oxodG is performed by HPLC using sequential injections of the reaction mixture onto the semi-preparative HPLC column. The mobile-phase flow rate is 5 mL/min. Verification of 8-oxodG (and dG) retention time is performed by preliminary injection of unlabelled standards (*see* **Note 5**). Eluting peaks are monitored by UV absorption at 245 and 254 nm, for 8-oxodG and dG, respectively.
2. Collect fractions at the appropriate retention time (*see* **Note 6**) and dry in a speedvac or freeze dryer. Dried material is then reconstituted by thorough vortexing in 1 mL water and reconstituted aliquots are then pooled and dried once more.
3. Dried material is reconstituted by thorough vortexing in 0.5 or 1 mL water.
4. Final quantification is performed by determining the concentration of 8-oxodG by UV absorbance at 245 nm, ε = 12,300 M^{-1} cm^{-1} (*see* **Note 7**).
5. Aliquot concentrated material and store at –80°C. Aliquots of concentrated material can be diluted, further aliquoted and stored at –80°C as desired. A working solution of 2.0 pmol/μL is appropriate for routine urinary 8-oxodG analysis.

3.3. Urine Collection and Creatinine Analysis

1. Collect spot urine samples from subject groups. Samples are usually early morning, first void, midstream urine samples (*see* **Note 8**). Samples can be stored at –20°C or –80°C until analysis, without significant degradation of the analyte (16).
2. Each urine should be analysed for creatinine concentration as a normalisation factor. This may be done using a commercially available colorimetric assay, based on the Jaffe alkaline picrate method, in a 96-well plate format (Metra® Creatinine, Quidel Corp., San Diego, CA, USA) (*see* **Note 9**).

3.4. Solid-Phase Extraction

1. To 1.2 mL urine, in an Eppendorf tube, add 6 μl of 2.0 pmol/μL [$^{15}N_5$]8-oxodG (i.e. 12 pmol) and 120 μL of 10% (v/v) formic acid, vortex mix (*see* **Notes 10** and **11**) **Fig. 20.1**.
2. Place acidified urine at 4–6°C for 1 h and then centrifuge at 14,000 rpm for 1 min in a bench-top microfuge.
3. Recover supernatant and dispose of any pellet.
4. Dilute, or make-up, urine with 20 mM formic acid to a final volume of 5.8 mL.

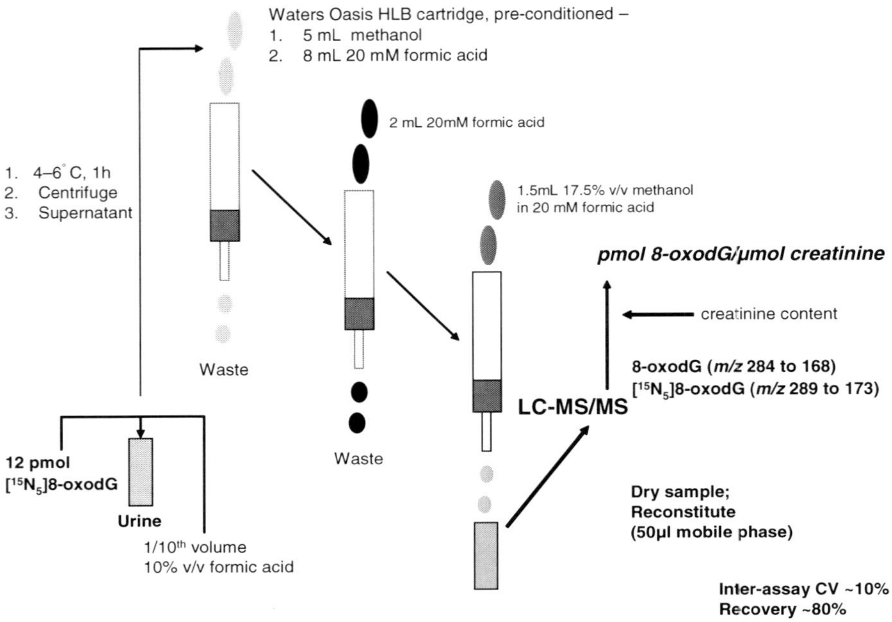

Fig. 20.1. Summary of urine treatment and solid-phase extraction procedure.

5. Use one SPE column per sample and dispose after use (*see* **Note 12**).
6. Pre-condition column packing material as follows: with vacuum applied, run liquid to waste: 5 mL methanol, followed by 8 mL of 20 mM formic acid (*see* **Note 13**).
7. Pass diluted urine through the column at a rate of ~2 mL/min, run to waste and dry column under vacuum.
8. Wash unbound material from the column with 2 mL of 20 mM formic acid, ~2 mL/min, run to waste and dry column under vacuum.
9. Elute 8-oxodG from the column by using 1.5 mL of 17.5% (v/v) methanol in 20 mM formic acid, collecting eluent into an appropriately labelled tube.
10. Either freeze eluent at this stage at –20°C, or –80°C, or proceed immediately to drying of samples in either a freeze dryer or a speedvac.

3.5. LC-MS/MS Analysis of Urine Extracts

1. The mass spectrometer is tuned using the standard 8-oxodG tuning solution (*see* **Note 14**).
2. Reconstitute dried urine extracts in 50 μL water with thorough vortexing.
3. Transfer reconstituted extracts into HPLC vials containing low volume inserts and inject 10 μL onto the analytical column.

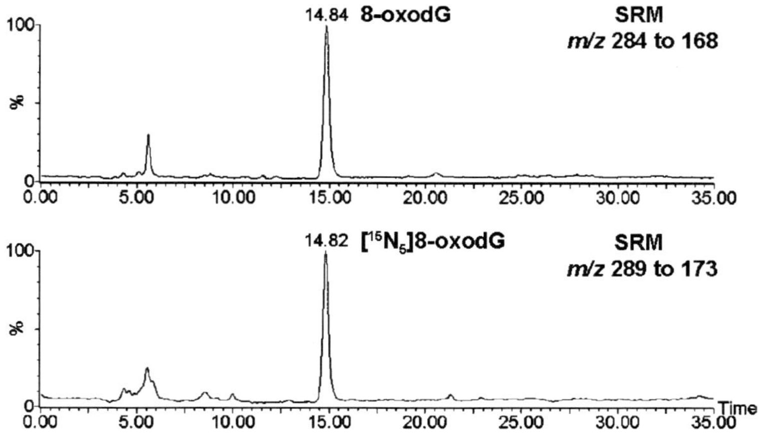

Fig. 20.2. Typical chromatograms, for the transitions *m/z* 284 to 168 and *m/z* 289 to 173 for 8-oxodG and the stable isotope internal standard [$^{15}N_5$]8-oxodG, respectively, obtained from solid-phase extract of healthy human urine, using the procedure outlined in this chapter.

4. The column is eluted isocratically with mobile phase at a flow rate of 120 μL/min.
5. Selected reaction monitoring (SRM) analysis is performed for the $[M+H]^+$ ion to oxidised base $[B+H_2]^+$ transitions of 8-oxodG (*m/z* 284 to 168) and the stable isotope internal standard [$^{15}N_5$]8-oxodG (*m/z* 289 to 173) (*see* **Note 15**).
6. Quantification of 8-oxodG in each urine sample is determined from the ratio of the peak area of 8-oxodG to that of the internal standard in the same sample:

$$Q_{\text{anal.}} = (A_{\text{anal.}}/A_{\text{istd.}}) \times Q_{\text{istd.}}$$

where $Q_{\text{anal.}}$ and $Q_{\text{istd.}}$ correspond to the amounts (pmol) of the analyte and the internal standard, respectively, and $A_{\text{anal.}}$ and $A_{\text{istd.}}$ correspond to the peak areas of the analyte the and internal standard, respectively. The quantity of 8-oxodG is then corrected for creatinine content yielding final values of picamoles of 8-oxodG/micromoles of creatinine. An example chromatogram is shown in **Fig. 20.2** (*see* **Note 16**).

4. Notes

1. Unless otherwise specified, throughout this procedure, 'water' refers to ultrapure water with resistance ≥18.2 MΩ and the content of organic matter <5 ppb.

2. A vacuum pump allows for better consistency and control of vacuum, compared to a water aspirator, but the latter can be used if a vacuum pump is not available.
3. The LC-MS/MS instrument used will depend on the individual laboratory performing this assay. The system used in method development, which we continue to use, consists of a Waters Alliance 2695 separation module connected to a Micromass Quattro Ultima Platinum (Waters Ltd., Manchester, UK) tandem quadrupole mass spectrometer with an electrospray interface. Specific details regarding the operation of each system are beyond the scope of this chapter.
4. Incubation at 37°C, compared to room temperature, has a negligible impact upon yield. Also, prolonging the incubation time does not appear to significantly modify the yield of 8-oxodG.
5. In addition to retention time, the identification of 8-oxodG is based upon the comparison of spectral properties [spectral overlay and absorbance ratios (250/260 nm; 280/260 nm) compared to the standard run under the HPLC purification conditions] to authentic, commercially obtained unlabelled standard. Further verification of stable isotope labelling, product identity and impurity of unlabelled material is done by mass spectrometry. It is important to verify that labelled material is in vast excess such that any unlabelled material, typically <2% of total material (depending on the exact purity of the starting material), does not contribute to levels of endogenously measured 8-oxodG in each sample during internal standardisation. Collectively, this information should assure the user of purity and identity of the internal standard.
6. The majority of the material remaining at the end of the reaction is [$^{15}N_5$]dG. We collect this unmodified [$^{15}N_5$]dG for recycling in subsequent syntheses.
7. Additionally, we determine the concentration using mass spectrometry. This is done by injecting an appropriately diluted sample of [$^{15}N_5$]8-oxodG onto the LC-MS/MS under the conditions used for urinary 8-oxodG analysis (*see* **Section 3.5**). Such a sample also contains a known amount of accurately quantified unlabelled 8-oxodG. Comparison of peak areas of labelled to unlabelled material allows quantification of the labelled standard.
8. Spot urine samples can be used from other times of the day usually without detection problems.
9. We generally use a hospital Routine Chemical Pathology Service; however, the ELISA assay quoted in the

protocol yields creatinine concentrations that compare very favourably with values obtained via an autoanalyser.

10. If urines have been thawed following a period of frozen storage, it is highly likely that material has precipitated. It is recommended that upon thawing, the urine sample is mixed and the volume used for analysis constitutes liquid containing a suspension of particulates.
11. Volume of urine – recent work in our laboratory indicates that the volume of urine used can be much less; we have tested down to 100 μL. Using lower volumes does not compromise sensitivity and generates corrected 8-oxodG values, comparable with those obtained from 1.2 mL (unpublished results). However, it is important to note that for the initial addition of 10% formic acid, the volume used is one-tenth the volume of urine. This information has implications for the analysis of smaller urine samples, such as those obtained from neonates or small laboratory animals.
12. In our experience. processing more than three cartridges at any one time leads to poor control of flow rate, but this issue is probably dependent on the pump and the manifold used.
13. The Oasis HLB cartridges are designed to function even if the column bed is allowed to dry out between steps (13). Leaving the column bed wet confers no advantage in terms of the final values obtained for urinary 8-oxodG.
14. In our laboratory the 8-oxodG tuning solution is introduced into the mass spectrometer by continuous infusion at a flow rate of 10 μL/min with a Harvard model 22 syringe pump (Harvard Apparatus Ltd., Edenbridge, UK).
15. LC-MS/MS instrument parameters for the system used in our laboratory – temperature of the electrospray source is maintained at 110°C and the desolvation temperature at 350°C. Nitrogen gas is used as the desolvation gas (650 L/h) and the cone gas is set to zero. The capillary voltage is 3.20 kV, and the cone and RF1 lens voltages are 42 and 30 V, respectively. The collision gas is argon (indicated cell pressure 2.0×10^{-3} mbar) and the collision energy is set at 12 eV. The dwell time is 200 ms and the resolution was two *m/z* units at peak base. Under these conditions, the limit of detection for the optimised analysis of urinary 8-oxodG by LC-MS/MS SRM on our system is 5 fmol on column (signal-to-noise ratio, S/N, = 4) for pure 8-oxodG standard.
16. Calculations should take any dilution or concentration factors into account.

e.g. urinary 8-oxodG level (pmol/mL/μmol creatinine) = [($A_{anal.}$/$A_{istd.}$) × 2.4 pmol × 5]/(1.2 mL/μmol creatinine).

In our experience <5% of samples analysed are likely to have peaks that interfere with either the endogenous 8-oxodG or the internal standard peaks. In the vast majority of cases where we have subjected these samples to re-extraction, the issue of interfering peaks has been resolved.

Acknowledgments

RS and PBF are supported by the UK Medical Research Council; development of this methodology in the laboratory of MDE and MSC was supported by technical support from Georgina Hall and Karendeep Sandhu, and using a University of Leicester miscellaneous income fund held by MSC. The authors of this paper are partners of ECNIS (Environmental Cancer Risk, Nutrition and Individual Susceptibility), a network of excellence operating within the European Union 6th Framework Program, Priority 5: 'Food Quality and Safety' (Contract No 513943).

The formation of a European-centred laboratory consortium, The European Standards Committee on Urinary (DNA) Lesion Analysis (ESCULA), is enabling an examination of methodology and issues surrounding the analysis of urinary 8-oxodG in the first instance. Further information about this consortium including activities and membership can be obtained by contacting Dr. M.S. Cooke (msc5@le.ac.uk); www.escula.org.

References

1. Kasai, H. (1997) Analysis of a form of oxidative DNA damage, 8-hydroxy-2'-deoxyguanosine, as a marker of cellular oxidative stress during carcinogenesis. *Mutat. Res.* **387**, 147–163.
2. Bogdanov, M.B., Beal, M.F., McCabe, D.R., Griffin, R.M., and Matson, W.R. (1999) A carbon column-based liquid chromatography electrochemical approach to routine 8-hydroxy-2'-deoxyguanosine measurements in urine and other biologic matrices: A one-year evaluation of methods. *Free Radic. Biol. Med.* **27**, 647–666.
3. Wu, L.L., Chiou, C.-C., Chang, P.-Y., and Wu, J.T. (2004) Urinary 8-OHdG: a marker of oxidative stress to DNA and a risk factor for cancer, atherosclerosis and diabetics. *Clin. Chim. Acta* **339**, 1–9.
4. Cooke, M.S., Evans, M.D., Dove, R., Rozalski, R., Gackowski, D., Siomek, A., Lunec, J., and Olinski, R. (2005) DNA repair is responsible for the presence of oxidatively damaged DNA lesions in urine. *Mutat. Res.* **574**, 58–66.
5. Siomek, A., Tujakowski, J., Gackowski, D., Rozalski, R., Foksinski, M., Dziaman, T., Roszkowski, K., and Olinski, R. (2006) Severe oxidatively damaged DNA after cisplatin treatment of cancer patients. *Int. J. Cancer* **119**, 2228–2230.
6. Cooke, M.S., Lunec, J., and Evans, M.D. (2002) Progress in the analysis of urinary oxidative DNA damage. *Free Radic. Biol. Med.* **33**, 1601–1614.
7. Haghdoost, S., Sjolander, L., Czene, S., and Harms-Ringdahl, M. (2006) The nucleotide

pool is a significant target for oxidative stress. *Free Radic. Biol. Med.* **41**, 620–626.

8. Bogdanov, M.B., Beal, M.F., McCabe, D.R., Griffin, R.M., and Matson, W.R. (1999) A carbon column-based liquid chromatography electrochemical approach to routine 8-hydroxy-2'-deoxyguanosine measurements in urine and other biologic matrices: A one-year evaluation of methods. *Free Radic. Biol. Med.* **27**, 647–666.
9. Rozalski, R., Siomek, A., Gackowski, D., Foksinski, M., Gran, C., Klungland, A., and Olinski, R. (2004) Diet is not responsible for the presence of several oxidatively damaged DNA lesions in mouse urine. *Free Radic. Res.* **11**, 1201–1205.
10. Ravanat, J.L., Guicherd, P., Tuce, Z., and Cadet, J. (1999) Simultaneous determination of five oxidative DNA lesions in human urine. *Chem. Res. Toxicol.* **12**, 802–808.
11. Hu, C.W., Wang, C.J., Chang, L.W., and Chao, M.R. (2006) Clinical-scale high-throughput analysis of urinary 8-oxo-7,8-dihydro-2'-deoxyguanosine by isotope-dilution liquid chromatography-tandem mass spectrometry with on-line solid-phase extraction. *Clin. Chem.* **52**, 1381–1388.
12. Weimann, A., Belling, D., and Poulsen, H.E. (2001) Measurement of 8-oxo-2'-deoxyguanosine and 8-oxo-2'-deoxyadenosine in DNA and human urine by high performance liquid chromatography-electrospray tandem mass spectrometry. *Free Radic. Biol. Med.* **30**, 757–764.
13. Lin, H.S., Jenner, A.M., Ong, C.N., Huang, S.H., Whiteman, M., and Halliwell, B. (2004) A high-throughput and sensitive methodology for the quantification of urinary 8-hydroxy-2'-deoxyguanosine: measurement with gas chromatography–mass spectrometry after single solid-phase extraction. *Biochem. J.* **380**, 541–548.
14. Cooke, M.S., Singh, R., Hall, G.K., Mistry, V., Duarte, T.L., Farmer, P.B., and Evans, M.D. (2006) Evaluation of enzyme-linked immunosorbent assay and liquid chromatography-tandem mass spectrometry methodology for the analysis of 8-oxo-7,8-dihydro-2'-deoxyguanosine in saliva and urine. *Free Radic. Biol. Med.* **41**, 1829–1836.
15. Singh, R., McEwan, M., Lamb, J.H., Santella, R.M., and Farmer, P.B. (2003) An improved liquid chromatography/tandem mass spectrometry method for the determination of 8-oxo-7,8-dihydro-2'-deoxyguanosine in DNA samples using immunoaffinity column purification. *Rapid Commun. Mass Spectrum.* **17**, 126–134.
16. Loft, S., Svoboda, P., Kasai, H., Tjonneland, A., Vogel, U., Moller, P., Overvad, K., and Raaschou-Nielsen, O. (2006). Prospective study of 8-oxo-7,8-dihydro-2'-deoxyguanosine excretion and the risk of lung cancer. *Carcinogenesis* **27**, 1245–1250.

Chapter 21

Oxidative Lipidomics of Apoptosis: Quantitative Assessment of Phospholipid Hydroperoxides in Cells and Tissues

Vladimir A. Tyurin, Yulia Y. Tyurina, Vladimir B. Ritov, Andriy Lysytsya, Andrew A. Amoscato, Patrick M. Kochanek, Ronald Hamilton, Steven T. DeKosky, Joel S. Greenberger, Hülya Bayir, and Valerian E. Kagan

Abstract

Oxidized phospholipids play essential roles in execution of mitochondrial stage of apoptosis and clearance of apoptotic cells. The identification and quantification of oxidized phospholipids generated during apoptosis can be successfully achieved by oxidative lipidomics. With this approach, diverse molecular species of phospholipids and their hydroperoxides are identified and characterized by soft-ionization mass-spectrometry techniques such as electrospray ionization (ESI). Quantitative assessment of lipid hydroperoxides is performed by fluorescence HPLC-based protocol. The protocol is based on separation of phospholipids using two-dimensional-high-performance thin-layer chromatography (2-D-HPTLC). Phospholipids are hydrolyzed using phospholipase A_2. The fatty acid hydroperoxides (FA-OOH) released is quantified by a fluorometric assay using Amplex red reagent and microperoxidase-11 (MP-11). Detection limit of this protocol is 1–2 pmol of lipid hydroperoxides. Lipid arrays vs. oxidized lipid arrays can be performed by comparing the abundance of phospholipids with the abundance of oxidized phospholipids. Using oxidative lipidomics approach we show that the pattern of phospholipid oxidation during apoptosis is nonrandom and does not follow their abundance in several types of cells undergoing apoptosis and a variety of disease states. This has important implications for evaluation of apoptosis in vivo. The anionic phospholipids, cardiolipin (CL) and phosphatidylserine (PS), are the preferred peroxidation substrates.

Key words: Phospholipid hydroperoxides, cardiolipin, phosphatidylserine, ESI-MS, fluorescence HPLC, apoptosis, cytochrome *c*, controlled cortical impact, gamma-irradiation, Alzheimer's disease.

1. Introduction

While lipids are relatively small biomolecules, their amphipathic nature makes them essential for compartmentalization and heterogeneous catalysis of life reactions in cells, tissues, and

R.M. Uppu et al. (eds.), *Free Radicals and Antioxidant Protocols*, Methods in Molecular Biology 610,
DOI 10.1007/978-1-60327-029-8_21, © Humana Press, a part of Springer Science+Business Media, LLC 1998, 2010

biofluids. The large diversity and metabolic conversions of lipids form the basis of biomembranes with their unique features of structural and functional flexibility and fluidity resulting in a broad range of adaptive mechanisms (including generation of a host of signaling molecules). These important features of lipids are largely due to the presence of double bonds in their hydrophobic moieties. This, in turn, makes lipids susceptible to oxidative modifications by molecular oxygen. Effective control of oxidative stress is one of the adaptive evolutionary achievements of aerobic life essential for the prevention and protection against toxicity of environmental chemical and biological hazards, degenerative diseases, and even aging (1). During oxidative stress, different types of molecules in cells and biofluids undergo peroxidation. One of the most common manifestations of oxidative stress is peroxidation of lipids in biomembranes and lipoproteins. Lipid peroxidation is considered as a deleterious process resulting in disruption of well-coordinated structural and functional organization of biomembranes. However, there is emerging understanding of new signaling roles that oxidized lipids may play in cell physiology.

Until recently, comprehensive analysis of different classes of lipids and their molecular species in cells, tissues, and biofluids was cumbersome and difficult if not impossible. However, with the advent of contemporary mass spectrometry and development of new soft-ionization techniques, identification and structural characterization of individual phospholipid molecules in complex mixtures became possible (2). These technological advancements designated the emergence of a new field of research and knowledge – lipidomics. The latter can be defined as "a branch of metabolomics that includes a systems-based study of all lipids, the molecules with which they interact, and their function within the cell" (2). Thus, lipidomics approach consists of analysis of the entire lipidome and lipid related metabolites, as well as characterization, and quantification of changes in the lipidome in response to internal and external stimuli such as alteration in nutritional status or under disease conditions (3). Application of principles and approaches of lipidomics to studies of oxidatively modified lipids yielded the area of oxidative lipidomics. Similar to lipidomics, the oxidative lipidomics includes identification and characterization of oxidatively modified lipids by MS, quantitative analysis of their content in cells, tissues, and biofluids as well as discovery of mechanisms and pathways through which they interact with other molecules and contribute to metabolism and physiological functions.

Soft-ionization techniques such as electrospray ionization (ESI) and matrix assisted laser desorption ionization (MALDI) can be successfully utilized for this purpose. However, vast diversity and relatively low quantity of peroxidized molecular species of phospholipids make MS analysis of oxidatively modified lipids technically challenging (4). In spite if this, we and others successfully utilized different MS protocols to detect and identify the major oxidized species of phospholipids of normal and injured cells and tissues (5–9). In our experience, after identification and characterization of individual molecular species of oxidatively modified phospholipids by MS, quantification of their content can be conveniently achieved by the assessment of hydroperoxides, the primary molecular products of phospholipid peroxidation, by high-performance liquid chromatography (HPLC). Comparison of the abundance of phospholipids with their respective hydroperoxides yields phospholipid/oxidized phospholipid arrays. This paper describes application of this methodology to analysis of phospholipid oxidative signaling in apoptosis.

Programmed cell death or apoptosis is viewed as an important adaptive evolutionary mechanism that allows eukaryotic cells in a multicellular organism to prevent proliferation of their dangerously modified genomes by self-elimination program (10). Mitochondria with their tool-set of pro-apoptotic factors play a central role in the execution of this program (11). Mitochondrial stages of apoptosis include the disruption of electron transport and enhanced generation of one-electron-reduced species of molecular oxygen (also called reactive oxygen species (ROS)) culminating in the release of cytochrome *c* (cyt *c*) and other dreadful pro-apoptotic factors from mitochondria into the cytosol (12). This results in mobilization and activation of cell's proteolytic enzymes ultimately leading to disassembly of essential bio-macromolecules into small fragments. Two important participants of this program are cyt *c* and cardiolipin (CL), specific components normally not found outside of mitochondria. Interactions between cyt *c* and CL within mitochondria and their role in triggering programmed cell death are not fully understood.

Recently, it has been established that binding of cyt *c* to anionic phospholipids – CL and PS – converts it into a peroxidase capable of utilizing these phospholipids as reaction substrates, ultimately leading to accumulation of their hydroperoxides (13, 14). Both oxidized phospholipids act as important signals in mitochondria and plasma membrane of apoptotic cells, respectively. While oxidized CL is required for the release of apoptotic factors into the cytosol, oxidized PS functions as a regulator of externalization of PS which is recognized by phagocytes as an "eat-me" signal and facilitates clearance of apoptotic cells (15–17). Not surprisingly, overexpression of mitochondrial antioxidant enzymes regulating phospholipid oxidation such as phospholipid hydroperoxide glutathione peroxidase inhibits release of cyt *c* from mitochondria likely by suppressing the oxidation of CL during apoptosis (18, 19). Similarly, lipid antioxidants effectively preventing PS oxidation inhibit PS externalization thus affecting phagocytosis of apoptotic cells (20). Therefore, understanding the mechanisms of CL and PS oxidation and analysis of their oxidation products is central to elucidating the events leading to execution of the mitochondria-dependent apoptosis and clearance of apoptotic cells. Recently, Girotti and coauthors (14–21) described a method for determination of phospholipid oxidation products (PL-OOH) based on separation of PL-OOH by HPTLC with their subsequent detection using staining with N, N, N', N'-tetramethyl-*p*-phenylenediamine (TPD). The detection limit of PL-OOH by this method is 100–200 pmol. While the protocol may be useful for analysis of relatively high concentrations of PL-OOH, its utility for the analysis of low endogenous levels of CL and PS oxidation products accumulating during execution of apoptotic program may be limited by insufficient sensitivity.

To characterize phospholipid oxidation generated during apoptosis we applied oxidative lipidomics approach that includes identification of diverse molecular species of phospholipids by ESI-MS analysis (22, 23) followed by quantitative assessments of lipid hydroperoxides using fluorescence HPLC-based protocol. The protocol is based on separation of phospholipids using 2-D-HPTLC. Phospholipids are hydrolyzed using phospholipase A_2 (PLA_2). The FA-OOH released are quantified by a fluorometric assay using Amplex red reagent (10-acetyl-3,7-dihydroxyphenoxazine) and MP-11. In this system, FA-OOH react with Amplex red reagent at a stoichiometry of 1:1 to generate a fluorescent product, resorufin (ex = 560 nm, em = 590 nm), analyzed by fluorescence HPLC. Using the protocol, 1–2 pmol of lipid hydroperoxides can be reliably detected in lipid extracts from different types of cells and tissues.

2. Materials

2.1. Reagents and Solvents

1. All organic solvents were of HPLC grade and obtained from Sigma Chemical Co (St. Louis, MO).
2. Unless otherwise indicated, all reagents were purchased from Sigma Chemical Co (St. Louis, MO).

2.2. Standards

1. 1,2-Dioleoyl-*sn*-glycero-3-phosphocholine (DOPC, PC), 1-palmitoyl-2-arachidonoyl-*sn*-glycero-3-phospho-L-serine (PAPS, PS), 1,1′,2,2′-tetraoleoylcardiolipin (TOCL), 1,1′, 2,2′-tetralinoleoyl-cardiolipin (TLCL), 1-palmitoyl-2-linoleoyl-*sn*-glycero-3-phosphoethanolamine (PE), 1,2-dioleoyl-*sn*-glycero-3-phosphoinositol (PI), *N*-stearoyl-D-*erythro*-sphingosylphosphorylcholine (Sphingomyelin, Sph) were purchased from Avanti Polar Lipids Inc (Alabaster, AL) and were of the highest purity available.

2. 9*Z*-Octadecenoic acid, 9*Z*, 12*Z*-octadecadienoic acid, 5*Z*, 8*Z*, 11*Z*, 14*Z*-eicosatetraenoic acid, 15*S*-hydroxy-5*Z*, 8*Z*, 11Z, 13*E*, 17*Z*-eicosapentaenoic acid, (15(*S*)-HEPE), 9*E*, 11*Z*-octadecadienic acid, 13-oxo-, (13-Oxo-ODE), (±)8 (9)-epoxy-5*Z*, 11*Z*, 14*Z*-eicosatrienoic acid, (8(9)-EpTrE), 9*S*-hydroperoxy-10*E*, 12*Z*-octa-decadienoic acid, (9(*S*)-HpODE), 13*S*-hydroperoxy-10*E*, 12*Z*-octa-decadienoic acid, (13(*S*)-HpODE), 15*S*-hydroperoxy-5*Z*, 8*Z*, 11*Z*, 13*E*-eicosatetraenoic acid, (15(*S*)-HpETE) were obtained from Cayman Chemical Co (Ann Arbor, MI) and were of the highest purity available.
3. Calibration solutions were prepared by mass, dissolved in HPLC grade ethanol, and stored under N_2 at -80°C prior to HPLC analysis.

2.3. Instruments and Equipment

1. *Mass spectrometry:*
   Mass spectrometry of phospholipids was performed by direct infusion into a high-throughput quadrupole linear ion-trap mass spectrometer Finnigan™ LXQ™ with the Xcalibur operating system (Thermo Electron, San Jose, CA).
   a) Finnigan™ LXQ™ quadrupole linear ion-trap mass spectrometer.
   b) Dell/Pentium®, Intel®, PC.
   c) Xcalibur operating system.
2. *HPLC System:*
   Analytical HPLC system from Shimadzu Scientific Instruments (Columbia, MD) was used.
   a) Shimadzu SCL-10A vp HPLC system controller.
   b) Shimadzu LC-10AT vp liquid chromatograph.
   c) Shimadzu SIL-10AD vp autosampler/injector.
   d) Shimadzu SPD M6A UV/VIS photodiode array.
   e) Shimadzu fluorescence detector RF-10Axl.
   f) Dell/Pentium®, Intel®, PC.
   g) Shimadzu EZstart v. 7.2 chromatography software.
3. *HPLC Columns:*
   a) Zorbax Eclipse XDB-C18, 4.6 × 150 mm, 5 μm (Agilent Technologies, Palo Alto, CA).
   b) Zorbax cartridge quard-columnXDB-C18, 4.6 × 20 mm, 5 μm.

2.4. Determination of Inorganic Lipid Phosphorus in Specific Classes of Phospholipids

1. 5 × 5 cm Whatman silica G HPTLC plates (Whatman, Clifton, NJ).
2. Chromatography chambers (e.g., 10 cm Latch-Lid Chromatotank, General Glassblowing Company, Inc., Richmond, CA).

3. Mobile phase solvent system 1: chloroform/methanol/28% ammonium hydroxide, 65/25/5 (v/v/v).
4. Mobile phase solvent system 2: chloroform/acetone/methanol/glacial acetic acid/water, 50/20/10/10/5 (v/v/v/v).
5. Nitrogen compressed, Grade 4.0 (Valley National Gases Inc., Wheeling, WV).
6. Cell culture disposable glass tubes (Fisher Scientific, Pittsburgh, PA).
7. Iodine crystals, chromatography chamber, and fume hood.
8. Vortex-2, Genie (Scientific Industries Inc., Bohemia, NY).
9. $HClO_4$, 70% w/v (Aldrich, Milwaukee, WI).
10. Heating apparatus to achieve 170–180°C (Troemner 501 hotplate with aluminum block test tube holders, Thorofare, NJ, or equivalent) and 90–100°C (for water bath container).
11. Sodium molybdate, 2.5% w/v (water solution).
12. Ascorbic acid, 10% w/v (water solution).
13. NaH_2PO_4 (water solution).
14. Tabletop centrifuge.
15. Shimadzu UV 160 V, spectrophotometer.

2.5. Reagents for Detection of Lipid Hydroperoxides of Specific Phospholipid Classes

1. 10-Acetyl-3, 7-dihydroxyphenoxazine, 10 mM, Amplex red reagent (Invitrogen, Carlsbad, CA).
2. Resorufin sodium salt, 32 μM solution in DMSO (Molecular Probes, Eugene, OR).
3. Phospholipase A_2 (PLA_2), porcine pancreas (EC 3.1.1.4).
4. Sodium dodecyl sulfate, ultra pure (SDS), 10 mM solution in ethanol (GibcoBRL, Grand Island, NY).
5. Butylated hydroxytoluene (BHT), 0.1 M.
6. Microperoxidase-11 (MP-11), water solution.
7. Hydrochloric acid, 1.0 M.
8. Ethanol.

3. Methods

3.1. Lipid Extraction and 2D-HPTLC Analysis

Total lipids were extracted from tissue samples by Folch procedure as described by Folch et al. (24). Methanol (4 mL) containing BHT (0.1 mg/mL) was added to the sample homogenate (1 mL) and mixed. Chloroform (8 mL) was then added and the mixture was stored for 2 h at 0°C under N_2. One milliliter of 0.1 M NaCl was added to the extract which was vortex mixed under N_2. After centrifugation for 5 min at 1,500×*g*, lower chloroform layer was collected and the upper aqueous methanol layer was mixed with equal volume of chloroform. After centrifugation for 5 min at 1,500×*g*, lower chloroform layer was collected and combined with chloroform phase after the first centrifugation and evaporated to dryness under a stream of N_2. Phospholipid classes in the extracts were separated by 2D-HPTLC on silica G plates as described by Kagan et al. (7). To bind transition metals from silica, prior to application and separation of phospholipids by HPTLC, plates were treated with methanol containing 1 mM EDTA, 100 μM DTPA. Then the plates were first developed with a solvent system consisting of chloroform/methanol/28% ammonium hydroxide (65/25/5, v/v/v). After the plates were dried with a forced N_2 blower to remove the solvent, they were developed in the second dimension with a solvent system consisting of chloroform/acetone/methanol/glacial acetic acid/water (50/20/10/10/5, v/v/v/v). The phospholipids were visualized by exposure to iodine vapors and identified by comparison with authentic phospholipid standards. For ESI-MS and Amplex red analysis the phospholipid spots on the silica plates were visualized by spraying the plates with deionized water. After this, the spots were scraped from the silica plates and phospholipids were extracted by chloroform/methanol/water (20/10/2, v/v/v). Extracted phospholipids were divided into aliquots for phosphorus, Amplex red, and ESI-MS analysis. Lipid phosphorus was determined by a micro-method (25).

3.2. Electrospray Ionization Tandem Mass Spectrometry

ESI-MS has emerged as a powerful tool for qualitative analysis of complex phospholipids (22, 23). Collision induced decomposition of both positive and negative molecular ion species yields rich information as to the polar head group of the phospholipid and the fatty acid substituents esterified to the glycerophospholipid backbone. ESI-MS analysis of phospholipids was performed by direct infusion into quadrupole linear ion-trap mass spectrometer Finnigan™ LXQ™ with the Xcalibur operating system (Thermo Electron, San Jose, CA). Samples collected after 2D-HPTLC separation were evaporated under N_2, resuspended in chloroform/methanol (1/2, v/v; 20 pmol/μL) and directly utilized for acquisition of negative-ion or positive-ion ESI mass spectra using a flow rate of 5 μL/min. The ESI probe was operated at a voltage differential of 5.0 kV in the negative- or positive-ion mode. Capillary temperature was maintained at 150°C. Tandem mass spectrometry (MS/MS analysis) of individual PL species was employed to determine the fatty acid composition. Using full range zoom (200–2,000 m/z) in positive- and negative-ion mode, the profile spectra were acquired. Two ion activation techniques were used for MS analysis: collision-induced dissociation (CID, $Q = 0.25$, low mass cut off at 28% of the precursor m/z)

and pulsed-Q dissociation technique (PQD), with $Q = 0.7$, and no low mass cut off for analysis of low molecular weight fragment ions (26).

3.3. Quantitation of Lipid Hydroperoxides

Lipid hydroperoxides were determined by fluorescence HPLC of resorufin formed in peroxidase-catalyzed reduction of specific PL-OOH with Amplex red. Phospholipids were hydrolyzed by porcine pancreatic PLA_2 (2 U/μL) in 25 mM phosphate buffer containing 1.0 mM Ca^{2+}, 0.5 mM EGTA, and 0.5 mM SDS (pH 8.0 at RT for 30 min), prior to exposure to reagents for the peroxidase reaction (MP-11/Amplex Red). This resulted in the production of lyso-phospholipids and release of hydroperoxide group-containing fatty acids (FA-OOH). While MP-11 can catalyze reduction of some PL-OOH without pretreatment with PLA_2, non-deesterified CL-OOH did not effectively react with MP-11/Amplex Red. For the peroxidase reaction, 50 μM Amplex red and MP-11 (1.0 μg/μL) were added to hydrolyzed lipids, and the samples were incubated at 4°C for 40 min. The reaction was started by addition of 1 μL of solution MP-11 (1.0 μg/μL). The reaction was terminated by addition of 100 μL of a stop reagent (solution of 10 mM HCl, 4 mM BHT in ethanol). The samples were centrifuged at 10,000×g for 5 min and the supernatant was used for HPLC analysis. Aliquots (5 μL) were injected into a C-18 reverse phase column (Eclipse XDB-C18, 5 μm, 150 × 4.6 mm). The column was eluted by mobile phase composed of 25 mM KH_2PO_4 (pH 7.0)/methanol (60/40, v/v) with 1 mL/min flow rate. The resorufin (an Amplex red oxidation product) fluorescence was measured at 590 nm after excitation at 560 nm. Shimadzu LC-100AT vp HPLC system equipped with fluorescence detector (model RF-10Axl) and autosampler (model SIL-10AD vp) was used.

4. Notes

MS Characterization of Molecular Species in Major Classes of Phospholipids with Emphasis on Polyunsaturated Substrates of Peroxidation

Two major tasks can be accomplished using ESI-MS analysis of phospholipids: (1) characterization of individual molecular species with particular emphasis on those that contain polyunsaturated fatty acid residues, hence represent likely substrates of lipid peroxidation and (2) identification of oxygenated molecular species that underwent oxidative modification. ESI-MS of anionic phospholipids such as PS, PI, and CL reveals molecular ion peaks readily observable in negative mode. A zwitterionic phospholipid, PC, is effectively detectable in positive mode, whereas another non-charged phospholipid, PE, could be determined equally well in either mode. Detection of phospholipids in positive ionization mode generally provides information about the phospholipid head groups, whereas tandem MS-MS analysis in negative mode is a source of structural information relevant to the identity and positional distribution at *sn*-1 and *sn*-2 of individual acyl chains.

Typical ESI mass spectra of different molecular species of several major classes of phospholipids from brain cortex are shown on **Fig. 21.1**. A negative-ion mode was employed for PE, PI, and CL, while spectra for PC were obtained using a positive-ion regimen. The molecular species of PE (**Fig. 21.1, A**) are represented by deprotonated ions $[M\text{-}H]^-$ with *m/z* 722; 747; 750; 766; 774; and 790. After fragmentation of the major molecular ion with *m/z* 766, two prominent ions with *m/z* 283 and 303 are obtained in negative ionization mode that correspond to stearic ($C_{18:0}$) and arachidonic ($C_{20:4}$) fatty acids, respectively (**Fig. 21.1, A, b**). Typical product ions of PE with *m/z* 140 and 196 derived from $[HPO_4CH_2CH_2NH_2]^+$ and $[CH_2C(OH)CH_2PO_4CH_2CH_2NH_2]^+$ were observed after fragmentation of molecular species of PE in a positive mode (data not shown).

The major product ion of PI under negative ionization mode (**Fig. 21.1, B, a**) is apparent as a peak $[M\text{-}H]^-$ with *m/z* 885. The negative ionization mass spectrum after fragmentation of PI (**Fig. 21.1, B, b**) showed the ion fragments with *m/z* 283 and *m/z* 303 that correspond to stearic ($C_{18:0}$) and arachidonic ($C_{20:4}$) fatty acids, respectively. Elimination of one of the fatty acids ($C_{18:0}$ or $C_{20:4}$) from the parent ion with *m/z* 885 resulted in the product ions at *m/z* 581 $[M\text{-}H\text{-}C_{20:4}]^-$ and *m/z* 601 $[M\text{-}H\text{-}C_{18:0}]^-$. Loss of the inositol group from the fragment with *m/z* 581 formed a product ion with *m/z* 419. Fragment of ion with *m/z* 259 corresponds to the polar head group (inositol phosphate) of PI. Loss of one and two molecules of water from this fragment yielded ions with *m/z* 241 and 223, respectively.

Molecular species of PC in positive-ion mode (**Fig. 21.1, C, a**) demonstrated the presence of the major molecular species with protonated ions $[M+H]^+$ with *m/z* 734, 760, and 782 and corresponding to multiple individual species containing $C_{16:0}/C_{16:0}$, $C_{16:0}/C_{18:1}$, and $C_{16:1}/C_{20:3}$ fatty acids, respectively. Several other molecular clusters of PC with polyunsaturated acyl species were present in relatively small quantities. A typical fragmentation spectrum in positive mode of the major parent ion with *m/z* 760 is presented in **Fig. 21.1**, panel **C, b**. The major fragment with *m/z* 184 is derived from the loss of the polar head group $[C_5H_{15}NPO_4]^+$. Additional information on PC comes from the fragment ions with *m/z* 478 and 504 of lysoPC derived from the loss of fatty acids: oleic ($C_{18:1}$) and palmitic ($C_{16:0}$). In the

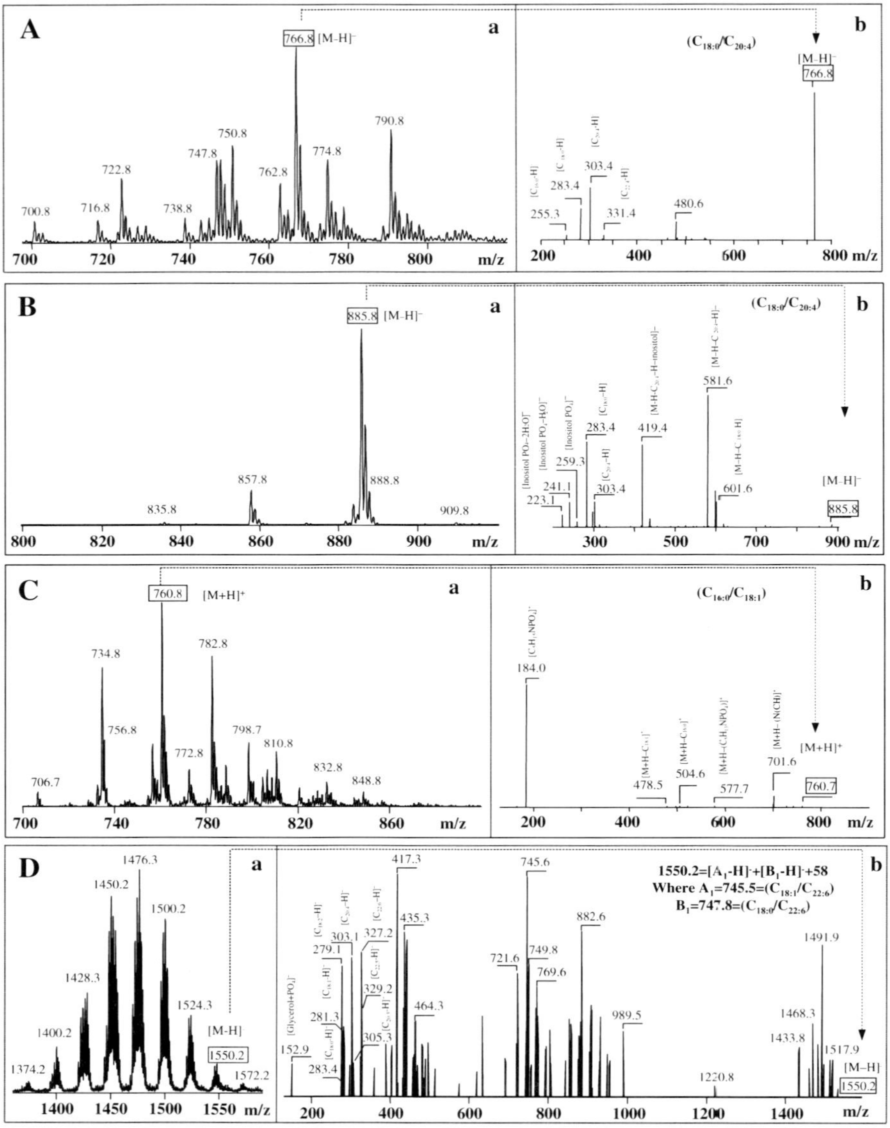

Fig. 21.1. Typical ESI mass spectra of molecular species of PE (**A**), PI (**B**), PC (**C**), and CL (**D**) obtained from rat brain cortical P2 fraction. Phospholipids separated by 2D-HPTLC were subjected to MS analysis by direct infusion into mass spectrometer. Typical MS/MS fragmentation experiments for the major molecular species of PE, PI, PC, and CL are shown on panels **Ab**, **Bb**, **Cb**, and **Db**, respectively. Note that MS spectra for PE, PI, CL were recorded in negative mode and for PC in positive mode.

negative ionization mode, most of the PC molecular species were observed as chloride adducts $[M+Cl]^-$ (data not shown).

Molecular species of CL were represented by at least nine different major clusters with a variety of fatty acid residues (from $C_{14:0}$ to $C_{22:6}$), including those containing polyunsaturated fatty acid residues, such as arachidonic ($C_{20:4}$) and docosahexaenoic ($C_{22:6}$) fatty acids, highly susceptible to oxidation

(**Fig. 21.1, D, a**). MS/MS fragmentation experiments with the CL molecular cluster containing docosahexaenoic ($C_{22:6}$) fatty acid with *m/z* 1550.2 revealed ions with *m/z* 279, 281, 283, 303, 305, 327, and 329 corresponding to several individual molecular CL species with the following fatty acid resi dues – $(C_{18:1})_1/(C_{18:0})_1/(C_{22:6})_2$; $(C_{18:0})_1/(C_{20:4})_2/(C_{22:5})_1$; $(C_{18:2})_1/(C_{20:3})_2/(C_{22:5})_1$; $(C_{18:2})_1/(C_{18:0})_1/(C_{22:5})_1/(C_{22:6})_1$ (**Fig. 21.1, D, b**). Complete structural characterization of major CL clusters was performed by using MS^n approach (multiple-stage fragmentation) on an ion-trap mass spectrometer. To characterize the structure of each CL species, the protocol described by Hsu and Turk was applied (27). Briefly, first, pairs of corresponding A/B or A/B+136 ions were identified in MS^2 spectra and then MS^3 was performed on each A or B ion to assign fatty acids and their positions. One major isomer was identified in CL molecular species of ion with *m/z* 1550.2 (1550.2 = $[A_1\text{–}H]^-$ – $[B_1\text{–}H]^-$+58, where A_1 = 745.5 corresponds to ($C_{18:1}/C_{22:6}$) and B_1 = 747.8 corresponds to ($C_{18:0}/C_{22:6}$)).

In contrast to multiple molecular species of CL and PE, a single major molecular ion of PS with *m/z* 834 was observed under negative ionization mode (**Fig. 21.2**). The main peak with *m/z* 747 formed after fragmentation of PS originated due to loss of the serine group. This fragmentation product ion together with the deprotonated ion (*m/z* 834) effectively determined the molecular mass of PS. There were also ions characteristic of the head group or its fragments (**Fig. 21.2, A, b**), as well as common fragments for phospholipids: *m/z* 79 for $[PO_3]^-$, *m/z* 153 for [glycerophosphate – H_2O-H]$^-$. Molecular fragments with *m/z* 283 and 327 correspond to carboxylate anions of stearic ($C_{18:0}$) and docosahexaenoic ($C_{22:6}$) fatty acids, respectively.

Based on these identification protocols, the distribution of different molecular species in major polyunsaturated phospholipid classes as well as characteristic fragments can be established (**Table 21.1**). Note that PE, CL, and PS contain molecular species with polyunsaturated fatty acid (PUFA) residues, particularly $C_{20:4}$, $C_{22:5}$, and $C_{22:6}$. These PUFA are known to be most susceptible to oxidative attack that is illustrated further.

MS Identification of Oxidized Molecular Species in Major Classes of Oxidizable Phospholipids

As indicated in the Introduction, ESI- and MALDI-MS can be also utilized for the detection and molecular characterization of individual species of phospholipids. Several examples below illustrate the applicability of ESI-MS for this purpose.

Using a model of acute brain injury that is commonly associated with the triggering of cell apoptosis and enhanced production of ROS (8, 28), we employed oxidative lipidomics approach for assessment of individual lipid oxidation products. ESI-MS analysis of PS in the brain cortex samples after traumatic brain

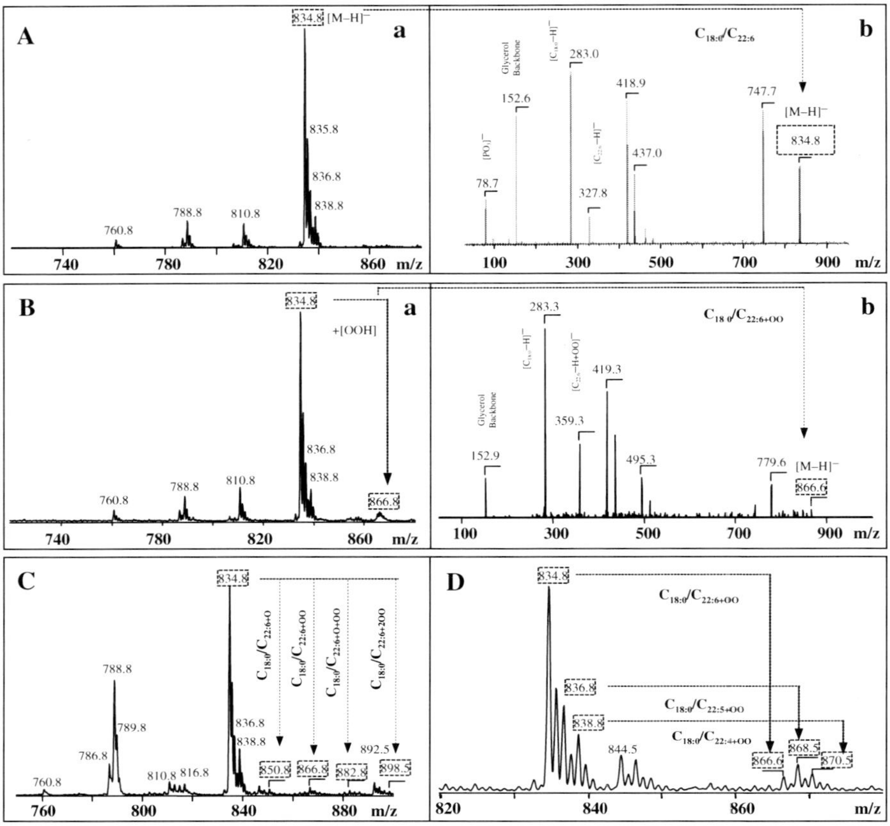

Fig. 21.2. Typical negative-ion ESI mass spectra of PS isolated from rat brain cortex (**A** – control and **B** – after CCl), (**C**) – midfrontal gyrus from human with Alzheimer's disease patients and (**D**) – from mouse gut after total body irradiation. Identification of individual oxidized molecular species of PS containing $C_{22:6+OOH}$. Tandem MS-MS experiments confirmed the structures of non-oxidized and oxidized PS species (panels **A, b; B, b**).

injury of rats showed the enhanced formation of PS molecular species with oxidized $C_{22:6}$, PS-OOH, with *m/z* 866 as compared with the control rats (**Fig. 21.2**). Detailed analysis of this peak by MS-MS demonstrated that the $[M–H]^-$ ion at *m/z* 866 corresponded to PS-OOH with the dominating product of ($C_{18:0}/C_{22:6}$+OO) and produced due to peroxidation of the PS molecular species with the ion at *m/z* 834 ($C_{18:0}/C_{22:6}$).

Oxidative damage to the brain has been long associated with the development of Alzheimer's disease (reviewed in (29)). Recently, it has been established that the loss of heme by the mitochondrial respiratory complex IV triggers the release of oxidants (e.g., H_2O_2); thus produced H_2O_2 may feed peroxidase activity of amyloid-β peptide–heme complexes and cause oxidative stress) in the brains of patients with Alzheimer's disease (30). By using ESI-MS, we were able to identify that PS containing $C_{22:6}$ species in brain samples from patients with Alzheimer's disease were converted into PS-OOH (**Fig. 21.2, C**). Tandem MS-MS

Table 21.1
Major phospholipid oxidizible molecular species from rat brain and mouse gut identified by ESI-MS approach

Phospholipids	m/z	Rat brain fatty acids	m/z	Mouse gut fatty acids
Phosphatidyl-ethanolamine	762.8	$C_{16:0}/C_{22:6}$	714.5	$C_{16:0}/C_{18:2}$
	766.8	$C_{18:0}/C_{20:4}$ and $C_{16:0}/C_{22:4}$	742.5	$C_{18:0}/C_{18:2}$
	790.8	$C_{18:0}/C_{22:6}$	766.7	$C_{18:0}/C_{20:4}$ and $C_{16:0}/C_{22:4}$
	794.7	$C_{18:0}/C_{22:4}$	768.6	$C_{18:0}/C_{20:3}$ and $C_{18:2}/C_{20:1}$
	722.8	$C_{16:0p}/C_{20:4}$	778.5	$C_{20:0}/C_{20:4}$
	747.8	$C_{18:1p}/C_{20:4}$ and $C_{16:0p}/C_{22:6}$	790.7	$C_{20:0}/C_{20:4}$
	750.8	$C_{16:0p}/C_{22:4}$ and $C_{18:0p}/C_{20:4}$	802.7	$C_{18:0}/C_{22:6}$
	747.8	$C_{18:0p}/C_{20:2}$	750.5	$C_{16:0p}/C_{22:4}$ and $C_{18:0p}/C_{20:4}$
	774.7	$C_{18:0p}/C_{22:6}$	776.5	$C_{18:0e}/C_{22:6}$ and $C_{20:0p}/C_{20:4}$
	778.8	$C_{18:0p}/C_{22:4}$	790.7	$C_{18:0p}/C_{22:6}$
	802.7	$C_{20:0p}/C_{22:6}$		
Phosphatidyl-inositol	857.8	$C_{16:0}/C_{20:4}$	833.7	$C_{16:0}/C_{18:2}$
	885.8	$C_{18:0}/C_{20:4}$	857.8	$C_{16:0}/C_{20:4}$
	887.8	$C_{18:0}/C_{20:3}$	861.8	$C_{18:1}/C_{18:2}$
	909.8	$C_{18:0}/C_{22:6}$	885.7	$C_{18:1}/C_{20:4}$
	909.7	$C_{18:0}/C_{22:6}$		
Phosphatidylserine	810.8	$C_{18:0}/C_{20:4}$ and $C_{16:0}/C_{22:4}$	810.7	$C_{18:0}/C_{20:4}$
	834.8	$C_{18:0}/C_{22:6}$	812.7	$C_{18:0}/C_{20:3}$ and $C_{20:1}/C_{18:2}$
	836.8	$C_{18:0}/C_{22:5}$	834.7	$C_{18:0}/C_{22:6}$
	838.8	$C_{18:0}/C_{22:4}$	836.7	$C_{18:1}/C_{22:4}$ and $C_{18:0}/C_{22:5}$
Phosphatidyl-choline	784.7	$C_{18:1}/C_{18:2}$	758.7	$C_{16:0}/C_{18:2}$
	806.7	$C_{16:0}/C_{22:6}$	782.7	$C_{16:0}/C_{20:4}$
	810.7	$C_{18:0}/C_{20:4}$	786.7	$C_{18:0}/C_{18:2}$
	812.7	$C_{18:0}/C_{20:2}$	806.7	$C_{16:0}/C_{22:6}$
	832.7	$C_{18:1}/C_{22:6}$	810.7	$C_{18:0}/C_{20:4}$
	838.7	$C_{18:0}/C_{22:4}$		
	832.7	$C_{18:0}/C_{22:6}$		
Cardiolipin	1424.6	$(C_{16:0})_1/(C_{16:1})_1/(C_{18:1})_1/(C_{20:4})_1$	1447.9	$(C_{18:2})_4$
	1426.4	$(C_{16:1})_1/(C_{18:2})_1/(C_{18:1})_2$	1450.0	$(C_{18:1})_1/(C_{18:2})_3$
	1448.5	$(C_{16:1})_1/(C_{18:2})_1/(C_{18:1})_1/(C_{20:4})_1$		
	1448.7	$(C_{18:2})_4$		

(continued)

Table 21.1 (continued)

Phospholipids	m/z	Rat brain fatty acids	m/z	Mouse gut fatty acids
	1450.5	$(C_{16:1})_1/(C_{18:2})_1/(C_{18:1})_1/(C_{20:3})_1$		
	1476.4	$(C_{18:1})_2/(C_{18:2})_1/(C_{20:4})_1$		
	1498.6	$(C_{18:2})_1/(C_{18:1})_1/(C_{20:4})_2$		
	1500.2	$(C_{18:1})_2/(C_{20:4})_2$		
	1502.6	$(C_{16:0})_1/(C_{18:1})_1/(C_{20:4})_1/(C_{22:4})_1$		
	1518.9	$(C_{16:1})_1/(C_{18:2})_1/(C_{22:6})_2$		
	1520.0	$(C_{18:2})_1/(C_{20:4})_3$		
	1525.2	$(C_{18:1})_2/(C_{20:4})_1/(C_{22:6})_1$		
	1528.6	$(C_{18:1})_2/(C_{20:4})_1/(C_{22:4})_1$		
	1550.2	$(C_{18:1})_1/(C_{18:0})_1/(C_{22:6})_2$		
	1550.2	$(C_{18:0})_1/(C_{20:4})_2/(C_{22:5})_1$		
	1550.2	$(C_{18:2})_1/(C_{20:3})_2/(C_{22:5})_1$		
	1550.2	$(C_{18:2})_1/(C_{18:0})_1/(C_{22:5})_1/(C_{22:6})_1$		
	1553.0	$(C_{18:0})_2/(C_{22:6})_2$		
	1570.2	$(C_{18:1})_1/(C_{20:4})_1/(C_{22:6})_2$		
	1582.7	$(C_{18:1})_1/(C_{20:2})_2/(C_{24:6})_1$		

experiments confirmed the structure of PS-OOH with oxidized $C_{22:6}$ (*m/z* 866). Additionally, other oxidized molecular products with *m/z* 850, 882, and 898 – formed from $C_{22:6}$ fatty acid, containing PS – corresponded to molecular clusters $(C_{18:0}/C_{22:6+O})$, $(C_{18:0}/C_{22:6+O+OO})$, and $(C_{18:0}/C_{22:6+2OO})$, respectively (**Table 21.2**)

In cellular membranes, ROS induced by ionizing irradiation may cause lipid peroxidation and the resulting cellular and/or nuclear damage can lead to apoptosis (31). Typical oxidative modification of PS extracted from mouse gut after the total body irradiation is presented in **Fig. 21.2, D**. Accumulation of oxidized PS species with *m/z* 866 $(C_{18:0}/C_{22:6+OO})$, 868 $(C_{18:0}/C_{22:5+OO})$, and 870 $(C_{18:0}/C_{22:4+OO})$ formed from molecular clusters of PS with *m/z* 834 $(C_{18:0}/C_{22:6})$, 836 $(C_{18:0}/C_{22:5})$, and 838 $(C_{18:0}/C_{22:4})$, respectively, was established (**Fig. 21.2, D**). All these examples clearly demonstrate that MS analysis can be used for *qualitative* identification and structural characterization of individual molecular species of major phospholipids involved in oxidative modifications catalyzed by different enzymatic and nonenzymatic mechanisms in cells and

Table 21.2
Major molecular species of PS in brains from patients with Alzheimer's disease

Molecular species	m/z [M–H]$^-$	Acyl chain composition
Diacyl species		*Acyl/Acyl*
34:1	760.8	$C_{16:0}/C_{18:1}$
36:2	786.8	$C_{18:1}/C_{18:1}$
36:1	788.8	$C_{18:0}/C_{18:1}$
38:4	810.8	$C_{18:0}/C_{20:4}$
40:6	834.8	$C_{18:0}/C_{22:6}$
40:5	836.8	$C_{18:0}/C_{22:5}$
40:4	838.8	$C_{18:0}/C_{22:4}$
40:3	840.8	$C_{18:0}/C_{22:3}$
40:0	846.5	$C_{18:0}/C_{22:0}$
40:6	850.8	$C_{18:0}/C_{22:6+O}$
40:6	866.8	$C_{18:0}/C_{22:6+OO}$
40:6	882.8	$C_{18:0}/C_{22:6+O+OO}$
40:6	898.5	$C_{18:0}/C_{22:6+2OO}$
42:7	860.8	$C_{20:1}/C_{22:6}$
42:6	862.8	$C_{18:0}/C_{24:6}$
42:5	864.8	$C_{18:0}/C_{24:5}$
44:10	882.8	$C_{22:4}/C_{22:6}$

Postmortem frozen tissue samples of cerebral cortex (midfrontal gyrus) from patients with Alzheimer's disease of Pittsburgh University UPMC Hospital were studied for PL content and composition.

tissues. Protocols for quantitative assessments of oxidized phospholipids by MS need further development.

Quantitative Assessments of Phospholipid Hydroperoxides in Complex Mixtures of Phospholipids and Major Classes of Phospholipids

While several techniques have been proposed to assess peroxidation of phospholipids (14, 32), their sensitivity is not always sufficient for the analyses of small amounts of these products in vivo or in cells, particularly as it relates to specific classes of phospholipids. Because PL-OOH are the primary molecular products of peroxidation, while the other numerous secondary products originate from the hydroperoxides and may be present in significantly lower concentrations (33), we reasoned that measurements of PL-OOH may be most representative of the peroxidation process as a whole. Therefore, our efforts were directed toward the development of an assay for PL-OOH. Quantitative determinations of the extent to which phospholipids undergo oxidation, can be performed by fluorescence HPLC. Our protocol is based on the use of Amplex red, a highly sensitive and chemically stable fluorogenic probe, which has been commonly utilized for the measurement of very low levels of H_2O_2 in peroxidase-mediated reactions (34). Here, we describe the use of Amplex red as a very sensitive and stable substrate for the peroxidase activity of MP-11 with high selectivity to FA-OOH.

Incubation of Amplex red with MP-11 in the presence of FA-OOH, 9*S* HpODE, reveals the formation of the oxidized Amplex red fluorescent product (**Fig. 21.3, A**), resorufin, with one major peak (R_f 3.5 min) after its HPLC separation (**Fig. 21.3 B**). The resorufin formation was not observed after incubation of Amplex red in phosphate buffer alone, or in the presence of 9*S*-HpODE alone or MP-11 alone (**Fig. 21.3, B, insert**). Fluorescence of resorufin was not changed when equimolar concentrations of resorufin and 9*S*-HpODE were incubated in the presence of different amounts of MP-11. Moreover, incubation of 9*S*-HpODE in the presence of Amplex red and different amounts of MP-11 resulted in increased formation of highly

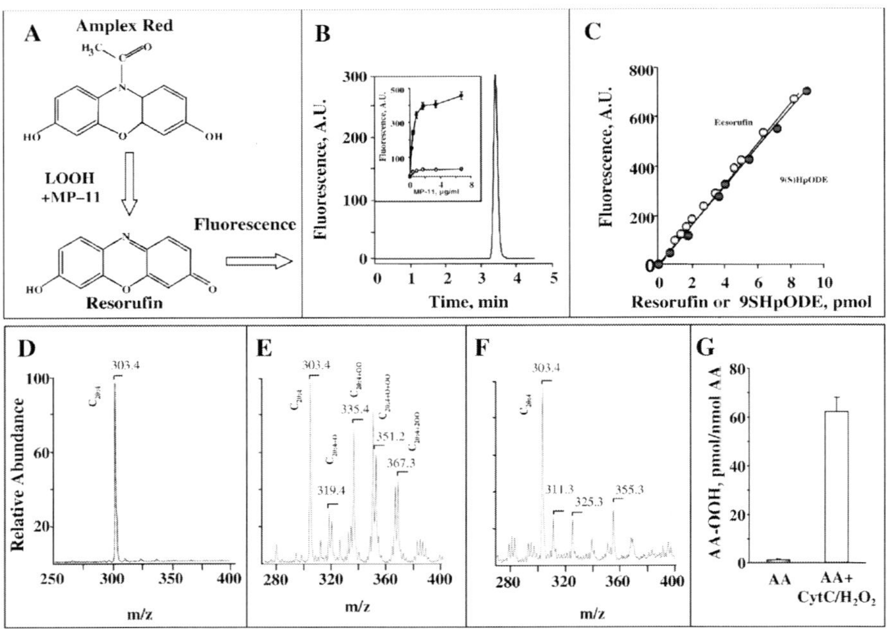

Fig. 21.3. Determination of FA-OOH using Amplex red protocol. Formation of resorufin from Amplex red in the presence of MP-11 and FA-OOH (**a**). Typical separation of the resorufin by HPLC (**b**). Insert shows fluorescent response of resorufin vs concentration of MP-11 in the presence (closed circus) and in the absence (opened circus) of FA-OOH (9*S* HpODE). Fluorescent response of resorufin vs concentration of resorufin or 9*S*-HpODE (**c**). Typical negative-ion ESI-MS of arachidonic acid (**d**), after its oxidation by cyt *c*/H_2O_2 (**e**), and oxidized arachidonic acid after reaction catalyzed by MP-11 in the presence of Amlex red (**f**). Quantification of arachidonic acid hydroperoxides catalyzed by cyt *c*/H_2O_2 (**g**). Reaction conditions: Arachidonic acid (2 mM) was incubated with or without cyt *c* (40 μM), H_2O_2 (2×500 μM), 100 μM DTPA in 50 mM phosphate buffer pH 7.4 at 37° C in for 30 min. Amount of arachidonic acid hydroperoxides was determined by Amlex red/MP-11 system (68.4 ± 5.9 LOOH pmol/nmol FA).

fluorescent resorufin (**Fig. 21.3, B, insert**). The fluorescence response was stable and no significant decrease of resorufin fluorescence was observed during prolonged incubations (**Fig. 21.3, B**, insert). Our results are consistent with published data that fluorescence of resorufin did not significantly change during the incubation of resorufin in the presence of H_2O_2 and horseradish peroxidase (34).

In contrast, to other protocols of FA-OOH determination (ferrous oxidation-xylenol orange, iodometric assays (35, 36)), our Amplex Red/MP-11 assay is free from complications due to potential oxidation of PUFA during the reaction. Incubation of Amplex red and MP-11 in the presence of 1,000-fold excess of non-oxidized arachidonic acid $C_{20:4}$ did not increase the resorufin fluorescence response. Amplex red, in the presence of enzyme MP-11, reacts with FA-OOH at a 1:1 stoichiometry to produce the highly fluorescent product, resorufin (**Fig. 21.3, C**). A linear relationship between 9*S*-HpODE concentration and fluorescence of oxidized Amplex red by MP-11 was consistently observed; the fluorescence response showed excellent correlation with the calibration using standard concentrations of resorufin. The limit of FA-OOH detection under conditions of our experiments was found to be 1–2 pmol. Moreover, the pool of FA-OOH was completely utilized after reaction with Amplex Red/MP-11, which was confirmed by MS (**Fig. 21.3 D, E, F**). Incubation of $C_{20:4}$ with cyt *c*/H_2O_2 resulted in the formation of $C_{20:4+OOH}$ (from 1.24 in control to 68.4 pmol FA-OOH per nmol fatty acid, **Fig. 21.3, G**). According to the ESI-MS data, non-oxidized $C_{20:4}$ had one product of molecular ion with a 303.3 *m/z* (**Fig. 21.1, D**). Incubation of $C_{20:4}$ with cyt *c*/H_2O_2 generated a series of $C_{20:4}$ oxidation products containing mono- $C_{20:4+OO}$ (335.3 *m/z*), and di-hydroperoxides $C_{20:4+2OO}$ (367.3 *m/z*), as well as other oxygenated species (**Fig. 21.3, E**), which were not observed in the MS spectrum after the reaction of oxidized $C_{20:4+OO}$ with Amplex Red/ MP-11 (**Fig. 21.3, F**).

Data presented in **Table 21.3** illustrate the resorufin formation catalyzed by Amplex Red/MP-11 from different fatty acid derivatives. One can see that resorufin was produced only in the presence of FA-OOH. This additionally confirms specific interactions of Amplex Red/MP-11 with FA-OOH.

Peroxidase activity of cyt *c* complexes with anionic phospholipids such as PS or CL can utilize their fatty acid residues as the reaction substrates, i.e., oxidize these phospholipids (7). Therefore, in a separate series of experiments, we employed the Amplex Red/MP-11 system to monitor cyt *c*-catalyzed production of PL-OOH in the presence of palmitoyl-arachidonoyl-PS (PAPS) or tetralinoleoyl-CL (TLCL) and H_2O_2. Incubation of cyt *c*/PAPS or cyt *c*/TLCL with H_2O_2 resulted in the accumulation of phospholpid oxidation products, which was also confirmed by

Table 21.3
Resorufin formation from Amplex red catalyzed by MP-11 in the presence of different fatty acid derivatives

Fatty acids	Resorufin pmol/nmol FA	Structures of fatty acids
9Z-Octadecenoic acid, Oleic acid	0.02 ± 0.02	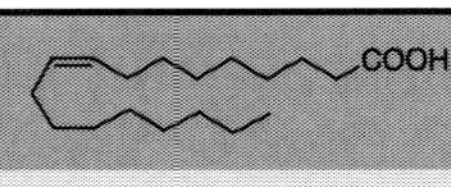
9Z, 12Z-Octadecadienoic acid, Linoleic acid	0.52 ± 0.11	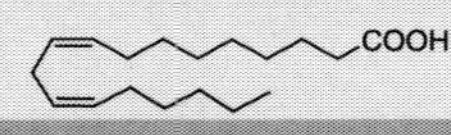
5Z, 8Z, 11Z, 14Z-Eicosatetraenoic acid, Aracidonic acid	1.24 ± 0.12	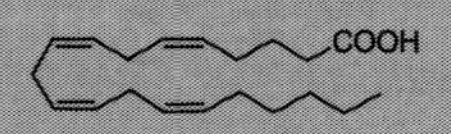
15S-Hydroxy-5Z, 8Z, 11Z, 13E, 17Z-eicosapentaenoic acid, 15(S)-HEPE	5.64 ± 7.35	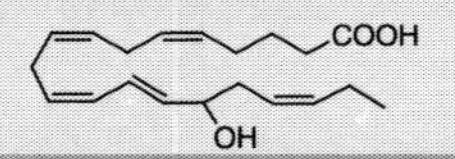
9E, 11Z -Octadecadienic acid, 13-oxo-, 13-Oxo-ODE	16.42 ± 8.87	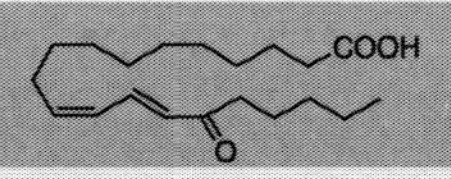
(+)8(9)-Epoxy-5Z, 11Z, 14Z-eicosatrienoic acid, 8(9)-EpTrE	1.99 ± 2.35	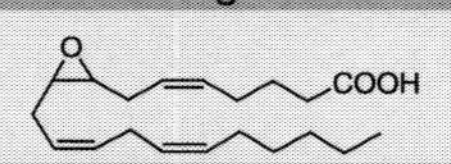
9S-Hydroperoxy-10E, 12Z-octa-decadienoic acid, 9(S)-HpODE	993.38 ± 29.72	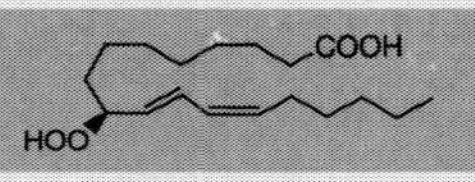
13S-Hydroperoxy-10E, 12Z-octa-decadienoic acid, 13(S)-HpODE	983.45 ± 39.74	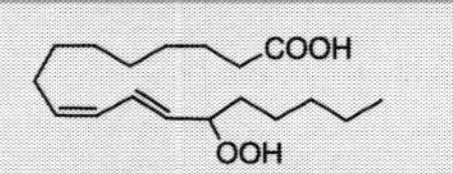
15S-Hydroperoxy-5Z, 8Z, 11Z, 13E-eicosatetraenoic acid, 15(S)-HpETE	996.45 ± 29.74	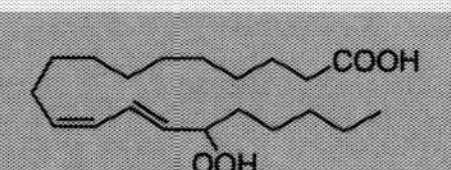

ESI-MS-analysis (data not shown). It was important to determine whether oxidized and non-oxidized species of PAPS (PAPSox) and TLCL (TLCLox) co-migrated during their separation by 2D-HPTLC. We found this to be the case as evidenced by the Amplex Red/MP-11 based assay as well as by ESI-MS analysis.

Because PS and CL molecules contain two and four fatty acid residues, respectively, the availability of MP-11 to oxidized acyl moieties in the respective phospholipid molecules could be different. To evaluate this, we performed experiments in which we treated PS and CL with PLA_2 and measured the formation of FA-OOH by the MP-11 catalyzed resorufin response. We found that PLA_2 treatment did not affect the formation of resorufin in PAPSox. In contrast, pretreatment of TLCLox with PLA_2 resulted in a marked additional increase of MP-11 catalyzed resorufin formation (up to 25–35%). Thus, treatment with PLA_2 should be uniformly used to reliably perform quantitative analyses of peroxidation in complex mixtures of phospholipids extracted from different cells and tissues.

Validation of the Amplex Red/MP-11/PLA_2 protocol confirmed its applicability for quantitative measurements of phospholipid peroxidation. Analyses of the contents of PL-OOH were performed in several cells lines exposed to oxidant and non-oxidant pro-apoptotic stimuli. In line with the MS measurements, we found that CL and PS underwent most robust oxidation, while other phospholipids such as PE, and PC were only slightly oxidized (**Fig. 21.4**). Comparison of the abundance of phospholipids with the abundance of oxidized phospholipids permits to present the results as lipid arrays vs. oxidized lipid arrays. Using different models of apoptosis in a variety of cell lines, we found that the pattern of phospholipid oxidation during apoptosis was nonrandom and did not follow their abundance in the analyzed samples. Considering the affinity of cyt *c* for anionic PS and CL and the emergence of its peroxidase activity, it was expected that cyt *c*/CL and cyt *c*/PS complexes are very effective catalysts of self-oxidation. Thus, cyt *c*/PS or cyt *c*/CL complexes may act

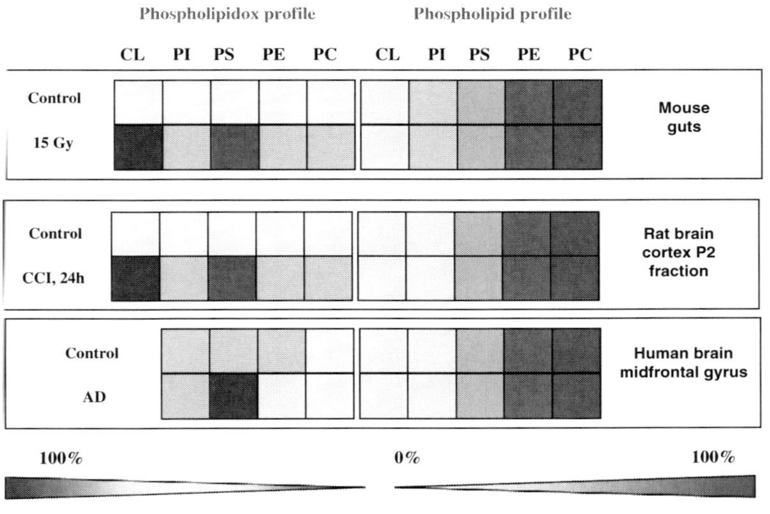

Fig. 21.4. Comparison of the abundance of major PL classes with their oxidation products. Profiles of phospholipids and PL-OOH in: (1) rat brain cortical P2 fractions after CCI, (2) human brain midfrontal gyrus with Alzheimer's disease, and (3) mouse gut after total body irradiation. Phospholipid content is expressed as percent of total phospholipids and shown in green-blue scale. Phospholipid hydroperoxides are presented as percent of phospholipids (pmol PL-OOH per nmol of phospholipid) and shown in yellow-brown scale. In a variety of apoptotic cells and the disease processes investigated, CL and PS underwent most robust oxidation compared with other phospholipids (PE, PI, and PC). One hundred percent corresponds to 110 $\pm$ 20 pmol; 71 $\pm$ 32 pmol, and 79 $\pm$ 20 pmol of PL-OOH per nmol of phospholipid in rat brain cortex, human brain (midfrontal gyrus), and mouse gut, respectively.

as CL-, or PS-specific oxygenases capable of oxidizing polyunsaturated molecular species of CL and PS in mitochondria and plasma membrane, respectively (7, 8, 13). The importance of these oxidative modifications of the phospholipid signaling in the release of pro-apoptotic factors from mitochondria into the cytosol and the externalization of PS and its appearance on the surface of apoptotic cells as an "eat-me" signal are described in several recent reviews (8, 9, 13, 37–39).

Acknowledgments

Supported by grants from NIH HL70755, NIH U19 AI068021, NIOSH OH008282, HD057587, AHA0535365N, Pennsylvania Department of Health SAP 4100027294, Human Frontier Science Program.

References

1. Vijg, J. and Suh, Y. (2005) Genetics of longevity and aging. *Annu. Rev. Med.* **56**, 193–212.
2. Watson, A.D. (2006) Thematic review series: systems biology approaches to metabolic and cardiovascular disorders. Lipidomics: a global approach to lipid analysis in biological systems. *J. Lipid Res.* **47**, 2101–2111.
3. Gaspar, M.L., Aregullin, M.A., Jesch, S.A., Nunez, L.R., Villa-García, M., and Henry, S.A. (2007) The emergence of yeast lipidomics. *Biochim. Biophys. Acta* **1771**, 241–254.
4. Marathe, G.K., Harrison, K.A., Murphy, R.C., Prescott, S.M., Zimmerman, G.A., and McIntyre, T.M. (2000) Bioactive phospholipid oxidation products. *Free Radic. Biol. Med.* **28**, 1762–1770.
5. DeLong, C.J., Baker, P.R., Samuel, M., Cui, Z., and Thomas, M.J. (2001) Molecular species composition of rat liver phospholipids by ESI-MS/MS: the effect of chromatography. *J. Lipid Res.* **42**, 1959–1968.
6. Podrez, E.A., Poliakov, E., Shen, Z., Zhang, R., Deng, Y., Sun, M., Finton, P.J., Shan, L. Febbraio, M., Hajjar, D.P., Silverstein, R.L., Hoff, H.F., Salomon, R.G., and Hazen, S.L. (2002) A novel family of atherogenic oxidized phospholipids promotes macrophage foam cell formation via the scavenger receptor CD36 and is enriched in atherosclerotic Lesions. *J. Biol. Chem.* **277**, 38517–38523.
7. Kagan, V.E., Tyurin, V.A., Jiang, J., Tyurina, Y.Y., Ritov, V.B., Amoscato, A.A., Osipov, A.N., Belikova, N.A., Kapralov, A.A., Kini, V., Vlasova, I.I., Zhao, Q., Zou, M., Di, P., Svistunenko, D.A., Kurnikov, I.V., and Borisenko, G.G. (2005) Cytochrome c acts as a cardiolipin oxygenase required for release of proapoptotic factors. *Nature Chem. Biol.* **4**, 223–232.
8. Kagan, V.E., Tyurina, Y.Y., Bayir, H., Chuh, C.T., Kapralov, A.A., Vlasova, I.I., Belikova, N.A., Tyurin, V.A., Amoscato, A., Epperly, M., Greenberger, J. DeKosky, S. Shvedova, A.A., and Jiang, J. (2006) The "pro-apoptotic genies" get out of mitochondria: Oxidative lipidomics and redox activity of cytochrome c/cardiolipin complexes. *Chem. Biol. Interact.* **163**, 15–28.
9. Bayir, H., Fadeel, B., Palladino, M.J., Witasp, E., Kurnikov, I.V., Tyurina, Y.Y., Tyurin, V.A., Amoscato, A.A., Jiang, J., Kochanek, P.M., DeKosky, S.T., Greenberger, J.S., Shvedova, A.A., and Kagan, V.E. (2006) Apoptotic interactions of cytochrome *c*: redox flirting with anionic phospholipids within and outside of mitochondria. *Biochim. Biophys. Acta* **1757**, 648–659.
10. Kujoth, G.C., Leeuwenburgh, C., and Prolla, T.A. (2006) Mitochondrial DNA mutations and apoptosis in mammalian aging. *Cancer Res.* **66**, 7386–7389.
11. Green, D.R. and Kroemer, G. (2004) The pathophysiology of mitochondrial cell death. *Science* **305**, 626–629.

12. Chandra, J., Samali, A., and Orrenius, S. (2000) Triggering and modulation of apoptosis by oxidative stress. *Free Radic. Biol. Med.* **29**, 323–333.
13. Kagan, V.E., Borisenko, G.G., Tyurina, Y.Y., Tyurin, V.A., Jiang, J., Potapovich, A.I., Kini, V., Amoscato, A.A., and Fujii, Y. (2004) Oxidative lipidomics of apoptosis: Redox catalytic interactions of cytochrome c with cardiolipin and phosphatidylserine. *Free Radic. Biol. Med.* **37**, 1963–1985.
14. Kriska, T., Korytowski, W., and Girotti, A.W. (2005) Role of mitochondrial cardiolipin peroxidation in apoptotic photokilling of 5-aminolevulinate-treated tumor cells. *Arch. Biochem. Biophys.* **433**, 435–446.
15. Kagan, V.E., Gleiss, B., Tyurina, Y.Y., Tyurin, V.A., Elenström-Magnusson, C., Liu, S.-X., Serinkan, F.B., Arroyo, A., Chandra, J., Orrenius, S., and Fadeel, B. (2002) A role for oxidative stress in apoptosis: Oxidation and externalization of phosphatidylserine is required for macrophage clearance of cells undergoing Fas-mediated apoptosis. *J. Immunol.* **169**, 487–499.
16. Tyurina, Y.Y., Kawai, K., Tyurin, V.A., Liu, S.X., Kagan, V.E., and Fabisiak, J.P. (2004a) The plasma membrane is the site of selective phosphatidylserine oxidation during apoptosis: role of cytochrome *c*. *Antioxid. Redox Signal* **6**, 209–225.
17. Wu, Y., Tibrewal, N., and Birge, R.B. (2006) Phosphatidylserine recognition by phagocytes: a view to a kill. *Trends Cell Biol.* **16**, 189–190.
18. Nomura, K, Imai, H, Koumura, T, Kobayashi, T., and Nakagawa, Y. (2000) Mitochondrial phospholipid hydroperoxide glutathione peroxidase inhibits the release of cytochrome c from mitochondria by suppressing the peroxidation of cardiolipin in hypoglycaemia-induced apoptosis. *Biochem. J.* **351**, 183–193.
19. Nakagawa, Y. (2004) Initiation of apoptotic signal by the peroxidation of cardiolipin of mitochondria. *Ann. NY Acad. Sci.* **1011**, 177–184.
20. Tyurina, Y.Y., Serinkan, F.B., Tyurin, V.A., Kini, V., Yalowich, J.C., Schroit, A.J., Fadeel, B., and Kagan, V.E. (2004b) Lipid antioxidant, etoposide, inhibits phosphatidylserine externalization and macrophage clearance of apoptotic cells by preventing phosphatidylserine oxidation. *J. Biol. Chem.* **279**, 6056–6064.
21. Kriska, T. and Girotti, A.W. (2004) Separation and quantitation of peroxidized phospholipids using high-performance thin-layer chromatography with tetramethyl-p-phenylenediamine detection. *Anal. Biochem.* **300**, 23–43.
22. Pulfer, M. and Murphy, R.C. (2003) Electrospray mass spectrometry of phospholipids. *Mass Spectrom. Rev.* **22**, 332–364.
23. Forrester, J.S., Milne, S.B., Ivanova, P.T., and Brown, H.A. (2004) Computational lipidomics: A multiplexed analysis of dynamic changes in membrane lipid composition during signal transduction. *Mol. Pharmacol.* **65**, 813–821.
24. Folch, J., Lees, M., and Sloan-Stanley, G.H. (1957) A simple method for isolation and purification of total lipids from animal tissue. *J. Biol. Chem.* **226**, 497–509.
25. Bottcher, C.J.F., Van Gent, C.M., and Pries, C. (1961) A rapid and sensitive sub-micro phosphorus determination. *Anal. Chim. Acta* **24**, 203–204.
26. Schwartz, J.C., Syka, J.E.P., and Quarmby, S.T., (2005) Improving the fundamentals of Msn on 2D linear ion traps: new ion activation and isolation techniques. *Proc. Am. Soc. MS.*, San Antonio, Texas.
27. Hsu, F.F. and Turk, J.J. (2006) Characterization of cardiolipin from Escherichia coli by electrospray ionization with multiple stage quadrupole ion-trap mass spectrometric analysis of [M - 2H + Na]- ions. *J. Am. Soc. Mass Spectrum.* **17**, 420–429.
28. Kochanek, P.M., Clark, R.S., Ruppel, R.A., Adelson, P.D., Bell, M.J., Whalen, M.J., Robertson, C.L., Satchell, M.A., Seidberg, N.A., Marion, D.W., and Jenkins, L.W. (2000) Biochemical, cellular, and molecular mechanisms in the evolution of secondary damage after severe traumatic brain injury in infants and children: lessons learned from the bedside. *Pediatr. Crit. Care Med.* **1**, 4–19.
29. Nunomura, A., Castellani, R.J., Zhu, X., Moreira, P.I., Perry, G., and Smith, M.A. (2006) Involvement of oxidative stress in Alzheimer disease. *J. Neuropathol. Exp. Neurol.* **65**, 631–641.
30. Atamna, H. and Boyle, K. (2006) Amyloid-beta peptide binds with heme to form a peroxidase: Relationship to the cytopathologies of Alzheimer's disease. *Proc. Natl. Acad. Sci. USA* **103**, 3381–3386.
31. Epperly, M.W., Osipov, A.N., Martin, I., Kawai, K.K., Borisenko, G.G., Tyurina, Y.Y., Jefferson, M., Bernarding, M., Greenberger, J.S., and Kagan, V.E. (2004) Ascorbate as a "redox senser" and protector against irradiation-induced oxidative stress in 32D CL3 hematopoietic cells and subclones overexpressing human manganese superoxide dis-

mutase. *Int. J. Rad. Oncol. Biol. Phys.* **58**, 851–861.

32. Mawatari, S. and Murakami, K. (1998) Analysis of membrane phospholipid peroxidation by isocratic high-performance liquid chromatography with ultraviolet detection. *Anal. Biochem.* **264**, 118–123.
33. Kagan, V.E. (1988) *Lipid Peroxidation in Biomembranes.* CRC Press, Boca Raton, Florida, pp. 1–181.
34. Zhou, M., Diwu, Z., Panchuk-Voloshina, N., and Haugland, R.P. (1997) A stable nonfluorescent derivative of resorufin for the fluorometric determination of trace hydrogen peroxide: applications in detecting the activity of phagocyte NADPH oxidase and other oxidases. *Anal. Biochem.* **253**, 162–168.
35. Gebicki, J.M., Collins, J., Baoutina, A., and Phair, P. (1996) The limitations of an iodometric aerobic assay for peroxides. *Anal. Biochem.* **240**, 235–241.
36. Dobarganes, M.C. and Velasco, J. (2002) Analysis of lipid hydroperoxides. *Eur. J. Lipid Sci. Technol.* **104**, 420–428.
37. Fadeel, B. (2004) Plasma membrane alterations during apoptosis: role in corpse clearance. *Antioxid. Redox Signal.* **6**, 269–275.
38. Kadl, A., Bochkov, V.N., Huber, J., and Leitinger, N. (2004) Apoptotic cells as sources for biologically active oxidized phospholipids. *Antioxid. Redox Signal.* **6**, 311–320.
39. Steinbrecher, U.P., Gomez-Munoz, A., and Duronio, V. (2004) Acid sphingomyelinase in macrophage apoptosis. *Curr. Opin. Lipid* **15**, 531–537.

Chapter 22

Simultaneous Analysis of Multiple Lipid Oxidation Products In Vivo by Liquid Chromatographic-Mass Spectrometry (LC-MS)

Huiyong Yin, Todd Davis, and Ned A. Porter

Abstract

Free radical-induced oxidation of polyunsaturated fatty acid (PUFAs) has been linked to a number of human diseases including atherosclerosis and neurodegenerative disorders. Oxidation of PUFAs generates hydroperoxides and cyclic peroxides that are reduced to lipid alcohol, such as hydroxyeicosatetraenoic acid (HETEs), and isoprostanes (IsoPs) respectively. The IsoPs are isomers of prostaglandins that are generated from autoxidation of arachidonic acid (C20:4). Quantification of F_2-IsoPs has been regarded as the "gold standard" to assess oxidative stress status in various human diseases. We herein report the protocol of analyzing HETEs and F_2-IsoPs using a triple quadrupole mass spectrometer coupled to reverse phase liquid chromatography. The selected reaction monitoring (SRM) mode selects the parent ion of interest in the first Quad (m/z 319 for HETE and m/z 353 for F_2-IsoPs) and fragments it in the second while an ion characteristic of the analyte of interest is monitored in the third Quad. This highly selective technique permits the simultaneous analysis of multiple oxidation products such as the HETEs and F_2-IsoPs. This LC-MS technique can be applied to study the free radical oxidation mechanism in vitro and assess the oxidative stress status in biological tissues and fluids.

Key words: Lipid peroxidation, free radicals, liquid chromatography – mass spectrometry (LC-MS), isoprostanes, oxidative stress, biomarkers.

Abbreviation: APCI, atmospheric pressure chemical ionization ; BHT, butylated hydroxytoluene; CID, collision induced dissociation; ESI-MS, electrospray mass spectrometry; GC, gas chromatography; HPLC, high performance liquid chromatography; IsoP, isoprostane; IPA, isopropanol; LC, liquid chromatography; MS, mass spectrometry; PG, prostaglandin; $PGF_{2\alpha}$, Prostaglandin $F_{2\alpha}$; HpETE, hydroperoxyeicosatetraenoate SRM, selective reaction monitoring.;

1. Introduction

Free radical-induced autoxidation of polyunsaturated fatty acids (PUFAs) has been implicated in numerous human disorders, including atherosclerosis and neurodegenerative diseases (1–4).

R.M. Uppu et al. (eds.), *Free Radicals and Antioxidant Protocols*, Methods in Molecular Biology 610,
DOI 10.1007/978-1-60327-029-8_22, © Humana Press, a part of Springer Science+Business Media, LLC 1998, 2010

A variety of techniques have been developed to measure free radical products in pathophysiological processes. These techniques include measurement of malondialdehyde (MDA) (5), conjugated dienes (6), short-chain alkanes, lipid hydroperoxides (7), and immunological assays (8), among others (9). These methods, however, suffer from inherent problems related to sensitivity and specificity, particularly when applied to in vivo situations. It has been proven, over the past decades, that quantification of a novel class of compounds termed isoprostanes by mass spectrometry coupled with gas chromatography (GC) represents a reliable and accurate approach to assess lipid oxidation and oxidative stress in vivo (10).

Lipid peroxidation gives rise to a host of products and the isoprostanes (IsoPs) are prostaglandin (PG)-like compounds that are formed from the peroxidation of arachidonic acid (11, 12). Formation of these compounds proceeds by successive 5-*exo* radical cyclizations to give bicyclic endoperoxides that are reduced to PGF_2-like compounds termed F_2-IsoPs (**Scheme 22.1**). A number of research groups including the group of Pryor at LSU have contributed to chemical studies involving the mechanism of formation of the isoprostanes (13). A number of IsoPs have been identified, such as those containing either F-type, E/D-type, or A/J-type prostane rings (10, 14). A novel aspect of IsoP formation is that, unlike cyclooxygenase-derived PGs, they are generated in situ esterified to phospholipids and cholesterol. In addition, they possess different stereochemical features compared to enzymatically derived PGs (15). Furthermore, some IsoPs exert potent biological activity by acting as ligands for either plasma membrane bound PG receptors or for nuclear receptors (15–18).

IsoP formation involves the initial hydrogen atom abstraction at either carbon 7, 10, or 13 of arachidonate,followed by molecular O_2 addition and radical cyclization to form unstable bicyclic endoperoxides (**Scheme 22.1a**) (19–21). A number of different parent IsoPs can be generated from these unstable endoperoxide precursors by either reduction or rearrangement. F_2-IsoPs are formed upon reduction of the bicyclic peroxides and hydroperoxide moieties in these molecules. Four regioisomeric F_2-IsoPs are formed according to this mechanism (**Scheme 22.1b**). Although five stereogenic centers are generated in the process of F_2-IsoP formation, only eight possible racemic diastereomers can be formed from one precursor peroxyl radical because the two hydroxyl groups on the prostane ring have to adapt a *cis* configuration. Therefore a total of 32 racemic F_2-IsoP stereoisomers can be generated theoretically since four sets of regioisomeric IsoPs are formed (10). The routine analysis of F_2-IsoPs by previously published gas chromatography (GC)/mass spectrometry (MS) methods quantifies all possible regioisomers and stereoiso-

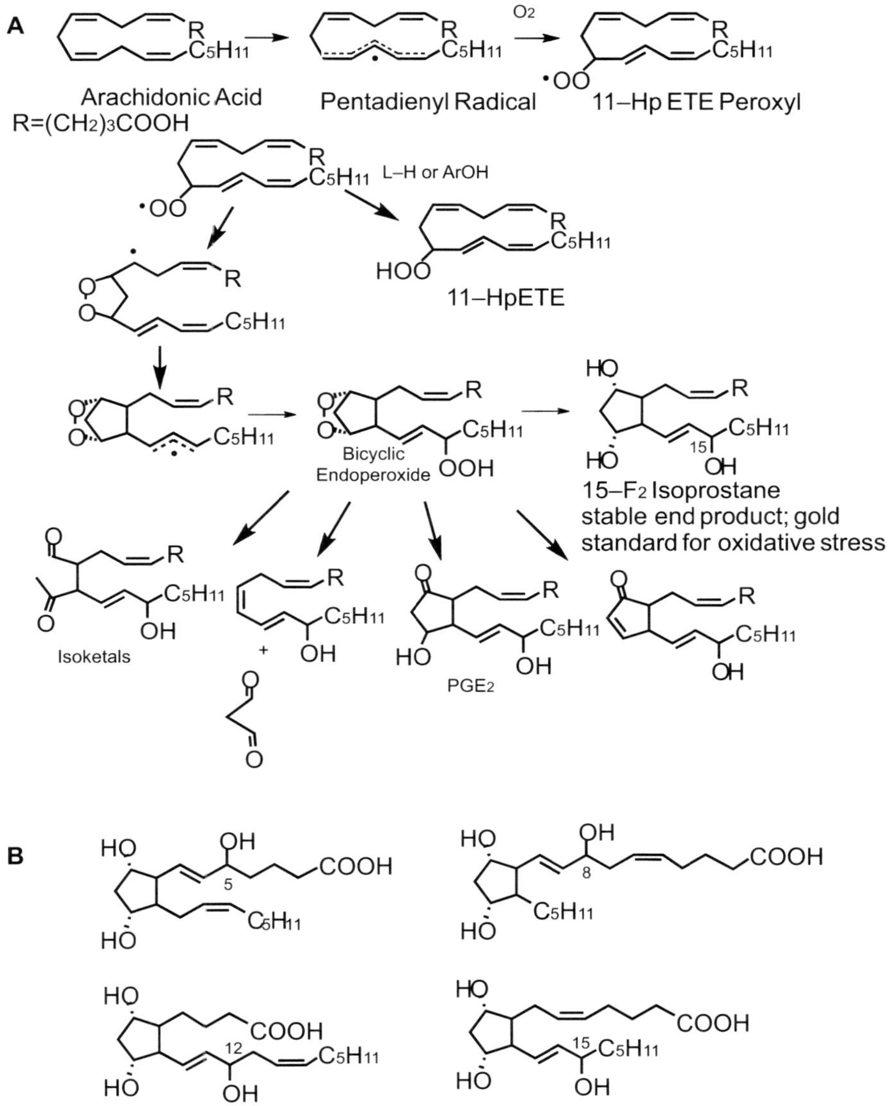

Fig. 21.4. **Scheme 22.1.**

mers. It may be of importance, however, to separate and quantify different F_2-IsoPs because they may have different biological activities and be formed and metabolized differently under human disease conditions which are linked to oxidative stress (8, 22, 23).

In addition to the formation of IsoPs, hydroxyeicosatetraenoic acids (HETEs) belong to another major class of oxidation products of arachidonate. They are formed from the reduction of hydroperoxyeicosatetraenoic acids (HpETEs) and the latter are generated from the hydrogen atom abstraction of peroxyl radicals. Six regioisomeric HETEs can be formed from the autoxidation of arachidonate, namely 5-, 8-, 9-, 11-, 12-, and 15-HETEs. The product distribution of arachidonates may indicate the concentration of hydrogen atom donors, such as α-tocopherol (Vit. E). Furthermore, some of the HETEs, such as 15-, 12-, and 5-HETEs, can be generated from the lipoxygenase pathways, and these lipid mediators have important biological activities (24).

Thus, simultaneous quantification of the multiple oxidation products of arachidonate, such as F_2-IsoPs and HETEs, is of significant importance to assess the oxidative stress status of the pathophysiological processes that are believed to be linked to oxidative stress. Separation and quantification of different regio- and diastereoisomers of F_2-IsoPs as well as the regioisomers of HETEs can best be accomplished by the use of LC-MS approaches. In this protocol (**Scheme 22.2**), reverse phase LC is coupled to electrospray (ESI) MS to analyze free acid F_2-IsoPs and HETEs. These two classes of major arachidonate oxidation products can be analyzed in a single LC-MS analysis. The four F_2-IsoP regioisomers and six HETEs can be distinguished based on their characteristic fragmentation in collision induced dissociation (CID) using the selective reaction monitoring (SRM) technique. For clarification of the origin of the HETEs, chiral LC can be coupled to the MS because nonenzymatic free radical reactions generate a racemic mixture whereas enzymatic processes, i.e., LOX enzymes, always give rise to only one enantiomer (25) (*see* **Note 1**).

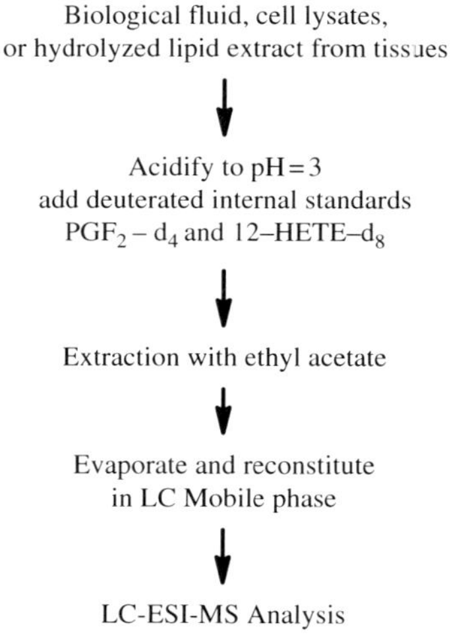

Fig. 21.4. **Scheme 22.2.**

2. Materials

2.1. Reagents

1. Arachidonic acid (Nu Chek Prep., Elysian, MI), >99+%.
2. 2,2′-azobis(4-methoxy-2,4-dimethylvaleronitrile) (MeO AMVN) (Wako Chemicals USA, Richmond, VA).
3. High-purity organic solvents: water, benzene, methanol, chloroform, ethyl acetate, acetonitrile, isopropanol (Burdick & Jackson, Muskegon, MI).
4. Butylated hydroxytoluene (BHT), Triphenylphosphine (TPP), Pentafluorobenzyl bromide (PFB-Br), *N,N*'-Diisopropylethylamine (DIPE), phosphomolybdic acid (Sigma-Aldrich Chemical Co., Milwaukee, WI).
5. Sodium Chloride, Potassium hydroxide, Sodium Sulfate (Na_2SO_4, anhydrous), HCl.
6. $PGF_{2\alpha}$-d_4, 12-HETE-d_8 (Cayman Chemical. Ann Arbor, MI).

2.2. Equipments

1. Blade homogenizer – PTA 10S generator (Brinkmann Instruments, Westbury, NY).
2. Tank of nitrogen gas.
3. Bench-top microcentrifuge (Fischer Scientific, Pittsburg, PA).
4. Water bath at 37°C.
5. LC Column: Phenomenex Luna C18 (2), ODS, 5μ, 2.1 mm × 15 cm or, 2.1 mm × 5 cm (Torrence, CA).
6. HPLC: Waters, Model 600E pump with a Waters 996 Photodiode array detector, autosampler, and Millenium32 chromatography software (Waters Corp, Milford, MA).
7. Mass spectrometer: Thermo Finnigan TSQ Quantum Ultra with Xcaliber software, version 1.3 (Thermo Electron, San Jose, CA).
8. Sep-Pak cartridges, C18 and Silica (Waters Associates, Milford, MA).
9. Disposable plastic syringes (10 mL).
10. Glass scintillation vials (20 mL).
11. Microcentrifuge tubes (2 mL).
12. Analytical evaporation units (such as Meyer N-Evap, Organomation Associates Inc., Berlin, MA).
13. Spin-X filter, pore size: 0.2μ (Corning, NY, USA).

3. Methods

3.1. Cautions for Handling and Storage of Biological Fluids and Tissues for Quantification of Lipid Oxidation Products

Since autoxidation of PUFAs in lipids of biological fluids and tissue can occur ex vivo, false results will be obtained if necessary steps in sample handling are not followed. Samples should be analyzed immediately or stored at –70°C (*see* **Note 2**). Tissue or biological fluids can be snap-frozen in liquid nitrogen and stored at –70°C. Furthermore, BHT and triphenyl phosphine (TPP) are added to the extraction solvent to minimize the ex vivo generation of oxidation products.

3.2. Extraction of Lipid Oxidation Products

The process to extract the lipid oxidation products from tissues or fluids is slightly different and thus is described in separate sections.

3.2.1. Extraction of Phospholipid-Bound Lipid Oxidation Products from Tissue

The method summarized here is applicable to all animal and human samples:

1. Weigh out 0.05–1.0 g of fresh or frozen tissue.
2. Add tissue to 20 mL of ice-cold Folch solution (chloroform: methanol, 2:1, v/v) containing 0.005% (w/v) BHT in a 40 mL glass centrifuge tube.
3. Homogenize tissue with blade homogenizer at full speed for 2 min to sufficiently grind the tissue.
4. Purge the air space in the centrifuge tube with nitrogen and cap the tube. Let the solution stand at room temperature for 30 min.; shake the tube every 10 min during this period of time.
5. Add 4 mL of aqueous sodium chloride 0.9% (w/v). Vortex or shake rigorously for 2 min.
6. Centrifuge on a bench-top instrument at 800×*g* for 10 min at room temperature.
7. After centrifugation, the top aqueous layer can be separated carefully by using a glass pipette and saved for analysis of oxidation products in free acid form. The center layer contains proteins and fibers of the tissue and can be removed and discarded. The bottom layer organic phase is evaporated using a rotavap. The residue is immediately processed for hydrolysis.

3.2.2. Extraction of Phospholipid-Bound Lipid Oxidation Products from Fluids

We have found that arachidonic acid in plasma is much more susceptible to oxidation and thus extra precaution is needed to prevent ex vivo oxidation when fluids are processed. TPP is added to the Folch solution in addition to BHT.

1. Add 1.0 mL of a biological fluid to 20 mL ice-cold Folch solution containing 0.005% (w/v) BHT and 5 mg of TPP in a 40 mL conical centrifuge tube.

2. Shake vigorously for 2 min.
3. Add 10 mL 0.043% (w/v) $MgCl_2$ and shake vigorously for 2 min.
4. Centrifuge 800×*g* for 10 min at room temperature.
5. Separate the two phases.

3.3. Hydrolysis of the Lipid Extract

1. To the lipid residue in a conical flask, add 1 mL of methanol containing 0.005% (w/v) BHT and 1 mL of aqueous KOH (15%, w/v).
2. After purging the flask with nitrogen, incubate the mixture at 37°C for 30 min.
3. Add 1–5 ng of deuterated internal standards, such as [2H] 15-F2α-IsoP (8-Iso PGF2α) and 12-HETE-d_8 after the mixture is cooled to room temperature.
4. Adjust the pH of the mixture to 3.0 by addition of 1.0 N HCl.
5. Add an equal volume of ethyl acetate to the reaction mixture and vortex rigorously.
6. Carefully separate the upper layer of organic phase and repeat step 5 mentioned above.
7. Combine the two organic phase and evaporate off the solvent.
8. Reconstitute the residue into 100μL of 80% (v/v) LC mobile phase A (2 mM NH_4Ac, pH 5.6) and 20% (v/v) B (CH_3CN: MeOH, 65:35, v/v).
9. Filter the solution by using Spin-X filter and it is ready for LC-ESI-MS analysis.

3.4. Analysis of Lipid Oxidation Products by LC-MS

HETEs and F_2-IsoPs can be analyzed by LC-MS using ESI-MS. Sensitivity of this protocol is in the lower pg range and thus can be applied to analyze most biological tissues and fluids (*see* **Note 3**).

For analysis of HETEs and F_2-IsoPs by ESI-MS, the experiment is conducted using a ThermoFinnigan TSQ Quantum Ultra equipped with a Finnigan Surveyor Autosampler Plus. Reverse phase HPLC is performed using a reverse phase Luna C18 (2) (2.1 mm×15 cm) column with a gradient summarized in **Table 22.1**. The MS is operated in the negative ion mode using the selective reaction monitoring (SRM) by monitoring the characteristic fragmentations as summarized in **Scheme 22.3**. MS parameters are optimized for the 15(*S*)-HETE-d_8 and $PGF_{2\alpha}$-d_4 and are as follows: auxiliary gas pressure was set at 20 psi, sheath gas pressure was 19 psi utilizing nitrogen for both. Discharge current is set at 20 eV and the vaporizer temperature is set at 370°C. Collision induced dissociation (CID) for the HETEs are

Table 22.1
Gradient for separation of F_2-IsoPs and HETEs on a C18 reverse phase column

Time (min)	A: 2 mM NH4Ac, pH 5.6 (%)	B: CH3CN:MeOH, 65:35 (%)	Flow (μl/min)
0	80	20	200
1	80	20	200
20	50	50	200
25	50	50	200
30	0	100	200
32	0	100	200
33	80	20	200
40	80	20	200

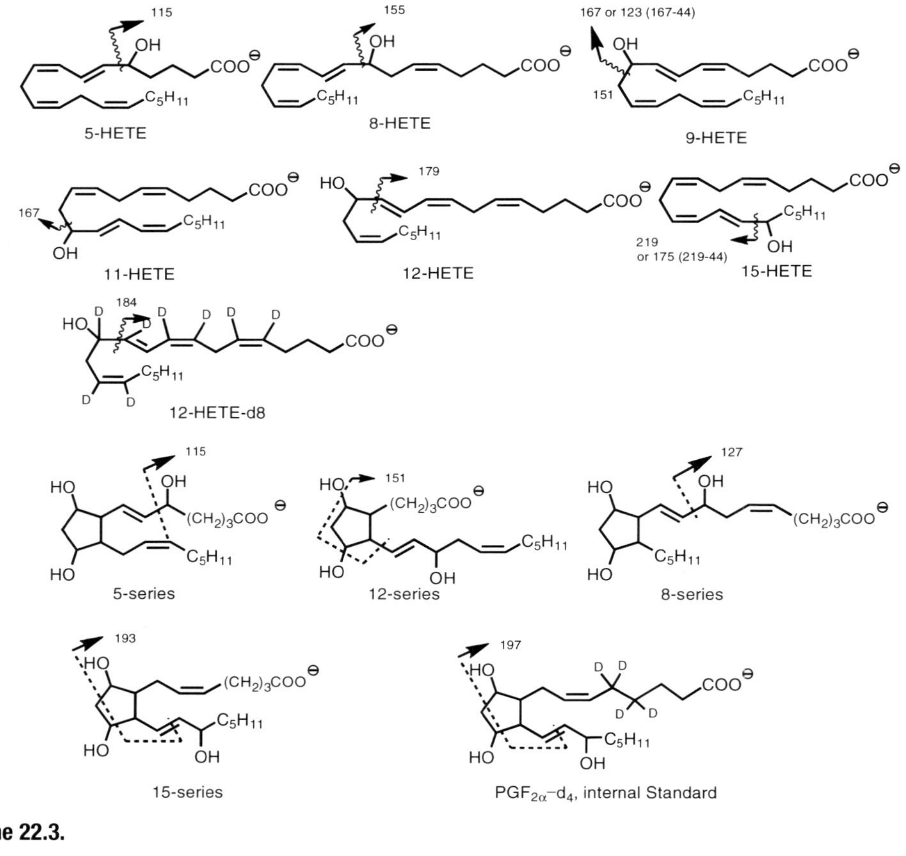

Scheme 22.3.

optimized for different HETEs and 30 eV for the four series of isoprostanes under 1.5 mTorr of argon. Spectra are displayed by averaging the scans across the individual chromatography peaks. Data acquisition and analysis is performed using Xcalibur software version 1.4 (San Jose, CA).

3.5. Results

Figure 22.1 shows a representative chromatogram of the LC-MS results obtained from the Folch extraction of rat liver after treatment with CCl_4 to induce oxidative stress (*see* **Notes 4** and **5**). **Figure 22.1a** is the SRM result of the four regioisomers of F_2-IsoPs. Among the four regioisomers, the 5- and 15-series are an order of magnitude more abundant than 8- and 12-series because the precursors that lead to 8- and 12-series can be further oxidized to form a novel series of oxidation products termed dioxolane-IsoPs whereas that of 5- and 15- can not (26). The diastereomers of 5- and 15-series F_2-IsoPs can be well separated using this method. The HETEs are shown in **Fig. 22.1b** and all six regioisomers can be analyzed by monitoring their respective characteristic fragmentation. 12-HETE-d_8 is used as an internal standard, and the response factors of different HETEs can be determined relative to this internal standard by analysis of a series of known mixtures.

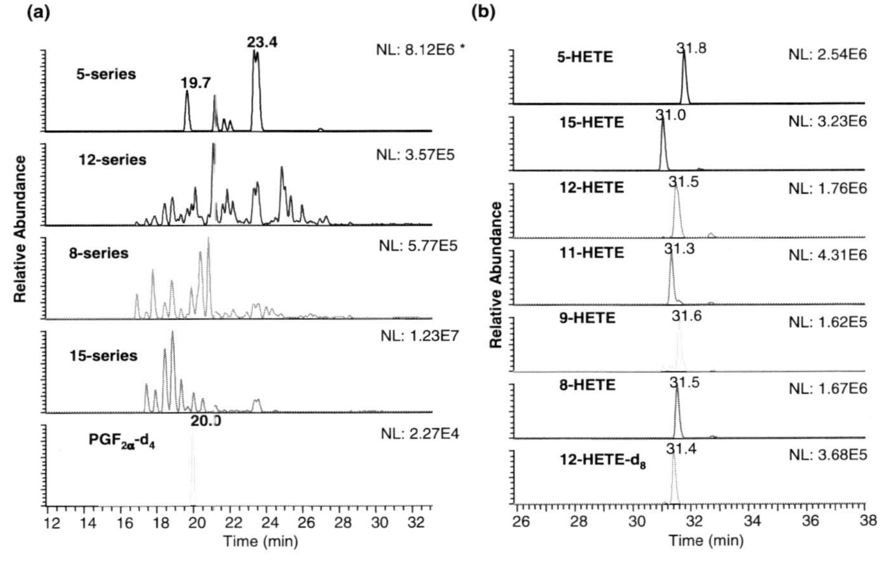

Fig. 22.1. SRM of F_2-IsoPs and HETEs from a rat liver extract after treatment with CCl_4. (a) SRM results of F_2-IsoPs, 5-series, SRM: 353 to 115 @ 30 eV; 12-series, SRM: 353 to 151 @ 30 eV; 8-series, SRM: 353 to 127 @ 30 eV; 15-series, SRM: 353 to 193@ 30 eV; Internal standard, $PGF_{2\alpha}$-d_4, SRM: 357 to 197 @ 30 eV. (b) SRM of HETE. 5-HETE, SRM: m/z 319 to 115 @ 16 eV; 15-HETE, SRM: m/z 319 to 175; 12-HETE, SRM: m/z 319 to 179 @ 15 eV; 11-HETE, SRM: m/z 319 to 167 @ 18 eV; 9-HETE, SRM: m/z 319 to 123 @ 22 eV; 8-HETE, SRM: 319 to 155; 12-HETE-d_8, SRM: 327 to 184 @ 15 eV. *, normalized intensity.

This chapter summarizes a protocol to simultaneously analyze multiple lipid oxidation products by reverse phase LC coupled to ESI-MS. Regioisomeric F_2-IsoPs and HETEs can be measured in a single analysis. The regioisomeric and diastereomeric distribution of the F_2-IsoPs and HETE may be important under different pathophysiological conditions where the formation and metabolism may be altered. This method offers more information for lipid oxidation than the conventional GC-MS method that quantifies the total amount of F_2-IsoPs. The sensitivity of this LC-MS method is somewhat poorer than that of GC-MS using electron capture negative ion chemical ionization method. But it is useful for the analysis of most biological tissues and fluids. Furthermore, the work-up is much simplified compared to the GC-MS methods, and higher throughput can be achieved.

4. Notes

1. Since 5-, 15-, and 12-HETE can be generated from LOX pathways, the origin of these HETEs can be definitively confirmed by chiral analysis.
2. As mentioned previously, polyunsaturated fatty acid can be oxidized ex vivo or during the work-up. Caution has to be followed to minimize the ex vivo oxidation. The tissue must be stored under –70°C or analyzed immediately, and BHT and TPP are essential to prevent the oxidation during work-up.
3. The levels of F_2-IsoPs in a number of diverse biological samples have been quantified using GC-MS. In most cases, the total amount of F_2-IsoPs has been reported in the literature. There are reports of the amounts of 8-Iso $PGF_{2\alpha}$ (or 15-F_{2t}-IsoP) by GC-MS. The separation of single isomers by GC is not nearly as good as using the LC-MS method described here. If better separation is needed for diastereomers, a different gradient can be used to achieve baseline separation (27).
4. For HETE quantification, reverse phase separation of the six regioisomers is not as good as using normal phase HPLC. And the normal phase separation can be readily coupled to atmospheric pressure chemical ionization (APCI) MS (21).
5. In the CID spectra of HETEs, there are overlapping fragmentations, such as 9- and 11-HETE have a common fragment of m/z 167. However, further loss of CO_2 from the fragmentation leading to m/z 123 seems more prominent in 9-HETE than in 11-HETE. Thus, SRM of m/z 319 to 123

is employed to monitor 9-HETE whereas *m/z* 319 to 167 is used for 11-HETE. Furthermore, the elution order of the six HETEs has been determined (28).

Acknowledgments

We thank Dr. David Hachey, Ms. Lisa Mannier, and Mrs. Betty Fox of the Mass Spectrometry Research Center of Vanderbilt University for assistance with the MS analysis. We acknowledge discussion with Dr. Jason Morrow in the Division of Clinical Pharmacology at Vanderbilt University. Financial support from NIEHS P01 ES013125, NIH grants DK 48831, RR00095, CA 77839, HL17921, GM15431, P30 ES00267 (HY, pilot grant), and NSF CHE 0412493 is gratefully acknowledged.

References

1. Funk, C.D. (2001) Prostaglandins and leukotrienes: Advances in eicosanoid biology. *Science* **294**, 1871–1875.
2. Chisolm, G.M. and Steinberg, D. (2000) The oxidative modification hypothesis of atherosclerosis: An Overview. *Free Radic. Biol. Med.* **28**, 1815–1826.
3. Montine, K.S., Quinn, J.F., Zhang, J., Fessel, J.P., Roberts, L.J., Jr., Morrow, J.D., and Montine, T.J. (2004) Isoprostanes and related products of lipid peroxidation in neurodegenerative diseases. *Chem. Phys. Lipids* **128**, 117–124.
4. Montine, T.J., Markesberry, W.R., Morrow, J.D., and Roberts, L.J., Jr. (1998) Cerebrospinal fluid F-2-isoprostane levels are increased in Alzheimer's disease. *Annals Neurol.* **44**, 410–413.
5. Kikugawa, K. (1997) Use and limitation of thiobarbituric acid (TBA) test for lipid peroxidation. *Recent Res. Devel. Lipids Res.* **1**, 73–96.
6. Kenar, J.A., Havrilla, C.M., Porter, N.A., Guyton, J.R., Brown, S.A., Klemp, K.R., and Selinger, E. (1996) Identification and quantification of the regioisomeric cholesteryl linoleate hydroperoxides in oxidized human low density lipoprotein and high density lipoprotein. *Chem. Res. Toxicol.* **9**, 737–744.
7. Jessup, W., Dean, R.T., and Gebicki, J.M. (1994) Iodometric determination of hydroperoxides in lipids and proteins. *Methods Enzymol.* **233**, 289–320.
8. Basu, S. (2004) Isoprostanes: Novel bioactive products of lipid peroxidation. *Free Radic. Res.* **38**, 102–105.
9. Halliwell, B.G. (1999) Free Radicals in Biology and Medicine, 3rd Edition, Oxford University Press, NY.
10. Morrow, J.D., Hill, E., Burk, R.F., Nammour, T.M., Badr, K.F., and Roberts, L.J., Jr. (1990) A series of prostaglandin-F_2-like compounds are produced in vivo in humans by a noncyclooxygenase, Free radical-catalyzed mechanism. *Proc. Natl. Acad. Sci. USA* **87**, 9383–9387.
11. Porter, N.A. (1986) Mechanisms for the autoxidation of polyunsaturated lipids. *Acc. Chem. Res.* **19**, 262–268.
12. Porter, N.A., Caldwell, S.E., and Mills, K.A. (1995) Mechanisms of free radical oxidation of unsaturated lipids. *Lipids* **30**, 277–290.
13. Pryor, W.A. and Stanley, J.P. (1975) Suggested mechanism for the production of malonaldehyde during the autoxidation of polyunsaturated fatty acids. Nonenzymic production of prostaglandin endoperoxides during autoxidation. *J. Org. Chem.* **40**, 3615–3617; see also Porter, N.A. and Funk, M.O. (1975) Peroxy radical cyclization as a model for prostaglandin biosynthesis. *J. Org. Chem.* **40**, 3614–3615.
14. Roberts, L.J., Jr. and Morrow, J.D. (2000) Measurement of F2-isoprostanes as an index of oxidative stress in vivo. *Free Radic. Biol. Med.* **28**, 505–513.

15. Morrow, J.D., Minton, T.A., and Roberts, L.J.I. (1992) The F_2-isoprostane, 8-epi-prostaglandin F_{2a}, a potent agonist of the vascular thromboxane/endoperoxide receptor, is a platelet thromboxane/endoperoxide receptor antagonist. *Prostaglandins* **42**, 155–163.
16. Pratico, D., Smyth, E.M., Violi, F., and FitzGerald, G.A. (1996) Local amplification of platelet function by 8-epi prostaglandin $F_{2\alpha}$ is not mediated by thromboxane receptor isoforms. *J. Biol. Chem.* **271**, 14916–14924.
17. Kunapuli, P., Lawson, J.A., Rokach, J.A., Meinkoth, J.L., and FitzGerald, G.A. (1998) Prostaglandin $F_{2\alpha}$ ($PGF_{2\alpha}$) and the isoprostane, 8,12-iso-isoprostane F2alpha -III, induce cardiomyocyte hypertrophy. Differential activation of downstream signaling pathways. *J. Biol. Chem.* **273**, 22442–22452.
18. Pratico, D., Lawson, J.A., and FitzGerald, G.A. (1995) Cyclooxygenase-dependent formation of the isoprostane, 8-epi prostaglandin F_2. *J. Biol. Chem.* **270**, 9800–9808.
19. Yin, H., and Porter, N.A. (2005) New insights regarding the autoxidation of polyunsaturated fatty acids. *Antiox. Redox Signal* 7, 170–184.
20. Yin, H., Havrilla, C.M., Morrow, J.D., and Porter, N.A. (2002) Formation of isoprostane bicyclic endoperoxides from the autoxidation of cholesteryl arachidonate. *J. Am. Chem. Soc.* **124**, 7745–7754.
21. Yin, H., Havrilla, C.M., Gao, L., Morrow, J.D., and Porter, N.A. (2003) Mechanisms for the formation of isoprostane endoperoxides from arachidonic acid: "Dioxetane" intermediate or beta-fragmentation of peroxyl radicals? *J. Biol. Chem.* **278**, 16720–16725.
22. Lawson, J.A., Rokach, J., and FitzGerald, G.A. (1999) Isoprostanes: Formation, analysis and use as indices of lipid peroxidation in vivo. *J. Biol. Chem.* **274**, 24441–24444.
23. Li, H., Lawson, J.A., Reilly, M., Adiyaman, M., Hwang, S.-W., Rokach, J., and FitzGerald, G.A. (1999) Quantitative high performance liquid chromatography/tandem mass spectrometric analysis of the four classes of F_2-isoprostanes in human urine. *Proc. Natl. Acad. Sci. USA* **96**, 13381–13386.
24. Kuhn, H. and O'Donnell, V.B. (2006) Inflammation and immune regulation by 12/15-lipoxygenases. *Progr. Lipid Res.* **45**, 334–356.
25. Schneider, C., Boeglin, W.E., and Brash, A.R. (2000) Enantiomeric separation of hydroxy eicosanoids by chiral column chromatography: Effect of the alcohol modifier. *Anal. Biochem.* **287**, 186–189.
26. Yin, H., Morrow, J.D., and Porter, N.A. (2004) Identification of a novel class of endoperoxides from arachidonate autoxidation. *J. Biol. Chem.* **279**, 3766–3776.
27. Yin, H., Porter, N.A., and Morrow, J.D. (2005) Separation and identification of F_2-isoprostane regioisomers and diastereomers by novel liquid chromatographic/mass spectrometric methods. *J. Chromatogr. B* **827**, 157–164.
28. Porter, N.A., Logan, J., and Kontoyiannidou, V. (1979) Preparation and purification of arachidonic acid hydroperoxides of biological importance. *J. Org. Chem.* **44**, 3177–3181.

Chapter 23

Lipoxygenase-Catalyzed Phospholipid Peroxidation: Preparation, Purification, and Characterization of Phosphatidylinositol Peroxides

E. Susan O'Connor Butler, Jessica N. Mazerik, Jason P. Cruff, Shariq I. Sherwani, Barbara K. Weis, Clay B. Marsh, Achuthan C. Raghavamenon, Rao M. Uppu, Harald H. O. Schmid, and Narasimham L. Parinandi

Abstract

The importance of understanding the mechanisms of modulation of cellular signaling cascades by the peroxidized membrane phospholipids (PLs) is well recognized. The enzyme-catalyzed peroxidation of PLs, as opposed to their oxidation by air and metal catalysis, is well controlled and rapid and yields well-defined PL peroxides which are highly desirable for biological studies. Therefore, here, we chose bovine liver phosphatidylinositol (PI), a crucial membrane PL which acts as the substrate for phospholipase C in cellular signal transduction, as a model membrane PL. We successfully generated the PI peroxides with soybean type-I lipoxygenase (LOX) in the presence of deoxycholate, which facilitates the LOX-mediated peroxidation of the polyunsaturated fatty acids esterified to the PL. The LOX-peroxidized PI, after enzymatic catalysis, was separated from the unoxidized PI in the reaction mixture by normal-phase, high-performance liquid chromatography (HPLC). The extent of LOX-mediated peroxidation of PI following HPLC purification was established by the analysis of lipid phosphorus, conjugated dienes by UV spectrophotometry, peroxides, and loss of fatty acids by gas chromatography. This study established the optimal conditions yielding ~46% of peroxidized PI from 300 μg of neat bovine liver PI that was peroxidized by soybean type-I LOX (50 μg) for 30 min in borate buffer (0.2 M, pH 9.0) containing 10 mM deoxycholate.

Key words: Lipoxygenase, phospholipid peroxidation, phosphatidylinositol peroxides, HPLC.

R.M. Uppu et al. (eds.), *Free Radicals and Antioxidant Protocols*, Methods in Molecular Biology 610,
DOI 10.1007/978-1-60327-029-8_23, © Humana Press, a part of Springer Science+Business Media, LLC 1998, 2010

1. Introduction

Free radical-catalyzed peroxidation of membrane lipids has been shown to be associated with a number of diseases among humans (1–4). In addition to the occurrence of free polyunsaturated fatty acid (PUFA) peroxidation in the biological milieu, the peroxidation of esterified PUFA at the *sn*-2 position of the intact phospholipid (PL) of biological membranes has long been known (1, 5). Needless to mention, several signal mediation properties of the peroxidized free PUFA have been demonstrated, but the role of peroxidized PLs (containing the esterified peroxidized PUFA) in cellular signal modulation is yet to be thoroughly understood. For such mechanistic studies, defined PL peroxide is an essential requirement, at least in model systems including the in vitro enzyme reactions and cell cultures.

Several methods of oxidation of PUFA-containing PLs such as prolonged air oxidation and transition metal-catalyzed oxidation have been widely employed in order to prepare the peroxidized PLs. However, these methods not only are time consuming but also yield a plethora of PL peroxidation products in an uncontrollable fashion. Moreover, the purification of defined peroxidized PLs generated from such methods of oxidation is difficult. Utilization of the enzyme-catalyzed peroxidation of intact PUFA in PLs offers a better approach to efficiently prepare well-defined PL peroxidation products for biological studies.

Lipoxygenases (LOXs) catalyze the oxidation of PUFA (linoleic, linolenic, arachidonic, and other higher fatty acids) by inserting molecular oxygen at *cis, cis*-1,4-pentadiene structure yielding a PUFA hydroperoxide with a conjugated diene rearrangement (6). Although soybean LOX readily utilizes free linoleic acid and other PUFAs to convert them effectively into their corresponding fatty acid hydroperoxides, in the presence of bile salts such as deoxycholate, the enzyme efficiently peroxidizes PUFA esterified to the PLs (7, 8) and converts them to PL hydroperoxides. Above all, the LOX-catalyzed peroxidation of PL-PUFAs is rapid and can be controlled to obtain defined PL peroxides.

Separation and purification of the PL hydroperoxides (PL-OOH) from the unoxidized PL in the reaction mixture are crucial. We have noticed that the thin-layer chromatography (TLC) technique for PL-OOH separation is not ideal as either the peroxidized fatty acyl (PUFA) chains may decompose during TLC separation or the reactive oxidized PLs may stick to silica on the plate, rendering the peroxidized PLs immobile and inseparable from the unoxidized PLs. However, it has been successfully demonstrated that the utilization of high-performance liquid chromatography

(HPLC) systems (both the normal and the reverse phases) can be used to separate, purify, characterize, and quantify the peroxidized PLs from their unoxidized counterparts in the mixture (9, 10).

Membrane PLs, in conjunction with phospholipases, play a pivotal role in a wide variety of cellular signaling cascades through the generation of bioactive lipid signal mediators (11). Phosphatidylinositol (PI) is an important component of the membrane phospholipid backbone, which is a substrate for the PI-specific phospholipase C (PI-PLC) (12). PI-PLC hydrolyzes PI producing the signal mediators such as diacylglycerol and 1,2-inositol phosphate, and the latter is then converted to inositol-1-phosphate (12, 13). These PI-specific PLC-generated metabolites of PI have been shown to regulate several important cellular functions through modulation of cellular signaling cascades (14). Oxidants including lipid peroxides have been shown to modulate phospholipase signaling events (15). As mammalian membrane PI is also rich in arachidonic acid, the most crucial PUFA of mammalian membrane PLs, it can undergo peroxidative modifications due to oxidative insults. Therefore, peroxidized PI or oxidatively modified PI can very well modulate the PI-specific PLC activity in mammalian cells and affect the downstream cellular signaling cascades.

Hence, there is a necessity to generate/prepare rapidly and purify well-defined PI peroxides for enzymatic, cellular, and signaling studies. To achieve our goal, here, we utilized soybean type-I LOX and generated the peroxidized PI from neat bovine liver PI. Optimal conditions were established to convert unoxidized (neat) PI to peroxidized PI in a reaction system containing deoxycholate for soybean type-I LOX to act on PI. The extent of LOX-mediated peroxidation of PI following normal-phase HPLC separation was established by determining the PL content, conjugated diene hydroperoxides by UV spectrophotometry, and loss of fatty acids by gas chromatography. The current method yielded ~46% of peroxidized PI from the starting unoxidized substrate. The initial substrate was peroxidized by soybean type-I LOX for 30 min in borate buffer (0.2 M, pH 9.0) containing 10 mM deoxycholate.

2. Materials

1. L-α-Phosphatidylinositol (PI) powder form (Avanti Polar Lipids, Inc., Alabaster, AL).
2. Whenever needed, PI powder was dissolved in chloroform/methanol (2/1, v/v), and immediately the extent of

oxidation of PI was assessed by UV spectrophotometry for conjugated dienes and by iodometric peroxide assay for peroxides.

3. Whenever the presence of peroxidized lipid was detected in the starting material, the phospholipid was not used for the study.
4. Deoxycholate (sodium salt) (Calbiochem, San Diego, CA).
5. Soybean type-I LOX (Sigma Chemical Company, St. Louis, MO).
6. All other reagents were of analytical grade. Hexane was of HPLC grade.

3. Methods

3.1. Lipoxygenase-Catalyzed Peroxidation of PI

1. The reaction medium was 0.2 M borate buffer (pH 9.0).
2. Prior to the oxidation of PI, the buffer was oxygenated by bubbling air for 1 h at room temperature.
3. A desired volume of PI from the stock solution was taken to dryness under a stream of nitrogen at 40°C and reconstituted in a final 1 mL volume of 0.2 M borate buffer containing 10 mM deoxycholate (prepared fresh; pH 9.0).
4. Peroxidation of PI was initiated at 37°C by the addition of the desired amount of freshly prepared soybean type-I LOX in the same reaction buffer.
5. In order to establish the optimal time of LOX-catalyzed peroxidation of PI, the reaction was carried out for different lengths of time (0, 5, 10, 20, and 30 min) at 37°C in the reaction medium containing 300 μg of PI and 50 μg of LOX.
6. The optimal enzyme concentration was established by including different concentrations (25, 50, 75, 125, 175, and 250 μg) of soybean type-I LOX to peroxidize PI (300 μg final) in a reaction carried out for 30 min at 37°C in the reaction buffer.
7. The optimal substrate concentration was established by using different concentrations (100–1,000 μg) of unoxidized neat PI for peroxidation by soybean type-I LOX (50 μg) in a reaction carried out for 30 min at 37°C.
8. Reactions were terminated by the addition of methanol, and the lipid was extracted with a final ratio of chloroform/methanol (2/1, v/v) according to Folch et al. (16).

9. All the lipid samples were stored at –80°C under a coat of argon until further analysis.

3.2. High-Performance Liquid Chromatography (HPLC)

1. Unoxidized and peroxidized PI species in the reaction mixture were separated by normal-phase (NP) HPLC according to Parinandi et al. (10).
2. An aliquot of PI (~0.3 mg) was taken to dryness under N_2 at 40°C, redissolved in ethanol/hexane (1/4, v/v), and analyzed by HPLC.
3. A Beckman HPLC system was used which included a model 421 controller, a model 340 organizer, two model 112 solvent delivery pumps, a model 160 UV detector at 214 nm, and a variable wavelength UV detector (Schoeffel Instrument Corp., USA) set at 235 nm.
4. The detectors were set at a range of 0.2 and 0.1 AUFS, respectively.
5. The normal phase column (Adsorbosphere Silica, 3 μm; 100 × 4.6 mm) was obtained from Alltech Associates (Deerfield, IL).
6. The two detectors were connected to a dual-channel recorder (Pharmacia, Sweden) and to a model SP4290 Spectra-Physics dual-channel integrator (San Jose, CA).
7. Lipids were eluted at room temperature with a flow rate of 2 mL/min.
8. Solvent A was hexane/ethanol/water (90/15/0.25, v/v/v) and solvent B was hexane/ethanol/water (90/60/2.8, v/v/v).
9. The gradient used was as follows: 0–20% B in 20 min; 20% B held for 10 min; 20–95% B in 5 min; 95% B held for 5 min; 95–0% B in 5 min; column equilibration in 100% A for 15 min.
10. The lipid fractions were collected and stored on ice.
11. They were taken to dryness under a stream of nitrogen at 40°C and redissolved in chloroform or other solvents for further analysis.

3.3. Conjugated Diene Assay

1. The extent of peroxidation in the unoxidized and peroxidized PI samples and in PI fractions separated by HPLC was determined by measuring the conjugated diene rearrangement in the unsaturation of PI according to Parinandi et al. (10).
2. Unoxidized PI, peroxidized PI, lipid peroxides, and individual HPLC fractions were taken to dryness under a stream of N_2 at 40°C and redissolved in cyclohexane, and the

absorbance at 233 nm was recorded on a Beckman DU-8 spectrophotometer against a cyclohexane blank.

3. Values were calculated using a molar extinction coefficient of 2.52×10^4 M^{-1}.

3.4. Iodometric–Spectrophotometric Analysis of Lipid Hydroperoxides

1. The hydroperoxide content of unoxidized, peroxidized, and HPLC fractions of PI was determined according to Parinandi et al. (10).
2. Samples were taken to dryness under a stream of nitrogen and redissolved quickly in 1 mL of acetic acid/chloroform (3/2, v/v), and 0.05 mL of potassium iodide solution (6 g/5 mL water) was added.
3. The tubes were stoppered, vortexed, and kept in the dark for 5 min.
4. Thereafter, 3 mL of cadmium acetate solution (0.5 g/100 mL water) was added, and the tubes were vortexed and centrifuged.
5. Absorbance of the upper phase was recorded at 353 nm against an appropriate blank.
6. Total hydroperoxides were calculated using molar extinction coefficient of cumene hydroperoxide (1.73×10^4 M^{-1}).

3.5. Thin-Layer Chromatography (TLC) and Lipid Phosphorus Determination

1. TLC was done on silica gel H plates using chloroform/methanol/water (65/25/4, v/v/v) as developing solvent, and fractions were detected either with ammonium molybdate spray (10) or by charring after spraying with 50% (v/v) H_2SO_4.
2. Fatty acid methyl esters, obtained by transesterification in methanolic NaOH (0.2 N) in the presence of methyl heptadecanoate, were separated on a column packed with SP-2330 (Supelco) in a Packard model 428 gas chromatograph connected to a Spectra-Physics integrator.
3. Lipid phosphorus was determined according to Bartlett (17).

3.6. Results

Initially, we investigated to identify an efficient and suitable oxidant to induce peroxidation of PI and other PUFA-containing PLs and to generate the peroxides of those PLs (*see* **Note 1**). We observed that the transition metal oxidants such as Fe^{2+}, Fe^{2+} + ascorbate, Cu^{2+}, and Cu^{2+} + H_2O_2, although induced peroxidation of PI, it was extremely difficult to generate and purify intact PI peroxides as the PI peroxides were highly labile in the metal-rich environment (data not shown). Also, in the metal-rich environment, the PI-hydroperoxide decomposed into reactive carbonyls, rendering the peroxidized PI immobile and inseparable on the TLC plate (*see* **Note 2**). Alternatively, we were successful in generating PI peroxides by oxidizing PI in air for prolonged periods of time (24–72 h). This method of peroxidation of PI was time consuming and also led to the formation of a wide variety (mixture) of PI peroxidation products including carbonyls and peroxides. Taking this into consideration, we avoided those methods of inducing PL peroxidation which could generate larger amounts of reactive carbonyls and less amounts of intact PL peroxides. Therefore, we sought after an enzyme capable of converting PUFA into hydroperoxides that could be conveniently utilized to generate PL peroxides.

In this study, we chose the well-characterized soybean type-I LOX to peroxidize the PUFA esters in the bovine liver PI. We had to include deoxycholate (10 mM final) in the reaction medium for the soybean LOX to act on the PUFA esters in PI (*see* **Note 3**). We established the optimal conditions to generate the PI peroxides in good yield with soybean type-I LOX using bovine liver PI as the substrate. Under condition 1, we kept the enzyme concentration constant (50 μg) and reaction time at 30 min and varied the substrate concentration (PI) from 100–1,000 μg. Under condition 2, we kept the substrate concentration constant (300 μg) and reaction time at 30 min and varied the enzyme concentration from 25–250 μg. In condition 3, we kept the substrate concentration constant (300 μg) and enzyme concentration constant (50 μg) but varied the reaction time from 0 to 30 min. Our initial experiments revealed the optimal conditions for soybean type-I LOX to efficiently convert the substrate (PI) into PI peroxides. From this, we established that 50 μg of soybean type-I LOX caused an optimal conversion of 300 μg of PI into PI peroxides in 0.2 M borate buffer (pH 9.0) containing 10 mM deoxycholate in a reaction time of 30 min at 37°C (*see* **Note 4**). The LOX-catalyzed formation of PI peroxides (positive to iodometric lipid hydroperoxide assay) was both qualitatively and quantitatively assayed by the conjugated diene UV spectrophotometry (**Fig. 23.1**). It was also revealed that inclusion of deoxycholate (10 mM) in the reaction medium did not cause an appreciable loss of extractable phospholipid (PI) following its peroxidation by LOX (**Table 23.1**). As the substrate used in this study (PI) contains nearly 16% of arachidonic acid (20:4), following the peroxidation of PI by LOX, the PUFA ester composition of the PL decreased dramatically with the increase in time of reaction (**Table 23.2**). At 30 min of reaction time, starting with arachidonic acid, all the PUFA esters in PI decreased. Thus, the loss of PUFA esters as measured by gas chromatography in the LOX-generated PI peroxide confirmed the peroxidation of the PL.

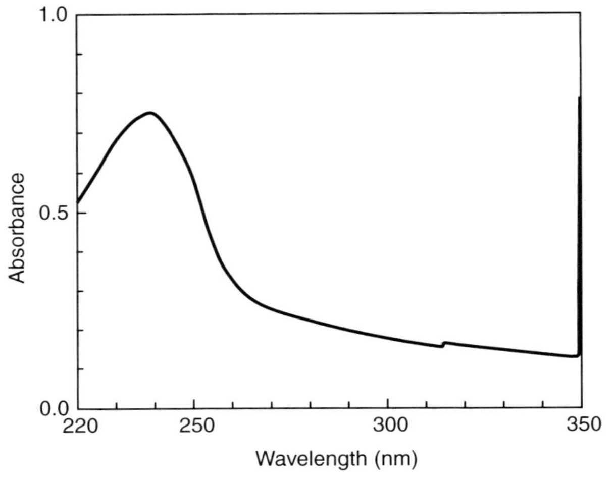

Fig. 23.1. Conjugated diene UV spectrum of LOX-generated peroxidized PI. Phosphatidylinositol (300 μg) was peroxidized by soybean type-I lipoxygenase (50 μg) for 30 min at 37°C in 0.2 M borate buffer (pH 9.0) containing 10 mM deoxycholate. At the end of the incubation period, PI was extracted, redissolved in cyclohexane, and the conjugated diene UV spectrum was recorded against equal amount of unoxidized neat PI as described in **Section 3**.

Table 23.1
Recovery of lipid phosphorus in PI after peroxidation by soybean lipoxygenase[a]

Assay condition	LP (nmol)	Absorbance at 235 nm
Prior to treatment with LOX	337.5	0.0
After treatment with deoxycholate (10 mM) unoxidized	305.0	0.0
After treatment with deoxycholate (10 mM) and lipoxygenase	316.0	2.5 (difference)

[a] PI (300 μg) was peroxidized by soybean type-1 LOX (50 βg) for 30 min at 37°C in 0.2 M borate buffer (pH 9.0) containing 10 mM deoxycholate. At the end of the reaction, the lipid was extracted and the amount of lipid phosphorous and conjugated diene absorbance at 235 nm were determined as described in **Section 3**.

Earlier, we utilized NP-HPLC and successfully separated cardiolipin peroxides from unoxidized cardiolipin (10). In the current study, we utilized the same NP-HPLC method to separate the LOX-generated PI peroxides from unoxidized PI. As shown in **Fig. 23.2**, the LOX-generated peroxides were clearly separated on the NP-HPLC system from the unoxidized PI with their

Table 23.2
Fatty acid composition of PI peroxidized by soybean lipoxygenase for various time periods[a]

Fatty acid	0 min μg	0 min %	2 min μg	2 min %	5 min μg	5 min %	10 min μg	10 min %	20 min μg	20 min %	30 min μg	30 min %
16:0	1.798	2.6	1.973	2.6	1.857	2.7	1.563	2.5	1.709	3.0	1.910	3.1
18:0	32.990	48.1	35.248	46.5	33.948	49.5	34.554	55.0	32.651	56.6	34.576	56.7
18:1	8.638	12.6	9.300	12.3	9.11	13.3	9.061	14.4	8.743	15.1	9.874	16.2
18:2	4.559	6.6	4.996	6.6	4.687	6.8	4.289	6.8	3.809	6.6	4.027	6.6
20:3ω6	6.679	9.7	7.796	10.3	6.275	9.2	5.176	8.2	4.309	7.5	4.379	7.2
20:4	10.882	15.9	12.511	16.5	9.495	13.8	6.442	10.3	5.200	9.0	5.009	8.2
24:1	1.486	2.2	2.008	2.6	1.650	2.4	1.083	1.7	0.831	1.4	0.895	1.5
22.5:ω3	1.557	2.3	1.996	2.6	1.545	2.3	0.643	1.0	0.484	0.8	0.361	0.6
Total	68.589		75.828		68.567		62.811		57.736		61.031	

[a] PI (300 μg) was peroxidized by soybean type-1 LOX (50 μg) for 30 min at 37°C in 0.2 M borate buffer (pH 9.0) containing 10 mM deoxycholate. At the end of the reaction, the lipid was extracted, subjected to alkaline methanolysis, and the fatty acid methyl esters were quantified by gas chromatography as described in **Section 3**.

characteristic conjugated diene peroxide UV absorption. Also, it was evident that with the increase in reaction time, the extent of formation of PI peroxides upon LOX catalysis increased. At 30 min of reaction, soybean type-I LOX converted 46% of the total PI (as measured by lipid phosphorus content) into PI peroxides (**Table 23.3**), which very well corresponded with their conjugated diene values (**Fig. 23.3**). The peroxidized PI fractions, which exhibited the presence of conjugated dienes, were also positive to the iodometric lipid hydroperoxide test (data not shown). Beginning with linoleic acid, a drastic and time-dependent loss of PUFA esters in the LOX-generated and NP-HPLC-separated PI peroxides was clearly evident from the gas chromatographic analysis of their methyl esters of fatty acids (**Table 23.4**). Beginning with 5 min of reaction of PI with LOX, higher PUFAs totally disappeared in the peroxidized PI fractions that were separated by NP-HPLC. Therefore, the loss of PUFA from the peroxidized PI HPLC fractions as revealed by the gas chromatographic analysis confirmed the peroxide nature of PI (*see* **Note 5**).

This study established the optimal conditions to generate ~46% of peroxidized PI from 300 μg of neat bovine liver PI that was catalyzed by soybean type-I LOX (50 μg) for 30 min in borate buffer (0.2 M, pH 9.0) containing 10 mM deoxycholate. The method can be used to generate PI peroxides for biochemical and biological studies. In addition, this method has the potential for use in quantification of PI peroxides both in vitro and in vivo.

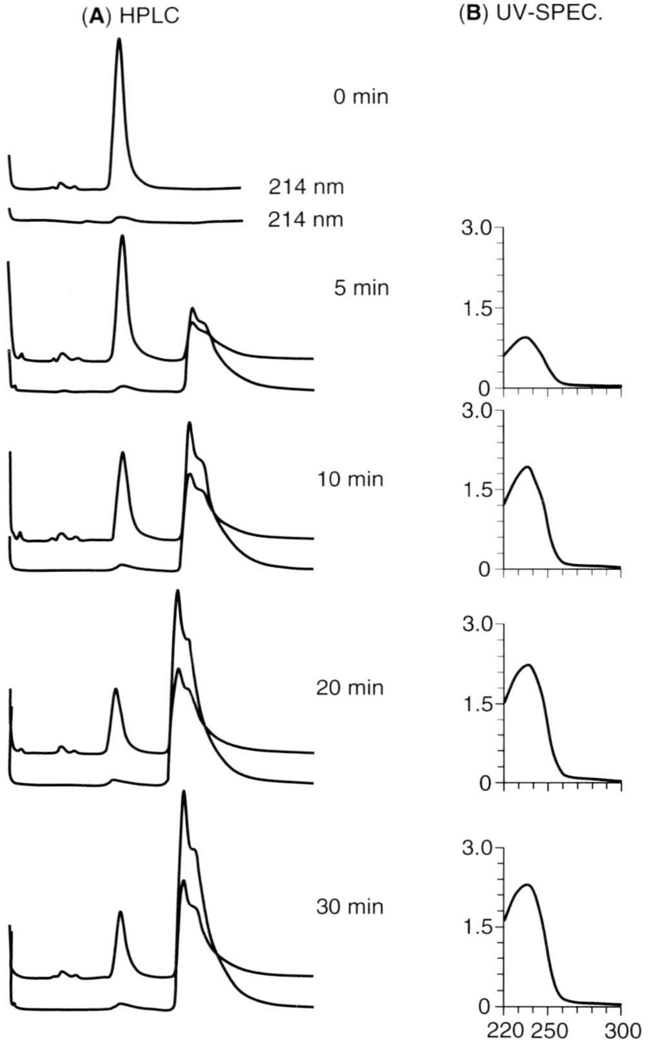

Fig. 23.2. Normal-phase (NP) HPLC separation of the peroxidized PI from unoxidized PI. Phosphatidylinositol (300 μg) was peroxidized by soybean type-I lipoxygenase (50 μg) for 0–30 min at 37ºC in 0.2 M borate buffer (pH 9.0) containing 10 mM deoxycholate. At the end of the incubation period, PI was extracted, subjected to NP-HPLC separation, and the unoxidized and peroxidized PI were separated (*panel A*) as described in **Section 3**. Two UV detectors were set at 214 nm (*top* tracing to track the PL) and 234 nm (*bottom* tracing to track the conjugated dienes of the peroxidized PL). *Panel B* illustrates the conjugated diene UV spectra of the individual corresponding peroxidized PI peaks in cyclohexane against cyclohexane blank.

Table 23.3
Quantitative determinations of lipoxygenase-peroxidized PI molecular species by normal-phase HPLC[a]

Period of incubation (min)	LP (nmol) Injected	Unoxidized fraction	Peroxidized fraction	Percentage of peroxidized PI formed
0	283.50	223.82	0.00	0.00
5	286.30	195.37	49.87	20.33
10	286.00	150.67	79.00	34.40
20	288.81	154.77	85.35	35.54
30	250.40	125.40	108.50	46.36

[a] PI (300 μg) was peroxidized by soybean type-1 LOX (50 μg) for 30 min at 37°C in 0.2 M borate buffer (pH 9.0) containing 10 mM deoxycholate. At the end of the reaction, the lipid was extracted and lipid phosphorous was determined in an aliquot used for HPLC separation and the unoxidized and peroxidized PI fractions were collected and their lipid phosphorus contents were determined as described in **Section 3**. The amount of peroxidized PI (%) at each time of reaction was calculated from the total lipid phosphorus injected for HPLC separation and the amount of lipid phosphorus recovered in the peroxidized PI fraction.

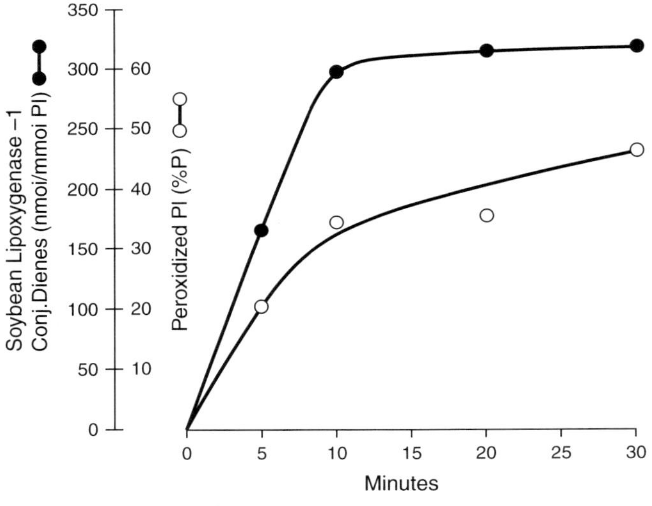

Fig. 23.3. Extent of lipoxygenase-catalyzed peroxidation of PI. Phosphatidylinositol (300 μg) was peroxidized by soybean type-I lipoxygenase (50 μg) for 0–30 min at 37°C in 0.2 M borate buffer (pH 9.0) containing 10 mM deoxycholate. At various intervals of time, the lipid present in the reaction mixture was extracted, and the amount of lipid phosphorus was determined in an aliquot used for NP-HPLC separation as described in **Section 3**. The lipid was then subjected to NP-HPLC separation, and the unoxidized and peroxidized PI fractions were collected. The amount of lipid phosphorus and the extent of peroxidation (formation of conjugated dienes) in each peroxidized PI fraction collected at each reaction time point were determined as described in **Section 3**.

Table 23.4
Fatty acid composition of PI peroxidized by soybean lipoxygenase for various time periods separated by HPLC[a]

	Unox (neat)		0 min		5 min				10 min				20 min				30 min			
	Unox		Unox		Unox		Ox		Unox		Ox		Unox		Ox		Unox		Ox	
Fatty	1A-1		2A-1		3A-1		3A-2		4A-1		4A-2		5A-1		5A-2		6A-1		6A-2	
acid	μg	%	μg	%	μg	%	μg	%	μg	%	μg	%	μg	%	μg	%	μg	%	μg	%
16:0	0.953	2.3	1.089	2.5	0.865	2.3	0.33	5.7	0.760	2.3	0.361	3.8	0.696	2.2	0.411	4.2	0.536	2.0	0.554	4.4
18:0	20.087	48.5	21.476	49.1	18.025	48.8	4.60	78.6	14.465	43.9	7.580	80.3	14.367	44.4	7.869	80.5	11.694	44.7	10.175	80.3
18:1	4.994	12.1	5.414	12.4	4.886	13.2	0.801	13.7	4.622	14.0	1.354	14.3	4.643	14.4	1.372	14.0	4.126	15.8	1.810	14.3
18:2	2.608	6.3	2.830	6.5	2.675	7.2	0.058	1.0	2.752	8.3	0.064	0.7	2.780	8.6	0.042	0.5	2.492	9.5	0.059	0.5
20:3ω6	4.164	10.1	4.277	9.8	3.704	10.0	0.028	0.5	3.866	11.7	0.038	0.4	3.662	11.3	0.033	0.3	2.977	11.4	0.034	0.3
20:4	6.621	16.0	6.737	15.4	5.271	14.3	0.037	0.6	5.061	15.4	0.048	0.5	4.727	14.6	0.045	0.5	3.353	12.9	0.040	0.3
24:1	1.014	2.4	0.943	2.2	0.833	2.3	0	0	0.859	2.6	0	0	0.911	2.8	0	0	0.654	2.5	0	0
22:5ω3	0.983	2.4	0.962	2.2	0.642	1.7	0	0	0.585	1.8	0	0	0.567	1.8	0	0	0.298	1.1	0	0
Total	41.424		43.728		36.901		5.855		32.970		9.445		32.353		9.778		25.150		12.672	

[a] PI (300 μg) was peroxidized by soybean type-1 LOX (50 μg) for 30 min at 37°C in 0.2 M borate buffer (pH 9.0) containing 10 mM deoxycholate. At the end of the reaction, the lipid was extracted, subjected to HPLC separation, unoxidized and peroxidized species of PI were separated and collected, and their fatty acid composition was determined by gas chromatography as described in **Section 3**. Unox: unoxidized; ox: peroxidized.

4. Notes

1. Commercial organic solvents (e.g., chloroform, ethanol, methanol) used to either dissolve PI or separate unoxidized and peroxidized PI species contain dissolved oxygen and traces of organic peroxides which will cause peroxidation of PI. In order to avoid the peroxidation of PI in organic solvents, it is recommended that distilled organic solvents be used to dissolve PI, separate the unoxidized and oxidized PI species on HPLC, and also to store those PI species. Both the unoxidized and the oxidized PI species should be stored at –80°C under argon coat. Since PI is rich in arachidonic acid, precautions should be taken to avoid or minimize its peroxidative degradation during handling, analysis, and storage.
2. Prior to soybean LOX-catalyzed peroxidation, a sample of the starting PI must be screened for any preformed lipid hydroperoxides by measuring the presence of conjugated dienes and carbonyls by UV spectrophotometry. A detectable absorption at 232–235 nm and at 270 nm is indicative of peroxidation of PI and the presence of carbonyl(s) in PI, respectively. Further, the presence of lipid hydroperoxides should be confirmed by the iodometric assay. Alternatively, the presence of carbonyls in PI can be detected by the dinitrophenylhydrazine (DNPH) test.
3. Soybean LOX-catalyzed peroxidation of PLs (e.g., PI) is enhanced by including deoxycholate in the reaction mixture. Deoxycholate acts as an emulsifier for PI facilitating the action of LOX on the PL. The choice of deoxycholate is very important such as using the commercially available ultrapure quality of deoxycholate. In our experience, highest quality of deoxycholate supplied by Calbiochem offered the optimum activity of soybean LOX in catalyzing the peroxidation of PI. Also, it is better to pass the solution of deoxycholate through chelax column to remove traces of heavy metals such as iron to avoid the degradation of lipid hydroperoxides and the formation of reactive carbonyls.
4. For unexplained reasons, the ideal reaction medium for soybean LOX action in vitro is the borate buffer. We also tried other buffers such as the phosphate, Tris, and HEPES buffers but the borate buffer appeared to cause the optimal formation of LOX-catalyzed formation of PI hydroperoxides. It is highly recommended that the borate buffer used to carry out the LOX-mediated peroxidation of PI in vitro

should be free of trace metals such as iron in order to avoid or minimize the degradation of PI peroxides into carbonyls. Therefore, borate buffer should be prepared with double-distilled water that is passed through chelax to remove the trace metal contaminants.

5. Upon purification of the PI hydroperoxides by HPLC, the hydroperoxide fractions should be taken to dryness under a stream of nitrogen at 30°C and stored in absolute ethanol or chloroform/ethanol (2/1, v/v) as opposed to chloroform/methanol (2/1, v/v). The use of chloroform alone as the solvent for storage degrades the PL (PI) and its hydroperoxides, so it should be avoided.

Acknowledgments

Support provided toward this work by the Hormel Institute of the University of Minnesota, the Dorothy M. Davis Heart and Lung Research Institute and the Division of Pulmonary, Allergy, Critical Care, and Sleep Medicine at the Ohio State University College of Medicine, and ARCH Program at Southern University-Baton Rouge (NIH grant number ES10018) is acknowledged.

References

1. Milne, G.L., Seal, J.R., Havrilla, C.M., Wijtmans, M., and Porter, N.A. (2005) Identification of products formed from phospholipids in free radical oxidation of human low density lipoproteins. *J. Lipid Res.* **46**, 307–319.
2. Bowler, R.P., Barnes, P.J., and Crapo, J.D. (2004) The role of oxidative stress in chronic obstructive pulmonary disease. *COPD* **1**, 255–277.
3. Hwang, E.S. and Kim, G.H. (2007) Biomarkers for oxidative stress status of DNA, lipids, and proteins in vitro and in vivo cancer research. *Toxicology* **229**, 1–10.
4. Szasz, T., Thakali K., Fink, G.D., and Watts, S.W. (2007) A comparison of arteries and veins in oxidative stress: Producers, destroyers, function, and disease. *Exp. Biol. Med.* **232**, 27–37.
5. Niki, E., Yoshida, Y., Saito, Y., and Noguchi, N. (2005) Lipid peroxidation: Mechanisms, inhibition, and biological effects. *Biochem. Biophys. Res. Commun.* **338**, 668–676.
6. Galliard, T. and Chan, H.W.-S. (1980) Lipoxygenases. In *The Biochemistry of Plants, A Comprehensive Treatise*, Vol. 4, Lipids: Structure and Function (P.K. Stumpf, Ed.), Academic Press, NY, pp. 131–161.
7. Brash, A.R., Ingram, C.D., and Harris, T.M. (1987) Analysis of a specific oxygenation reaction of soybean lipoxygenase-1 with fatty acids esterified in phospholipids. *Biochemistry* **26**, 5465–5471.
8. Began, G., Sudharshan, E., and Appu Rao, A.G. (1999) Change in the positional specificity of lipoxygenase 1 due to insertion of fatty acids into phosphatidylcholine deoxycholate mixed micelles. *Biochemistry* **38**, 13920–13927.
9. Terao, J. and Matsushita, S. (1987) Application of HPLC for the determination of phospholipid hydroperoxides in food and biological systems. *Free Radic. Biol. Med.* **3**, 345–347.
10. Parinandi, N.L., Weis, B.K., and Schmid, H.H.O. (1988) Assay of cardiolipin peroxidation by high-performance liquid chromatography. *Chem. Phys. Lipids* **49**, 215–220.
11. Varadharaj, S., Steinhour, E., Hunter, M.G., Watkins, T., Baran, C.P., Magalang, U., Kuppusamy, P., Zweier, J.L., Marsh, C.B.,

Natarajan, V., and Parinandi, N.L. (2006) Vitamin C-induced activation of phospholipase D in lung microvascular endothelial cells: Regulation by MAP kinases. *Cell Signal.* **18**, 1396–1407.

12. Shao, C., Shi, X., Wehbi, H., Zambonelli, C., Head, J. F., Seaton, B.A., and Roberts, M.F. (2007) Dimer structure of an interfacially impaired phosphatidylinositol-specific phospholipase C. *J. Biol. Chem.* **282**, 9228–9235.
13. Cocco, L., Faenza, I., Fiume, R., Billi, A.M., Gilmour, R.S., and Manzoli, F.A. (2006) Phosphoinositide-specific phospholipase C (PI-PLC) 1 and nuclear lipid-dependent signaling. *Biochim. Biophys. Acta* **1761**, 509–521.
14. Katan, M. (1998) Families of phosphoinositide-specific phospholipase C: Structure and function. *Biochim. Biophys. Acta* **1436**, 5–17.
15. Cummings, R., Parinandi, N., Wang, L., Usatyuk, P., and Natarajan, V. (2002) Phospholipase D/phosphatidic acid signal transduction: Role and physiological significance in lung. *Mol. Cell Biochem.* **234–235**, 99–109.
16. Folch, J., Lees, M., and Sloane Stanley, G.H. (1957) A simple method for the isolation and purification of total lipids from animal tissue. *J. Biol. Chem.* **226**, 497–509.
17. Bartlett, G.R. (1959) Phosphorus assay in column chromatography. *J. Biol. Chem.* **234**, 466–468.

Chapter 24

Oxidized Low-Density Lipoprotein

Sampath Parthasarathy, Achuthan Raghavamenon, Mahdi Omar Garelnabi, and Nalini Santanam

Abstract

Oxidized low-density lipoprotein (Ox-LDL) has been studied for over 25 years. Numerous pro- and anti-atherogenic properties have been attributed to Ox-LDL. Yet, Ox-LDL has neither been defined nor characterized, as its components and composition change depending on its source, method of preparation, storage, and use. It contains unoxidized and oxidized fatty acid derivatives both in the ester and free forms, their decomposition products, cholesterol and its oxidized products, proteins with oxidized amino acids and cross-links, and polypeptides with varying extents of covalent modification with lipid oxidation products, and many others. It seems to exist in vivo in some form not yet fully characterized. Until its pathophysiological significance, and how it is generated in vivo are determined, the nature of its true identity will be only of classical interest. In this review, its components, their biological actions and methods of preparation will be discussed.

Key words: Atherosclerosis, oxidative stress, lipid peroxides, antioxidants, aldehydes.

1. Introduction

The concept that oxidative stress and the oxidation of low-density lipoprotein (LDL) might play a role in atherosclerosis originated over 25 years ago (1–3). The concept originated from a simple observation that in vitro incubation of macrophages with oxidized LDL and not with native LDL led to cholesterol ester accumulation (2, 4). The oxidation of LDL is a complex process during which both the protein and the lipids undergo oxidative changes and form complex products. For Example, non-enzymatic oxidative changes in amino acids as well as proteolysis and cross-links of apoprotein B (apo B) occur that result in extensive alteration in

R.M. Uppu et al. (eds.), *Free Radicals and Antioxidant Protocols*, Methods in Molecular Biology 610,
DOI 10.1007/978-1-60327-029-8_24, © Humana Press, a part of Springer Science+Business Media, LLC 1998, 2010

the protein composition and structure (5). In addition, the peroxidized lipids decompose generating both free and core aldehydes and ketones that covalently modify ε-amino groups of lysine residues of the protein moiety (6). The latter not only generates Schiff's bases, thus modifying charges on the amino acids, but also results in both intra- and intermolecular cross-links between proteolyzed apo B. No wonder the LDL that is damaged to such an extent is avidly scavenged and degraded by macrophages. In fact, the oxidative modification is not unique to LDL as other lipoproteins such as very low-density lipoprotein (VLDL), beta very low-density lipoprotein (β-VLDL), and even high-density lipoprotein (HDL) have been suggested to undergo similar oxidative changes (7–18) with accompanying changes in their pro- or anti-atherosclerotic behavior (19–23).

What is Ox-LDL? The definition depends on the purpose of the question. The original concept defined Ox-LDL as "oxidatively modified LDL" that contained protein components that were "modified" presumably by aldehyde products creating net negative charges that were essential for its interaction and uptake by macrophages (3, 24). However, during the past 25 years myriads of protein-independent pro-atherogenic effects have been attributed to the lipid components of oxidized LDL that one has to redefine oxidized LDL. In order to define oxidized LDL, one needs to know what components are usually associated with it.

Table 24.1 is a list of products that are generated during in vitro oxidation of LDL.

The formation of all of the above products or the changes in the properties of circulating LDL are not guaranteed during the oxidation of LDL as many are secondary products of oxidation and their formation might depend on the type of oxidant, the extent of oxidation, and the presence or absence of other agents such as redox metals. Also, some products, such as malondialdehyde, readily diffuse out of Ox-LDL. More importantly, the fatty acid profile plays a major role in the formation of many of the products and of Ox-LDL (25–27). Generally, polyunsaturated fatty acids favor the oxidation of LDL while monounsaturated fatty acids are less conducive to its formation (28, 29). In addition, the protein component of the LDL (apo B) might also determine how the oxidation is propagated within the particle. Specific amino acids might generate and propagate oxidation (30–34).

In light of these, Ox-LDL can be defined as follows: A particle derived from circulating LDL that may have peroxides or their degradation products generated within the LDL molecule or elsewhere in the body associated with the particle. This would include minimally oxidized LDL that may have lipid peroxides or their degradation products (for Example, oxovaleryl PtdCho) but no apoprotein changes, MDA modified particles with MDA

Table 24.1
List of lipid/protein oxidation products generated during the oxidation of LDL

A. Fatty acid oxidation products:	
(i)	Free and esterified fatty acid peroxides, such as 13-hydroperoxylinoleic acid (13-HPODE)
(ii)	Free and esterified fatty acid hydroxides, such as 13-hydroxylinoleic acid (13-HODE)
(iii)	Prostaglandin-like products, such as isoprostanes in free and esterified forms
(iv)	Aldehydes, such as malondialdehyde (MDA), 4-hydroxy nonenal, and hexanal
(v)	Core aldehydes that contain esterified lipid backbone, such as oxovaleryl phosphatidylcholine
(vi)	Pentane and other hydrocarbons
B. Lipid derived products:	
(i)	Lysophosphatidylcholine
(ii)	Cholesterol oxidation products, such as 7 keto cholesterol
(iii)	Internally modified phosphatidyl ethanolamine/serine products
C. Protein oxidation products:	
(i)	Protein carbonyls
(ii)	Non-enzymatic proteolyzed fragments
(iii)	Modified cysteine, cystine, histidine, methionine, lysine, arginine, tryptophan, and tyrosine
(iv)	Protein cross-links due to tyrosine cross-links as well as due to bifunctional aldehydes
(v)	Lipid–protein adducts which could be classified as ceroids (lipofuscins)
(vi)	Many of the above changes as well as conformational changes might lead to antigenicity
D. Other changes:	
(i)	Increased buoyant density
(ii)	Increased negative charge
(iii)	Loss of characteristic yellow color (human)
(iv)	Loss of enzyme activities associated with LDL

arising from platelets or elsewhere, and others. However, to this date, LDL particles with oxidized apo B amino acids without lipid changes have not been described. **Figure 24.1** describes some of the potential Ox-LDL particles that may be of relevance.

Although not depicted in the picture, major changes in apo B might occur in addition to aldehyde modification and proteolysis, depending on the type of oxidant used. As a result, countless antigenic epitopes are possible (35, 36). As mentioned before, these changes are not unique to LDL and other lipoprotein might undergo similar changes (37, 38). However, the author has not seen any proteolysis of HDL apoproteins as a result of oxidation but has noted aggregation changes.

In summary, Ox-LDL may contain a specific oxidation product (*see* **Table 24.2**). The simplest Example may be lipoxygenase-treated LDL. The lipoprotein subjected to such treatment may contain varying amounts of phospholipid and cholesterol ester

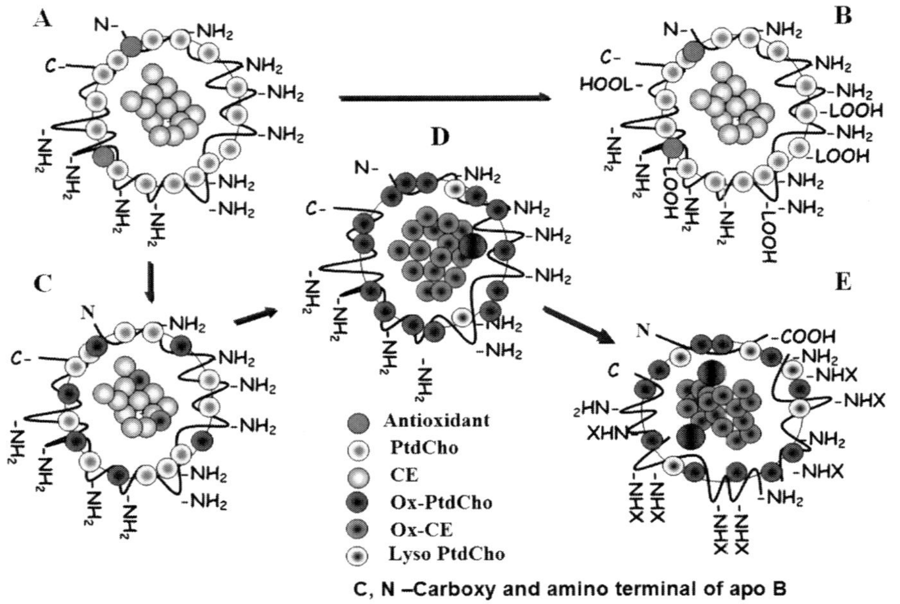

Fig. 24.1. Forms of oxidized low-density lipoprotein (reproduced from Parthasarathy et al. (157)). (**a**)Unoxidized native LDL with amino groups of lysine residues of apo B and representative lipids. (**b**) Lipid peroxides generated elsewhere associated with such LDL. (**c**) LDL lipids might get oxidized resulting in the generation of cholesterol ester and phospholipid peroxides. (**d**) Such LDL might undergo extensive oxidation leading to protein changes. (e) Extensive protein changes and l pid decomposition might hallmark the end stages of oxidation.

Table 24.2
Suggested mechanisms for the oxidation of LDL

Mechanism	References
1. Lipoxygenase reaction	(47–53)
2. Copper and ceruloplasmin-mediated oxidation	(54–58)
3. Iron-mediated oxidation	(54, 59, 60)
4. Peroxidase-mediated oxidation including myeloperoxidase and heme	(61–63)
5. Peroxynitrite-mediated oxidation	(64–66)
6. Thiol-dependent oxidation	(67–70)
7. Xanthine oxidase, NADPH oxidase, and other superoxide generators	(71–78)
8. AAPH or other means of radical generation including cytochromes	(79–81)

hydroperoxides. Ox-LDL may also contain limited amounts of a variety of degraded oxidized lipid products. "Minimally modified LDL" as described by Berliner, Fogelman and associates may qualify for such a criteria (39, 40). There is also Ox-LDL that is recognized by macrophages (41, 42). While it was believed for a long time that protein changes are essential for macrophage recognition, at least some studies seem to suggest that oxidized lipids might mediate the whole Ox-LDL uptake (43–45). There are, however, Ox-LDL preparations that do not share the properties of conventionally oxidized lipoproteins. For Example, LDL oxidized by photooxidation in the presence of dyes, such as Rose Bengal, showed massive increase in electrophoretic mobility as a result of oxidation of histidine residues (46). However, other than a loss of phospholipase A2 activity, it did not bear any resemblance to oxidized LDL.

Ox-LDL, thus, might represent the elephant that is described by blind men. It is without description and yet distinct; it is a complex mixture of chemical entities and yet they are measurable, and individual investigator is responsible for setting forth the definition that suits a set of criteria. These definitions should be the basis for experimental validation and scientific reproducibility.

Not all the oxidation mechanisms are comparable or lead to similar products even under in vitro conditions. For Example, peroxidase-mediated oxidations require co-oxidants such as H_2O_2 or lipid peroxides. Some damage the proteins more readily than others. For Example, treatment of LDL with 2,2′-azobis(2-amidinopropane) dihydrochloride (AAPH), a radical generator or peroxynitrite, Results in more protein oxidation than lipid peroxidation (65, 80, 82, 83). Similarly, peroxidase-catalyzed oxidation generate very little aldehyde products as compared to metal-catalyzed oxidations (84). Lipoxygenase reactions which are exclusively intracellular might require additional reactions or transfers before or after lipoxygenase reactions occur (85, 86). To our knowledge, no one has characterized oxidized LDL isolated from animals or humans to the extent that mechanistic insights could be derived, although there are numerous claims of the presence of an electronegative LDL (presumably minimally oxidized LDL) in human and animal plasma (87–95).

While all these reactions are oxidative in nature, they are not uniformly amenable to inhibition by traditional antioxidants. Vitamin E or simple phenols such as tyrosine or estradiol actually enhance peroxidase-mediated oxidation of LDL (96–98). This has been taken as evidence for an intermediate prooxidant role for antioxidants. Similarly, many thiols actually enhance the oxidation of LDL depending on the peroxide content of LDL (99).

The detection of Ox-LDL, at least in vitro, has been quite easy, although earlier work, before the discovery of the involvement of oxidative process, revolved around the increase in net negative charge and increase in buoyant density. The presence and amounts of MDA has always been used as a yard stick for minimally or terminally oxidized LDL (100–103), although numerous publications concluded a short-term oxidation always resulted in a "mildly" oxidized LDL as opposed to long-term oxidations

(104–107). Yet, enzymatic (lipoxygenase, peroxidase) reactions and treatment with peroxynitrite that generated lower amounts of MDA were always considered fully oxidized LDL due the ability of macrophages to engulf such LDL.

The measurement of lipid oxidation has been a great boon, not only to the understanding of the process of LDL oxidation but also in providing numerous serendipitous discoveries and methodologies. The formation of conjugated dienes during lipid peroxidation was successfully exploited to generate the "lag time" concept by Esterbauer and associates (108) that is still used as a yard stick for measuring the oxidizability of LDL. Similarly, the discovery of isoprostanes by Roberts, Morrow and others (109) also created tremendous excitement and opportunity to look for such LDL in vivo (110).

Since the discovery of oxidized LDL, over 5,000 articles have appeared on the topic with over 400 articles appearing every year during the past decade (**Fig. 24.2**). Combined with epidemiological evidence and success in several animal trials in a number of species using a variety of antioxidants, the hypothesis appeared to be on solid ground until recently. This euphoria of initial success led to clinical trials that were planned mostly to be the "first" to validate the hypothesis. Obviously vitamin E was the chosen antioxidant as the expectation of the general public and scientists was that a natural antioxidant (such as vitamin E) would have less undesirable effects. Accordingly a number of clinical trials were performed (111–114) using vitamin E at pharmacological doses. Some of these trials were piggyback studies on cancer and other diseases (115, 116). These trials with vitamin E have not been overwhelmingly supportive of the hypothesis. Understanding the

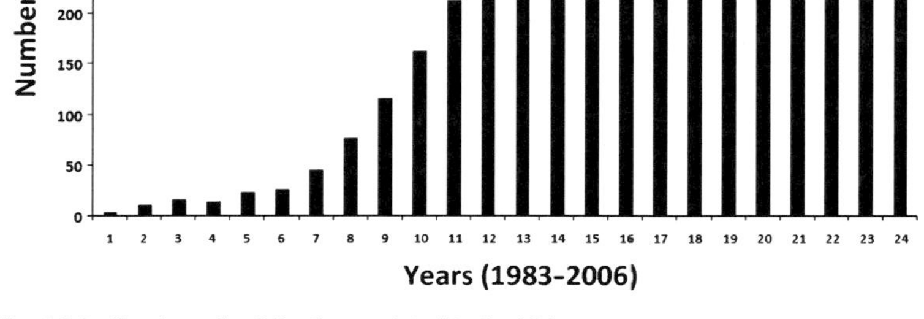

Fig. 24.2. Number of publications related to Ox-LDL.

sequence of events that led to the hypothesis, the molecular basis of antioxidant action, the potential interaction with other drugs, and the stage of the disease process to which oxidation was suggested to contribute, dosage, the transport mechanisms involved in the release and utilization of vitamin E in the body, its interaction(s) with cytochrome systems, and its function as a prooxidant in peroxidase-mediated oxidations played little role in the design of the study (117, 118). More importantly, little if any attention was paid to factors that would affect and control oxidation in vivo. There have been many reviews written dissecting out the trials that "disproved" the oxidation hypothesis (119–122).

On the other hand, more and more recent publications have described the effects of oxidized lipids on atherosclerosis and associated risk factors (123–130). The prevalence of antibodies that recognize epitopes of Ox-LDL seem to correlate well with the disease process (35, 95, 131–145). These antibodies also detect such LDL in the plasma of subjects who have higher risk associated with atherosclerosis (35, 139, 145–153). Oxidative enzymes such as Myeloperoxidase seem to predict the vulnerable population and seem to correlate with the severity of the disease (154–156). Considering that vitamin E might actually enhance peroxidase-catalyzed oxidations (98), the clinical trials might have actually endorsed the hypothesis rather than shooting it down. Perhaps, when inhibitors of specific oxidases are available, additional trials might be warranted to induce the development of anti-atherosclerotic pharmacological agents.

Acknowledgment

This work was supported by funding from National Institutes of Health, HL-069038 and HL-74239 (SP) and HL74239 (NS).

References

1. Steinbrecher, U.P., Parthasarathy, S., Leake, D.S., Witztum, J.L., and Steinberg, D. (1984) Modification of low density lipoprotein by endothelial cells involves lipid peroxidation and degradation of low density lipoprotein phospholipids. *Proc. Natl. Acad. Sci. USA* **81**, 3883–3887.
2. Quinn, M.T., Parthasarathy, S., Fong, L.G., and Steinberg, D. (1987) Oxidatively modified low density lipoproteins: a potential role in recruitment and retention of monocyte/macrophages during atherogenesis. *Proc. Natl. Acad. Sci. USA* **84**, 2995–2998.
3. Steinberg, D., Parthasarathy, S., Carew, T.E., Khoo, J.C., and Witztum, J.L. (1989) Beyond cholesterol. Modifications of low-density lipoprotein that increase its atherogenicity [see comments]. *N. Engl. J. Med.* **320**, 915–924.
4. Parthasarathy, S., Quinn, M.T., and Steinberg, D. (1988) Is oxidized low density lipoprotein involved in the recruitment and retention of monocyte/macrophages

in the artery wall during the initiation of atherosclerosis? Basic Life Sci. **49**, 375–380.
5. Fong, L.G., Parthasarathy, S., Witztum, J.L., and Steinberg, D. (1987) Nonenzymatic oxidative cleavage of peptide bonds in apoprotein B-100. *J. Lipid Res.* **28**, 1466–1477.
6 Fruebis, J., Parthasarathy, S., and Steinberg, D. (1992) Evidence for a concerted reaction between lipid hydroperoxides and polypeptides. *Proc. Natl. Acad. Sci. USA* **89**, 10588–10592.
7. Parthasarathy, S., Quinn, M.T., Schwenke, D.C., Carew, T.E., and Steinberg, D. (1989) Oxidative modification of beta-very low density lipoprotein. Potential role in monocyte recruitment and foam cell formation. *Arteriosclerosis* **9**, 398–404.
8. de Rijke, Y.B., Hessels, E.M., and van Berkel, T.J. (1992) Recognition sites on rat liver cells for oxidatively modified beta-very low density lipoproteins. *Arterioscler. Thromb.* **12**, 41–49.
9. Bowry, V.W., Stanley, K.K., and Stocker, R. (1992) High density lipoprotein is the major carrier of lipid hydroperoxides in human blood plasma from fasting donors. *Proc. Natl. Acad. Sci. USA* **89**, 10316–10320.
10. Bradamante, S., Barenghi, L., Giudici, G.A., and Vergani, C. (1992) Free radicals promote modifications in plasma high-density lipoprotein: nuclear magnetic resonance analysis. *Free Radic. Biol. Med.* **12**, 193–203.
11. Bonnefont-Rousselot, D., Khalil, A., Delattre, J., Jore, D., and Gardes-Albert, M. (1997) Oxidation of human high-density lipoproteins by OH and OH/O(-)2 free radicals. *Radiat. Res.* **147**, 721–728.
12. Greilberger, J. and Jurgens, G. (1998) Oxidation of high-density lipoprotein HDL3 leads to exposure of apo-AI and apo-AII epitopes and to formation of aldehyde protein adducts, and influences binding of oxidized low-density lipoprotein to type I and type III collagen in vitro. *Biochem. J.* **331**, 185–191.
13. Bergt, C., Oram, J.F., and Heinecke, J.W. (2003) Oxidized HDL: the paradox-idation of lipoproteins. *Arterioscler. Thromb. Vasc. Biol.* **23**, 1488–1490.
14. Asztalos, B.F. (2004) High-density lipoprotein metabolism and progression of atherosclerosis: new insights from the HDL Atherosclerosis Treatment Study. *Curr. Opin. Cardiol.* **19**, 385–391.
15. Navab, M., Ananthramaiah, G.M., Reddy, S.T., Van Lenten, B.J., Ansell, B.J., Fonarow, G.C., Vahabzadeh, K., Hama, S., Hough, G., Kamranpour, N., et al. (2004) The oxidation hypothesis of atherogenesis: the role of oxidized phospholipids and HDL. *J. Lipid Res.* **45**, 993–1007.
16. Stojanovic, N., Krilov, D., and Herak, J.N. (2006) Slow oxidation of high density lipoproteins as studied by EPR spectroscopy. *Free Radic. Res.* **40**, 135–140.
17. Malle, E., Marsche, G., Panzenboeck, U., and Sattler, W. (2006) Myeloperoxidase-mediated oxidation of high-density lipoproteins: fingerprints of newly recognized potential proatherogenic lipoproteins. *Arch. Biochem. Biophys.* **445**, 245–255.
18. Ferretti, G., Bacchetti, T., Negre-Salvayre, A., Salvayre, R., Dousset, N., and Curatola, G. (2006) Structural modifications of HDL and functional consequences. *Atherosclerosis* **184**, 1–7.
19. Kervinen, K., Horkko, S., Beltz, W.F., and Kesaniemi, A. (1995) Modification of VLDL apoprotein B by acetaldehyde alters apoprotein B metabolism. *Alcohol* **12**, 189–194.
20. Nagano, Y., Arai, H., and Kita, T. (1991) High density lipoprotein loses its effect to stimulate efflux of cholesterol from foam cells after oxidative modification. *Proc. Natl. Acad. Sci. USA* **88**, 6457–6461.
21. Ghiselli, G., Giorgini, L., Gelati, M., and Musanti, R. (1992) Oxidatively modified HDLs are potent inhibitors of cholesterol biosynthesis in human skin fibroblasts. *Arterioscler. Thromb.* **12**, 929–935.
22. Van Lenten, B.J., Wagner, A.C., Nayak, D.P., Hama, S., Navab, M., and Fogelman, A.M. (2001) High-density lipoprotein loses its anti-inflammatory properties during acute influenza a infection. *Circulation* **103**, 2283–2288.
23. Jaouad, L., Milochevitch, C., and Khalil, A. (2003) PON1 paraoxonase activity is reduced during HDL oxidation and is an indicator of HDL antioxidant capacity. *Free Radic. Res.* **37**, 77–83.
24. Parthasarathy, S. (1994) *Modified Lipoproteins in the Pathogenesis of Atherosclerosis.* R.G. Landes Co., Austin, TX, 152 pp.
25. Lada, A.T. and Rudel, L.L. (2004) Associations of low density lipoprotein particle composition with atherogenicity. *Curr. Opin. Lipidol.* **15**, 19–24.
26. Moreno, J.J. and Mitjavila, M.T. (2003) The degree of unsaturation of dietary fatty acids and the development of atherosclerosis (review). *J. Nutr. Biochem.* **14**, 182–195.
27. Kratz, M., Cullen, P., Kannenberg, F., Kassner, A., Fobker, M., Abuja, P.M., Assmann,

G., and Wahrburg, U. (2002) Effects of dietary fatty acids on the composition and oxidizability of low-density lipoprotein. *Eur. J. Clin. Nutr.* **56**, 72–81.

28. Reaven, P., Parthasarathy, S., Grasse, B.J., Miller, E., Steinberg, D., and Witztum, J.L. (1993) Effects of oleate-rich and linoleate-rich diets on the susceptibility of low density lipoprotein to oxidative modification in mildly hypercholesterolemic subjects. *J. Clin. Invest.* **91**, 668–676.
29. Lada, A.T. and Rudel, L.L. (2003) Dietary monounsaturated versus polyunsaturated fatty acids: which is really better for protection from coronary heart disease? *Curr. Opin. Lipidol.* **14**, 41–46.
30. Giessauf, A., van Wickern, B., Simat, T., Steinhart, H., and Esterbauer, H. (1996) Formation of N-formylkynurenine suggests the involvement of apolipoprotein B-100 centered tryptophan radicals in the initiation of LDL lipid peroxidation. *FEBS Lett.* **389**, 136–140.
31. Miller, Y.I., Felikman, Y., and Shaklai, N. (1996) Hemoglobin induced apolipoprotein B crosslinking in low-density lipoprotein peroxidation. *Arch. Biochem. Biophys.* **326**, 252–260.
32. Alaiz, M., Beppu, M., Ohishi, K., and Kikugawa, K. (1994) Modification of delipidated apoprotein B of low density lipoprotein by lipid oxidation products in relation to macrophage scavenger receptor binding. *Biol. Pharm. Bull.* **17**, 51–57.
33. Picard, S., Parthasarathy, S., Fruebis, J., and Witztum, J.L. (1992) Aminoguanidine inhibits oxidative modification of low density lipoprotein protein and the subsequent increase in uptake by macrophage scavenger receptors. *Proc. Natl. Acad. Sci. USA* 89, 6876–6880.
34. Van Antwerpen, P., Legssyer, I., Zouaoui Boudjeltia, K., Babar, S., Moreau, P., Moguilevsky, N., Vanhaeverbeek, M.,Ducobu, J., and Neve, J. (2006) Captopril inhibits the oxidative modification of apolipoprotein B-100 caused by myeloperoxidase in a comparative in vitro assay of angiotensin converting enzyme inhibitors. *Eur. J. Pharmacol.* **537**, 31–36.
35. Salonen, J.T., Yla-Herttuala, S., Yamamoto, R., Butler, S., Korpela, H., Salonen, R., Nyyssonen,K., Palinski, W., and Witztum, J.L. (1992) Autoantibody against oxidised LDL and progression of carotid atherosclerosis [see comments]. *Lancet* **339**, 883–887.
36. Palinski, W., Horkko, S., Miller, E., Steinbrecher, U.P., Powell, H.C., Curtiss, L.K., and Witztum, J.L. (1996) Cloning of monoclonal autoantibodies to epitopes of oxidized lipoproteins from apolipoprotein E-deficient mice. Demonstration of epitopes of oxidized low density lipoprotein in human plasma. *J. Clin. Invest.* **98**, 800–814.
37. Nakajima, T., Sakagishi, Y., Katahira, T., Nagata, A., Kuwae, T., Nakamura, H., Inoue, I., Takahashi, K., Katayama, S., and Komoda, T. (1995) Characterization of a specific monoclonal antibody 9F5-3a and the development of assay system for oxidized HDL. *Biochem. Biophys. Res. Commun.* **217**, 407–411.
38. Shao, B., O'Brien, K.D., McDonald, T.O., Fu, X., Oram, J.F., Uchida, K., and Heinecke, J.W. (2005) Acrolein modifies apolipoprotein A-I in the human artery wall. *Ann. NY Acad. Sci.* **1043**, 396–403.
39. Cushing, S.D., Berliner, J.A., Valente, A.J., Territo, M.C., Navab, M., Parhami, F., Gerrity, R., Schwartz, C.J., and Fogelman, A.M. (1990) Minimally modified low density lipoprotein induces monocyte chemotactic protein 1 in human endothelial cells and smooth muscle cells. *Proc. Natl. Acad. Sci. USA* **87**, 5134–5138.
40. Liao, F., Berliner, J.A., Mehrabian, M., Navab, M., Demer, L.L., Lusis, A.J., and Fogelman, A.M. (1991) Minimally modified low density lipoprotein is biologically active in vivo in mice [published erratum appeared in J. Clin. Invest. 88, 721, 1991]. *J. Clin. Invest.* **87**, 2253–2257.
41. Parthasarathy, S., Fong, L.G., Otero, D., and Steinberg, D. (1987) Recognition of solubilized apoproteins from delipidated, oxidized low density lipoprotein (LDL) by the acetyl-LDL receptor. *Proc. Natl. Acad. Sci. USA* **84**, 537–540.
42. Sparrow, C.P., Parthasarathy, S., and Steinberg, D. (1989) A macrophage receptor that recognizes oxidized low density lipoprotein but not acetylated low density lipoprotein. *J. Biol. Chem.* **264**, 2599–2604.
43. Morel, D.W., Hessler, J.R., and Chisolm, G.M. (1983) Low density lipoprotein cytotoxicity induced by free radical peroxidation of lipid. *J. Lipid. Res.* **24**, 1070–1076.
44. Hoff, H.F., O'Neil, J., Chisolm, G.M.D., Cole, T.B., Quehenberger, O., Esterbauer, H., and Jurgens, G. (1989) Modification of low density lipoprotein with 4-hydroxynonenal induces uptake by macrophages. *Arteriosclerosis* **9**, 538–549.
45. Tertov, V.V., Kaplun, V.V., Dvoryantsev, S.N., and Orekhov, A.N. (1995) Apolipoprotein B-bound lipids as a marker for evaluation of low density lipoprotein oxi-

dation in vivo. *Biochem. Biophys. Res. Commun.* **214**, 608–613.

46. Parthasarathy, S., and Barnett, J. (1990) Phospholipase A2 activity of low density lipoprotein: evidence for an intrinsic phospholipase A2 activity of apoprotein B-100. *Proc. Natl. Acad. Sci. USA* **87**, 9741–9745.
47. Sparrow, C.P., Parthasarathy, S., and Steinberg, D. (1988) Enzymatic modification of low density lipoprotein by purified lipoxygenase plus phospholipase A2 mimics cell-mediated oxidative modification. *J. Lipid Res.* **29**, 745–753.
48. Parthasarathy, S., Wieland, E., and Steinberg, D. (1989) A role for endothelial cell lipoxygenase in the oxidative modification of low density lipoprotein. *Proc. Natl. Acad. Sci. USA* **86**, 1046–1050.
49. Cathcart, M.K., McNally, A.K., and Chisolm, G.M. (1991) Lipoxygenase-mediated transformation of human low density lipoprotein to an oxidized and cytotoxic complex. *J. Lipid Res.* **32**, 63–70.
50. Rankin, S.M., Parthasarathy, S., and Steinberg, D. (1991) Evidence for a dominant role of lipoxygenase(s) in the oxidation of LDL by mouse peritoneal macrophages. *J. Lipid Res.* **32**, 449–456.
51. Sigal, E., Laughton, C.W., and Mulkins, M.A. (1994) Oxidation, lipoxygenase, and atherogenesis. *Ann. NY Acad. Sci.* **714**, 211–224.
52. Kuhn, H., Belkner, J., Suzuki, H., and Yamamoto, S. (1994) Oxidative modification of human lipoproteins by lipoxygenases of different positional specificities. *J. Lipid Res.* **35**, 1749–1759.
53. Kuhn, H., and Chan, L. (1997) The role of 15-lipoxygenase in atherogenesis: pro- and antiatherogenic actions. *Curr. Opin. Lipidol.* **8**, 111–117.
54. Heinecke, J.W., Rosen, H., and Chait, A. (1984) Iron and copper promote modification of low density lipoprotein by human arterial smooth muscle cells in culture. *J. Clin. Invest.* **74**, 1890–1894.
55. Quehenberger, O., Jurgens, G., Zadravec, S., and Esterbauer, H. (1988) Oxidation of human low density lipoprotein initiated by copper (II) chloride. *Basic Life Sci.* **49**, 387–390.
56. Parthasarathy, S., Fong, L.G., Quinn, M.T., and Steinberg, D. (1990) Oxidative modification of LDL: comparison between cell-mediated and copper-mediated modification. *Eur. Heart. J.* **11** Suppl E, 83–87.
57. Ehrenwald, E., Chisolm, G.M., and Fox, P.L. (1994) Intact human ceruloplasmin oxidatively modifies low density lipoprotein. *J. Clin. Invest.* **93**, 1493–1501.
58. Lamb, D.J. and Leake, D.S. (1994) Acidic pH enables ceruloplasmin to catalyse the modification of low- density lipoprotein. *FEBS Lett.* **338**, 122–126.
59. Balla, G., Jacob, H.S., Eaton, J.W., Belcher, J.D., and Vercellotti, G.M. (1991) Hemin: a possible physiological mediator of low density lipoprotein oxidation and endothelial injury. *Arterioscler. Thromb. Vasc. Biol.* **11**, 1700–1711.
60. Sakurai, T., Kimura, S., Nakano, M., and Kimura, H. (1991) Oxidative modification of glycated low density lipoprotein in the presence of iron. *Biochem. Biophys. Res. Commun.* **177**, 433–439.
61. Wieland, E., Parthasarathy, S., and Steinberg, D. (1993) Peroxidase-dependent metal-independent oxidation of low density lipoprotein in vitro: a model for in vivo oxidation? *Proc. Natl. Acad. Sci. USA* **90**, 5929–5933.
62. Savenkova, M.L., Mueller, D.M., and Heinecke, J.W. (1994) Tyrosyl radical generated by myeloperoxidase is a physiological catalyst for the initiation of lipid peroxidation in low density lipoprotein. *J. Biol. Chem.* **269**, 20394–20400.
63. Natella, F., Nardini, M., Ursini, F., and Scaccini, C. (1998) Oxidative modification of human low-density lipoprotein by horseradish peroxidase in the absence of hydrogen peroxide. *Free Radic. Res.* **29**, 427–434.
64. Darley-Usmar, V.M., Hogg, N., O'Leary, V.J., Wilson, M.T., and Moncada, S. (1992) The simultaneous generation of superoxide and nitric oxide can initiate lipid peroxidation in human low density lipoprotein. *Free Radic. Res. Commun.* **17**, 9–20.
65. Graham, A., Hogg, N., Kalyanaraman, B., O'Leary, V., Darley-Usmar, V., and Moncada, S. (1993) Peroxynitrite modification of low-density lipoprotein leads to recognition by the macrophage scavenger receptor. *FEBS Lett.* **330**, 181–185.
66. Panasenko, O.M., Briviba, K., Klotz, L.O., and Sies, H. (1997) Oxidative modification and nitration of human low-density lipoproteins by the reaction of hypochlorous acid with nitrite. *Arch. Biochem. Biophys.* **343**, 254–259.
67. Parthasarathy, S. (1987) Oxidation of low-density lipoprotein by thiol compounds leads to its recognition by the acetyl LDL receptor. *Biochim. Biophys. Acta.* **917**, 337–340.

68. Heinecke, J.W., Kawamura, M., Suzuki, L., and Chait, A. (1993) Oxidation of low density lipoprotein by thiols: superoxide-dependent and -independent mechanisms. *J. Lipid Res.* **34**, 2051–2061.
69. Sparrow, C.P. and Olszewski, J. (1993) Cellular oxidation of low density lipoprotein is caused by thiol production in media containing transition metal ions. *J. Lipid Res.* **34**, 1219–1228.
70. Wood, J.L. and Graham, A. (1995) The role of thiols in oxidation of low-density lipoprotein by macrophages. *Biochem. Soc. Trans.* **23**, 242S.
71. Lynch, S.M. and Frei, B. (1993) Mechanisms of copper- and iron-dependent oxidative modification of human low density lipoprotein. *J. Lipid Res.* **34**, 1745–1753.
72. Napoli, C., Ambrosio, G., Palumbo, G., Elia, P.P., and Chiariello, M. (1991) Human low-density lipoproteins are peroxidized by free radicals via chain reactions triggered by the superoxide radical. *Cardiologia* **36**, 527–532.
73. Jessup, W., Simpson, J.A., and Dean, R.T. (1993) Does superoxide radical have a role in macrophage-mediated oxidative modification of LDL? Atherosclerosis **99**, 107–120.
74. Aviram, M., Rosenblat, M., Etzioni, A., and Levy, R. (1996) Activation of NADPH oxidase required for macrophage-mediated oxidation of low-density lipoprotein. *Metabolism* **45**, 1069–1079.
75. Montgomery, R.R., Nathan, C.F., and Cohn, Z.A. (1986) Effects of reagent and cell-generated hydrogen peroxide on the properties of low density lipoprotein. *Proc. Natl. Acad. Sci. USA* **83**, 6631–6635.
76. Hiramatsu, K., Rosen, H., Heinecke, J.W., Wolfbauer, G., and Chait, A. (1987) Superoxide initiates oxidation of low density lipoprotein by human monocytes. *Arteriosclerosis.* **7**, 55–60.
77. Steinbrecher, U.P. (1988) Role of superoxide in endothelial-cell modification of low-density lipoproteins. *Biochim. Biophys. Acta* **959**, 20–30.
78. Cathcart, M.K., McNally, A.K., Morel, D.W., and Chisolm, G.M.D. (1989) Superoxide anion participation in human monocyte-mediated oxidation of low-density lipoprotein and conversion of low-density lipoprotein to a cytotoxin. *J. Immunol.* **142**, 1963–1969.
79. Kawabe, Y., Cynshi, O., Takashima, Y., Suzuki, T., Ohba, Y., and Kodama, T. (1994) Oxidation-induced aggregation of rabbit low-density lipoprotein by azo initiator. *Arch. Biochem. Biophys.* **310**, 489–496.
80. Noguchi, N., Gotoh, N., and Niki, E. (1994) Effects of ebselen and probucol on oxidative modifications of lipid and protein of low density lipoprotein induced by free radicals. *Biochim. Biophys. Acta* **1213**, 176–182.
81. Leonhardt, W., Bergmann, R., and Hanefeld, M. (1997) Initiation of LDL oxidation by copper ions or AAPH yields different kinetic parameters which are correlated [letter]. *Clin. Chim. Acta* **259**, 195–197.
82. Kim, J.G., Sabbagh, F., Santanam, N., Wilcox, J.N., Medford, R.M., and Parthasarathy, S. (1997) Generation of a polyclonal antibody against lipid peroxide-modified proteins. *Free Radic. Biol. Med.* **23**, 251–259.
83. Dinis, T.C., Santosa, C.L., and Almeida, L.M. (2002) The apoprotein is the preferential target for peroxynitrite-induced LDL damage protection by dietary phenolic acids. *Free Radic. Res.* **36**, 531–543.
84. Heinecke, J.W. (1997) Pathways for oxidation of low density lipoprotein by myeloperoxidase: tyrosyl radical, reactive aldehydes, hypochlorous acid and molecular chlorine. *Biofactors* **6**, 145–155.
85. Cornicelli, J.A. and Trivedi, B.K. (1999) 15-Lipoxygenase and its inhibition: a novel therapeutic target for vascular disease. *Curr. Pharm. Des.* **5**, 11–20.
86. Kuhn, H., Romisch, I., and Belkner, J. (2005) The role of lipoxygenase-isoforms in atherogenesis. *Mol. Nutr. Food Res.* **49**, 1014–1029.
87. Gugliucci Creriche, A. and Stahl, A.J. (1993) Glycation and oxidation of human low density lipoproteins reduces heparin binding and modifies charge. *Scand. J. Clin. Lab Invest.* **53**, 125–132.
88. Sanchez-Quesada, J.L., Perez, A., Caixas, A., Ordonmez-Llanos, J., Carreras, G., Payes, A., Gonzalez-Sastre, F., and de Leiva, A. (1996) Electronegative low density lipoprotein subform is increased in patients with short-duration IDDM and is closely related to glycaemic control. *Diabetologia* **39**, 1469–1476.
89. Demuth, K., Myara, I., Chappey, B., Vedie, B., Pech-Amsellem, M.A., Haberland, M.E., and Moatti, N. (1996) A cytotoxic electronegative LDL subfraction is present in human plasma. *Arterioscler. Thromb. Vasc. Biol.* **16**, 773–783.
90. Dai, L., Zhang, Z., Winyard, P.G., Gaffney, K., Jones, H., Blake, D.R., and Morris, C.J. (1997) A modified form of low-density lipoprotein with increased electronegative

charge is present in rheumatoid arthritis synovial fluid. *Free Radic. Biol. Med.* **22**, 705–710.

91. Moro, E., Zambon, C., Pianetti, S., Cazzolato, G., Pais, M., and Bittolo Bon, G. (1998) Electronegative low density lipoprotein subform (LDL-) is increased in type 2 (non-insulin-dependent) microalbuminuric diabetic patients and is closely associated with LDL susceptibility to oxidation. *Acta Diabetol.* **35**, 161–164.
92. Benitez, S., Perez, A., Sanchez-Quesada, J.L., Wagner, A.M., Rigla, M., Arcelus, R., Jorba, O., and Ordonez-Llanos, J. (2007) Electronegative low-density lipoprotein subfraction from type 2 diabetic subjects is proatherogenic and unrelated to glycemic control. *Diabetes. Metab. Res. Rev.* **23**, 26–34.
93. Sanchez-Quesada, J.L., Camacho, M., Anton, R., Benitez, S., Vila, L., and Ordonez-Llanos, J. (2003) Electronegative LDL of FH subjects: chemical characterization and induction of chemokine release from human endothelial cells. *Atherosclerosis* **166**, 261–270.
94. Parasassi, T., Bittolo-Bon, G., Brunelli, R., Cazzolato, G., Krasnowska, E.K., Mei, G., Sevanian, A., and Ursini, F. (2001) Loss of apoB-100 secondary structure and conformation in hydroperoxide rich, electronegative LDL(-). *Free Radic. Biol. Med.* **31**, 82–89.
95. Barros, M.R., Bertolami, M.C., Abdalla, D.S., and Ferreira, W.P. (2006) Identification of mildly oxidized low-density lipoprotein (electronegative LDL) and its autoantibodies IgG in children and adolescents hypercholesterolemic offsprings. *Atherosclerosis* **184**, 103–107.
96. Yamamoto, K. and Niki, E. (1988) Interaction of alpha-tocopherol with iron: antioxidant and prooxidant effects of alpha-tocopherol in the oxidation of lipids in aqueous dispersions in the presence of iron. *Biochim. Biophys. Acta* **958**, 19–23.
97. Bowry, V.W., Ingold, K.U., and Stocker, R. (1992) Vitamin E in human low-density lipoprotein. When and how this antioxidant becomes a pro-oxidant. *Biochem. J.* **288**, 341–344.
98. Santanam, N. and Parthasarathy, S. (1995) Paradoxical actions of antioxidants in the oxidation of low density lipoprotein by peroxidases. *J. Clin. Invest.* **95**, 2594–2600.
99. Santanam, N., and Parthasarathy, S. (1995) Cellular cysteine generation does not contribute to the initiation of LDL oxidation. *J. Lipid Res.* **36**, 2203–2211.
100. Palinski, W., Rosenfeld, M.E., Yla-Herttuala, S., Gurtner, G.C., Socher, S.S., Butler, S.W., Parthasarathy, S., Carew, T.E., Steinberg, D., and Witztum, J.L. (1989) Low density lipoprotein undergoes oxidative modification in vivo. *Proc. Natl. Acad. Sci. USA* **86**, 1372–1376.
101. Aviram, M. (1990) Malondialdehyde affects the physico-chemical and biological characteristics of oxidized low density lipoprotein. *Atherosclerosis* **84**, 141–143.
102. Lecomte, E., Artur, Y., Chancerelle, Y., Herbeth, B., Galteau, M.M., Jeandel, C., and Siest, G. (1993) Malondialdehyde adducts to, and fragmentation of, apolipoprotein B from human plasma. *Clin. Chim. Acta* **218**, 39–46.
103. Requena, J.R., Fu, M.X., Ahmed, M.U., Jenkins, A.J., Lyons, T.J., Baynes, J.W., and Thorpe, S.R. (1997) Quantification of malondialdehyde and 4-hydroxynonenal adducts to lysine residues in native and oxidized human low-density lipoprotein. *Biochem. J.* **322**, 317–325.
104. Lyons, T.J., Li, W., Wells-Knecht, M.C., and Jokl, R. (1994) Toxicity of mildly modified low-density lipoproteins to cultured retinal capillary endothelial cells and pericytes. *Diabetes* **43**, 1090–1095.
105. Scaccini, C., and Jialal, I. (1994) LDL modification by activated polymorphonuclear leukocytes: a cellular model of mild oxidative stress. *Free Radic. Biol. Med.* **16**, 49–55.
106. Sigari, F., Lee, C., Witztum, J.L., and Reaven, P.D. (1997) Fibroblasts that overexpress 15-lipoxygenase generate bioactive and minimally modified LDL. *Arterioscler. Thromb. Vasc. Biol.* **17**, 3639–3645.
107. Kennedy, S., Fournet-Bourguignon, M.P., Breugnot, C., Castedo-Delrieu, M., Lesage, L., Reure, H., Briant, C., Leonce, S., Vilaine, J.P., and Vanhoutte, P.M. (2003) Cells derived from regenerated endothelium of the porcine coronary artery contain more oxidized forms of apolipoprotein-B-100 without a modification in the uptake of oxidized LDL. *J. Vasc. Res.* **40**, 389–398.
108. Esterbauer, H., Striegl, G., Puhl, H., and Rotheneder, M. (1989) Continuous monitoring of in vitro oxidation of human low density lipoprotein. *Free Radic. Res. Commun.* **6**, 67–75.
109. Moore, K.P., Darley-Usmar, V., Morrow, J., and Roberts, L.J., 2nd (1995) Formation of F2-isoprostanes during oxidation of human low-density lipoprotein and plasma by peroxynitrite. *Circ. Res.* **77**, 335–341.

110. Moore, K., and Roberts, L.J., 2nd (1998) Measurement of lipid peroxidation. *Free Radic. Res.* **28**, 659–671.
111. Gaziano, J.M. (1994) Antioxidant vitamins and coronary artery disease risk. *Am. J. Med.* 97, 18S–21S; Discussion 22S–28S.
112. Rapola, J.M., Virtamo, J., Ripatti, S., Huttunen, J.K., Albanes, D., Taylor, P.R., and Heinonen, O.P. (1997) Randomised trial of alpha-tocopherol and beta-carotene supplements on incidence of major coronary events in men with previous myocardial infraction [see comments]. *Lancet* **349**, 1715–1720.
113. Porkkala-Sarataho, E.K., Nyyssonen, M.K., Kaikkonen, J.E., Poulsen, H.E., Hayn, E.M., Salonen, R.M., and Salonen, J.T. (1998) A randomized, single-blind, placebo-controlled trial of the effects of 200 mg alpha-tocopherol on the oxidation resistance of atherogenic lipoproteins. *Am. J. Clin. Nutr.* **68**, 1034–1041.
114. McKechnie, R., Rubenfire, M., and Mosca, L. (2002). Antioxidant nutrient supplementation and brachial reactivity in patients with coronary artery disease. *J. Lab Clin. Med.* **139**, 133–139.
115. Gey, K.F. (1993) Prospects for the prevention of free radical disease, regarding cancer and cardiovascular disease. *Br. Med. Bull.* **49**, 679–699.
116. Blot, W.J., Li, J.Y., Taylor, P. Guo, W., Dawsey, S., Wang, G.Q., Yang, C.S., Zheng, S.F., Gail, M., Li, G.Y., et al. (1993) Nutrition intervention trials in Linxian, China: supplementation with specific vitamin/mineral combinations, cancer incidence, and disease- specific mortality in the general population. *J. Natl. Cancer Inst.* **85**, 1483–1492.
117. Steinberg, D. (1995) Clinical trials of antioxidants in atherosclerosis: are we doing the right thing? *Lancet* **346**, 36–38.
118. Parthasarathy, S., Khan-Merchant, N., Penumetcha, M., Khan, B.V., and Santanam, N. (2001) Did the antioxidant trials fail to validate the oxidation hypothesis? *Curr. Atheroscler. Rep.* **3**, 392–398.
119. Heinecke, J.W. (2001) Is the emperor wearing clothes? Clinical trials of vitamin E and the LDL oxidation hypothesis. *Arterioscler. Thromb. Vasc. Biol.* **21**, 1261–1264.
120. Yusoff, K. (2002) Vitamin E in cardiovascular disease: has the die been cast? *Asia Pac. J. Clin. Nutr.* 7, S443–447.
121. Heinecke, J.W. (2003) Clinical trials of vitamin E in coronary artery disease: is it time to reconsider the low-density lipoprotein oxidation hypothesis? *Curr. Atheroscler. Rep.* **5**, 83–87.
122. Ferns, G.A. and Lamb, D.J. (2004) What does the lipoprotein oxidation phenomenon mean? *Biochem. Soc. Trans.* **32**, 160–163.
123. Mowri, H., Chinen, K., Ohkuma, S., and Takano, T. (1986) Peroxidized lipids isolated by HPLC from atherosclerotic aorta. *Biochem. Int.* **12**, 347–352.
124. Carpenter, K.L., Wilkins, G.M., Fussell, B., Ballantine, J.A., Taylor, S.E., Mitchinson, M.J., and Leake, D.S. (1994) Production of oxidized lipids during modification of low-density lipoprotein by macrophages or copper. *Biochem. J.* **304**, 625–633.
125. Folcik, V.A., Nivar-Aristy, R.A., Krajewski, L.P., and Cathcart, M.K. (1995) Lipoxygenase contributes to the oxidation of lipids in human atherosclerotic plaques. *J. Clin. Invest.* **96**, 504–510.
126. Staprans, I., Rapp, J.H., Pan, X.M., Hardman, D.A., and Feingold, K.R. (1996) Oxidized lipids in the diet accelerate the development of fatty streaks in cholesterol-fed rabbits. *Arterioscler. Thromb. Vasc. Biol.* **16**, 533–538.
127. Berliner, J., Leitinger, N., Watson, A., Huber, J., Fogelman, A., and Navab, M. (1997) Oxidized lipids in atherogenesis: formation, destruction and action. *Thromb. Haemost.* **78**, 195–199.
128. Parthasarathy, S., Santanam, N., Ramachandran, S., and Meilhac, O. (2000) Potential role of oxidized lipids and lipoproteins in antioxidant defense [In Process Citation]. *Free Radic. Res.* **33**, 197–215.
129. Navab, M., Hama, S.Y., Reddy, S.T., Ng, C.J., Van Lenten, B.J., Laks, H., and Fogelman, A.M. (2002) Oxidized lipids as mediators of coronary heart disease. *Curr. Opin. Lipidol.* **13**, 363–372.
130. Birukov, K.G. (2006) Oxidized lipids: the two faces of vascular inflammation. *Curr. Atheroscler. Rep.* **8**, 223–231.
131. Salmon, S., Maziere, C., Theron, L., Beucler, I., Ayrault-Jarrier, M., Goldstein, S., and Polonovski, J. (1987) Immunological detection of low-density lipoproteins modified by malondialdehyde in vitro or in vivo. *Biochim. Biophys. Acta.* **920**, 215–220.
132. Parums, D.V., Brown, D.L., and Mitchinson, M.J. (1990) Serum antibodies to oxidized low-density lipoprotein and ceroid in chronic periaortitis. *Arch. Pathol. Lab. Med.* **114**, 383–387.
133. Virella, G., Virella, I., Leman, R.B., Pryor, M.B., and Lopes-Virella, M.F. (1993) Anti-oxidized low-density lipoprotein antibodies in patients with coronary heart disease and

normal healthy volunteers. *Int. J. Clin. Lab Res.* **23**, 95–101.

134. Holvoet, P., Perez, G., Zhao, Z., Brouwers, E., Bernar, H., and Collen, D. (1995) Malondialdehyde-modified low density lipoproteins in patients with atherosclerotic disease. *J. Clin. Invest.* **95**, 2611–2619.
135. Festa, A., Kopp, H.P., Schernthaner, G., and Menzel, E.J. (1998) Autoantibodies to oxidised low density lipoproteins in IDDM are inversely related to metabolic control and microvascular complications. *Diabetologia* **41**, 350–356.
136. Lehtimaki, T., Lehtinen, S., Solakivi, T., Nikkila, M., Jaakkola, O., Jokela, H., Yla-Herttuala, S., Luoma, J.S., Koivula, T., and Nikkari, T. (1999) Autoantibodies against oxidized low density lipoprotein in patients with angiographically verified coronary artery disease [In Process Citation]. *Arterioscler. Thromb. Vasc. Biol.* **19**, 23–27.
137. Frostegard, J., Wu, R. Lemne, C., Thulin, T., Witztum, J.L., and de Faire, U. (2003) Circulating oxidized low-density lipoprotein is increased in hypertension. *Clin. Sci. (London)* **105**, 615–620.
138. Herrick, A.L., Illingworth, K.J., Hollis, S., Gomez-Zumaquero, J.M., and Tinahones, F.J. (2001) Antibodies against oxidized low-density lipoproteins in systemic sclerosis. *Rheumatology (Oxford)* **40**, 401–405.
139. Tanaga, K., Bujo, H., Inoue, M., Mikami, K., Kotani, K., Takahashi, K., Kanno, T., and Saito, Y. (2002) Increased circulating malondialdehyde-modified LDL levels in patients with coronary artery diseases and their association with peak sizes of LDL particles. *Arterioscler. Thromb. Vasc. Biol.* **22**, 662–666.
140. Meraviglia, M.V., Maggi, E., Bellomo, G., Cursi, M., Fanelli, G., and Minicucci, F. (2002) Autoantibodies against oxidatively modified lipoproteins and progression of carotid restenosis after carotid endarterectomy. *Stroke* **33**, 1139–1141.
141. Hsu, R.M., Devaraj, S., and Jialal, I. (2002) Autoantibodies to oxidized low-density lipoprotein in patients with type 2 diabetes mellitus. *Clin. Chim. Acta* **317**, 145–150.
142. Tsimikas, S., Bergmark, C., Beyer, R.W., Patel, R., Pattison, J., Miller, E., Juliano, J., and Witztum, J.L. (2003) Temporal increases in plasma markers of oxidized low-density lipoprotein strongly reflect the presence of acute coronary syndromes. *J. Am. Coll. Cardiol.* **41**, 360–370.
143. Wang, J., Qiang, H., Zhang, C., Liu, X., Chen, D., and Wang, S. (2003) Detection of IgG-bound lipoprotein(a) immune complexes in patients with coronary heart disease. *Clin. Chim. Acta* **327**, 115–122.
144. Koskenmies, S., Vaarala, O., Widen, E., Kere, J., Palosuo, T., and Julkunen, H. (2004) The association of antibodies to cardiolipin, beta 2-glycoprotein I, prothrombin, and oxidized low-density lipoprotein with thrombosis in 292 patients with familial and sporadic systemic lupus erythematosus. *Scand. J. Rheumatol.* **33**, 246–252.
145. Luoma, J.S., Kareinen, A., Narvanen, O., Viitanen, L., Laakso, M., and Yla-Herttuala, S. (2005) Autoantibodies against oxidized LDL are associated with severe chest pain attacks in patients with coronary heart disease. *Free Radic. Biol. Med.* **39**, 1660–1665.
146. Maggi, E., Finardi, G., Poli, M., Bollati, P., Filipponi, M., Stefano, P.L., Paolini, G., Grossi, A., Clot, P., Albano, E., et al. (1993) Specificity of autoantibodies against oxidized LDL as an additional marker for atherosclerotic risk. *Coron. Artery Dis.* **4**, 1119–1122.
147. Bui, M.N., Sack, M.N., Moutsatsos, G., Lu, D.Y., Katz, P., McCown, R., Breall, J.A., and Rackley, C.E. (1996) Autoantibody titers to oxidized low-density lipoprotein in patients with coronary atherosclerosis. *Am. Heart J.* **131**, 663–667.
148. Uusitupa, M.I., Niskanen, L., Luoma, J., Vilja, P., Mercuri, M., Rauramaa, R., and Yla-Herttuala, S. (1996) Autoantibodies against oxidized LDL do not predict atherosclerotic vascular disease in non-insulin-dependent diabetes mellitus. *Arterioscler. Thromb. Vasc. Biol.* **16**, 1236–1242.
149. van de Vijver, L.P., Steyger, R., van Poppel, G., Boer, J.M., Kruijssen, D.A., Seidell, J.C., and Princen, H.M. (1996) Autoantibodies against MDA-LDL in subjects with severe and minor atherosclerosis and healthy population controls. *Atherosclerosis* **122**, 245–253.
150. Kim, J.G., Taylor, W.R., and Parthasarathy, S. (1999) Demonstration of the presence of lipid peroxide-modified proteins in human atherosclerotic lesions using a novel lipid peroxide-modified anti-peptide antibody. *Atherosclerosis* **143**, 335–340.
151. Tsai, W.C., Li, Y.H., Chao, T.H., and Chen, J.H. (2002) Relation between antibody against oxidized low-density lipoprotein and extent of coronary atherosclerosis. *J. Formos. Med. Assoc.* **101**, 681–684.
152. Radulescu, L., Stancu, C., and Antohe, F. (2004) Antibodies against human oxidized

low-density lipoprotein (LDL) as markers for human plasma modified lipoproteins. *Med. Sci. Monit.* **10**, BR207–214.

153. Tsimikas, S. (2006) Oxidized low-density lipoprotein biomarkers in atherosclerosis. *Curr. Atheroscler. Rep.* **8**, 55–61.
154. Moguilevsky, N., Zouaoui Boudjeltia, K., Babar, S., Delree, P., Legssyer, I., Carpentier, Y., Vanhaeverbeek, M., and Ducobu, J. (2004) Monoclonal antibodies against LDL progressively oxidized by myeloperoxidase react with ApoB-100 protein moiety and human atherosclerotic lesions. *Biochem. Biophys. Res. Commun.* **323**, 1223–1228.
155. Nambi, V. (2005) The use of myeloperoxidase as a risk marker for atherosclerosis. *Curr. Atheroscler. Rep.* 7, 127–131.
156. Yamaguchi, Y., Yoshikawa, N., Kagota, S., Nakamura, K., Haginaka, J., and Kunitomo, M. (2006) Elevated circulating levels of markers of oxidative-nitrative stress and inflammation in a genetic rat model of metabolic syndrome. *Nitric Oxide* **15**, 380–386.
157. Parthasarathy, S. et al. (1999) *J. Lipid Res.* **40,** 2143–2157.

Chapter 25

Detection and Localization of Markers of Oxidative Stress by In Situ Methods: Application in the Study of Alzheimer Disease

Paula I. Moreira, Lawrence M. Sayre, Xiongwei Zhu, Akihiko Nunomura, Mark A. Smith, and George Perry

Abstract

Oxidative stress is a key factor involved in the development and progression of Alzheimer disease (AD), and it is well documented that free radical oxidative damage, particularly of neuronal lipids, proteins, nucleic acids, and sugars, is extensive in brains of AD patients. The complex chemistry of peroxynitrite has been the subject of intense study and is now evident that there are two principal pathways for protein modification: the first one involves homolytic hydroxyl radical-like chemistry that results in protein-based carbonyls and the second involves electrophilic nitration of vulnerable side chains, in particular the electron-rich aromatic rings of Tyr and Trp. In the presence of buffering bicarbonate, peroxynitrite forms a CO_2 adduct, which augments its reactivity. Formation of 3-nitrotyrosine by this route has become the classical protein marker specifically for the presence of peroxynitrite. Protein-based carbonyls can be detected by two methods: (i) derivatization with 2,4-dinitrophenylhydrazine (DNPH) and detection of the protein-bound hydrazones using an enzyme-linked anti-2,4-dinitrophenyl antibody and (ii) derivatization with biotin-hydrazide and detection of the protein-bound acyl hydrazone with enzyme-linked avidin or streptavidin. Glycation of proteins by reducing sugars (Maillard reaction) results in a profile of time-dependent adduct evolution rendering susceptibility to oxidative elaboration. In addition, oxidative stress can result in oxidized sugar derivatives which can subsequently modify protein through a process known as glycoxidation. Of more general importance, oxidative stress results in lipid peroxidation and the production of a range of electrophilic and mostly bifunctional aldehydes that modify numerous proteins. The more important protein modifications are referred to as advanced glycation end products (AGEs) and advanced lipoxidation end products (ALEs). Protein modification can result in both non-cross-link and cross-link AGEs and ALEs, the latter arising from the potential bifunctional reactivity, such as that of the lipid-derived modifiers 4-hydroxy-2-nonenal (HNE) and malondialdehyde (MDA). Oxidative damage to nucleic acids results in base modification, substitutions, and deletions. Among the most common modifications, 8-hydroxyguanosine (8OHG) is considered a signature of oxidative damage to nucleic acid.

Cells are not passive to increased oxygen radical production but rather upregulate protective responses. In neurodegenerative diseases, heme oxygenase-1 (HO-1) induction is coincident with the formation of neurofibrillary tangles. This enzyme thatconverts heme, a prooxidant, to biliverdin/bilirubin (antioxi-

R.M. Uppu et al. (eds.), *Free Radicals and Antioxidant Protocols*, Methods in Molecular Biology 610,
DOI 10.1007/978-1-60327-029-8_25, © Humana Press, a part of Springer Science+Business Media, LLC 1998, 2010

dants) and free iron has been considered an antioxidant enzyme. But seen in the context of arresting apoptosis, HO-1 and tau may play a role in maintaining the neurons free from the apoptotic signal (cytochrome *c*), since tau has strong iron-binding sites. Given the importance of iron as a catalyst for the generation of reactive oxygen species, changes in proteins associated with iron homeostasis can be used as an index of cellular responses. One such class of proteins is the iron regulatory proteins (IRPs) that respond to cellular iron concentrations by regulating the translation of proteins involved in iron uptake, storage, and utilization. Therefore, IRPs are considered to be the central control components of cellular iron concentration.

Key words: Advanced glycation end products, advanced lipoxidation end products, glycation, glycoxidation, heme oxygenase-1, 8-hydroxyguanosine, 4-hydroxy-2-nonenal, iron regulatory proteins, malondialdehyde, 3-nitrotyrosine, protein carbonyls.

1. Introduction

Oxidative damage is relevant in the brain (1) for several reasons: (1) it is a post-mitotic tissue with a high energy demand; (2) it is exposed to high oxygen concentration and utilizes about one-fifth of the oxygen consumed by the body; (3) it contains relatively poor concentrations of antioxidants and related enzymes; (4) it is rich in polyunsaturated fatty acids that are more prone to oxidation; and (5) it is rich in iron content, which accumulates as a function of age and can be a source of potent catalyst for oxidative species formation.

Oxidative stress is characterized by a compromise in the physiological removal of reactive oxygen species (ROS) and/or an increase in their production that results from irreversible damage to biomacromolecules. Oxidative damage of biomacromolecules has been described in the brains of Alzheimer disease (AD) patients: (1) DNA and RNA oxidation is marked by increased levels of 8-hydroxy-2-deoxyguanosine (8OHdG) and 8-hydroxyguanosine (8OHG) (2–4); (2) protein oxidation is marked by elevated levels of protein carbonyl and nitration of tyrosine residues (5, 6). Moreover, cross-linking of proteins by oxidative processes may contribute to the resistance of the lesions, senile plaques, and neurofibrillary tangles (NFT) to intracellular and extracellular removal despite being extensively ubiquitinated (7). This resistance of NFT to proteolysis might play an important role in the progression of AD (8); (3) lipid peroxidation is marked by high levels of thiobarbituric acid-reactive substances (TBARS), malondialdehyde (MDA), 4-hydroxy-2-nonenal (HNE), and isoprostanes and altered phospholipid composition (9, 10); and (4) modification to sugars is marked by increased glycation and glycoxidation (11, 12). While intracellular oxidative balance is tightly regulated, the activation of multiple signaling pathways and the upregulation of compensatory mechanisms would be an anticipated outcome in neurons from AD patients.

The importance of in situ methods over bulk analysis cannot be overplayed when considering the structural and cellular complexity of the nervous system. First, the affected neurons comprise a small percentage of the tissue, and second, it is unknown how other cells such as glial cells are affected by the disease. For instance, while neurons may show one type of damage or response, the glial cells may show a compensatory response unrelated to or opposing responses to that of neurons. Further, bulk analysis of oxidative damage will always be hampered by long-lived proteins modified during physiological aging. Extracellular matrix proteins of major vessels provide a record of long-term oxidative insult and therefore are sensitive markers of pathological changes. These properties that render vessels as ideal monitors for aging limit their sensitivity during the detection of disease-specific changes. While isolation of neurons or other cell fractions can address the issue of specificity, limitation exists for poorly represented cell types with the concerns of vascular contamination. Vascular contamination of insoluble proteins can interfere because even glass bead columns or collagenase treatments seldom quantitatively remove all vascular elements if the initial isolation procedure does not exclude them.

This chapter provides details of in situ detection methods that can be used for the qualitative and semi-quantitative measurement of various oxidative stress markers in tissue sections obtained at autopsy. While oxidative stress-related damage detected implicates general or specific cytotoxicity, it is not the purpose of this chapter to suggest that the appearance of certain markers correlates with a particular mechanism of toxicity. Indeed, as discussed in the last section, any such correlation requires a consideration of whether the damage apparent represents a primary or a secondary event in the cascade of biochemical processes leading to neuronal death.

2. Materials

2.1. Pathological Tissue

Liquid nitrogen/isopentane, methacarn (60% methanol, 30% chloroform, 10% acetic acid), and paraffin.

2.2. Direct Histochemical Methods

2.2.1. 2,4-Dinitrophenylhydrazine Reactivity of Protein-Bound Versus Protein-Based Carbonyls

2,4-Dinitrophenylhydrazine (DNPH), pyruvate 2,4-dinitrophenylhydrazone, enzyme-linked anti-2,4-dinitrophenyl antibody, biotin-hydrazide, enzyme-linked avidin or enzyme-linked streptavidin, xylene, ethanol, methanol, sodium chloride (NaCl), Tris-buffered saline (TBS; 50 mM Tris–HCl, 150 mM NaCl, pH 7.6), normal goat serum (NGS), rat monoclonal antibody (LO-DNP-2; Zymed Laboratories, San Francisco, CA), rat IgG (Boehringer-Mannheim, Indianapolis, IN), anti-rat peroxidase–antiperoxidase complex (ICN Pharmaceuticals, Costa Mesa, CA), hydrogen peroxide (H_2O_2), 3,3′-diaminobenzidine (DAB; Sigma, St. Louis, MO), sodium borohydride ($NaBH_4$), and sodium cyanoborohydride ($NaBH_3CN$).

2.2.2. Redox-Active Transition Metals; Cytochemical Detection of Iron(II)/(III)

Xylene, ethanol, potassium ferrocyanide [$K_4Fe(CN)_6{\bullet}3H_2O$], potassium ferricyanide $K_3Fe(CN)_6$], HCl, DAB, Tris–HCl (50 mM, pH 7.6), H_2O_2, and ultrapure water.

2.2.3. Metal Chelation/Binding

Deferoxamine (DFX) or diethylenetriaminepentaacetic acid (DTPA), ethylenediaminetetraacetic acid (EDTA), ferrocyanide, ferricyanide, H_2O_2, DAB, iron citrate ($FeC_6H_8O_7$), iron chloride ($FeCl_3$), and copper(II) chloride ($CuCl_2$).

2.3. Immunocytochemical Methods

2.3.1 Immunocytochemistry Using a Peroxidase-Coupled Secondary Antibody

Xylene, ethanol, H_2O_2, methanol, NGS, TBS, DAB, goat anti-mouse antibody, goat anti-rabbit antibody, mouse- and rabbit-specific peroxidase–antiperoxidase complexes, sodium hydroxide (NaOH), trypsin, Tris, NaCl, calcium chloride ($CaCl_2$), citrate buffer, guanidine, formic acid (HCOOH), antigen retrieval solution (Biogenix Laboratories), proteinase K (Boehringer Mannheim), and phosphate buffer (PBS; pH 7.4).

2.3.2. Nitrotyrosine Antibody for Determination of Protein Tyrosine Nitration by Peroxynitrite

Affinity-purified rabbit antiserum or mouse monoclonal antibodies (7A2 or 1A6) raised to nitrated keyhole limpet hemocyanin, NGS, NaCl, Tris, HCl, nitrated bovine serum albumin (BSA-NO_2), nitrated Gly-Tyr-Ala peptide, sodium hydrosulfite ($Na_2S_2O_4$), and carbonate buffer.

2.3.3. Antibodies to Advanced Glycation End Products

Keyhole limpet hemocyanin (KLH), pentosidine–KLH conjugate, Freund's complete adjuvant, Freund's incomplete adjuvant, mouse monoclonal pyrraline antibodies raised to keyhole limpet hemocyanin–caproyl pyrraline conjugate, free pentosidine, and pyrraline haptens.

2.3.4. Antibodies to Lipid Peroxidation End Products

Antiserum raised to keyhole limpet hemocyanin modified by 4-oxononanal (ON-KLH), antiserum raised to keyhole limpet hemocyanin conjugated with preformed 6-(2-pentylpyrrol-1-yl)caproic acid (PPC-KLH), DAB, H_2O_2, Tris–HCl, Congo red, BSA-derived antigens (ON-BSA, PPC-BSA).

2.3.5. Antibodies to DNA Damage

Anti-8OHG (QED Bioscience, San Diego, CA) and 1F7 (Trevigen, Gaithersburg, MD), NGS, proteinase K, DAB, H_2O_2, Tris–HCl, PBS, DNase (Boehringer Mannheim), S1 DNase (Boehringer Mannheim), and RNase (Boehringer Mannheim).

2.3.6. Antibodies to Cellular Response Factor, Heme Oxygenase-1

Rabbit antisera to heme oxygenase-1 (HO-1; Stressgen Biotechnologies, Victoria, Canada) and purified HO-1 protein (Stressgen Biotechnologies, Victoria, Canada).

2.3.7. Antibodies to the Cellular Response Factor; Iron-Response Element

Rabbit polyclonal antibodies to iron-response protein-2 (IRP-2) and purified IRP-2 protein.

3. Methods

3.1. Pathological Tissue

Tissue can either be flash frozen in liquid nitrogen/isopentane and cryostat sections prepared at 20 nm or be fixed in methacarn, paraffin embedded, and sections cut at 6 μm (*see* **Note 1**).

3.2. Direct Histochemical Methods

3.2.1. 2,4-Dinitrophenylhydrazine Reactivity of Protein-Bound Versus Protein-Based Carbonyls

Two related methods for the detection of protein-based carbonyls are (i) derivatization with DNPH and detection of the protein-bound hydrazones using an enzyme-linked anti-2,4-dinitrophenyl antibody and (ii) derivatization with biotin-hydrazide and detection of the protein-bound acyl hydrazone with enzyme-linked avidin or enzyme-linked streptavidin (5, 13) (*see* **Note 2**).

After deparaffinization in xylene and rehydration through graded ethanol for methacarn-fixed material or brief fixation in ice-cold acetone for cryostat material, sections are covered with 0.1–0.001% DNPH in 2 N HCl. All incubations are conducted at room temperature, unless otherwise indicated, in a plastic box in which water-saturated paper towels are placed in the bottom over which the slides are placed on a platform. After 1 h incubation, sections are exhaustively rinsed in TBS followed by incubation for 30 min in 10% NGS to block non-specific binding sites. After rinsing with 1% NGS/TBS, a rat monoclonal antibody (LO-DNP-2) to DNP is diluted at 1:100 in 1% NGS/TBS and incubated with the sections at 4°C for 16 h. Sections are then rinsed with 1% NGS/TBS, followed by incubation with goat antiserum to rat IgG diluted at 1:50 with 1% NGS/TBS for 30 min. After rinsing in 1% NGS/TBS, rat peroxidase–antiperoxidase complex (1:250) in the same buffer is incubated with the section at room temperature for 1 h, after which it is rinsed with 1% NGS/TBS. Peroxidase activity is localized by development with 0.015% H_2O_2 in 50 mM Tris–HCl, pH 7.6, and 0.75 mg/mL DAB for 5–10 min (*see* **Note 3**). The development of the sections is directly monitored for maximum contrast under the 10× objective of a Zeiss Axioskop 20 microscope.

Chemical and immunochemical controls are used to define carbonyl-specific binding. Chemical reduction of free carbonyls and Schiff bases is performed by incubating sections with 25 mM

$NaBH_4$ in 80% methanol for 30 min before incubation with DNPH. Selective reduction of Schiff bases as opposed to free carbonyls can be achieved by incubation with 50 mM sodium cyanoborohydride in 0.1 M phosphate buffer, pH 6.0, for 1 h. Immunochemical specificity is demonstrated by omission of the DNPH treatment or the antibody to DNP. Immunoabsorption of the antibody to DNP is performed by incubating the antibody (1:100) with 5 mM pyruvate 2,4-dinitrophenylhydrazone (stock 1 mg/mL in ethanol) at 4°C for 16 h and comparing the resulting immunoreactivity with unabsorbed antibody that had been similarly treated with ethanol.

3.2.2. Cytochemical Detection of Iron(II)/(III) and In Situ Oxidation

This technique incorporates three key modifications of the established histological method (14, 15) that greatly increase the detection sensitivity of this technique: (i) increasing the concentrations of potassium ferrocyanide (7%) and hydrochloric acid (3%); (ii) lengthening the incubation time and/or increasing the temperature (15 h at room temperature or 1 h at 37°C); and (iii) using methacarn versus formalin-fixed tissue. The last modification is perhaps the most critical to iron detection in human tissue where formaldehyde-based fixation for extended periods is routine. Indeed, we found that even a brief post-fixation with formaldehyde (3.7%, 5–60 min at room temperature) reduced labeling in a time-dependent manner (16). The tesults with methacarn fixation are similar to those found in sections made from frozen blocks, which were neither fixed nor embedded, suggesting that the iron localization is not a result of tissue fixation or embedding. This method takes advantage of forming the intensely blue mixed valence Fe(II)/Fe(III)–cyanide complex, which can be observed directly as a pale blue stain on tissue sections or can be enhanced by taking advantage of the complex to catalyze the H_2O_2-dependent oxidation of DAB, giving an insoluble brown precipitate localized to bound iron. For the latter case, the tissue sections are deparaffinized in xylene, rehydrated through graded ethanol, and then incubated for 15 h in 7% potassium ferrocyanide (for iron(III) detection) or 7% potassium ferricyanide (for iron(II) detection) in aqueous HCl (3%) and subsequently incubated with a solution containing 0.75 mg/mL DAB tetrahydrochloride in 50 mM Tris–HCl, pH 7.6, and 0.015% H_2O_2 for 5–10 min.

3.2.3. Metal Chelation/Binding

Redox-active metals bound adventitiously to proteins in tissue sections can be removed by exposure to up to 0.1 M DFX or DTPA at room temperature for a period of time depending on the affinity of the tissue sites. For example, whereas for 0.1 M DFX, 20 min was sufficient to remove ferrocyanide-detectable iron from certain sites, a 15 h incubation was required to completely remove iron bound to others and the use of 0.1 M EDTA for 15 h was only partly effective in removing iron (16). Observed differential efficiency of the removal of H_2O_2-dependent DAB oxidizing activity by, e.g., DFX vs. DTPA provides information on the nature of endogenous metals present, since DFX and DTPA are somewhat selective for iron and copper, respectively. The fact that following chelation of iron, incubation of AD tissue sections with ferrocyanide or ferricyanide did not result in H_2O_2-dependent oxidation of DAB on the tissue indicated that formation of the mixed valence complex represents real tissue-bound iron and is not an artifact of the use of an iron-containing reagent.

Potential sites for binding of redox-active transition metals in tissue sections or attempted rebinding of metals following chelation-dependent removal of endogenous metals can be carried out, for example, by incubating with 0.01 mM iron(III) citrate plus 0.01 mM iron(II) chloride or with 0.01 mM copper(II) chloride at room temperature for 3 h, followed by detection (*see* **Note 5**).

3.2.4. In Situ Oxidation of Redox-Active Transition Metals

Protein-bound redox-active transition metals can be directly demonstrated by incubation of sections with 3% H_2O_2 and 0.75 mg/mL DAB in 50 mM Tris–HCl, pH 7.6. This protocol does not involve pre-formation of a mixed valence iron complex, and so it does not directly identify the nature of the tissue-bound metal (17). However, iron and copper are the metals most likely to be bound adventitiously in redox-active form (*see* **Note 5**).

3.3. Immunocytochemical Methods

3.3.1. Immunocytochemistry Using a Peroxidase-Coupled Secondary Antibody

Following deparaffinization with xylene, sections are rehydrated through graded ethanol (100, 95, 70, 50, and 30%). Endogenous peroxidase activity in the tissue is inactivated by incubation with 3% H_2O_2 in methanol for 20 min at room temperature and non-specific binding sites are blocked in a 30-min incubation at room temperature with 10% NGS in TBS. Immunostaining is accomplished by the peroxidase–antiperoxidase technique using DAB as co-substrate chromogen (18). Adjacent sections can be immunostained with antisera to known proteins and to the location of specific structures.

There are a great variety of protocols available for the pretreatment of tissue sections to facilitate antigen retrieval (i.e., exposure of hidden epitopes at the protein–protein interface). For example, in previous studies we, and others, have used pretreatments with (i) NaOH (0.2 M, 15 min, room temperature); (ii) trypsin (400 μg/mL in 0.05 M Tris, 0.3 M NaCl, 0.02 M $CaCl_2$, pH 7.6 for 10 min, room temperature); (iii) microwave oven or pressure cooker in citrate buffer (100 mM); (iv) guanidine (8 M for 15 min, room temperature); (v) formic acid (70% for 15 min, room temperature); (vi) antigen retrieval solution; or (vii) proteinase K [10 μg/mL in PBS (pH 7.4), 40 min at 37°C].

3.3.2. Nitrotyrosine Antibody for Determination of Protein Tyrosine Nitration by Peroxynitrite

Affinity-purified rabbit antiserum or mouse monoclonal antibodies (7A2 or 1A6), raised to nitrated keyhole limpet hemocyanin (19), are used at a 1/100 dilution in 1% NGS, 150 mM NaCl, 50 mM Tris–HCl, pH 7.6.

Controls consist of the following: (i) omission of the primary antibody; (ii) absorption of the antibody with 50 μM nitrated BSA or 15 μM nitrated Gly-Tyr-Ala peptide at 4°C overnight prior to application to the section; and (iii) chemical reduction of nitrotyrosine by treating sections with 15 mM sodium hydrosulfite (dithionite) in 50 mM carbonate buffer, pH 8.0, for 15 min at room temperature (20) prior to immunostaining (*see* **Note 6**).

3.3.3. Antibodies to Advanced Glycation End Products

Polyclonal antibodies against pentosidine were raised in rabbits. Briefly, 250 μg of pentosidine–KLH conjugate prepared by the carbodiimide coupling technique (21) was emulsified with an equal volume of Freund's complete adjuvant and injected intradermally and intramuscularly into New Zealand Wistar rabbits. At 2–3 week intervals, booster injections of pentosidine–KLH emulsified in Freund's incomplete adjuvant were made. Serum was obtained 7–10 days post-injection and antibody titer assessed by ELISA. When antibody levels plateau, the rabbits were exsanguinated by cardiac puncture, and the collected serum was stored at –80°C until required. The production of monoclonal antibodies against pyrraline has been described previously (22).

Labeling was by the peroxidase–antiperoxidase technique described earlier. Adsorption experiments can be performed in parallel by incubation of primary anti-pentosidine or anti-pyrraline antibody with either free pentosidine or pyrraline hapten (10 nM) for 3 h at 37°C. Cross adsorption of pentosidine antisera with free pyrraline and pyrraline antibodies with free pentosidine was also performed as a control against artifactual adsorption. The preparation of pyrraline and pentosidine has been described previously (22, 23).

3.3.4. Antibodies to Lipid Peroxidation End Products

Two different antisera to HNE–pyrrole have been used, one raised to ON-KLH (24, 25), which generates as the major product the same 2-pentylpyrrole on lysine ε-amino groups formed (in low yield) upon direct exposure to HNE, but which also gives some side products. The other antiserum was raised to PPC-KLH, which generates the pure pyrrole epitope, but with a longer tether than when the pyrrole is formed from HNE (24). Thus, the coincidence of labeling with both antisera was deemed to be most convincing of specific immunocytochemical recognition of the HNE-derived pyrrole.

HNE pyrrole antisera were used at a dilution of 1:10–1,000. We found a 1:100 dilution to be optimal and this dilution was used throughout for immunostaining with the exception of the immunoadsorption experiment, see below. Immunostaining utilized the peroxidase–antiperoxidase procedure and was developed under a microscope by using 0.75 mg/mL DAB in 0.015% H_2O_2, 50 mM Tris–HCl, pH 7.6. The sections were dehydrated through ethanol and xylene solutions and then mounted in Permount (Fisher). The location of aberrant protein aggregates such as NFT and senile plaques can be determined following immunostaining by counterstaining the sections with Congo red and viewing them under crossed polarized light. Antisera specific to the protein constituents of aberrant aggregates can be used on adjacent sections in order to determine the microscopic localization of the lipoxidation-dependent adduct.

The specificity of both HNE–pyrrole antibodies was confirmed by omitting the primary antibody or by performing an adsorption. The two antisera to HNE–pyrrole were diluted at 1:150 and then incubated with 0.1 mg/mL of ON-BSA or PPC-BSA (9) for 16 h at 4°C prior to immunocytochemistry. The results of adsorbed immunostaining were compared with adjacent sections labeled with 1:150 dilution of unabsorbed antibody.

3.3.5. DNA Damage

Two mouse monoclonal antibodies to 8OHG have been used (4): anti-8OHG (1:250) and 1F7 (1:30; Trevigen) (26), which recognize RNA- as well as DNA-derived 8OHG. For 1F7, sections were treated with proteinase K prior to incubation with 10% NGS.

Immunostaining was developed by the peroxidase–antiperoxidase procedure (18) using 0.75 mg/mL DAB as co-substrate in 0.015% H_2O_2, 50 mM Tris–HCl, pH 7.6.

The specificity of both antibodies to 8OHG was confirmed by comparison with sections in which (i) the primary antibody was omitted or (ii) absorption with purified deoxyribo-8OHG (0.24 mg/mL) was compared to deoxyribo-G (4). Following the proteinase K treatment, additional sections were pretreated with DNase I (10 U/μL, PBS, 1 h at 37°C), S1 DNase (10 U/μL, PBS, 1 h at 37°C), or RNase (5 μg /μL, PBS, 1 h at 37°C) prior to incubation with 8OHG antibody.

3.3.6. Antibodies to the Cellular Response Factor, Heme Oxygenase-1

Rabbit antisera to HO-1 are available from Stressgen Biotechnologies Corporation. In addition, the MAP system described by Posnett et al. (27) was employed for the production of a rabbit antibody against a peptide corresponding to residues 1–30 from the sequence reported for rat HO-1 (28). Details of the preparation and characterization of the rabbit antibody to HO-1 peptide as well as its ability to react with human HO-1 have been described (29). Use of this antibody required a trypsin pretreatment of the section (see antigen retrieval on section 2.3) prior to the immunocytochemical protocol described above (*see* **Note** 7). Adsorption experiments were performed on anti-HO-1 to confirm the specificity of antibody binding. The immunostaining protocol was repeated, except here using adsorbed antisera in parallel. Adsorbed antisera were generated by incubation of primary antisera with purified HO-1 protein diluted to a final concentration of 10 μg/mL for 3 h at 37°C.

3.3.7. Antibodies to the Cellular Response Factor, Iron-Response Element

Rabbit polyclonal antibodies to IRP-2 raised to a 75 amino acid peptide representing the N-terminal sequence unique to IRP-2 (30) or IRP-1 (31) can be used as previously described.

Adsorption experiments were performed with the antibody to IRP-2 to confirm the specificity of antibody binding (32). The immunostaining protocol was repeated, except here using adsorbed antisera in parallel. Adsorbed antiserum was generated by incubation of primary antisera with purified IRP-2 protein (30) diluted to a final concentration of 100 μg/mL overnight at 4°C. Adsorption of antibodies to tau with IRP-2 protein was also performed as a control against artifactual adsorption.

3.4. Application of In Situ Methods to AD Tissue Sections to Visualize Senile Plaques and NFT

The methods outlined in this chapter have been applied in an effort to localize markers of oxidative damage in AD. We and others (11, 33, 34) determined that advanced glycation end product (AGE) modifications are present in both NFT and senile plaques in AD. These studies employed not only antibodies to the specific structures pentosidine and pyrraline but also less defined AGE antibodies raised to carrier proteins treated with reducing sugars for long periods of time. AGE modification is likely an early in vivo event since both diffuse senile plaques (11) and paired helical filaments (33, 34), considered two of the earliest pathological changes in AD (35, 36), were stained by the anti-AGE antibodies.

Immunostaining using antibodies to lipoxidation-derived modifications has exhibited more distinctive patterns. Whereas some antibodies (e.g., to MDA adducts (33) or to ill-defined HNE adducts) display variable staining of both NFT and senile plaques, antibodies to characterized HNE adducts are exclusively localized to neuronal cell bodies and neurofibrillary pathology (9, 25). In fact, antibodies to the HNE–lysine-derived pyrrole, an advanced lipoxidation end product (ALE), were found to stain not only intraneuronal and extraneuronal NFT but also apparently normal hippocampal neurons in AD but not in controls (9). Counterstaining with Congo red on the same sections or antisera to tau (37) on the adjacent serial sections was used to assess relationships to NFT. The same profile of staining seen for HNE–pyrrole was seen by using antibodies to markers of direct protein oxidation, including nitrotyrosine and protein-based carbonyls (5, 6, 13). Again, the heightened sensitivity of immunochemical follow-up subsequent to derivatization of protein-based carbonyls with 2,4-dinitrophenylhydrazine permitted the in situ detection of carbonyl reactivity not only within NFT but also within vulnerable neurons in AD (5, 13). Two key objectives in the development of multiple in situ immunochemical markers for oxidative stress damage are the following: (i) to obtain clues to the temporal aspects of pathogenesis and (ii) to help track down the sources of oxidative damage. In the former case, it is of interest to determine whether, and at what time point, there is a biochemical response to oxidative damage revealed by induction of stress proteins such as HO-1. We found that HO-1 is associated with neurofibrillary pathology (38) at the same (early) stage of degeneration as is revealed by the antibody (Alz50) to abnormal conformation of tau. As far as sources of oxidative stress are concerned, the modified Perl's stain for protein-bound iron showed iron accumulation in both NFT and senile plaques. In the absence of the formation of the mixed valence iron complex through employment of hexacyanoiron (II/III), direct utilization of the H_2O_2-dependent DAB oxidation protocol localized endogenously bound redox-active transition metals at the same sites on NFT and senile plaques. Overall, the panel of in situ methods for localizing oxidative stress damage in AD is permitting an assessment of the spatio-temporal aspects to neuronal degeneration that define the disease process. One question that remains to be answered has to do with better characterization of the cellular response to AD lesions since certain oxidative stress markers show up only in neurofibrillary pathology, whereas others are associated with both NFT and senile plaques.

3.5. Quantification and Statistical Analyses

The intensity of immunoreaction on tissue sections can be evaluated by measuring the optical density with a suitable imaging system (4, 39) (e.g., Zeiss Axiocam and Axiovision software). All measurements are done under the same optical and light conditions as well as using an electronic shading correction to compensate for any un-evenness that might be present in the illumination. Statistical analysis for the differences in the corrected OD value among subgroups is by the analysis of variance (ANOVA) using StatView program (Abacus Concepts, Inc., Berkeley, CA). Fisher's protected least significant difference can be applied in the post hoc analysis.

4. Notes

1. It is essential that the modifications analyzed be both stable and not produced as a result of postmortem or other time-dependent post-experimental changes. The latter is a

particular concern if initial lipid oxidative adducts are analyzed. One way we have addressed this issue is to analyze and compare results of localizing early lipoxidation adducts versus ALE. Material fixed in glutaraldehyde or even paraformaldehyde is problematic since it precludes the analysis of lipoxidative and other modifications that share the property of carbonyl condensation-dependent cross-links. The latter concern is the most significant limitation in practice since the majority of pathological tissue available for retrospective study is fixed in formalin. If formalin-fixed tissue must be used, it is best to focus on highly defined products, unlikely to share homology with formalin or nucleic acid modifications.

2. Although we have utilized only the DNPH method in situ, the biotin-hydrazide method should perform similarly and may be preferable in cases where there appears to be excessive non-specific binding of DNPH. In either case, it is important to recognize that protein-bound carbonyls can represent either oxidized side chains or the univalent adduction of a bifunctional glycoxidation or lipoxidation product such that the adduct still contains a free carbonyl group. Modification of unoxidized lysine amino groups by monofunctional lipid- or sugar-derived carbonyl compounds will not interfere because DNPH or biotin-hydrazide will displace the carbonyl compound into bulk solution. In these cases, one can learn about the nature of the lysine modification through direct LC-MS analysis of the DNP derivatives in solution. If it is important to localize and/or quantify total protein-associated carbonyls (both protein-based carbonyls and carbonyl-modified lysines), one approach is to carry out an in situ determination of protein-bound tritium using [^{3}H]-$NaBH_4$ with detection by fluorography.
3. When the peroxidase–antiperoxidase method is used for visualization of bound DNP antibody, we have typically utilized a 20-min pre-incubation of tissue sections with 3% H_2O_2 in methanol to prevent artifacts of staining arising from endogenous peroxidase activity in the tissue (18). Although H_2O_2 itself, or synergistically with tissue-bound transition metals, might create oxidative modifications through hydroxyl radical formation (e.g., carbonyl residues), deletion of this step had no effect upon subsequent DNPH labeling conducted in our laboratory, though it did convert Fe(II) to Fe(III) (16). Therefore, a control that omits the H_2O_2 treatment step should be performed with each experimental paradigm. Alternatively, utilization of a secondary antibody coupled to alkaline phosphatase,

with *p*-nitrophenyl phosphate as chromogen, obviates the necessity of the prior H_2O_2 treatment.

4. In these experiments, it is important to recognize that DAB is supplied as the tetrahydrochloride and that DAB free base is only sparingly soluble in water. Thus, it is crucial that the tissue incubation solution remains at a low enough pH that the DAB stays in solution. In this manner, the insoluble brown DAB oxidation product formed at tissue sites of redox activity is readily visualized without interference from precipitated unoxidized DAB.
5. By altering the parameters of metal chelation and/or rebinding, one can obtain valuable information regarding the avidity and affinity of various metal sequestration sites.
6. These procedures are performed in parallel with the antisera to known markers as controls against artifacts of inactivation of either primary or secondary antibodies from the use of sodium hydrosulfite-reduced sections and against non-specific adsorption with nitrated BSA or nitrated Gly-Tyr-Ala.
7. For immunostaining with HO-1 antisera (Stressgen), we found that methacarn-fixed tissue was superior to formalin-fixed tissue, requiring shorter incubation times and lower titers of antisera.

General Note: In assessing markers of oxidative damage, there are a number of potential complications that investigators should be aware of and, if possible, control against processing artifacts since oxidative processes continue after death and therefore it is imperative to minimize the time before fixation. In this respect, it is important to realize that iron, a potent catalyst of oxidative chemistry, is frequently liberated following death. Additionally, one needs to pay particular attention to the fixation protocol employed since oxidation-related modifications could either be destroyed or created by the fixative. In this regard, we routinely use fixation in methacarn, which is relatively inert and appears to optimize labeling by immunocytochemical and histochemical techniques. The variability of oxidative changes also serves as an important control: by destroying or altering the oxidative modification with specific reagents, one can readily assert whether a particular technique is selective. Examples are reduction of free carbonyls with sodium borohydride, reduction of nitrotyrosine with sodium dithionite, and enzymatic removal of oxidized nucleic acids with DNase or RNase.

Finally, perhaps the most important aspect that one has to tackle is that oxidative modifications are fundamental aspects of both aging and disease. Indeed, cell death, both by necrosis and apoptosis, involves alterations in redox chemistry. Therefore, if one is looking at a pathological process, it is highly unlikely that one would not detect oxidative changes. In order to put such changes into context, it is important to understand the relevance of these changes with respect to other detrimental events. Thus, it is extremely important to empirically determine the conditions required for the detection of selective changes relative to the appropriate control. The definition of "selective" depends on the goal of the study. For example, if one is interested in observing immunocytochemical evidence for oxidative damage in a particular age-related neurodegenerative disease, then the antibody response should be titrated to a level such that age-matched control tissue exhibits changes at the background threshold level. On the other hand, for conclusions regarding the effect of aging itself, the immunocytochemistry should be titrated such that young controls exhibit only background levels. Overall, it is important that the researcher takes into consideration the various factors thought, at the time, to underlie the changes of interest, in order to permit an informed evaluation of the experimental observations. This will pave way for a proper perspective in evaluating whether oxidative damage represents a primary or a secondary phenomenon.

Acknowledgments

Work in the authors' laboratories is supported by the National Institutes of Health, the Alzheimer's Association, and Philip Morris USA and Philip Morris International.

References

1. Halliwell, B. (1989) Protection against tissue damage in vivo by desferrioxamine: what is its mechanism of action? *Free Radic. Biol. Med.* 7, 645–651.
2. Nunomura, A., Perry, G., Aliev, G., Hirai, K., Takeda, A., Balraj, E.K., Jones, P.K., Ghanbari, H., Wataya, T., Shimohama, S., Chiba, S., Atwood, C.S., Petersen, R.B., and Smith, M.A. (2001) Oxidative damage is the earliest event in Alzheimer disease. *J. Neuropath. Exp. Neurol.* **60**, 759–767.
3. Nunomura, A., Perry, G., Pappolla, M.A., Friedland, R.P., Hirai, K., Chiba, S., and Smith, M.A. (2000) Neuronal oxidative stress precedes amyloid-beta deposition in Down syndrome. *J. Neuropath. Exp. Neurol.* **59**, 1011–1017.
4. Nunomura, A., Perry, G., Pappolla, M.A., Wade, R., Hirai, K., Chiba, S., and Smith, M.A. (1999) RNA oxidation is a prominent feature of vulnerable neurons in Alzheimer's disease. *J. Neurosci.* **19**, 1959–1964.
5. Smith, M.A., Perry, G., Richey, P.L., Sayre, L.M., Anderson, V.E., Beal, M.F., and Kowall, N. (1996) Oxidative damage in Alzheimer's. *Nature* **382**, 120–121.
6. Smith, M.A., Richey Harris, P.L., Sayre, L.M., Beckman, J.S., and Perry, G. (1997) Widespread peroxynitrite-mediated damage in Alzheimer's disease. *J. Neurosci.* **17**, 2653–2657.
7. Cras, P., Smith, M.A., Richey, P.L., Siedlak, S.L., Mulvihill, P., and Perry, G. (1995) Extracellular neurofibrillary tangles reflect neuronal loss and provide further evidence of extensive protein cross-linking in Alzheimer disease. *Acta Neuropathol.* **89**, 291–295.
8. Smith, M.A. (1998) Alzheimer disease. *Int. Rev. Neurobiol.* **42**, 1–54.
9. Sayre, L.M., Zelasko, D.A., Harris, P.L., Perry, G., Salomon, R.G., and Smith, M.A.

(1997) 4-Hydroxynonenal-derived advanced lipid peroxidation end products are increased in Alzheimer's disease. *J. Neurochem.* **68**, 2092–2097.

10. Zhu, X., Raina, A.K., Lee, H.G., Casadesus, G., Smith, M.A., and Perry, G. (2004) Oxidative stress signalling in Alzheimer's disease. *Brain Res.* **1000**, 32–39.
11. Smith, M.A., Taneda, S., Richey, P.L., Miyata, S., Yan, S.D., Stern, D., Sayre, L.M., Monnier, V.M., and Perry, G. (1994) Advanced Maillard reaction end products are associated with Alzheimer disease pathology. *Proc. Nat. Acad. Sci. USA* **91**, 5710–5714.
12. Smith, M.A., Rudnicka-Nawrot, M., Richey, P.L., Praprotnik, D., Mulvihill, P., Miller, C.A., Sayre, L.M., and Perry, G. (1995) Carbonyl-related posttranslational modification of neurofilament protein in the neurofibrillary pathology of Alzheimer's disease. *J. Neurochem.* **64**, 2660–2666.
13. Smith, M.A., Sayre, L.M., Anderson, V.E., Harris, P.L., Beal, M.F., Kowall, N., and Perry, G. (1998) Cytochemical demonstration of oxidative damage in Alzheimer disease by immunochemical enhancement of the carbonyl reaction with 2,4-dinitrophenylhydrazine. *J. Histochem. Cytochem.* **46**, 731–735.
14. Perls, M. (1867) Nachweis von Eisenoxyd in gewissen Pigmenten. *Virchows Arch. [Pathol. Anat.]* **39**, 42–48.
15. Virchow, R. (1847) Die Pathologischen pigmente. *Virch. Arch. [Pathol. Anat.]* **1**, 379–407.
16. Smith, M.A., Harris, P.L., Sayre, L.M., and Perry, G. (1997) Iron accumulation in Alzheimer disease is a source of redox-generated free radicals. *Proc. Nat. Acad. Sci. USA* **94**, 9866–9868.
17. Sayre, L.M., Perry, G., and Smith, M.A. (1999) In situ methods for detection and localization of markers of oxidative stress: application in neurodegenerative disorders. *Methods Enzymol.* **309**, 133–152.
18. Sternberger, L.A. (Ed.) (1986) Immunocytochemistry, Wiley, NY.
19. Beckman, J.S. (1996) Oxidative damage and tyrosine nitration from peroxynitrite. *Chem. Res. Toxicol.* **9**, 836–844.
20. Cuatrecasas, P., Fuchs, S., and Anfinsen, C.B. (1968) The tyrosyl residues at the active site of staphylococcal nuclease. Modifications by tetranitromethane. *J. Biol. Chem.* **243**, 4787–4798.
21. Yamada, H., Imoto, T., Fujita, K., Okazaki, K., and Motomura, M. (1981) Selective modification of aspartic acid-101 in lysozyme by carbodiimide reaction. *Biochemistry* **20**, 4836–4842.
22. Miyata, S. and Monnier, V. (1992) Immunohistochemical detection of advanced glycosylation end products in diabetic tissues using monoclonal antibody to pyrraline. *J. Clin. Invest.* **89**, 1102–1112.
23. Sell, D.R. and Monnier, V.M. (1990) End-stage renal disease and diabetes catalyze the formation of a pentose-derived crosslink from aging human collagen. *J. Clin. Invest.* **85**, 380–384.
24. Sayre, L.M., Sha, W., Xu, G., Kaur, K., Nadkarni, D., Subbanagounder, G., and Salomon, R.G. (1996) Immunochemical evidence supporting 2-pentylpyrrole formation on proteins exposed to 4-hydroxy-2-nonenal. *Chem. Res. Toxicol.* **9**, 1194–1201.
25. Montine, K.S., Kim, P.J., Olson, S.J., Markesbery, W.R., and Montine, T.J. (1997) 4-hydroxy-2-nonenal pyrrole adducts in human neurodegenerative disease. *J. Neuropathol. Exp. Neurol.* **56**, 866–871.
26. Yin, B., Whyatt, R.M., Perera, F.P., Randall, M.C., Cooper, T.B., and Santella, R.M. (1995) Determination of 8-hydroxydeoxyguanosine by an immunoaffinity chromatography-monoclonal antibody-based ELISA. *Free Radic. Biol. Med.* **18**, 1023–1032.
27. Posnett, D.N., McGrath, H., and Tam, J.P. (1988) A novel method for producing anti-peptide antibodies. Production of site-specific antibodies to the T cell antigen receptor beta-chain. *J. Biol. Chem.* **263**, 1719–1725.
28. Shibahara, S., Muller, R., Taguchi, H., and Yoshida, T. (1985) Cloning and expression of cDNA for rat heme oxygenase. *Proc. Nat. Acad. Sci. USA* **82**, 7865–7869.
29. Kutty, R.K., Nagineni, C.N., Kutty, G., Hooks, J.J., Chader, G.J., and Wiggert, B. (1994) Increased expression of heme oxygenase-1 in human retinal pigment epithelial cells by transforming growth factor-beta. *J. Cell. Physiol.* **159**, 371–378.
30. Rouault, T.A., Tang, C.K., Kaptain, S., Burgess, W.H., Haile, D.J., Samaniego, F., McBride, O.W., Harford, J.B., and Klausner, R.D. (1990) Cloning of the cDNA encoding an RNA regulatory protein–the human iron-responsive element-binding protein. *Proc. Nat. Acad. Sci. USA* **87**, 7958–7962.
31. Pantopoulos, K., Gray, N.K., and Hentze, M.W. (1995) Differential regulation of two related RNA-binding proteins, iron regulatory protein (IRP) and IRPB. *RNA* **1**, 155–163.

32. Smith, M.A., Wehr, K., Harris, P.L., Siedlak, S.L., Connor, J.R., and Perry, G. (1998) Abnormal localization of iron regulatory protein in Alzheimer's disease. *Brain Res.* **788**, 232–236.
33. Yan, S.D., Chen, X., Schmidt, A.M., Brett, J., Godman, G., Zou, Y.S., Scott, C.W., Caputo, C., Frappier, T., and Smith, M.A. (1994) Glycated tau protein in Alzheimer disease: a mechanism for induction of oxidant stress. *Proc. Nat. Acad. Sci. USA* **91**, 7787–7791.
34. Ledesma, M.D., Bonay, P., Colaco, C., and Avila, J. (1994) Analysis of microtubule-associated protein tau glycation in paired helical filaments. *J. Biol. Chem.* **269**, 21614–21619.
35. Perry, G. and Smith, M.A. (1993) Senile plaques and neurofibrillary tangles: what role do they play in Alzheimer disease. *Clin. Neurosci.* **1**, 199–203.
36. Trojanowski, J.Q., Schmidt, M.L., Shin, R.-W., Bramblett, G.T., Goedert, M., and Lee, V.M.Y. (1993) PHF-tau (A68): from pathological marker to potential mediator of neuronal dysfunction and degeneration in Alzheimer's disease. *Clin. Neurosci.* **1**, 184–191.
37. Perry, G., Kawai, M., Tabaton, M., Onorato, M., Mulvihill, P., Richey, P., Morandi, A., Connolly, J.A., and Gambetti, P. (1991) Neuropil threads of Alzheimer's disease show a marked alteration of the normal cytoskeleton. *J. Neurosci.* **11**, 1748–1755.
38. Smith, M.A., Kutty, R.K., Richey, P.L., Yan, S.D., Stern, D., Chader, G.J., Wiggert, B., Petersen, R.B., and Perry, G. (1994) Heme oxygenase-1 is associated with the neurofibrillary pathology of Alzheimer's disease. *Am. J. Pathol.* **145**, 42–47.
39. Ghanbari, H.A., Ghanbari, K., Harris, P.L., Jones, P.K., Kubat, Z., Castellani, R.J., Wolozin, B.L., Smith, M.A., and Perry, G. (2004) Oxidative damage in cultured human olfactory neurons from Alzheimer's disease patients. *Aging Cell* **3**, 41–44.

Chapter 26

Enzyme Immunoassay of Isoprostanes

Denis M. Callewaert and Charles Sloan

Abstract

Isoprostanes are a group of prostaglandin-like compounds that are derived in vivo primarily by free radical-mediated peroxidation of arachidonic acid within phospholipids. The resulting isoprostane moieties are rapidly hydrolyzed, metabolized, and excreted. It is now well recognized that isoprostane levels, most commonly those of $15F_{2t}$-isoprostane (formerly denoted 8-iso-prostaglandin $F_{2\alpha}$) in blood or urine, are useful biomarkers for the assessment of oxidative stress, which is a major risk factor for many diseases. Isoprostanes were initially quantified by GC/MS following multiple extraction and purification steps. However, to facilitate routine cost-effective analysis of these biomarkers, immunoassays for isoprostanes were developed. Given the large number of isoprostanes, isoprostane metabolites, and isoprostane conjugates in vivo, the potential for their ex vivo generation from arachidonic acid, and differences among protocols for GC/MS and immunoassay of these biomarkers, it is not surprising that discrepancies arise in values obtained using different methods. When appropriate precautions are taken during sample collection, preparation, and analysis, enzyme immunoassays for $15F_{2t}$-isoprostane can be used for the reproducible and reliable assessment of oxidative stress in experimental animals or human subjects.

Key words: Isoprostanes, immunoassay, ELISA, oxidative stress, biomarker, reactive oxygen species.

1. Introduction

In contrast to specific eicosanoids that are generated enzymatically, the attack of reactive oxygen species on free or esterified arachidonic acid gives rise to a series of isoprostane (IsoP) isomers (1). Based on the initial availability of one synthetic isoprostane isomer [15-F_{2t}-isoprostane (15-F_{2t}-IsoP)] that could serve as a standard, Roberts and Morrow developed a GC/MS method

R.M. Uppu et al. (eds.), *Free Radicals and Antioxidant Protocols*, Methods in Molecular Biology 610,
DOI 10.1007/978-1-60327-029-8_26, © Humana Press, a part of Springer Science+Business Media, LLC 1998, 2010

(2) for the quantification of 15-F_{2t}-IsoP and have extensively demonstrated its utility as a biomarker for systemic oxidative stress (3–5). More recently, LC/MS (6) and immunoassay methods (7–9) for isoprostane quantification have been developed. Some investigators have reported good correlation among these methods (7, 8), whereas others have not (9). Given the large number of isoprostanes, the fact that isoprostanes are rapidly and extensively metabolized in humans, and differences among laboratories with regard to sample processing (whether for GC/MS or immunoassay) and variations in antibody specificities, it is not surprising that results for isoprostane levels in biological fluids obtained using different sample preparation and/or analysis methods do not always yield identical results. However, since even relatively small changes in isoprostane levels may be very significant and have been reported to be a significant risk indicator for cardiovascular disease (10), it is critical that all components of isoprostane assays – from sample collection to data analysis – be controlled to ensure the reproducibility and reliability of isoprostane assays. This chapter focuses on the quantification of IsoP using a commercial liquid phase competitive enzyme-linked immunosorbent assay (ELISA) in which the free IsoP in a test sample and a constant amount of enzyme-labeled 15-F_{2t}-IsoP compete for IgG anti-15-F_{2t}-IsoP adsorbed onto the surface of a microplate. Although immunoassay of isoprostanes in plasma (11, 12) and urine (7, 8) samples have been shown to be reliable indicators of oxidative stress, they differ with regard to (a) susceptibility to artifactual generation of isoprostanes ex vivo and (b) the extent and nature of conjugated forms, as well as other factors. Therefore, specific requirements for quantitative analysis of isoprostane levels in each of these fluids are provided.

2. Materials

2.1. Sample Collection and Storage

2.1.1. Plasma

1. Antioxidant anticoagulant solution: 3.8% (w/v) trisodium citrate solution containing 140 μM indomethacin and 226 μM butylated hydroxytoluene (BHT) (*see* **Note 1**). Indomethacin and BHT inhibit isoprostane production via cyclooxygenase and ROS pathways, respectively (13).
2. BHT stock solution: dissolve 5 mg of BHT per 100 mL of methanol to obtain a stock solution of 226 μM. (*see* **Note 2**)

2.1.2. Urine

2.2. Sample Processing

2.2.1. Plasma

1. Folch reagent: prepared by mixing $CHCl_3$ with methanol at a ratio of 2:1 (v/v). This reagent also contains BHT and triphenylphosphine (TPP) at final concentrations of 2.26 mM and 1.91 mM, respectively (*see* **Note 3**).

2. Magnesium chloride, 2. 1 mM (*see* **Note 3**).
3. Potassium hydroxide, 2.67 mM (*see* **Note 3**).
4. Water, pH 3.0: reagent grade water (*see* **Note 1**) is adjusted to pH 3.0 by the addition of dilute hydrochloric acid.
5. Nitrogen gas (analytical grade).
6. Solid phase extraction columns: immediately prior to use, wash Sep-Pak C_{18} columns (Waters; product WAT043395) with 5 mL of ethanol followed by 5 mL of water (pH 3.0). Sep-pak silica columns (Waters; product WAT043400) are prewashed with 5 mL of methanol followed by 5 mL of ethyl acetate immediately prior to use.
7. Reagents for solid phase extraction: reagent grade ethanol, heptane, sodium sulfate, and ethyl acetate are required. Ethyl acetate/heptane (1/1, v/v) is prepared by mixing equal volumes of ethyl acetate and heptane. Approximately 10 mL of this solution is required for solid phase extraction of each sample. Ethyl acetate/methanol (1/1, v/v) is prepared by mixing equal volumes of ethyl acetate and methanol. Approximately, 5 mL of this reagent is required for solid phase extraction of each sample.
8. Equipment: a 20-position extraction manifold (Waters; product WAT200609) and a suitable vacuum pump are recommended to facilitate solid phase extractions.

2.2.2. Urine

1. Dilution buffer: a proprietary reagent containing nonionic detergents and proteins that essentially eliminates interference in IsoP immunoassays of urine specimens is commercially available (Oxford Biomedical Research; *see* **Note 4**).
2. Glucuronidase: *Helix pomatia* β-glucuronidase (Sigma; product G1512) is suspended in water at a concentration of 250,000 units/mL (*see* **Note 5**).

2.3. Enzyme Immunoassay (see Note 6)

2.3.1. Immunoassay of IsoP in Plasma Samples

1. IsoP immunoassay kits (Oxford Biomedical Research; product EA84) include high-binding 96 strip well microplates (NUNC) are available precoated with the purified IgG fraction of goat anti-IsoP and with non-specific binding site blocked with casein in Tris-buffered saline, 15-F_{2t}-IsoP standard (dissolved in ethanol), horseradish peroxidase (HRP) conjugate of 15-F_{2t}-IsoP (15-F_{2t}-IsoP-HRP), wash and dilution buffers, and a solution of 3, 3′, 5, 5′-tetramethylbenzidine (TMB). TMB is one of the substrates commonly used for colorimetric assay of HRP.
2. In addition to reagents supplied with commercial IsoP immunoassay kits, precision pipettes (range: 5–1,000 μL), disposable tips, and a 96-well plate reader are required.

Washing/aspiration devices are not required but recommended if large numbers of assays are routinely analyzed.

3. 3.0 N sulfuric acid is used to inactivate HRP and thereby stop color development at 450 nm in fixed point assays.

2.3.2. Immunoassay of IsoP in Urine Samples

1. Immunoassay kit: in addition to general IsoP immunoassay kits (mentioned above) which can be used for any biological specimens (i.e., following solid phase extraction), a commercial kit optimized for quantification of IsoP in urine specimens is commercially available (Oxford Biomedical Research; product EA85).
2. Creatinine assay: it is recommended that urine samples are normalized for creatinine content or other comparable biomarker. This allows reasonable sample-to-sample comparison of urinary IsoP values. A microplate creatinine assay kit specifically designed for this purpose is available (Oxford Biomedical Research; product CR01).

2.3.3. Data Analysis

The four-parameter fit sine wave algorithm offers the most precise method to fit data to a standard curve. Most modern microplate readers are equipped with software that automatically transforms the standard curve and calculates values for unknowns using the four-parameter fit method. Alternatively, a stand-alone software package such as GraphPad® (Prism) can be used to fit raw data using the four-parameter fit.

3. Methods

Isoprostanes (IsoPs) are a family of arachidonic acid derivatives that are rapidly generated and extensively metabolized in humans (14–17). The most widely used GC/MS method for IsoP quantification involves two solid phase extraction steps followed by thin layer chromatography. Although intended for the quantification of a single isomer, of 15-F_{2t}-IsoP, it is now known that four different isoprostanes contribute to the $m/z = 569$ peak used for "15-F_{2t}-IsoP" quantification by GC/MS (*18*). Given the large number of isoprostanes and their metabolites, it is not unsurprising that polyclonal antibodies generated against 15-F_{2t}-IsoP exhibit some cross-reactivity with other isoprostanes and/or isoprostane metabolites, particularly metabolites of 15-F_{2t}-IsoP. One study demonstrated that different antibodies differ in the specificity for epitopes on isoprostanes (17). Isoprostane metabolism, and the potential impact of inter-individual differences in metabolism on the use of IsoP as a biomarker for oxidative stress, has not been systematically studied.

Since ROS react with arachidonic acid esterified in phospholipids, it is important to perform the Folch reaction to release isoprostanes esterified in phospholipids prior to quantification of the total isoprostane level in plasma samples. Similarly, since a significant fraction of the 15-F_{2t}-IsoP, and presumable other isoprostanes and their metabolites, are excreted as glucuronic acid conjugates (19), it is important to pretreat urine specimens with glucuronidase in order to quantify the total isoprostane level in urine (**Fig. 26.1**).

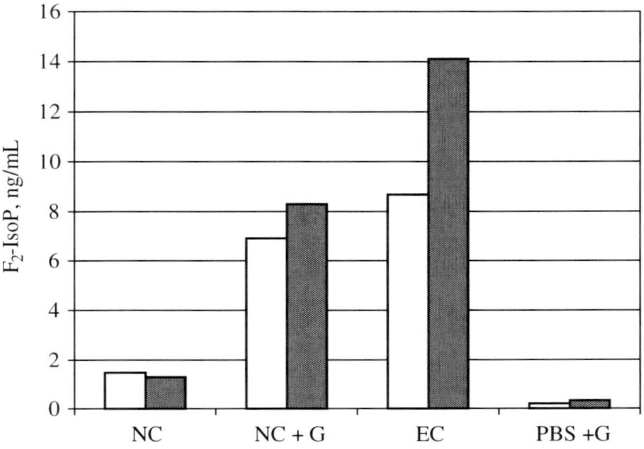

Fig. 26.1. Effect of pretreatment with β-glucuronidase on the concentration of F2-isoprostanes in pooled human urine samples. NC is a normal control urine sample; EC is a pooled urine sample that has been spiked with 10 ng/mL of 15F_{2t}-IsoP. Gray bars represent values obtained by ELISA using the protocols described herein. Open bars represent values obtained for the same specimens by GC/MS following the standard extraction protocols used for that procedure (2). +G are values obtained following pretreatment with β-glucuronidase. PBS represents values obtained for a control sample of phosphate-buffered saline preincubated with the same quantity of β-glucuronidase.

3.1. Sample Collection and Storage

3.1.1. Collection and Preservation of Plasma Samples

1. Peripheral blood is collected into polyethylene tubes containing an antioxidant–anticoagulant solution at a 9:1 ratio (*see* **Note 7**).
2. After 30 min of incubation at 4°C, which enables complete inhibition of cyclooxygenase, samples are centrifuged at 1,120×*g* for 15 min at 4°C.
3. Plasma fractions are then transferred to a separate set of polypropylene screw-cap tubes and stored at –80°C until extraction and analysis.

3.1.2. Collection and Preservation of Urine Samples

1. Either spot or 24-h urine samples can be collected. The samples are stored in small aliquots at –80°C.

2. Thimerosal (final concentration: 0.02% (w/v); retards microbial growth) and BHT are added as needed (*see* **Note 9**). In general, these reagents are not needed because there cannot be a significant ex vivo generation of isoprostanes in urine samples (2, 7).

3.2. Sample Processing

3.2.1. Folch Extraction of Esterified Isoprostanes in Plasma

1. In a 50 mL-capacity conical centrifuge tube, mix 1 mL of plasma with 20 mL of ice-cold Folch reagent (*see* **Note 8**). Vortex at high speed for 1 min (*see* **Note 10**).
2. Add 10 mL of ice-cold, 2.1 mM magnesium chloride and vortex for 1 min.
3. Centrifuge the tubes at $3{,}000\times g$ for 3 min at 4°C to facilitate the separation of phases.
4. Remove the tubes from the centrifuge with minimal or no agitation of phases. There will be three noticeable phases. The top aqueous phase is discarded by aspiration with a pipette. Insert a Pasteur pipette through the middle, buff-colored phase and transfer the bottom organic phase to a clean 50 mL-capacity conical centrifuge tube.
5. Dry the organic solvent(s) in the centrifuge tube under a gentle stream of nitrogen.
6. Resuspend the residue in 1.0 mL of methanol containing 2.26 mM BHT. Vortex as needed.
7. Add 2.0 mL of 2.67 mM KOH, mix briefly by vortexing, and then incubate at 37°C for 30 min in a water bath.
8. Add 18 mL of water at pH 3.0 to the centrifuge tube. Mix briefly by vortexing. Check the pH of the solution. If necessary, readjust the pH to 3.0 using 0.1 N HCl or 0.1 N NaOH. The sample is now ready for solid phase extraction.

3.2.2. Solid Phase Extraction

1. Sep-Pak C_{18} extraction: for each sample to be extracted, affix one column in the extraction manifold (*see* **Note 11**). Label and place one waste collection tube in the manifold chamber for each column (*see* **Note 12**).
2. With the stopcocks for each column in the closed position, apply 3–5 psi vacuum and check the apparatus for leaks.
3. With the vacuum left on, open the stopcock for each respective column and prewash with 5 mL of absolute ethanol. Monitor the passage of the solvent and, when meniscus reaches the sorbent layer, close the stopcock.
4. Add 5 mL of water (pH 3.0) to each of the columns and open the stopcock for a second prewash. Monitor the passage of solvent as suggested above and turn off the stopcock when the meniscus reaches the sorbent layer.

5. With the stopcock still in the closed position, load samples onto the columns. For each sample, open the stopcock slowly and maintain the flow rate ca. 1 mL/min. Allow the sample to flow through until the meniscus reaches the sorbent layer and then close the stopcock.
6. Wash each column with 10 mL of pH 3.0 water, closing the stopcock each time when the solution meniscus reaches the sorbent layer.
7. Wash each column with 10 mL of heptane, again closing the stopcock when the solution meniscus reaches the sorbent layer.
8. Remove the waste collection tubes from the extraction manifold and replace them with sample collection tubes.
9. Elute each sample with 10 mL of ethyl acetate/heptane (1/1, v/v) at a flow rate of 1 mL per minute until the column is dry.
10. Add approximately 1 g of solid sodium sulfate to each eluent to remove any residual water content. After 2 min, decant the solution into a clean vial.
11. Silica Sep-Pak extraction: replace the Sep-pak (C_{18}) columns in the extraction manifold with Silica Sep-Pak columns and new waste collection tubes. Check the apparatus and adjust the vacuum as per step 2 of this section.
12. Following the same process as described in steps 3 and 4, prewash each column with 5 mL of absolute methanol followed by 5 mL of ethyl acetate.
13. With the stopcock still in the closed position load the sample onto the column. Open the stopcock slowly to accommodate a flow rate of 1 mL per minute. Allow the sample to flow through until the solution meniscus reaches the sorbent layer then close the stopcock.
14. Wash column with 5 mL of methanol followed by 5 mL of ethyl acetate, closing the stopcock when the solution meniscus reaches the sorbent layer.
15. Remove the waste collection tubes from the extraction manifold and replace them with sample collection tubes.
16. Elute each sample with 5 mL of ethyl acetate/methanol (1/1, v/v) at a flow rate of 1 mL per minute until the column is dry.
17. Evaporate the ethyl acetate/methanol (1/1, v/v) from each sample under a stream of nitrogen. Once dry, the samples can be stored at –80°C or redissolved for analysis.

3.2.3. Liberation of Isoprostanes from Glucuronic Acid Conjugates in Urine

Add 80 μL of *H. pomatia* β-glucuronidase solution to 1.0 mL urine sample, mix well, and incubate the mixture for 1 h at 37°C.

3.3. Enzyme Immunoassay

3.3.1. Extracted Plasma Samples

1. Immediately prior to immunoassay, the dried material extracted from 1.0 mL of each plasma sample should be carefully resuspended in 0.5 mL of dilution buffer (*see* **Note 13**). After removing aliquots for immunoassay, freeze the remainder of each sample to allow for additional assays or reassay at multiple dilutions (*see* **Note 14**).
2. Following the instructions provided with the immunoassay kit, prepare a series of dilutions of the stock 15-F_{2t}-IsoP standard provided (*see* **Note 15**).
3. Prepare the specified dilution of the stock 15-F_{2t}-IsoP–HRP conjugate (*see* **Note 16**).
4. Remove the microplate that has been precoated with the IgG fraction of goat 15-F_{2t}-IsoP from the desiccated foil pouch.
5. Add 100 μL each of the diluted 15-F_{2t}-IsoP standard and resuspended extracts from the plasma samples to the wells. A recommended arrangement for standards and samples is shown in **Fig. 26.2**. For the determination of B_0, the maximal color development in the absence of any unconjugated 15-F_{2t}-IsoP, 100 μL of dilution buffer is added rather than sample or standard. Two separate wells are used to measure reagent blank (RB) readings. For these wells, 100 μL of the

	1	2	3	4	5	6	7	8	9	10	11	12
A	B0	B0	U1	U1	U9	U9	U17	U17	U25	U25	U33	U33
B	S1	S1	U2	U2	U10	U10	U18	U18	U26	U26	U34	U34
C	S2	S2	U3	U3	U11	U11	U19	U19	U27	U27	U35	U35
D	S3	S3	U4	U4	U12	U12	U20	U20	U28	U28	U36	U36
E	S4	S4	U5	U5	U13	U13	U21	U21	U29	U29	U37	U37
F	S5	S5	U6	U6	U14	U14	U22	U22	U30	U30	U38	U38
G	S6	S6	U7	U7	U15	U15	U23	U23	U31	U31	U39	U39
H	S7	S7	U8	U8	U16	U16	U24	U24	U32	U32	RB	RB

Fig. 26.2. Recommended template for the analysis of 39 samples in duplicate on a standard 96-well immunoassay plate. S1 through S7 are standard duplicates; B_0 is used to determine maximal absorbance in the absence of free isoprostane. RB is the reagent blank that does not contain 15-F_{2t}-HRP conjugate. U1 through U39 are the unknown samples and dilutions of unknowns assayed in duplicate.

dilution buffer is added, and dilution buffer is also added at a later step rather than 15-F_{2t}-IsoP -HRP conjugate.

6. Add 100 μL of diluted 15-F_{2t}-IsoP-HRP conjugate to each well and allow plate to stand at room temperature for 2 h.
7. Empty the contents of the plate by inversion. Pat dry upside-down on a lint-free towel.
8. Remove any unbound 15-F_{2t}-IsoP-HRP conjugate by adding 300 μL of wash buffer to each well, wait 2 min, then invert to empty the contents, and pat dry. Repeat this wash step two more times. Alternatively, an automated microplate washing instrument may be employed for this step.
9. Add 200 μL of TMB substrate solution to each well and allow the plate to incubate at room temperature for 30 min.
10. Add 50 μL of 3.0 N sulfuric acid to each well to stop the HRP-catalyzed reaction.
11. Read the absorbance for each well at 450 nm (*see* **Note 17**).

3.3.2. Unextracted Urine Samples

The immunoassay procedure is essentially the same as the one described for the extracted serum samples. However, unextracted urine samples must first be diluted at least 4-fold with an enhanced dilution buffer containing a mixture of proteins and detergents that has been formulated to minimize interference due to nonspecific binding. Dilutions of the standard are also prepared in this buffer.

3.4. Data Analysis

3.4.1. Blank Subtraction and Construction of a Standard Curve

1. Average the reagent blank (RB) absorbance values and subtract this average from the absorbance value for every other well (*see* **Note 18**).
2. Average the values obtained for each set of replicates for each dilution of the standard. Divide the average for each by that obtained for B0 and multiply by 100 to obtain $\%B_0$ values.
3. Graph the $\%B_0$ values (linear *y*-axis) vs. the concentration of 15-F_{2t}-IsoP in each dilution of the standard (logarithmic *x*-axis) to obtain a standard curve. **Figure 26.3** shows a typical standard curve.
4. Average the replicates of each unknown and divide by the average B_0 value to obtain $\%B_0$, then determine corresponding concentration using the standard curve.

3.4.2. Use of the Four-Parameter Fit to Obtain Reliable Results

1. The relatively narrow standard curve range ($\times{\sim}0.05$–2.0 concentration units for 20–80% B/B_0 (B = signal, B_0 = maximum signal, adjusted to full scale range) defines the portion of the standard curve (**Fig. 26.3**) that has the most reliable concentration dependence. To obtain the most

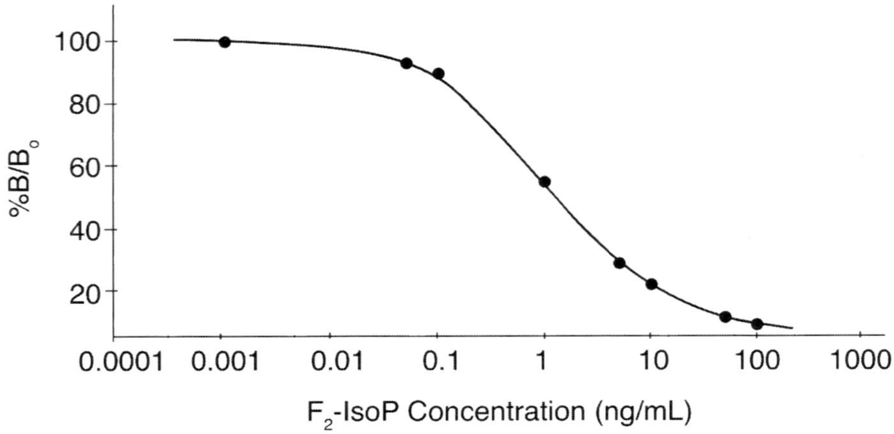

Fig. 26.3. A typical standard curve obtained for the competitive binding of several dilutions of 15-F_{2t}-IsoP standard and 15-F_{2t}-IsoP-HRP conjugate with anti-15-F_{2t}-IsoP adsorbed on the wells of a microplate. As the concentration of free 15-F_{2t}-IsoP is increased, less 15-F_{2t}-IsoP-HRP is bound to the adsorbed antibody resulting in a lower signal. The colorimetric signal, expressed as the percentage of the maximum signal in the absence of free 15-F_{2t}-IsoP, i.e., %B/B0, is inversely and logarithmically related to the concentration of 15-F_{2t}-IsoP in the assay. The concentrations of 15-F_{2t}-IsoP in biological specimens using this method should be restricted to values for %B/B0 ranging from 20 to 80%, 0.2–10 ng/mL for this standard curve.

reliable results, raw data should be fit to a plot of the signal intensity (%B/B_0 vs. the log of the concentration of 15F_{2t}-IsoP standard using the four-parameter equation (20) for which a = the maximum signal = B_0, d = the minimum signal, b = a slope-related term at the inflection point, and c = the concentration at the inflection point (50% B/B_0).

$$y = \frac{(a-d)}{1+\left(\frac{x}{c}\right)} + d$$

2. Multiple dilutions should be performed prior to analysis and only data obtained for only the 20–80% B/B_0 portion of the standard curve reported (*see* **Note 19**).

3.4.3. Normalization of Urinary Isoprostane Levels

To allow for the meaningful comparison of urinary isoprostane levels among subjects, creatinine levels must be determined for urine sample and isoprostane values obtained for each sample must be normalized and expressed as the isoprostane level per milligram of creatinine in the urine specimen (*see* **Notes 20** and **21**).

4. Notes

1. Unless stated otherwise, water used in all the experiments has resistivity of 18.2 MΩ cm and the total organic content is less than 5 ppb.

2. Ex vivo generation of isoprostanes is not an issue when analyzing urine samples (3). However, the samples should be refrigerated and/or an antimicrobial agent is added to resist microbial growth.
3. These reagents may be prepared and stored at room temperature but should be chilled to 4°C prior to use.
4. Urine samples are diluted 4-fold (or higher) using this proprietary buffer. If undiluted samples are to be used, they must be subjected to solid phase extraction following the protocol described for plasma samples.
5. Glucuronidase is reconstituted just prior to use. Aliquots of the reconstituted enzyme should be stored at –80°C for future use.
6. These protocols are intended for quantification of the relative levels of isoprostanes in plasma or urine samples in conjunction with commercial standards manufactured by Oxford Biomedical Research. The assays are based on competition between isoprostanes and their metabolites (IsoP) in the samples that are immunoreactive with a polyclonal antibody generated against 15-F_{2t}-IsoP and a 15-F_{2t}-IsoP-HRP conjugate. IsoP immunoassays have also been developed using 15-F_{2t}-IsoP-alkaline phosphatase or 15-F_{2t}-IsoP-acetylcholinesterase conjugates. Each of these systems have unique characteristics regarding sensitivity and stability. In general, based on relatively high stability and resistance to inactivation by a wide range of agents, HRP is the most widely employed enzyme.
7. In order to minimize ex vivo generation of isoprostanes from either the free or the esterified arachidonic acid in blood, it is crucial to inhibit cyclooxygenase activity (a minor source of isoprostanes) and add an antioxidant as soon as possible. The isoprostanes should be extracted immediately or else the plasma samples must be stored at –80°C until extraction.
8. Folch extraction (21) should be performed on ice or at 4°C to minimize ex vivo generation of isoprostanes from arachidonic acid.
9. These additives are not required since there is no significant ex vivo generation of isoprostanes in urine samples (2, 7).
10. This mixture will turn milky white and has a tendency to foam after vortexing.
11. Affinity chromatography on immobilized anti-15-F_{2t}-IsoP antibody has been used as an alternative for isoprostane extraction prior to GC/MS analysis or immunoassay (18). Although this approach is practical, one must be careful

to make certain that the immunoaffinity column has the capacity to extract all of the 15-F_{2t}-IsoP in the sample. Furthermore, given the fact that a low pH is required for sample elution and that the immunoaffinity columns potentially loose capacity for binding the ligand(s) of interest, as far as solid phase extraction goes, it is preferable to use such columns only once. Regardless of the sample preparation procedure employed, it is necessary that one determines and corrects for the efficiency of recovery of isoprostanes by immunoaffinity columns. A suitable approach is the addition of a known quantity (e.g., 5 ng) of authentic 15-F_{2t}-IsoP (spiking) to few plasma samples, followed by extraction and immunoassay of both spiked and unspiked samples to compute the typical percentage of the 15-F_{2t}-IsoP recovered.

12. It is possible that the solid phase columns do not accommodate the entire volume of the sample to be extracted. In such cases, adapters that allow accommodation of higher loading volumes can be used. The extraction column under no circumstances is allowed to run dry during sample applications.
13. These instructions are meant for investigators who use the entire immunoassay kit (all of the wells at one time). If portions of the kit are to be used, it is advisable to prepare smaller quantities and save the remaining stock for use at a later time. When using the kit in part, adhere to the following: All unused components should be stored at 4°C. Unused portions of the microplate should be kept in a ziplock bag with desiccant and store at 4°C. The 15-F_{2t}-IsoP-HRP conjugate is most stable at the stock concentration as provided; therefore, dilute only the appropriate amount of this stock and store the remaining stock solution for subsequent uses. Generate a standard curve each time the assay is performed. Two vials of standard are provided for added convenience and use.
14. Since the absorbance values obtained in competitive enzyme immunoassays are inversely and logarithmically related to the concentration of the analyte, the most precise determinations will be obtained for samples in which the isoprostane concentration falls near the inflection point of the standard curve. Therefore, if the value obtained with a given sample is not within the 20–80% B/B_0 portion of the standard curve, it is advisable to perform additional immunoassays using multiple dilutions of the sample.
15. Pipette tips should be pre-wetted or rinsed prior to dispensing reagents. To do this, adjust pipette to the volume

desired and draw up reagent of choice into the tip 2–3 times before dispensing.

16. To ensure recovery of any liquid that may have been trapped in the cap or on walls of the vials in which the immunoassay reagents are provided, briefly centrifuge the vials before withdrawing specified aliquots.
17. Alternatively, if 3.0 N sulfuric acid is not added to stop the reaction, the plate may be read at 650 nm.
18. Most modern microplate readers can be easily programmed to perform this task automatically.
19. The IsoP concentrations of several human urine samples were determined by immunoassay and by GC/MS following solid phase extraction of separate aliquots. Although the two methods quantify different IsoPs and IsoP metabolites, a correlation coefficient (r^2) of >0.8 is typically obtained. (*See* **Fig. 26.4**).
20. This is valid except for subjects with significant renal deterioration
21. Unless one is using a method that clearly detects only a single isomer, regardless of the detection method employed, results should be expressed as (isoprostanes) rather than of F2-isoprostanes (15-F_{2t}-IsoP).

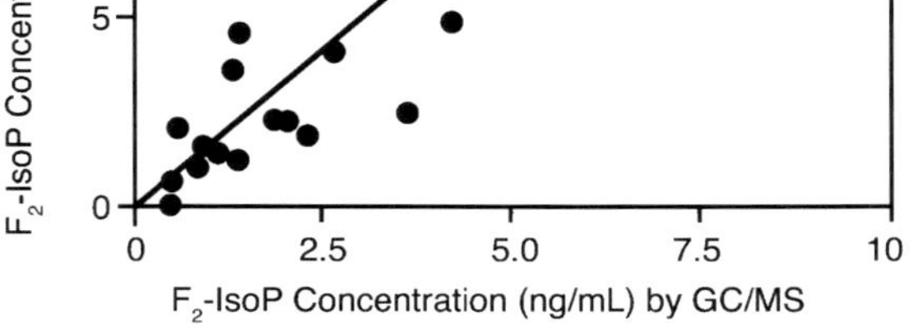

Fig. 26.4. When immunoassays are performed under appropriate conditions, values obtained for F_{2t}-IsoP by immunoassay correlate well ($r^2 > 0.8$) with values obtained by the standard GC/MS method (2). Four F2-isoprostanes contribute to the GC/MS peak, whereas the immunoassay method quantifies a set of F2-isoprostanes and their metabolites that are dependent on the specificity of the antibody employed. The reproducibly lower correlation that is observed for some samples is attributed to interindividual variations in IsoP metabolism.

References

1. Morrow, J.D., Harris, T., and Roberts, L.J. (1990) Noncyclooxygenase oxidative formation of a series of novel prostaglandins: analytical ramifications for measurement of eicosanoids. *Anal. Biochem.* **14**, 1–10.
2. Morrow, J.D. and Roberts, L.J. (1999) Mass spectrometric quantification of F2-isoprostanes in biological fluids and tissues as measure of oxidant stress. *Meth. Enzymol.* **300**, 3–12.
3. Morrow, J.D., Hill, K., Burke, R.F., Nammour, T.M., Badr, K.F., and Roberts, L.J. (1990) A series of prostaglandin F2-like compounds are produced in vivo in humans by a non-cyclooxygenase, free radical-catalyzed mechanism. *Proc. Natl. Acad. Sci. USA* **87**, 9383–9387.
4. Morrow, J.D. and Roberts, L.J. (1997) The isoprostanes: unique bioactive products of lipid peroxidation. *Prog. Lipid Res.* **36**, 1–21.
5. Morrow, J.D., Awad, J.A., Boss, H.J., Blair, I.A., and Roberts, L.J. (1992) Non-cyclooxygenase-derived prostanoids (F2-isoprostanes) are formed in situ on phospholipids. *Proc. Natl. Acad. Sci. USA* **89**, 10721–10725.
6. Liang, Y., Wei P., Duke, R.W., Reaven, P.D., Harman, S.M., Cutler, R.G., and Heward, C.B. (2003) Quantification of 8-iso-prostaglandin-F2α and 2,3-dinor-8-iso-prostaglandin-F2α in human urine using liquid chromatography-tandem mass spectrometry. *Free Radic. Biol. Med.* **34**, 409–418.
7. Helmersson, J. and Basu, S. (1999) F2-isoprostane excretion rate and diurnal variation in human urine. *Prostaglandins Leukot. Essent. Fatty Acids* **61**, 203–205.
8. Sasaki, D.M., Yuan, Y., Gikas, K., Taber, D., Morrow, J.D., Roberts, L.J., and Callewaert, D.M. (1999) An immunometric ELISA for 15-F2t-Isoprostane, an urinary biomarker for oxidant stress. *Free Radic. Biol. Med.* **27**, S43.
9. Proudfoot, J., Barden, A., Mori, T.A., Burke, V., Croft, K.D., Beilin, L.J., and Puddey, I.B. (1999) Measurement of urinary F2 isoprostanes as markers of in vivo lipid peroxidation – A comparison of enzyme immunoassay with gas chromatography/mass spectrometry. *Anal. Biochem.* **272**, 209–215.
10. Schwedlhelm, E., Bartling, A., Lenzen, H., Tsikas, D., Maas, R., Brummer, J., Gutzki, F-M., Chem, I., Berger, J., Frolioch, J.C., and Boger, R.H. (2004) Urinary 8-iso-prostaglandin F2α as a risk marker in patients with coronary heart disease: a matched case-control study. *Circulation* **109**, 843–848.
11. Basu, S. (1998) Radioimmunoassay of 8-iso-prostaglandin F2α: an index for oxidative injury via free radical catalyzed lipid peroxidation. *Prostaglandins Leukot. Essent. Fatty Acids* **58**, 319–325.
12. Helmersson, J., Arnlov, J., Vessby, B., Larsson, A., Alfthan, G., and Basu, S. (2005) Serum selenium predicts levels of F2-isoprostanes and prostaglandin F2α in a 27 year follow-up study of Swedish men. *Free Radic. Res.* **39**, 763–770.
13. Sampson, M.J., Gopaul, N., Davies, I.R., Hughes, D.A., and Carrier, M.J. (2002) Plasma F2 Isoprostanes. Direct evidence of increased free radical damage during acute hyperglycemia in type 2 diabetes. *Diabetes Care* **25**, 537–541.
14. Roberts, L.J., Moore, K.P., Zackert, W.E., Oates, J.A., and Morrow, J.D. (1996) Identification of the major urinary metabolite of the F2-isoprostane 8-iso-prostaglandin F2α in humans. *J. Biol. Chem.* **271**, 20617–20620.
15. Awad, J.A., Morrow, J.D., Takahashi, K., and Roberts, L.J. (1993) Identification of non-cyclooxygenase-derived prostanoid (F2-isoprostane) metabolites in human urine and plasma. *J. Biol. Chem.* **268**, 4161–4169.
16. Basu, S. (1998) Metabolism of 8-iso-prostaglandin F2α. *FEBS Lett.* **428**, 32–36.
17. Chiabrando, C., Valagussa, A., Rivalta, C., Durand, T., Guy, A., Zuccato, E., Villa, P., Rossi, J.C., and Fanelli, R. (1999) Identification and measurement of endogenous beta-oxidation metabolites of 8-epi-Prostaglandin F2α. *J. Biol. Chem.* **274**, 1313–1319.
18. Tsikas, D., Schwedhelm, E., Suchy, M.T., Niemann, J., Gutzki, F.M., Erpenbeck, V.J., Hohlfeld, J.M., Surdacki, A., and Frolich, J.C. (2003) Divergence in urinary 8-iso-PGF2α. (iP F2α-III, 15-F2t-IsoP) levels from gas chromatography-tandem mass spectrometry quantification after thin-layer chromatography and immunoaffinity column chromatography reveals heterogeneity of 8-iso-PG F2 Possible methodological, mechanistic and clinical implications. *J. Chromatogr. B Analyt. Technol. Biomed. Life Sci.* **794**, 237–255.
19. Callewaert, D.M., McGowen, R., Sloan, C., Godschalk, K., Basu, S., Morrow, J., and Gupta, S.V. (2005) Isoprostane Glucuronides in Human Urine and the

Evaluation of Oxidative Stress. *Free Radic. Biol. Med.* **39**, S110.
20. Maciel, R.J. (1985) Standard curve fitting in immunodiagnostics: A primer. *J. Clin. Immunoassay* **8**, 98–106.
21. LeBaron, F.N. and Folch, J. (1959) The effect of pH and salt concentration on aqueous extraction of brain proteins and lipoproteins. *J. Neurochem.* **4**, 1–8.

Chapter 27

Application of Membrane Extraction with Sorbent Interface for Breath Analysis

Victor Ma, Heather Lord, Melissa Morley, and Janusz Pawliszyn

Abstract

The detection of volatile organic compounds (VOCs) in human breath may be useful for routine clinical diagnosis of several diseases in a non-invasive manner. Traditional methods of breath analysis have some technical limitations. Membrane extraction with a sorbent interface (MESI), however, has many advantages over current methods, including good selectivity and sensitivity, and is well suited for breath analysis.

The aim of this project was to develop a simple and reproducible sampling device and method based on the MESI system for breath analysis. The feasibility and validity of the MESI system was tested with real human breath samples. The use of breath CO_2 as an internal standard for the analysis of breath VOCs is an effective method to solve the difficulties associated with variations in the target analyte concentrations in a sample, which are attributed to mass losses and different breathing patterns of different subjects. In this study, the concentration of breath acetone was successfully expressed normalized to CO_2 as in the alveolar air. Breath acetone of healthy males and females profiled at different times of the day was plotted using the MESI system, and results are consistent with literature. This technique can be used for monitoring breath acetone concentrations of diabetic patients and for applications with other biomarker monitoring.

Key words: Membrane extraction with sorbent interface (MESI), CO_2 sensor, internal standard, carboxen, breath analysis, breath acetone, on-line monitoring.

1. Introduction

Breath analysis may be viewed as a headspace extraction of the VOCs present in the blood present in lungs. Breath testing for VOCs is intrinsically safe and non-invasive which offers an alternative approach to the early diagnosis or evaluation of several

R.M. Uppu et al. (eds.), *Free Radicals and Antioxidant Protocols*, Methods in Molecular Biology 610,
DOI 10.1007/978-1-60327-029-8_27,

common disorders including lung cancer and heart disease through endogenous biomarkers (1).

To date, there are approximately 3,000 VOCs that have been detected at least once in human breath, although most breath samples contain around 200 VOCs, which are present in picomolar concentrations (2). Breath testing challenges scientists in two ways. First, it is technically very difficult to analyze breath VOCs that are present in picomolar concentrations. Second, the source and biochemical significance of most breath VOCs are still unknown, making it difficult to recognize abnormalities in a patient's breath (3, 4).

Membrane extraction with a sorbent interface (MESI), a new analytical method invented by J. Pawliszyn, has been developed to allow for rapid routine analysis and long-term continuous monitoring of VOCs in various environmental matrices (5, 6). The analytical system consists of a sampling chamber, a membrane extraction module, a sorbent trap, and a gas chromatograph equipped with a flame ionization detector (FID). It has been used for on-line monitoring of human expired breath, specifically the last 250 mL portion since this is most valuable for analytical purposes. The sampling chamber consists of a tube fitted with one-way valves at both ends to allow the last portion of exhaled air to enter during expiration, but not to exit or allow mixing with ambient air. The last portion of breath is held in the chamber at the end of expiration. The organic components in the trapped breath are transferred to the stripping gas by permeation through the membrane in the extraction module and concentrated in the sorbent trap before introduction as a sharp plug on the front of the chromatographic column. Sampling can also be done off-line with the extraction membrane module and sorbent trap which is beneficial for field sampling since the sorbent trap can later be connected to a GC and desorbed in a separate step.

MESI offers a high degree of selectivity and enrichment power. The method is simple, effective, solvent-free, and easy to automate. It minimizes the loss of analytes by coupling the membrane extraction module directly to a capillary gas chromatograph, thus bypassing the use of the GC injector. The application of hydrophobic membranes prevents moisture from entering the carrier gas and the GC column. The sensitivity of the system is significantly enhanced with the cryogenic sorbent trap.

For continuous monitoring applications, membrane extraction is an attractive approach. Permeation through a membrane is a specific extraction process where the sorption into and desorption out of the extraction phase occur simultaneously. The sample (donor phase) is in contact with one side of the membrane where extraction into the membrane material occurs, while permeated analytes are removed by the stripping phase (acceptor). For membrane extraction with good flow (agitation) conditions at both the acceptor and the donor sites, as well as efficient stripping, the rate of mass transport through the membrane is controlled by the diffusion of analytes through the membrane material. The concentration gradient, which facilitates transport across the membrane, is formed by the difference in analyte concentrations between the sample side ($K_{es}C_s$) and the stripping phase, which is close to zero for high flow rates associated with the stripping phase (6). **Figure 27.1** is a schematic representing the concentration gradient of a sample by membrane extraction.

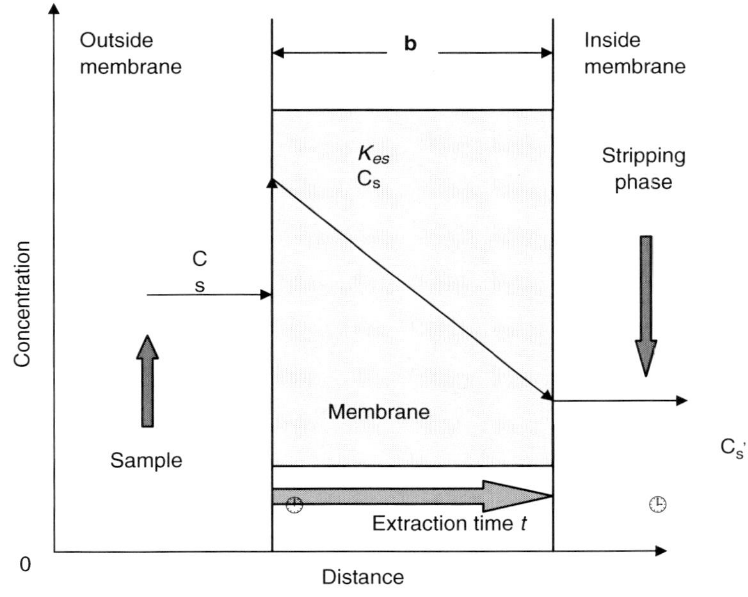

Fig. 27.1. Concentration gradient of the sample by membrane extraction.

The mass transfer rate through the membrane, n/t, can be estimated at steady-state conditions using Eq. [**1**]:

$$\frac{n}{t} = \frac{B_2 A D_e K_{es} C_s}{b} \qquad [1]$$

where b: the thickness of the membrane

n: the extraction amount

B_2: a geometric factor defined by the shape of the membrane

A: the surface area of membrane

D_e: the diffusion coefficient in the membrane material

K_{es}: the membrane material/sample matrix distribution constant

t: the extraction time

The concentration of an unknown can be calculated by converting Eq. [**1**] to Eq. [2].

$$C_s = \frac{bn}{B_2 A D_e K_{es} t} \quad [2]$$

In order to model the mass transport, the gradation in fluid motion and convection of molecules in the space surrounding the membrane can be simplified as a boundary layer zone (6). The thickness of the boundary layer (δ) is determined by both the rate of convection (agitation) in the sample and the diffusion coefficient of the analyte. The analytes are transferred through the membrane by the following seven-step mechanism: (1) convection and diffusion through the sample to the boundary layer outside the membrane surface; (2) diffusion through the boundary layer to the outer surface of the membrane; (3) partitioning between the sample and the outer surface of the membrane; (4) diffusion through the membrane; (5) partitioning between the inner surface of the membrane and the stripping gas; (6) diffusion through the boundary layer from the inner surface of the membrane to the bulk of mobile phase; and (7) diffusion and convection away from the membrane with the mobile phase to the sorbent interface. A schematic of membrane extraction process is shown in **Fig. 27.2**.

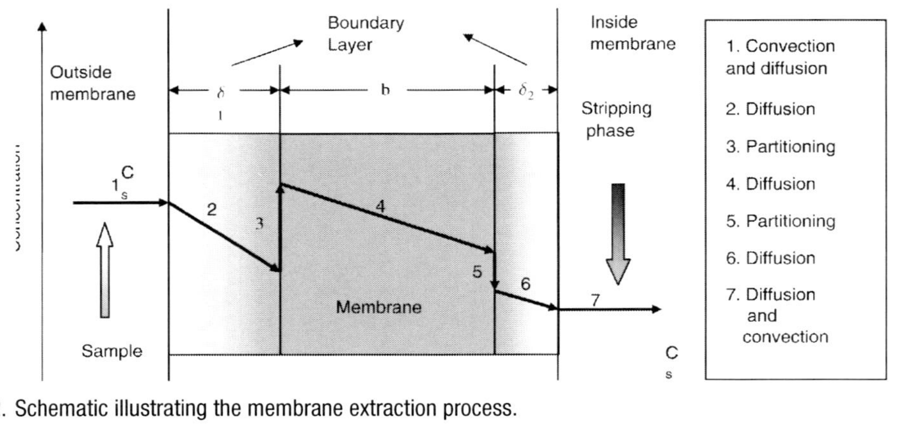

Fig. 27.2. Schematic illustrating the membrane extraction process.

The MESI system consists of four major sections: (1) The membrane extraction module; (2) the cryofocusing trap and thermal desorption sorbent interface; (3) the separation and detection system (GC/FID or GC/MS); and (4) the computer control and data acquisition system. A schematic diagram of the MESI-GC system is illustrated in **Fig. 27.3**.

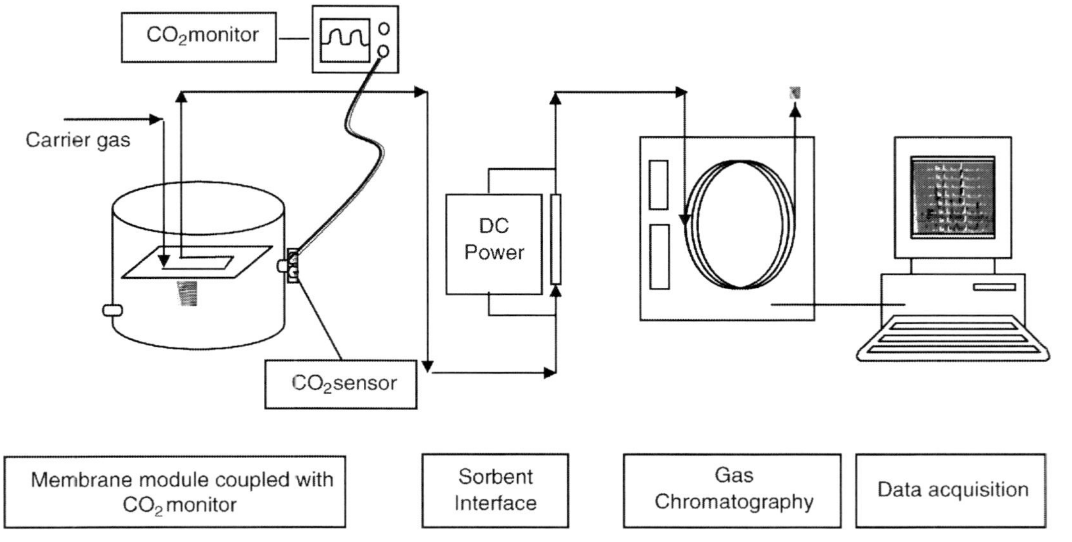

Fig. 27.3. Schematic of the MESI-GC system.

2. Materials

Helium (5.0 ultra high purity, 99.999%), nitrogen (5.0 ultra high purity, 99.999%), and hydrogen (5.0 Polymer, 99.999%) were purchased from Praxair (Kitchener, ON, Canada). Acetone (HPLC grade, ≥99.5%) was purchased from Merck (Germany). Isoprene (99%) was purchased from Sigma-Aldrich (Oakville, ON, Canada). Ethylene (5.0 ultra high purity, 99.999%) was purchased from Praxair (Kitchener, ON, Canada). The plexiglas tubular sampler (124 mL) and the glass vial sampler (250 mL) were custom-made at the University of Waterloo (Waterloo, ON, Canada). Flat sheet silicone polycarbonate membranes SSP-M213 (0.0005″) were purchased from Special Silicone Products, Inc. (Ballston Spa, NY, USA). Tenax TA 80/100 mesh, Carboxen 569 20/45 mesh, and Carboxen 1000 60/80 mesh were purchased from Supelco (Bellefonte, PA., USA). Rtx-U Plot column (30 m × 0.25 mm i.d. × 1.40 μm d.f.), Hydroguard MXT guard columns and transfer lines (0.28 and 0.53 mm i.d.), coiled Silcosteel tubing (0.53 mm i.d.), and gastight syringes (Hamilton, 1.0 mL) were obtained from Restek (Bellefonte, PA, USA). A two-stage Peltier cooler was purchased from Melcor (Trenton, NJ, USA). The gas chromatograph coupled with FID (Chrompack CP-9002) was supplied by Varian (Walnut Creek, CA, USA). A DC power supply (HP Harrison 6427B) from Hewlett Packard (Palo Alto, CA, USA), an electronic thermometer (Fluke 53II) from Fluke Corporation (Everett, WA, USA), and an electronic flow meter (ADM 2000 Intelligent flow meter) from

J & W Scientific (Folsom, CA, USA) were also used. The membrane module, the power supply for the cooler (S&D – 066), the temperature controller for the DC power supply (S&D – 070), and the heating timer (S&D – 073) were custom-made by the Science Shop of the University of Waterloo (Waterloo, ON, Canada). An NICO CO_2 monitor and CAPNOSTAT CO_2 sensor (Respironics Novametrix, LLC) were also used.

2.1. Membrane Module

The flat sheet membrane module was constructed as illustrated in **Fig. 27.4**. A flat sheet silicone polycarbonate membrane (0.0005″, SSP-M213) was mounted between two Teflon spacers as shown. The upper Teflon spacer has two holes that match the holes in the upper steel plate and permit the passage of the carrier gas. The lower Teflon spacer is slightly thinner and has a U-shaped channel cut into it to match with the channel cut into the lower steel plate.

During operation, the pressure of the carrier gas causes the membrane to balloon into the U-shaped channel in the lower Teflon spacer. The carrier gas travels along the channel, and the dimensions of the channel in the lower Teflon spacer determine the overall dimensions of the channel on the receiving side of the membrane. A wire mesh was affixed to the lower steel plate to support the membrane and prevent it from ballooning out of the bottom of the module during sampling. The module was sealed by evenly tightening the 12 machine bolts that pass through the

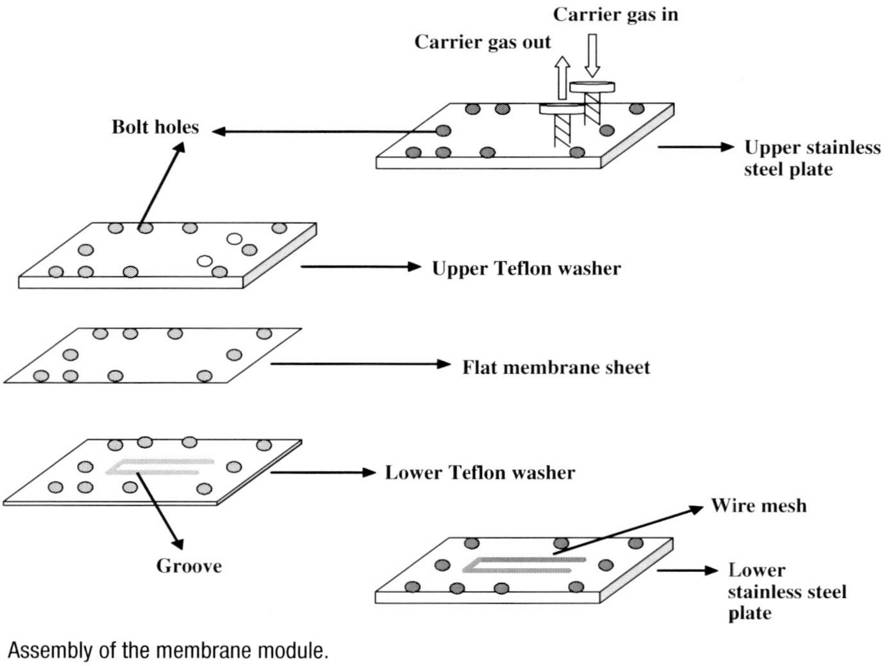

Fig. 27.4. Assembly of the membrane module.

module and compress the two steel plates together. To ensure the module was gastight, it was immersed in water or methanol while the carrier gas was passed through the system, followed by thorough drying prior to sampling.

2.2. Sorbent Interface

Pre-concentration is necessary because most breath VOCs are present in the nmol/L to pmol/L range. As discussed in the introduction, the sorbent trap is most frequently used as an efficient method for pre-concentration.

2.2.1. Sorbent Trap System

Analytes stripped at the membrane are transported by the carrier gas to the trap, where they are retained until a sufficient amount has been trapped for thermal desorption and analysis. **Figure 27.5** illustrates the sorbent interface system.

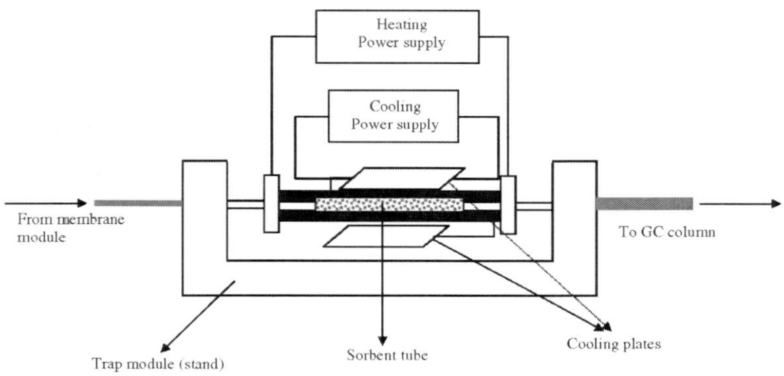

Fig. 27.5. Schematic diagram of the sorbent interface system.

Trapping time is optimized so that the amount trapped is within the linear response range of the detector but does not result in breakthrough in the trap. The breakthrough time is determined as the time after which no additional analyte is trapped with increasing trapping time. During sampling, the trap is cooled with a two-stage Peltier cooler (Melcor Corp., Trenton, NJ, USA) to improve trap capacity and reduce breakthrough. The trap is desorbed at pre-determined times by applying a capacitive discharge voltage from a custom-made power supply (Science Shops, University of Waterloo). After thermal desorption, analytes are carried in a narrow band to the top of the separation column mounted in the gas chromatograph.

2.2.2. Adsorbent Materials

Adsorbents collect analytes on their surface by physically trapping or chemically reacting with the analytes. The porosity of the sorbent is critical as it determines the surface area of the sorbent, thus determining the sorbent capacity. The most commonly used sorbents for breath analysis are those made of organic polymers (e.g., Tenax TA), activated charcoal, graphitized carbon, and carbon molecular sieves.

As a result of the different boiling points of breath VOCs, adsorbents in sorbent traps have to be selected carefully to avoid breakthrough and memory effects. Organic polymers, such as Tenax TA, are least affected by high water contents in the samples but have low breakthrough volumes. In contrast, carbon molecular sieves and graphitized carbon have high breakthrough volumes for breath VOCs; however, during thermal desorption, these adsorbents exhibit serious memory effects. Memory and breakthrough effects can be minimized by using multibed sorbent traps (1).

2.2.3. Sorbent Tube Packing

The sorbent tube system is a crucial component of trace analysis, and all of the experimental results are based on the reliability of the sorbent trap. However, packing the thermal desorption tube is a highly skilled and time-consuming operation. Loosely packed tubes may cause sorbent movement and air space formation, resulting in sampling errors. Conversely, over compression of the sorbent bed will result in high tube impedance, which may result in inefficient thermal desorption.

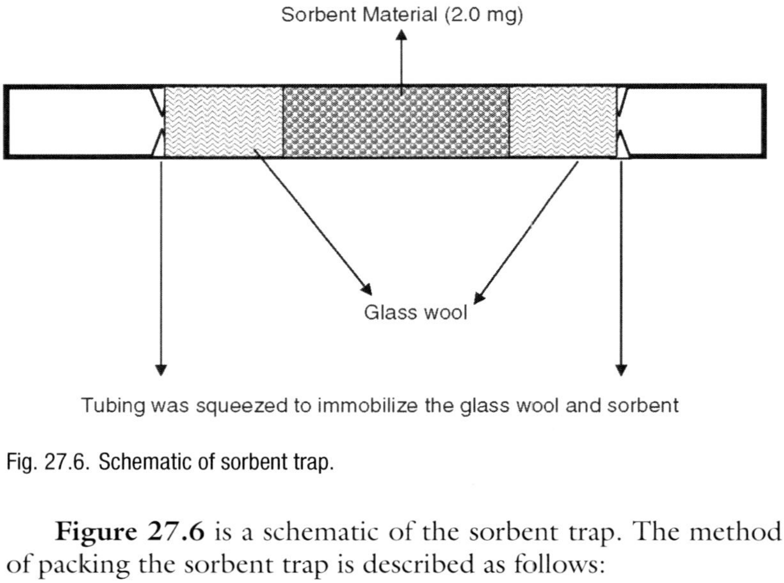

Fig. 27.6. Schematic of sorbent trap.

Figure 27.6 is a schematic of the sorbent trap. The method of packing the sorbent trap is described as follows:

- Materials Needed: silcosteel-treated tubing (55 mm length, 0.53 mm id, 1.59 mm od); trap sorbent: Carboxen569 20/45 mesh 1.0 mg; Carboxen1000 45/60 mesh 1.0 mg; and deactivated glass wool
- Packing Procedure:

A section of silcosteel-treated tubing (55 mm length) is cut and cleaned with methanol. Glass wool is immobilized by inserting it into one end of the tube and partially crimping the tube

2 cm from the end. Carboxen 569 (1.0 mg) is weighed into a 1.5 mL vial. The end of the tube with the glass wool is connected with a vacuum source and the vacuum is used to draw the sorbent from the vial into the tube. Carboxen 1000 (1.0 mg) is then weighed into a 1.5 mL vial. The trap is then further packed with small size sorbent particles. The larger Carboxen particles, having a diameter only slightly smaller than the inner diameter of the tubing, are packed first and ultimately located at the beginning of the sorbent bed. The smaller particles placed at the end of the trap slightly restrict the gas flow toward the membrane module during the heating pulse. Glass wool is placed at the end of the sorbent bed. The silcosteel-treated tubing is crimped slightly at the end of the final glass wool packing to help immobilize the packing material. The tube is crimped more severely at the beginning of the tube than at the end. With this configuration, when a heating pulse is applied, the carrier gas inside the trap, following the less restricted path, will mainly flow toward the column, pushing the analytes in this direction. The packed sorbent tube is conditioned for 2 h (220°C) with nitrogen as the stripping gas. Finally, the sorbent is connected with the MESI system, acetone is spiked into the sampling chamber, extraction time is set for 20 min, a heating pulse is applied every 15 min, and the performance of the sorbent trap is tested.

2.2.4. Breath Sampling Chamber

The helium carrier/stripping gas flows at a rate of 3.0 mL/min and passes over the receiving side of the membrane to strip the analytes as they diffuse through the membrane. The membrane module is mounted into an aerosol-holding chamber equipped with one-way valves (Boehringer Ingelheim Canada Ltd., Burlington, ON, Canada). A schematic representation of the assembled sampling chamber with the membrane module contained in the Aero-chamber is shown in **Fig. 27.7**.

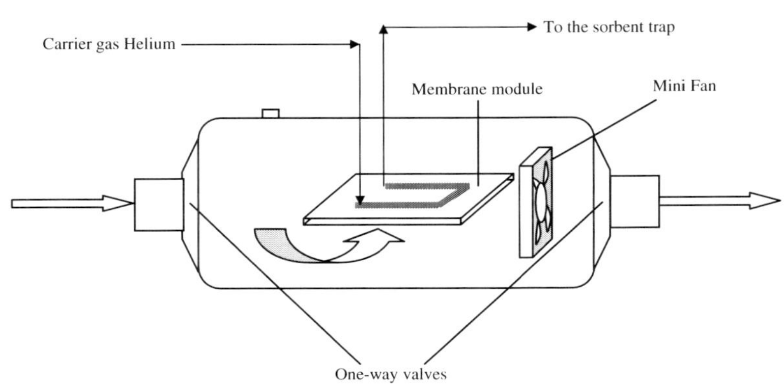

Fig. 27.7. Schematic diagram of the breath sampling chamber.

For sampling, the subject is instructed to expire as much breath as possible into the device. The sampling chamber holds a volume of 250 mL and has a one-way valve mounted at both the inlet and the outlet. Thus, only the last 250 mL of breath is sampled in the chamber. The flow-through design minimizes mixing and no concentration or dilution occurs in the chamber. The sampling chamber also possesses a septum-plugged hole to allow for parallel solid-phase microextraction (SPME) sampling for validation purposes.

The breath sampling chamber is designed to fulfill the following requirements:

1. *Subject comfort*: The subject breathes into a wide-bore tube (approximately 1 in. diameter), which presents very little resistance to expiration. The person providing the breath sample experiences no discomfort and encounters virtually no resistance while breathing into the device.
2. *Subject safety*: Using a paper disposable mouthpiece with inlet and outlet valves, the subject inhales room air and exhales into the instrument, so that there is no potential hazard of exposure to infection.
3. *Alveolar sampling*: The breath sample is drawn principally from alveolar breath, not from dead space breath in which no VOC interchange has occurred.
4. *Chemical contamination*: The device incorporates no structural components such as volatile adhesives, which might outgas VOCs and contaminate the sample. The breath sample is ducted through stainless steel tubing; the plastic component is manufactured with a low content of plasticizer.
5. *Water interfering*: Breath is saturated with water; the membrane significantly prevents water vapor from permeating through.
6. *Concentration of the sample*: The membrane selectively concentrates VOCs contained in alveolar breath, while allowing the nitrogen, oxygen, and carbon dioxide in the breath to escape.
7. *User convenience*: The breath sampling chamber is simple to use and does not require a high degree of operator training.

2.2.5. CO_2 Sensor

A schematic representing a normal capnogram is shown in **Fig. 27.8** (1). Phase I is the first expiratory stage. Gas sampled during this phase represents anatomical dead space and would typically not contain CO_2 and endogenous VOCs. Phase II reflects CO_2 appearance and a steep upstroke of CO_2 tension in the normal capnogram. Gas sampled during this phase typically contains a mixture of alveolar and dead space air. Phase III reflects a minimal increase in the CO_2 concentration, due to alve-

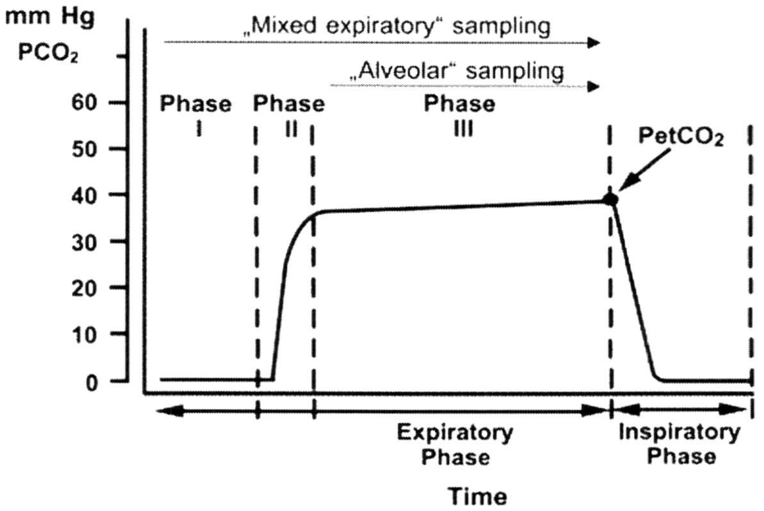

Fig. 27.8. Schematic of a normal capnogram Adapted from reference 1, used with permission from Elsevier.

olar emptying in the lungs. This phase is referred to as the alveolar plateau or expiratory plateau. $PetCO_2$ is the terminal portion of exhaled carbon dioxide, which reveals the actual end tidal concentration ($ETCO_2$).

The measure of CO_2 concentration is important because CO_2 can be used as an internal standard to calibrate the breath VOCs for different breath patterns and against the dilution of dead space air.

The breath CO_2 monitor used in these studies (NICO monitor) was generously provided by Respironics Novametrix, LLC. It is a medical device that measures various respiratory monitoring parameters, including CO_2 elimination and alveolar minute ventilation. The NICO monitor uses the CAPNOSTAT CO_2 sensor to measure CO_2 by the infrared absorption technique. The principle is based on the fact that CO_2 molecules absorb infrared (IR) light energy of specific wavelengths, with the amount of energy absorbed being directly related to the CO_2 concentration. When an IR beam is passed through a gas sample containing CO_2, an electronic signal from the photo detector (which measures the remaining light energy) can be obtained. This signal is then compared to the energy of the IR source and calibrated to accurately reflect the CO_2 concentration.

Infrared absorption methods for gas measurement can be sensitive and selective as well as provide a continuous, accurate, precise, and rapid response that is not saturated nor damaged by concentrations of the target gases, such as CO_2. Infrared energy is safe because it causes no permanent changes in the molecules. The absorption of IR light increases the molecular rotation and vibration, which increases the temperature and pressure, but only

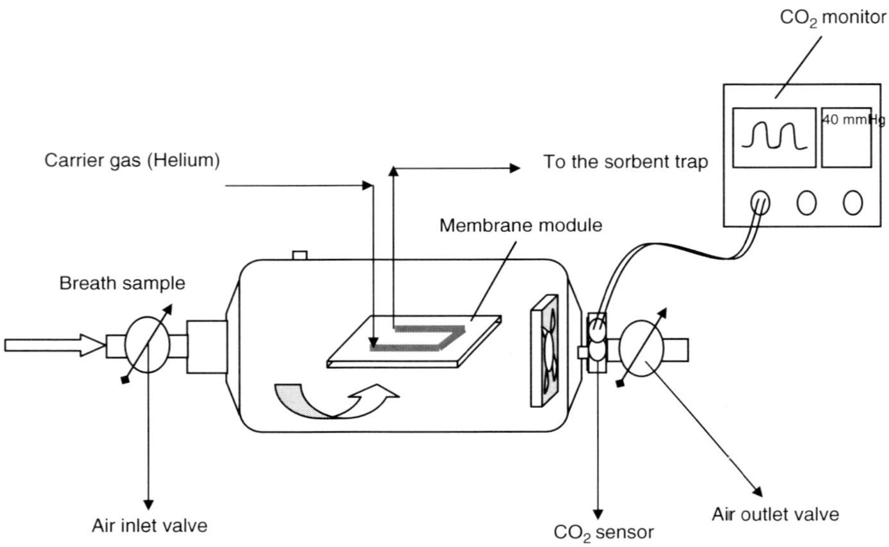

Fig. 27.9. Schematic of MESI-GC coupled with CO_2 monitoring system.

modestly. CO_2 has a strong absorption band near the IR region (4.26 μm).

Figure 27.9 is a schematic of the MESI system coupled with a CO_2 sensor. The CAPNOSTAT CO_2 sensor is mounted at the output port of the sampling chamber.

3. Methods

Helium was used as the carrier gas and stripping gas at a flow rate of 5.0 mL/min. The GC/FID oven temperature was isothermal at 100°C. The trap cooling temperature ranged from –15 to +20.5°C. Desorption heating parameters were 150–200°C for 60 s for acetone and isoprene. The Carboxen sorbent trap was conditioned on-line at temperatures up to 180°C by several 60 s desorptions in pure nitrogen gas until a sufficiently low contaminant background was achieved. When trace quantities are to be determined, the background pattern becomes important.

For method development, a standard gas containing pentane and/or acetone was used, generated by permeation tubes. An electric pressure control was set in front of the pentane permeation tube, so a constant flow of gas passed through the permeation tube chamber. This way a constant concentration of pentane gas was generated and used an internal standard. A similar design was also used with an acetone permeation tube. Different flow rates of acetone were acquired by introducing a split valve in front of the sampling chamber.

After setting up the experimental system, a stable base line was obtained. The membrane module was immersed in a N_2 environment and the background of the system was very clean. There was no background noise when a heating pulse was applied to the trap module. These results show that the membrane and the trap were clean and the instrument worked properly.

4. Notes

1. *Breath profile*: Real breath samples were tested using the MESI system. Two breath samples, one from a male and one from a female were analyzed (**Fig. 27.10**). Each trace shows four separate analyses, with a pair of significant peaks in each. The first peak is higher than the second. These two major peaks are acetone and isoprene, which are two of the most abundant compounds in human breath.

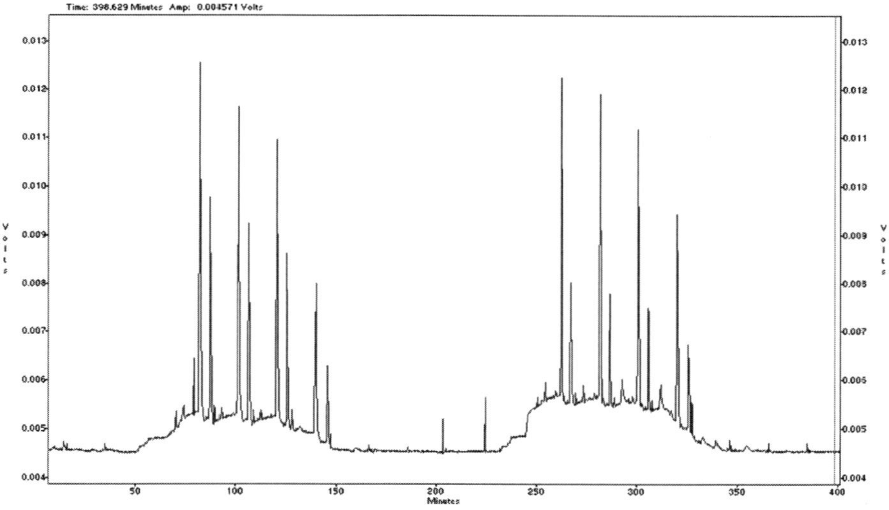

Fig. 27.10. Breath profiles obtained from two healthy individuals.

Figures **27.11** and **27.12** are the breath profiles of a male and a female non-smoker. It can be seen that both subjects' breath sample profiles contain acetone and isoprene. The peak heights and peak areas are all different, which indicates different concentrations of the breath VOCs.

2. *Breath acetone profile*: Breath acetone was quantitatively analyzed by MESI-GC coupled with the CO_2 monitoring

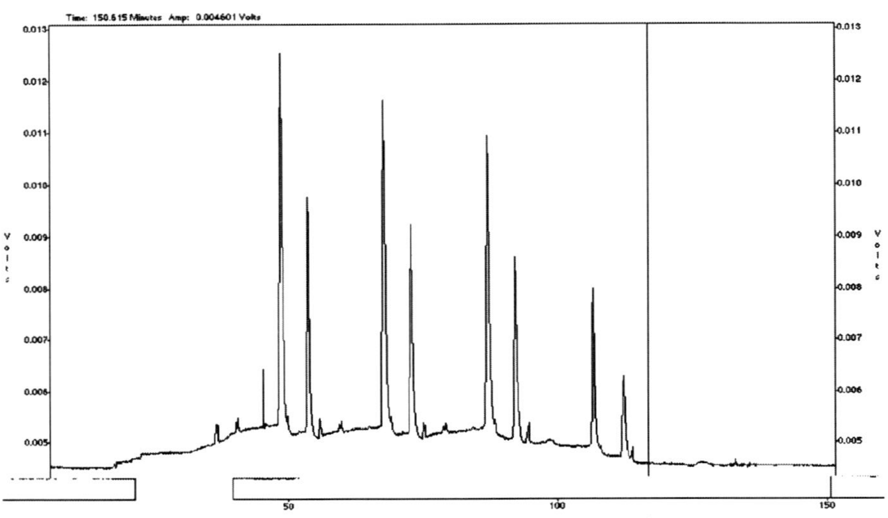

Fig. 27.11. Breath profile of a male non-smoker.

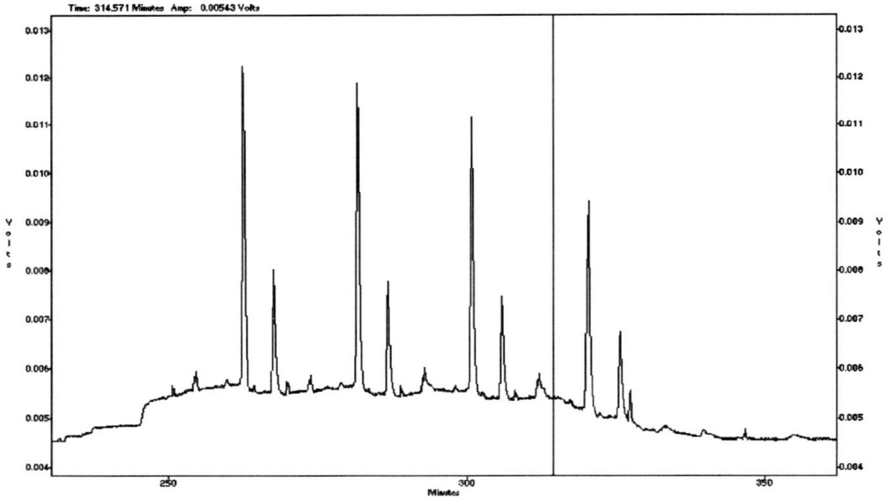

Fig. 27.12. Breath profile of a female non-smoker.

system. A subject was asked to inhale and then exhale into the sample chamber. At the same time, the end tidal pressure of CO_2 was monitored. The extraction was initiated once the breath sample was collected in the sample chamber. The extraction took 20 min, followed by a heating pulse that was applied to the sorbent trap for desorption. The acetone chromatogram was obtained using the FID. Different levels of breath acetone were compared for the different end tidal pressures of CO_2. **Figure 27.13** is a breath acetone profile

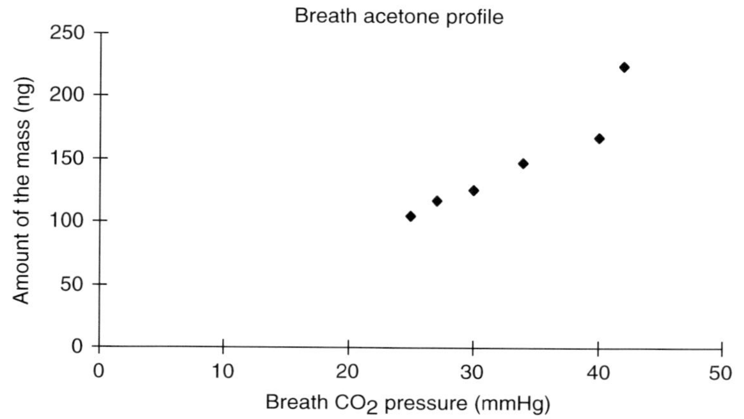

Fig. 27.13. Breath acetone profile of health male at different $ETCO_2$ pressure.

of a healthy male at different $ETCO_2$ pressures (end tidal CO_2 pressure).

From **Fig. 27.13**, a linear correlation between acetone and the $ETCO_2$ pressure is demonstrated. This linear relationship illustrates that CO_2 can be used as an internal standard to monitor breath VOCs at different breathing patterns. The ratio of the amount of breath acetone (ng) and the concentration of CO_2 is shown in **Fig. 27.14** .

$$\frac{\text{Amount of breath acetone}}{\text{ET } CO_2\text{pressure}} = C$$

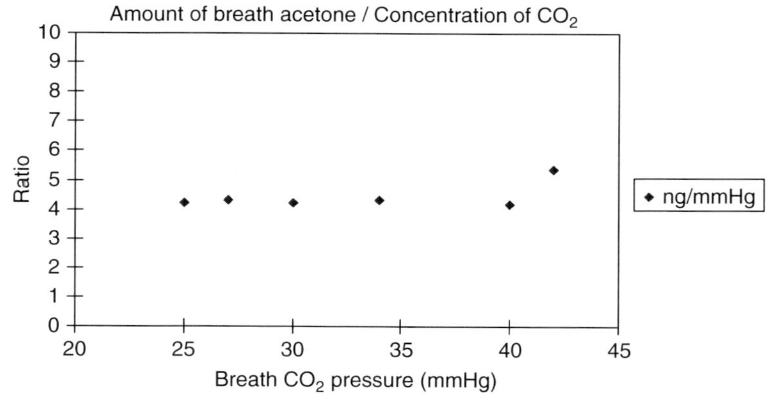

Fig. 27.14. Ratio of the amount of breath acetone to the concentration of CO_2.

The ratio of the amount of breath acetone (ng) and the concentration of CO_2 is a constant value. After using CO_2 as an internal standard, normalization was validated with analytes at higher concentration (acetone or isoprene) and the technique was applied to the analysis of other breath VOCs.

3. *Breath acetone profile of male and female at different times*: Acetone is formed by decarboxylation of acetoacetate, which is derived from lipolysis or lipid peroxidation. Ketone bodies like acetone are oxidized via the Krebs cycle in peripheral tissue (7). Ketone bodies in blood (including acetoacetate and β-hydroxybutyrate) increase in ketonemic subjects during times of fasting or starving or during dieting. Breath acetone concentrations are increased in patients with uncontrolled diabetes. As acetone is produced by spontaneous decarboxylation of acetoacetate, it is impossible to quantify the fraction that arises from lipid peroxidation.

A healthy male subject was asked to inhale and then exhale air into the sample chamber through the inlet nozzle. Each time the subject exhaled, he was asked to exhale as much as possible until the maximum $ETCO_2$ pressure was reached (which was 43 mmHg). The breath was collected at 8 a.m. (after breakfast), 1 p.m. (before lunch), and 8 p.m. (after dinner). **Figure 27.15** illustrates the male breath acetone profile at different times of the day.

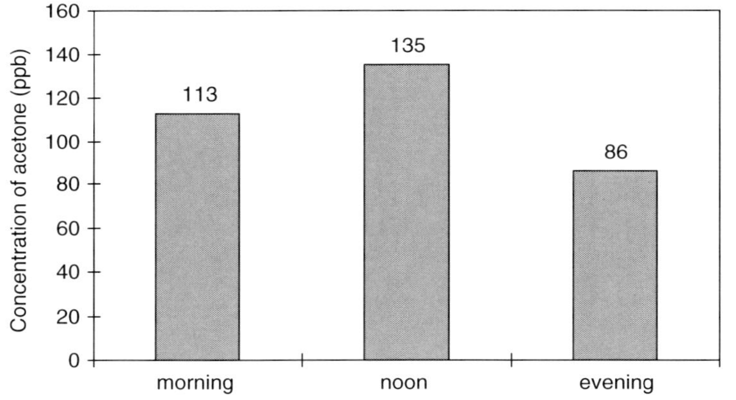

Fig. 27.15. Male breath acetone profile at different times of the day.

From **Fig. 27.15**, it is evident that the concentration of breath acetone increased when the subject had not yet eaten, and decreased after dinner, which is consistent with findings in the literature related to the use of acetone as a diet biomarker.

A similar experiment was also conducted with a female subject. A healthy female subject was asked to inhale and then exhale air into the sample chamber through the inlet nozzle. Each time the subject exhaled, she was asked to exhale as much as possible until the maximum $ETCO_2$ pressure was reached (which was 44 mmHg). The breath was collected at 10 a.m. (after breakfast),

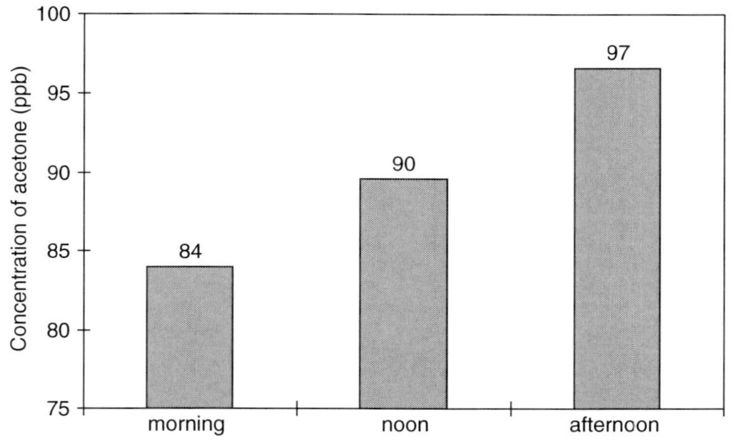

Fig. 27.16. Female breath acetone profile at different times of the day.

1 p.m. (before lunch), and 5 p.m. (before dinner). **Figure 27.16** provides a diagram of the female breath acetone profile at different times of the day.

As illustrated in **Fig. 27.16**, the acetone concentration in the female subject increased during the day, following the morning fasting. **Figure 27.17** provides a comparison of the breath acetone profiles of a male and a female subject at different times of the day.

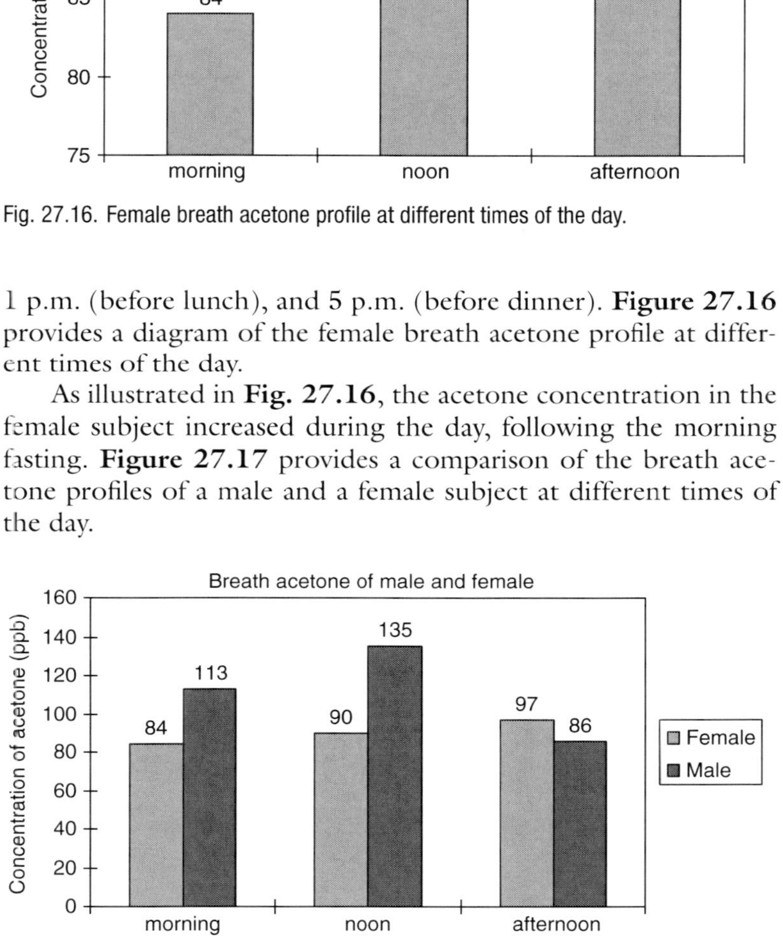

Fig. 27.17. Comparison of breath acetone of a male and a female at different times of the day.

Each person's breath was tested at different diet conditions, and the subject was asked to breathe as much as possible (the $ETCO_2$ reached was about 45 mmHg). The breath sample was collected 3 h after breakfast and 1 h after lunch.

From **Fig. 27.18**, we observe that the five subjects possess different acetone levels in their breath. From this result, it is evident that the acetone level in human breath samples decreases after food is consumed. The results show that membrane

Table 27.1
Comparison of breath acetone monitoring data for five individuals

Subject	ETCO$_2$ (mmHg)	Age	M/F	Acetone peak area Before	Acetone peak area After	Acetone concentration (ppb) Before	Acetone concentration (ppb) After
1	44	33	F	87,981	66,510	43.26	32.70
2	43	24	M	75,544	67,271	37.14	33.08
3	42	36	F	74,222	61,797	36.49	30.38
4	46	36	M	174,702	142,068	85.90	69.85
5	44	24	F	196,519	170,670	96.63	83.92

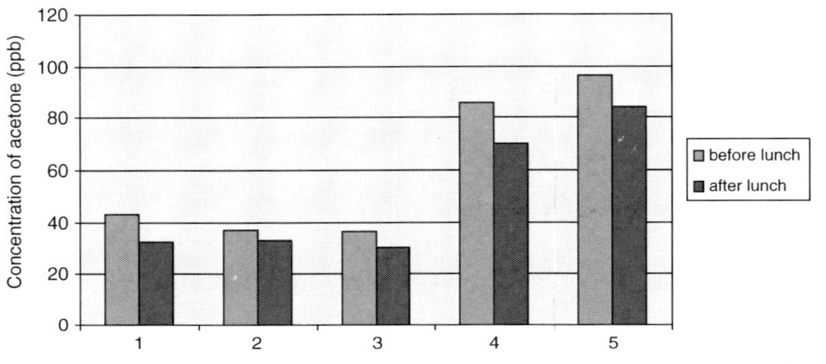

Fig. 27.18. Breath acetone monitoring for five volunteers.

extraction with a sorbent interface (MESI) coupled with a CO_2 monitoring system is a simple, rapid, sensitive, and solvent-free method for the determination of low concentrations of acetone in human breath and the analysis of breath acetone in subjects can be used for diet monitoring as well as a supplementary tool for the diagnosis of diabetes.

References

1. Miekisch, W. and Schubert, J. K. (2004) Diagnostic potential of breath analysis – focus on volatile organic compounds. *Clin. Chim. Acta* **347**, 25–39.
2. Nandor, M. (Ed.) (2003) *Disease Markers in Exhaled Breath*, Marcel Dekker Inc., NY.
3. Phillips, M., Herrera, J., Krishnan, S., Zain, M., Greenberg, J., and Cataneo, R.N. (1999) Variation in volatile organic compounds in the breath of normal humans. *J. Chromatogr. B* **729**, 75–88.
4. Phillips, M. (1997) Method for the collection and assay of volatile organic compounds in breath. *Anal. Biochem.* **247**, 272–278.
5. Luo, Y.Z., Adams, M., and Pawliszyn, J. (1997) Aqueous sample direct extraction and analysis by membrane extraction with a sorbent interface. *Analyst* **122**, 1461–1469.
6. Pawliszyn, J. (Ed.) (2002) *Sampling and Sample Preparation for Field and Laboratory*. Elsevier Science, Amsterdam.
7. Jonsson, J.A. and Mathiasson, L. (2000) Membrane-based techniques for sample enrichment. *J. Chromagr. A* **902**, 205–225.

Subject Index

R.M. Uppu et al. (eds.), *Free Radicals and Antioxidant Protocols*, Methods in Molecular Biology 610,
DOI 10.1007/978-1-60327-029-8, © Humana Press, a part of Springer Science+Business Media, LLC 1998, 2010

F

G

H

I

J

K

L

M

Q

R

S

T

U

V

W

X

Y

Z